TABLE OF INTEGRALS

Basic Forms

1. $\displaystyle\int u\, dv = uv - \int v\, du$

2. $\displaystyle\int u^n\, du = \frac{1}{n+1} u^{n+1} + C, \quad n \neq -1$

3. $\displaystyle\int \frac{du}{u} = \ln |u| + C$

4. $\displaystyle\int e^u\, du = e^u + C$

5. $\displaystyle\int a^u\, du = \frac{1}{\ln a} a^u + C$

6. $\displaystyle\int \sin u\, du = -\cos u + C$

7. $\displaystyle\int \cos u\, du = \sin u + C$

8. $\displaystyle\int \sec^2 u\, du = \tan u + C$

9. $\displaystyle\int \csc^2 u\, du = -\cot u + C$

10. $\displaystyle\int \sec u \tan u\, du = \sec u + C$

11. $\displaystyle\int \csc u \cot u\, du = -\csc u + C$

12. $\displaystyle\int \tan u\, du = \ln |\sec u| + C$

13. $\displaystyle\int \cot u\, du = \ln |\sin u| + C$

14. $\displaystyle\int \sec u\, du = \ln |\sec u + \tan u| + C$

15. $\displaystyle\int \csc u\, du = \ln |\csc u - \cot u| + C$

16. $\displaystyle\int \frac{du}{\sqrt{a^2 - u^2}} = \arcsin \frac{u}{a} + C$

17. $\displaystyle\int \frac{du}{a^2 + u^2} = \frac{1}{a} \arctan \frac{u}{a} + C$

18. $\displaystyle\int \frac{du}{u\sqrt{u^2 - a^2}} = \frac{1}{a} \operatorname{arcsec} \frac{u}{a} + C$

19. $\displaystyle\int \frac{du}{a^2 - u^2} = \frac{1}{2a} \ln \left|\frac{u+a}{u-a}\right| + C$

20. $\displaystyle\int \frac{du}{u^2 - a^2} = \frac{1}{2a} \ln \left|\frac{u-a}{u+a}\right| + C$

Forms Involving $\sqrt{a^2 + u^2}$

21. $\displaystyle\int \sqrt{a^2 + u^2}\, du = \frac{u}{2} \sqrt{a^2 + u^2} + \frac{a^2}{2} \ln |u + \sqrt{a^2 + u^2}| + C$

22. $\displaystyle\int \frac{\sqrt{a^2 + u^2}}{u}\, du = \sqrt{a^2 + u^2} - a \ln \left|\frac{a + \sqrt{a^2 + u^2}}{u}\right| + C$

23. $\displaystyle\int \frac{\sqrt{a^2 + u^2}}{u^2}\, du = -\frac{\sqrt{a^2 + u^2}}{u} + \ln |u + \sqrt{a^2 + u^2}| + C$

24. $\displaystyle\int \frac{u^2\, du}{\sqrt{a^2 + u^2}} = \frac{u}{2} \sqrt{a^2 + u^2} - \frac{a^2}{2} \ln |u + \sqrt{a^2 + u^2}| + C$

25. $\displaystyle\int \frac{du}{\sqrt{a^2 + u^2}} = \ln |u + \sqrt{a^2 + u^2}| + C$

26. $\displaystyle\int \frac{du}{u\sqrt{a^2 + u^2}} = -\frac{1}{a} \ln \left|\frac{\sqrt{a^2 + u^2} + u}{u}\right| + C$

27. $\displaystyle\int \frac{du}{u^2 \sqrt{a^2 + u^2}} = -\frac{\sqrt{a^2 + u^2}}{a^2 u} + C$

28. $\displaystyle\int \frac{du}{(a^2 + u^2)^{3/2}} = \frac{u}{a^2 \sqrt{a^2 + u^2}} + C$

29. $\displaystyle\int u^2 \sqrt{a^2 + u^2}\, du = \frac{u}{8}(a^2 + 2u^2) \sqrt{a^2 + u^2} - \frac{a^4}{8} \ln |u + \sqrt{a^2 + u^2}| + C$

(continued on next page)

Forms Involving $\sqrt{a^2 - u^2}$

30. $\int \sqrt{a^2 - u^2}\, du = \frac{u}{2}\sqrt{a^2 - u^2} + \frac{a^2}{2} \arcsin \frac{u}{a} + C$

31. $\int \frac{\sqrt{a^2 - u^2}}{u^2}\, du = -\frac{1}{u}\sqrt{a^2 - u^2} - \arcsin \frac{u}{a} + C$

32. $\int u^2 \sqrt{a^2 - u^2}\, du = \frac{u}{8}(2u^2 - a^2)\sqrt{a^2 - u^2} + \frac{a^4}{8} \arcsin \frac{u}{a} + C$

33. $\int \frac{\sqrt{a^2 - u^2}}{u}\, du = \sqrt{a^2 - u^2} - a \ln \left| \frac{a + \sqrt{a^2 - u^2}}{u} \right| + C$

34. $\int \frac{u^2\, du}{\sqrt{a^2 - u^2}} = -\frac{u}{2}\sqrt{a^2 - u^2} + \frac{a^2}{2} \arcsin \frac{u}{a} + C$

35. $\int \frac{du}{u\sqrt{a^2 - u^2}} = -\frac{1}{a} \ln \left| \frac{a + \sqrt{a^2 - u^2}}{u} \right| + C$

36. $\int \frac{du}{u^2 \sqrt{a^2 - u^2}} = -\frac{1}{a^2 u}\sqrt{a^2 - u^2} + C$

37. $\int \frac{du}{(a^2 - u^2)^{3/2}} = \frac{u}{a^2 \sqrt{a^2 - u^2}} + C$

38. $\int (a^2 - u^2)^{3/2}\, du = -\frac{u}{8}(2u^2 - 5a^2)\sqrt{a^2 - u^2} + \frac{3a^4}{8} \arcsin \frac{u}{a} + C$

Forms Involving $\sqrt{u^2 - a^2}$

39. $\int \sqrt{u^2 - a^2}\, du = \frac{u}{2}\sqrt{u^2 - a^2} - \frac{a^2}{2} \ln |u + \sqrt{u^2 = a^2}| + C$

40. $\int u^2 \sqrt{u^2 - a^2}\, du = \frac{u}{8}(2u^2 - a^2)\sqrt{u^2 - a^2} - \frac{a^4}{8} \ln |u + \sqrt{u^2 - a^2}| + C$

41. $\int \frac{\sqrt{u^2 - a^2}}{u^2}\, du = -\frac{\sqrt{u^2 - a^2}}{u} + \ln |u + \sqrt{u^2 - a^2}| + C$

42. $\int \frac{u^2\, du}{\sqrt{u^2 - a^2}} = \frac{u}{2}\sqrt{u^2 - a^2} + \frac{a^2}{2} \ln |u + \sqrt{u^2 - a^2}| + C$

43. $\int \frac{\sqrt{u^2 - a^2}}{u}\, du = \sqrt{u^2 - a^2} - a \arccos \frac{a}{u} + C$

44. $\int \frac{du}{\sqrt{u^2 - a^2}} = \ln |u + \sqrt{u^2 - a^2}| + C$

45. $\int \frac{du}{u^2 \sqrt{u^2 - a^2}} = \frac{\sqrt{u^2 - a^2}}{a^2 u} + C$

46. $\int \frac{du}{(u^2 - a^2)^{3/2}} = -\frac{u}{a^2 \sqrt{u^2 - a^2}} + C$

Forms Involving $a + bu$

47. $\int \frac{u\, du}{a + bu} = \frac{1}{b^2}(a + bu - a \ln |a + bu|) + C$

48. $\int u\sqrt{a + bu}\, du = \frac{2}{15b^2}(3bu - 2a)(a + bu)^{3/2} + C$

49. $\int \frac{u^2\, du}{a + bu} = \frac{1}{2b^3}[(a + bu)^2 - 4a(a + bu) + 2a^2 \ln |a + bu|] + C$

50. $\int \frac{du}{u(a + bu)} = \frac{1}{a} \ln \left| \frac{u}{a + bu} \right| + C$

51. $\int \frac{du}{u^2(a + bu)} = -\frac{1}{au} + \frac{b}{a^2} \ln \left| \frac{a + bu}{u} \right| + C$

52. $\int \frac{u\, du}{(a + bu)^2} = \frac{a}{b^2(a + bu)} + \frac{1}{b^2} \ln |a + bu| + C$

53. $\int \frac{du}{u(a + bu)^2} = \frac{1}{a(a + bu)} - \frac{1}{a^2} \ln \left| \frac{a + bu}{u} \right| + C$

54. $\int \frac{u^2\, du}{(a + bu)^2} = \frac{1}{b^3}\left(a + bu - \frac{a^2}{a + bu} - 2a \ln |a + bu|\right) + C$

55. $\int \frac{u\, du}{\sqrt{a + bu}} = \frac{2}{3b^2}(bu - 2a)\sqrt{a + bu} + C$

56. $\int \frac{u^2\, du}{\sqrt{a + bu}} = \frac{2}{15b^3}(8a^2 + 3b^2 u^2 - 4abu)\sqrt{a + bu} + C$

(continued on back page)

THE OREGON STATE UNIVERSITY
CALCULUS CONNECTIONS PROJECT

Calculus of Several Variables

Thomas P. Dick

Charles M. Patton

PWS Publishing Company

I(T)P *An International Thomson Publishing Company*
Boston • Albany • Bonn • Cincinnati • Detroit • London • Madrid • Melbourne • Mexico City • New York • Paris • San Francisco • Singapore • Tokyo • Toronto • Washington

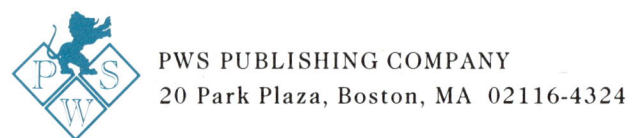

PWS PUBLISHING COMPANY
20 Park Plaza, Boston, MA 02116-4324

Copyright © 1995 by PWS Publishing Company, a division of International Thomson Publishing Inc. All rights reserved. No part of this book may be reproduced, stored in a retrieval system, or transcribed, in any form or by any means -- electronic, mechanical, photocopying, recording, or otherwise -- without the prior written permission of PWS Publishing Company.

International Thomson Publishing

The trademark ITP is used under license

For more information, contact:

PWS Publishing Co.
20 Park Plaza
Boston, MA 02116

International Thomson Publishing Europe
Berkshire House 168-173
High Holborn
London WC1V 7AA
England

Thomas Nelson Australia
102 Dodds Street
South Melbourne, 3205
Victoria, Australia

Nelson Canada
1120 Birchmount Road
Scarborough, Ontario
Canada M1K 5G4

International Thomson Editores
Campos Eliseos 385, Piso 7
Col. Polanco
11560 Mexico D.F., Mexico

International Thomson Publishing GmbH
Konigswinterer Strasse 418
53227 Bonn, Germany

International Thomson Publishing Asia
221 Henderson Road
#05-10 Henderson Building
Singapore 0315

International Thomson Publishing Japan
Hirakawacho Kyowa Building, 31
2-2-1 Hirakawacho
Chiyoda-ku, Tokyo 102
Japan

Library of Congress Cataloging-in-Publication Data

Dick, Thomas P.
    Calculus of several variables / Thomas P. Dick, Charles M. Patton.
       p.   cm.
    At head of title: The Oregon State University, Calculus Connections Project
    Includes index.
    ISBN 0-534-94716-6
    1. Calculus.   I. Patton, Charles M. II. Oregon State University.
Calculus Connections Project. III. Title.
    QA303.D595   1995
515'.84--dc20   94-39282
                                                        CIP

Sponsoring Editor Steve Quigley
Editorial Assistant John Ward
Marketing Manager Marianne C. P. Rutter
Production Editor Helen Walden
Manufacturing Coordinator Ellen Glisker
Text Printer/Binder Courier / Westford
Cover Printer Henry N. Sawyer Company
Cover Image Stephen Hunt

Printed and bound in the United States of America.
94 95 96 97 98 - 10 9 8 7 6 5 4 3 2 1

Contents

10 Infinite Processes -- Sequences and Series   619

- **10.1** SEQUENCES   619
- **10.2** SERIES   636
- **10.3** CONVERGENCE AND DIVERGENCE OF SERIES   646
- **10.4** MORE TESTS OF CONVERGENCE -- CONDITIONAL AND ABSOLUTE CONVERGENCE   656
- **10.5** POWER SERIES   668
- **10.6** ITERATION   679

11 Fundamentals of Vectors   693

- **11.1** REPRESENTATIONS OF VECTORS   694
- **11.2** ALGEBRA OF VECTORS   701
- **11.3** DOT PRODUCTS OF VECTORS   708
- **11.4** MATRICES   723
- **11.5** GEOMETRIC APPLICATIONS OF VECTORS IN R^3 732
- **11.6** EQUATIONS OF LINES AND PLANES IN R^3 740

12 Calculus of Curves   749

- **12.1** POSITION FUNCTIONS AND GRAPHICAL REPRESENTATIONS   749
- **12.2** DIFFERENTIAL CALCULUS OF CURVES   758
- **12.3** PROPERTIES OF DERIVATIVES   768
- **12.4** INTEGRATION AND ARC LENGTH   775
- **12.5** NORMAL VECTORS TO A CURVE   785
- **12.6** CURVATURE   792

13 Fundamentals of Multivariable Functions   801

- **13.1** MULTIVARIABLE FUNCTIONS - EXAMPLES AND TERMINOLOGY   802
- **13.2** VISUALIZING AND INTERPRETING MULTIVARIABLE FUNCTIONS   815
- **13.3** LINEAR AND QUADRATIC MULTIVARIABLE FUNCTIONS   829

14 Differential Calculus of Multivariable Functions   845

- **14.1** PARTIAL DERIVATIVES   846
- **14.2** THE TOTAL DERIVATIVE   854
- **14.3** THE GRADIENT   864
- **14.4** HIGHER ORDER PARTIAL DERIVATIVES   878
- **14.5** FINDING EXTREMA OF MULTIVARIABLE FUNCTIONS -- OPTIMIZATION   885
- **14.6** THE METHOD OF LAGRANGE MULTIPLIERS -- EXTREMA UNDER CONSTRAINTS   895

15 *Integral Calculus of Multivariable Functions*   903

- **15.1** MULTIPLE INTEGRALS   904
- **15.2** DOUBLE INTEGRALS OVER MORE GENERAL REGIONS   915
- **15.3** DOUBLE INTEGRALS IN POLAR COORDINATES   927
- **15.4** TRIPLE INTEGRALS OVER MORE GENERAL REGIONS   932
- **15.5** CYLINDRICAL AND SPHERICAL COORDINATES   937
- **15.6** NUMERICAL TECHNIQUES FOR MULTIPLE INTEGRALS   948

16 *Vector Analysis*   959

- **16.1** DERIVATIVES OF VECTOR FIELDS - DIVERGENCE AND CURL   960
- **16.2** TOTAL DERIVATIVE OF A VECTORED FIELD --JACOBIANS   972
- **16.3** LINE INTEGRALS   985
- **16.4** PARAMETRIZED SURFACES   998
- **16.5** SURFACE INTEGRALS   1008

17 *Fundamental Theorems of Vector Calculus*   1021

- **17.1** CONSERVATIVE FIELDS AND POTENTIALS   1021
- **17.2** STOKES' THEOREM   1033
- **17.3** GREEN'S THEOREM   1043
- **17.4** THE DIVERGENCE THEOREM -- GAUSS' THEOREM   1049

Appendices   A 1

- **A.1** TRIANGLE TRIGONOMETRY   A 2
- **A.2** TECHNIQUES OF INTEGRATION   A 10
- **A.3** METHOD OF PARTIAL FRACTIONS   A 21
- **A.4** POLAR COORDINATES   A 29
- **A.5** COMPLEX NUMBERS   A 46
- **A.6** TAYLOR'S THEOREM   A 56

DIFFERENTIATION PRACTICE   A 60
INTEGRATION PRACTICE   A 62
USEFUL FACTS AND FORMULAS   A 63
FORMULAS FROM ALGEBRA   A 66
TRIGONOMETRIC IDENTITIES   A 68
INVERSE TRIGONOMETRIC FUNCTIONS   A 70
FORMULAS FROM ANALYTIC GEOMETRY   A 71

Answers To Selected Exercises   A 115

Index   I 1

Preface to the Instructor

Curriculum revision is generally a process of gradual evolution of scope and sequence, occasionally punctuated by calls for more fundamental changes in content or delivery. The launching of Sputnik precipitated such a call for reform in mathematics education in the 1950s. We now find ourselves in the midst of a new period of widespread revitalization efforts in mathematics curriculum and instruction. A forward-looking vision of the entire K-12 mathematics curriculum is outlined in the *Curriculum and Evaluation Standards* of the National Council of Teachers of Mathematics. The Mathematical Sciences Education Board has made an eloquent and urgent case for revitalizing mathematics instruction at all levels in preparation for our country's future workforce needs in *Everybody Counts*. Both of these influential documents recognize the emergence of sophisticated computer and calculator technology as redefining the tools of mathematics education.

Calculus occupies a particularly critical position in mathematics education as the gateway to advanced training in most scientific and technical fields. It is fitting that calculus should receive particular attention as we prepare for the needs of the twenty-first century. The Sloan Conference (Tulane, 1986) and the Calculus for a New Century Conference (Washington, 1987) sounded the call for reform in the calculus curriculum. Now the entire introductory course in calculus is being reexamined under the closest scrutiny that it has received in several years. Through a special funding initiative, the National Science Foundation has made resources available for a variety of calculus curriculum revision efforts to be tried and implemented. The Calculus Connections Project is one of these NSF funded efforts, and this book is a major result of the project.

MAJOR THEMES

The text does not differ radically from a traditional calculus text in terms of major topics. This is as it should be - calculus reform will not change the importance and vitality of the major ideas of calculus, and any wholesale departure from those ideas should be viewed with great skepticism. What *is* possible is a fresh approach to these important ideas in light of the availability of modern technology. In particular, the technology can invite us to change or adopt new emphases in instruction.

Making intelligent use of technology

Computer algebra systems, spreadsheets, and graphing calculators are just a few of the readily available technological tools providing students with new windows of understanding and new opportunities for applying calculus. However, technology should not be viewed as a panacea for calculus instruction. This book seeks to take advantage of these new tools, while at the same time alerting the student to their inherent limitations and the care that must be taken to use technology wisely.

While being technology-aware, the text itself does not assume the availability of any particular machine or software. To do so would invite immediate obsolescence and ignore how quickly technology advances. Rather, the text adopts a language appropriate for the kinds of numerical, graphical, and symbolic capabilities that are found (and will continue to be found) on a wide variety of computer software packages and sophisticated calculators. For example, the language of "zooming in" on the graph of a function is powerfully suggestive without the need for listing specific keystrokes or syntax.

Technology can provide students new opportunities for understanding calculus, but it must be used with care. Numerical computations performed by a machine are subject to magnitude and precision limitations. For example, the calculation of difference quotients is naturally prone to cancellation errors. Machine-generated graphs can also provide misleading information, since graphs consist of discrete collections of pixels whose locations are computed numerically. Symbolic algebra results need to be interpreted in context. Helping students understand the limitations of technology is a major goal of this text. Students are reminded of the care that must be taken to make intelligent use of technology without becoming a victim of its pitfalls.

Since no specific hardware or software is assumed, an instructor will need to judge the appropriateness of any particular activity in light of the technology available. However, the exercises are designed to be compatible with a very wide variety of available software and hardware. A graphing calculator will be adequate for most of the activities.

Multiple representation approach to functions

The most important concept in all of mathematics is that of *function*, and the function concept is central in calculus. The idea of a function as a process accepting inputs and returning outputs can be captured in a variety

PREFACE TO THE INSTRUCTOR

of representations: numerically as a table of input-output pairs, graphically as a plot of outputs vs. inputs, and symbolically as a formula describing or modeling the input-output process. The interpretations of the core calculus topics of limits and continuity, differentiation, and integration all have different flavors when approached through different representations. The connections we forge among them enrich our personal concept of function.

All too often, students leave the calculus course with an impoverished mental image of function formed in a context dominated by symbolic forms. This book seeks to take a more balanced three-fold approach to functions. With each new topic or result, an explicit effort is made to interpret the meaning and consequences in a numerical, graphical, and symbolic context. Such an approach does not require technology, but the availability of an appropriate device allows us greater access to numerical and graphical representations, while at the same time reducing the need for heavy emphasis of rote "by hand" symbol manipulation skills.

Visualization and approximation

Two themes that become increasingly important with the availability of technology are visualization and approximation. The ability to obtain a machine-generated graph as a first step instead of a last one can completely turn around our approach to a variety of calculus topics. Graphical interpretation skills become primary. In particular, graphing can be used as a powerful problem-solving aid, both in estimating and in monitoring the reasonableness of results obtained numerically or symbolically. Whenever possible, explicit mention is made of the visual interpretation of definitions, theorems, and example solutions, often with direct reference to machine-generated graphs.

Much of calculus grew out of problems of approximation, and many of the key concepts of calculus are best understood as limits of approximations. Numerical tools make once exorbitantly tedious calculations into viable computational estimation strategies. Accordingly, approximation and estimation techniques are given a high priority throughout the text.

Overview of the materials

Chapter 10 opens with a discussion of long division as a familiar example of an infinite process that can produce an infinite sequence of approximations. Several examples of sequences, including recursive and iterative sequences, are examined. Root-finding methods, including the bisection method and Newton's method, are also included as examples of iterative techniques yielding sequences of approximations. The Archimedean property of real numbers and Zeno's paradox are used to motivate the idea of a series. A series is then defined as the limit of a sequence of partial sums. Tests of convergence include the Nth term test, comparison and limit comparison tests, the integral test, the alternating series test, and the root and ratio tests. Absolute and conditional convergence are contrasted. After a discussion of power series, including interval and radius of convergence, this chapter concludes with a closer look at iterative methods in general.

Chapter 11 covers the fundamentals of vectors, including a section on matrices. The distinction between vector and scalar quantities is emphasized, and the basics of vector arithmetic are treated from both analytic and geometric viewpoints. Most of the discussion focuses on vectors in two and three dimensions, but generalizations to vectors in n dimensions are also indicated where appropriate. The section on dot products also introduces the notions of norms and vector components. Following the section on matrix algebra and determinants, the cross product and triple scalar products are introduced. The chapter concludes with applications of vectors to analytic geometry, and the final section discusses parametric equations for lines and planes.

Chapter 12 introduces the notion of a vector-valued function of a single variable. Such a function can generate the curve traced out by a moving object when the independent variable is considered as representing time and the output as representing the position of the object. The first section takes care to emphasize the distinction between a curve (a locus of points in the plane or space) and the many possible parametrizations (position functions) of that curve. While few graphing calculators include built-in capabilities for displaying curves in space, any graphing calculator can be used effectively for displaying three views (projections) of a space curve. Limits, continuity, and differentiability of position functions are treated in the next section, and the derivative is interpreted in both geometric (tangent) and physical (velocity) contexts. Following a section on the key properties of the derivative of a position function, integration of position functions is introduced and applied to finding the arc length of a curve. The next section includes material on the principal unit normal vector, the osculating plane, the unit binormal vector, and torsion. The chapter concludes with a section on curvature.

PREFACE TO THE INSTRUCTOR

Chapter 13 shifts the emphasis from vector-valued functions of one variable to scalar-valued functions of several variables. The first section provides several examples of functions and relations of several variables, including the important class of quadric surfaces. Techniques for visualizing and interpreting multivariable functions are the subject of the next section. It should be noted that cross-sections or "slices" of the graph of a function of any number of variables can be obtained with a graphing calculator by "freezing" all of the independent variables but one and graphing the resulting function of one variable. Contour plots of level curves and wireframe plots provide other visual aids for functions of several variables. The chapter concludes with a detailed analysis of linear and quadratic functions of two variables.

Chapter 14 treats the differential calculus of multivariable functions. Partial derivatives are interpreted both geometrically in terms of the slopes of the "slices" of a function's graph and physically in terms of the instantaneous rate of change of a function when only one variable is allowed to change. The total derivative (of a C^1-differentiable function is then defined and interpreted in terms of the approximately locally linear behavior of the function. In particular, the total derivative is used to find the best linear approximation of a function of several variables. The gradient is then introduced and applied to finding directional derivatives, direction of fastest change, the rate of change along a path, and the equation of a tangent plane to the graph of a function. Higher order derivatives are used to develop the best quadratic approximation to a function of several variables, and this in turn is used to motivate the second derivative test for functions of two variables. The chapter concludes with applications to finding the extrema of functions, both over closed domains and under constraints (using the method of Lagrange multipliers).

Chapter 15 treats the integral calculus of multivariable functions. Multiple integrals are interpreted both geometrically and physically, first over rectangular regions of integration, and then over more general regions. Integration using polar coordinates over plane regions, and using cylindrical or spherical coordinates over regions of space are also discussed. The chapter concludes with a section on numerical techniques of multiple integration, including Monte Carlo methods.

Chapter 16 introduces the notion of a vector field and discusses the fundamentals of vector analysis. Velocity flows and force fields provide two important physical examples of vector fields. The first section defines the divergence and curl of a vector field and interprets their meaning in physical contexts. The corresponding ideas of incompressible or irrotational vector fields are also discussed. The section concludes with a discussion of the Laplacian. The total derivative of a vector field is defined and Jacobian matrices are applied to change of variables in multiple integration. Chapter 16 then turns to a detailed discussion of line integrals over parametrized curves and surface integrals over parametrized surfaces.

Chapter 17 is the final chapter of the book, and fittingly provides the major "punchlines" of vector calculus--- the fundamental theorems. The relationship between conservative force fields and potential functions is developed and results in a fundamental theorem of calculus for line integrals. The fundamental theorems of Stokes, Green, and Gauss are the centerpieces of each of the final sections of the chapter.

The appendices provide review material on trigonometry as well as additional material on techniques of integration (including the method of partial fractions), polar coordinates, complex numbers, and Taylor's formula. The appendices conclude with additional practice exercises for differentiation and integration. Besides short answers to almost all the odd-numbered exercises, several useful formulas and tables can be found in the end pages of the book.

Corrections, comments, and criticisms of the materials are welcomed, and can be directed to the authors.

ANCILLARIES

The **Instructor's Resource Manual** provides answers to all the exercises, a pool of test items for each chapter (with answers), and additional commentary on goals and philosophy of the book, suggestions for pacing, and section-by-section notes to aid instructors using the materials.

The **Student's Resource Manual** is a supplement that provides detailed answers to selected exercises, as well as programs for graphing calculators that should prove useful in the sections on numerical techniques. If your school makes use of a computer algebra system in teaching calculus, or if students have access to such software -- Mathematica, Maple, Derive, or Theorist -- other resources are available from PWS Publishing Company. **Notebooks** have been prepared to accompany this book for each of these computer algebra systems. Each includes examples of step-by-step worked exercises from the text. Contact your bookstore for more information.

Preface to the Student

Books are written to be read. Yes, that is true even of mathematics books! We strongly encourage you to read the chapter introductions and the explanations in each section carefully, and to follow closely the discussion of examples. Perhaps you are accustomed to skipping to the exercises of a mathematics textbook first, and then searching back for an example that is a "clone" of the problem at hand that you can mimic. Certainly, this book has many examples and exercises to illustrate and help you practice your calculus skills. But there are also many problems in this text that ask you to reflect on and explain in your own words some of the important ideas of calculus. Other problems are designed to force you to think about these ideas in new ways. You may feel frustrated at times, but keep in mind that the effort you make to really understand the ideas in calculus will give you an ownership of them that will last long after you forget some of the specific technical details.

USING TECHNOLOGY TO STUDY CALCULUS

Some of the technology made possible by calculus includes devices such as graphing calculators and computers. We live in an exciting age where these powerful computational tools enable us to perform complex numerical and symbolic computations and provide tremendous graphics capabilities at our fingertips. In turn, we now have both new ways to understand the ideas of calculus and new opportunities to apply calculus.

However, even the most powerful technology is of little use if we do not know how it can and cannot be applied. We recognize that new technological tools are available and this book was written with the *intelligent* use of those tools in mind. The use of technology to study functions is not without its pitfalls. To use calculators and computers intelligently requires a knowledge of their limitations. Solving important mathematical problems will always require the inspiration, recognition, and application of the right idea at the right time.

If you have access to one of the computer algebra systems now being used to teach calculus at many schools -- Mathematica, Maple, Derive, or Theorist -- you may be interested in another problem-solving aid provided by the publisher. The *Notebooks* prepared to accompany this book are data disks comprised of worked examples and exercises from the text. The examples on each disk show, step-by-step, how to use a particular computer algebra system to solve selected problems from the text. (Contact your bookstore for more information.)

In calculus and other branches of mathematics, you will often encounter problems for which there is no specific recipe to solve them. Even in these instances, there are a variety of strategies you can use to make progress toward a solution. The next section gives you some hints from a master problem solver. We hope you find them useful.

GENERAL HINTS FOR SOLVING PROBLEMS: POLYA'S FOUR STEPS

George Polya (1887-1985) was considered by many as the greatest teacher of mathematical problem solving. In his work *How To Solve It*, Polya discusses in detail many aspects of the problem-solving process. He provides several useful general strategies (or heuristics) for mathematical problem solving. Here are the four basic steps Polya outlined in the problem-solving process.

POLYA'S FOUR STEPS IN PROBLEM SOLVING

1. **UNDERSTAND THE PROBLEM**
2. **DEVISE A PLAN**
3. **CARRY OUT THE PLAN**
4. **LOOK BACK**

Let's elaborate on these problem-solving steps.

1. UNDERSTAND THE PROBLEM

This means *understand what the problem is asking for*. While that may seem obvious, there are many times when we dive into a problem and waste a lot of time and effort that could have been saved by a few extra moments of reflection at the beginning. Ask yourself these questions: Do I understand all the terminology? What is given? What is the goal? Am I required to find something or to prove something? Is there enough information? Is there extraneous information? Have I seen a similar problem before? Rewriting the problem in your own words, drawing a figure, trying some examples are all ways to clarify a problem statement.

The full power of algebra and calculus can be unleashed if we can model a problem situation as a function or as an equation or inequality. We may be able to introduce a coordinate system for the purposes of graphing. The act of identifying and labeling variable quantities in and of itself may clarify aspects of the problem to us.

2. DEVISE A PLAN

Devise a plan of action for the problem. If you don't know where to begin, then try a general problem-solving heuristic or strategy. Three very useful heuristics include:

Trial and Error. At worst, you may get a better feel for the constraints of the problem situation. At best, you may stumble on the answer directly. Trial and error doesn't necessarily mean blind guesswork; our early guesses can help guide us in making better guesses. Making a list of the results of our trials may reveal a pattern or relationship. Mathematics is sometimes called the science or art of finding patterns.

Try a Simpler Problem. If the original problem seems too complex or confusing, try simplifying it first and solving that version. The solution to the simpler problem may give insights on how to solve the original problem. Exactly how do you make a problem simpler? Some of the ways include: substituting a smaller number in place of a larger one given in the problem; substituting a specific numerical value for an unknown constant or parameter (0 or 1 are often good substitution choices); making up a related problem that involves fewer dimensions or unknowns; and adding or dropping some of the problem constraints.

Try Extreme or Special Cases. We may get a special understanding from examining the problem situation in extreme or special cases. For example, if a problem involved the *elliptical* orbits of planets, we might benefit by considering the special case of a *circular* orbit. Substituting extreme values for an unknown variable can also give us useful information. For example, a question involving lines in the plane can be examined for the special cases of horizontal (zero slope) and vertical (undefined slope) lines. Making a list of special cases may also reveal a pattern or relationship.

3. CARRY OUT THE PLAN

Carry out your plan of action. Implement the strategy you've chosen until the problem is solved or until a new course of action is suggested. Give yourself a reasonable period of time to solve the problem. Monitor yourself.

If you feel that you've embarked on a dead-end road then consider a change of strategy. Don't be afraid of starting all over. Many times a fresh start and a new strategy lead to success. You can have a flash of insight when you least expect it!

4. LOOK BACK

Check your answer to see if it really satisfies the requirements of your problem. Looking back means more than just checking your answer, though. Also look at your method of solution. Can you see another way of coming up with the answer? Can you see how your method could be used on other problems? Look forward to how you might generalize or extend your solution.

Calculus arose in response to the need to solve certain problems, and to understand the what and why of calculus requires understanding how, when, and where calculus can be used to solve problems. Calculus provides some very powerful tools for calculating quantities related to change. The derivative provides a means of measuring rates of change, and the integral provides a means of measuring accumulated change. This book is devoted to helping you understand these fundamental ideas so that you can successfully apply calculus to solving problems.

Your study of calculus can be an exciting intellectual adventure. Good luck on your journey!

Acknowledgments

The Calculus Connections Project has been made possible with the support of the National Science Foundation, Oregon State University and the Lasells Stewart Foundation, the Hewlett-Packard Corporation, and PWS Publishing Company.

This book was prepared using Donald Knuth's TeX with Textures (Blue Sky Research) and the AMS-TeX - Version 2.0 macro package (American Mathematical Society). Thanks to Marilyn Wallace, Donna Kent, D'Anne Hammond, and Colleen Dick for their contributions to the technical typesetting of the text. The illustrations were produced using MacPaint and MacDraw II (Claris), Adobe Illustrator, PSMathgraphs II (Maryann software), and Grapher 881 (thanks to Steve Scarborough). Thanks to George Dick for his painstaking preparation and revisions of the illustrations.

A project such as this owes thanks to many people. First and foremost, we wish to thank Dr. Dianne Hart for her exemplary work throughout the life of this project, including the preparation of supporting materials for instructors, coordination of so many of the project's instructional activities, and most importantly, for her research on students' use of multiple representations and technology. Special thanks are also due to Howard L. Wilson of Oregon State University, for his exceptional instructional and in-service work with the Calculus Connections project. We would also like to acknowledge our appreciation for the efforts and support of many others:

To the mathematics department at Oregon State University, for its encouragement of the laboratory approach to calculus, and to all the faculty and graduate teaching assistants involved in the experimental calculus program at Oregon State University.

To Dianne Hart and Marie Franzosa, for their work on the instructional materials for the textbook. In particular, their many helpful suggestions on the exercises and the preparation and checking of the answers is greatly appreciated.

To past and present advisors and consultants to the project: Bert Waits (Ohio State University), Franklin Demana (Ohio State University), Gregory D. Foley (Sam Houston State University), Thomas Tucker (Colgate University), Robert Moore (University of Washington), William Wickes (Hewlett-Packard), John Kenelly (Clemson University), Don LaTorre (Clemson University), and Jeanette Palmiter (Portland State University).

Many people reviewed and/or pilot tested the preliminary edition drafts and revisions. We wish to acknowledge them for their many useful suggestions and constructive criticisms.

Nacer Abrouk	Rose-Hulman Institute of Technology
Nancy Baggs	University of Colorado - Colorado Springs
Maureen A. Bardwell	Westfield State College
Barry Bergman	Clackamas County Community College
Marcelle Bessman	Frostburg State University
E. E. Burniston	North Carolina State University
Herb Brown	SUNY at Albany
Dan Chiddix	Ricks College
Mary Louise Collette	Mount St. Mary's College
Debra Crawford	Riverside High School
Carol Crawford	United States Naval Academy
Deborah Crocker	Miami University (Ohio)
Catherine Curtis	Mt. Hood Community College
Gerald Daniels	Mitchell High School
Wade Ellis	West Valley College
Kevin Fitzpatrick	Greenwich High School
William Francis	Michigan Technological University
Dewey Furness	Ricks College
Charles Geldaker	Lakeridge High School
Anthony J. Giovannitti	University of Southern Mississippi
Ron Goolsby	Winthrop University
Mary Ann Gore	Warner Robins High School
Samuel Gough	East Meklenburg High School
Karen Graham	University of New Hampshire
Murli Gupta	George Washington University
James Hall	Westminster College
Donnie Hallstone	Green River Community College
Betty Hawkins	Shoreline Community College
Warren Hickman	Westminster College
Mark Howell	Gonzaga College High School
Howard Iseri	Mansfield University
Gary S. Itzkowitz	Glassboro State College
William Kiele	United States Air Force Academy
Elaine Klett	Brookdale Community College
Paul Latiolais	Portland State University
Tom R. Lucas	University of North Carolina - Charlotte
Lewis Lum	University of Portland
Mary Martin	Winthrop College
Marian McCain	Centennial High School
Joan McCarter	Arizona State University
Richard Metzler	University of New Mexico
Dennis Mick	Carroll College
Teresa Michnowicz	Jersey City State College
Laura Moore-Mueller	Green River Community College
Lawrence Morgan	Montgomery County Community College
Stephen Murdock	Tulsa Junior College
Mel Noble	Olympic High School
Michele Olsen	College of the Redwoods
John Oman	University of Wisconsin - Oshkosh
Robert Piziak	Baylor University

ACKNOWLEDGMENTS

Priscilla Putman-Haindl	Jersey City State College
William Raddatz	Linfield College
Carla Randall	Lake Oswego High School
Laurel Rogers	University of Colorado - Colorado Springs
Audrey Rose	Tulsa Junior College
Donald Rossi	DeAnza College
David Royster	University of North Carolina - Charlotte
G.T. Springer	Alamo Heights High School
Larry Sternberger	Tulsa Junior College
Ted Sundstrom	Grand Valley State University
W. Todd Timmons	Westark Community College
Sandra Vrem	College of the Redwoods
John Whitesitt	Southern Oregon State College
Kei Yasuda	Lane Community College

To the hundreds of other instructors and the thousands of students who used the published preliminary edition, for their valuable feedback. Special thanks to Lewis Lum, William Kiele, and James Hall, for their very detailed reviews of the book, and for the improvements in exposition and problems they suggested to us.

To the editorial and production staff at PWS Publishing Company, for their support and expertise in helping disseminate the work of the project: Leslie Bondaryk, David Dietz, Steve Quigley, Barbara Lovenvirth, Helen Walden, and John Ward.

Finally, to Leslie and Colleen, Daniel, Jean, Connor, Eamon, and Eleanor, for enduring the authors throughout this project with love and support, we dedicate this book.

Thomas P. Dick
Charles M. Patton

Corvallis, Oregon

About the Authors

Calculus of a Single Variable is the product of five years of intensive experimentation, writing, and revision by Drs. Thomas Dick and Charles Patton. The textbook is the outgrowth of the authors' interest in the teaching and learning of calculus, and how emerging technologies can be intelligently used to enhance students' understanding of calculus.

Thomas Dick is the director of the Calculus Connections Project, a program funded by the National Science Foundation as part of a major effort to revise the calculus curriculum in preparation for the needs of a new century. He earned his Ph.D. at the University of New Hampshire, where he served as the coordinator of the UNH Calculus Testing Center for two years before joining the mathematics faculty of Oregon State University. In 1988 he received the Carter Award for outstanding and inspirational undergraduate teaching in Oregon State University's College of Science. He has been widely involved in high school and college teacher education, particularly in calculus and the use of technology in the mathematics classroom. He has served both as a reader and as a consultant for the Advanced Placement Program in Mathematics, and has directed several workshops and institutes for teachers and instructors. Dr. Dick is on the editorial board of the Journal of Computers in Mathematics and Science Teaching, and has been appointed a member of the Committee on Research in Undergraduate Mathematics Education by the American Mathematical Society. He has written extensively on the use of technology in the mathematics classroom, including papers for *The Computing Teacher*, the Mathematical Association of America's *Calculus as a Laboratory Course*, and the National Council of Teachers of Mathematics 1992 Yearbook on *Calculators in Mathematics Education*.

Charles Patton earned his Ph.D. at the State University of New York at Stonybrook, studying also at the Mathematical Institute, Oxford, and Institute des Hautes Etudes Scientifique, France. His doctoral work was in the mathematical foundations of relativity and quantum mechanics. He was awarded an American Mathematical Society Postdoctoral Research Fellowship and has been a member of the Institute for Advanced Study, Princeton. After teaching mathematics at the University of Utah, Dr. Patton joined the Calculator Research and Development team for Hewlett-Packard, where he has been instrumental in the first implementation of a computer algebra system on hand-held computers. He has been awarded several patents for his work related to computational systems for computer algebra. Dr. Patton has numerous publications on representation theory and twistor theory in such journals as the *Transactions* and the *Bulletin of the American Mathematical Society*.

10

Infinite Processes—Sequences and Series

Approximations and infinite processes are two of the distinguishing features of calculus, and in many ways the ideas of approximation and infinite processes go hand in hand.

The idea of an infinite process is a recurring central theme in calculus. The derivative can be thought of as the limiting value of an infinite sequence of secant line slopes. The definite integral is the limiting value of an infinite sequence of Riemann sums. In this chapter, we study infinite processes and sequences in more detail.

With any sequence of real numbers is associated a *series*, which can be thought of as the "sum" of all the terms in the sequence. Note that a decimal number can be thought of as a series. For example,

$$\frac{1}{3} = 0.3 + 0.03 + 0.003 + 0.0003 + \cdots$$

Sequences can often be described iteratively. You have seen two examples of iterative processes already: the bisection method (Chapter 2) and Newton's method (Chapter 4) for approximating roots of functions.

A polynomial is a sum of terms that are multiples of powers of x. Just as a series can be thought of as the "sum" of infinitely many real numbers, a *power series* can be thought of as a "polynomial" with infinitely many terms. You will see that power series can be used to define functions. The limit of the infinite sequence of Taylor polynomial approximations is called a *Taylor series*.

10.1 SEQUENCES

To get some feel for the ideas involved in sequences and series, let's look at a familiar example. Suppose that you want to express the fraction 22/7 as a

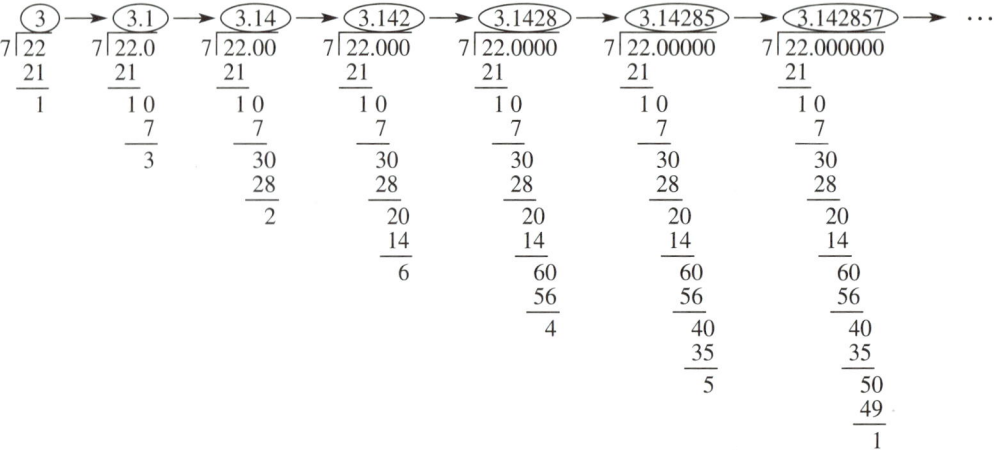

Figure 10.1 Long division as an infinite process.

decimal. As we perform the long division of 22 by 7 as shown in Figure 10.1, we get a sequence of partial results

$$a_1 = 3, \ a_2 = 3.1, \ a_3 = 3.14, \ a_4 = 3.142, \ a_5 = 3.1428, \ldots.$$

Since 22/7 is reduced to lowest terms and its denominator 7 is not a factor of any power of 10, we know that this long division process will never end. We say that the decimal representation of 22/7 does not terminate. We can think of the partial results as forming an infinite sequence of decimal approximations, each one closer to the exact value 22/7.

When you first encountered such a long division problem years ago, the realization that the process would never end may have troubled you, since it means that it is impossible to write down the "complete" decimal expansion of 22/7. Nevertheless, this infinite sequence of results somehow does have a definite "end," in the sense that it represents 22/7. We could say that the real number 22/7 is the *limit* of the sequence. Similarly, the real number $\sqrt{2}$ is the limit of the sequence

$$b_1 = 1, \ b_2 = 1.4, \ b_3 = 1.41, \ b_4 = 1.414, \ b_5 = 1.4142, \ldots.$$

You can generally think of a *sequence* as an *infinitely long list* of numbers (or any other objects, for that matter). We call the individual numbers in a sequence the **terms**, and we identify them by their position in the list. For example, in the sequence

$$a_1 = 3, \ a_2 = 3.1, \ a_3 = 3.14, \ a_4 = 3.142, \ a_5 = 3.1428, \ \ldots,$$

10.1 SEQUENCES

we call a_1 the first term, a_2 the second term, a_3 the third term, and so on. In general, a_n is called the nth term of the sequence. The subscript n used to label the terms of the sequence is sometimes called the **index**.

Technically speaking, a sequence of real numbers can be considered a function from the set of positive integers $\mathbb{N} = \{1, 2, 3, \ldots\}$ to the set of real numbers $\mathbb{R}$. In other words, the numbers in the list

$$a_1, \ a_2, \ a_3, \ \ldots, \ a_n, \ \ldots$$

can be viewed as the outputs

$$f(1), \ f(2), \ f(3), \ \ldots, \ f(n), \ \ldots$$

of a function $$f : \mathbb{N} \to \mathbb{R}.$$

> The notation for a sequence is $\{a_n\}_{n=1}^{\infty}$.

Examples of sequences

The **harmonic sequence** consists of the reciprocals of the positive integers in order:

$$1, \ \frac{1}{2}, \ \frac{1}{3}, \ \frac{1}{4}, \ \ldots, \ \frac{1}{n}, \ \ldots.$$

The positive integers themselves form a sequence: $1, 2, 3, 4, \ldots$.

The values appearing as terms in a sequence need not all be distinct. The sequence

$$0, \ 1, \ 2, \ 0, \ 1, \ 2, \ 0, \ 1, \ 2, \ \ldots$$

has infinitely many terms (as does any sequence), but only three distinct values. It's quite possible to have a **constant sequence** such as

$$-\frac{5}{7}, \ -\frac{5}{7}, \ -\frac{5}{7}, \ -\frac{5}{7}, \ \ldots,$$

with only one value appearing.

If the signs of the terms in a sequence alternate between positive and negative, we call it an **alternating sequence**. The **alternating harmonic sequence** is

$$1, \ -\frac{1}{2}, \ \frac{1}{3}, \ -\frac{1}{4}, \ \frac{1}{5}, \ -\frac{1}{6}, \ \ldots$$

Describing sequences

As we mentioned above, a sequence of real numbers can be thought of as a function assigning a real number a_n to each positive integer index n. Certainly, it's impossible to write down all the terms in a sequence. However, we could say that we "know" a sequence if we have some means by which we can generate the value of any term in the sequence.

The most convenient way this could be accomplished is if we have a *formula*

$$a_n = f(n)$$

providing the value a_n in terms of n. In this case, we can write

$$\{f(n)\}_{n=1}^{\infty}.$$

EXAMPLE 1 Find a formula describing the harmonic sequence

$$1, \frac{1}{2}, \frac{1}{3}, \frac{1}{4}, \ldots$$

Solution   If we use a_1 to represent the first term, we can note that

$$a_1 = 1, \ a_2 = \frac{1}{2}, \ a_3 = \frac{1}{3}, \ a_4 = \frac{1}{4}, \ \ldots$$

In general, $a_n = f(n) = \frac{1}{n}$, so we can write the harmonic sequence

$$\{\frac{1}{n}\}_{n=1}^{\infty}.$$ ∎

EXAMPLE 2 Find a formula describing the alternating harmonic sequence

$$1, -\frac{1}{2}, \frac{1}{3}, -\frac{1}{4}, \frac{1}{5}, -\frac{1}{6}, \ldots$$

Solution   Note that $a_1, a_3, a_5, \ldots$ are all positive and $a_2, a_4, a_6, \ldots$ are all negative. We need some means of describing these alternating signs. The factor $(-1)^{n+1}$ works well for this purpose. When n is odd ($n = 1, 3, 5, 7, \ldots$), then $n+1$ is even and $(-1)^{n+1} = 1$. When n is even ($n = 2, 4, 6, 8, \ldots$), then $n+1$ is odd and $(-1)^{n+1} = -1$. In general, $a_n = f(n) = \frac{(-1)^{n+1}}{n}$, and we can write the alternating harmonic sequence as

$$\{\frac{(-1)^{n+1}}{n}\}_{n=1}^{\infty}.$$ ∎

10.1 SEQUENCES

EXAMPLE 3 Find a formula describing the constant sequence

$$-\frac{5}{7}, -\frac{5}{7}, -\frac{5}{7}, \ldots$$

Solution Every term $a_n = -5/7$, so we can simply write

$$\left\{-\frac{5}{7}\right\}_{n=1}^{\infty}.$$

Occasionally, it is convenient to have the index n start out with a value other than $n = 1$. In particular, $n = 0$ is often a starting index value for a sequence. For example, the sequence of powers of two

$$1, 2, 4, 8, 16, 32, \ldots,$$

can be written as

$$\{2^n\}_{n=0}^{\infty},$$

since $2^0 = 1$, $2^1 = 2$, $2^2 = 4$, and so on.

If you are trying to find a closed-form formula for a sequence, you certainly should check that the formula generates the correct values for some specific index values n.

▶ ▶ ▶ **We should be careful to distinguish between a sequence and an infinite set. A sequence has a definite ordering of its terms, while a set is simply a collection of objects with no particular ordering.**

Arithmetic sequences

One important type of sequence is an **arithmetic sequence**. An arithmetic sequence is distinguished by the fact that the *difference of any two consecutive terms is constant*. In other words, given any term, we obtain the next term by adding a specific constant. The following are examples of arithmetic sequences:

$$1, 2, 3, 4, 5, \ldots \quad \text{(constant difference} = 1)$$
$$1, 3, 5, 7, 9, \ldots \quad \text{(constant difference} = 2)$$
$$17, -22, -27, -32, \ldots \quad \text{(constant difference} = -5).$$

An arithmetic sequence can be spotted by examining the difference between consecutive terms to see if it is constant. We can find a formula describing any arithmetic sequence, provided that we know the value of the initial term and the constant difference. It is convenient to start with the index $n = 0$ so that a_0 is the initial term. If d is the constant difference, then a formula for the arithmetic sequence is

$$\{a_0 + dn\}_{n=0}^{\infty}.$$

The formula $a_0 + dn$ generates

$$a_0,\ a_0 + d,\ a_0 + 2d,\ a_0 + 3d,\ \ldots$$

for $n = 0, 1, 2, 3, \ldots$.

EXAMPLE 4 Find formulas for each of the arithmetic sequences described above.

Solution For the first sequence $1, 2, 3, \ldots$, we have $a_0 = 1$ and $d = 1$, so we can write

$$\{1 + n\}_{n=0}^{\infty}.$$

For the second sequence $1, 3, 5, 7, 9, \ldots$, we have $a_0 = 1$ and $d = 2$, so we can write

$$\{1 + 2n\}_{n=0}^{\infty}.$$

For the third sequence $-17, -22, -27, -32, \ldots$, we have $a_0 = -17$ and $d = -5$, so we can write

$$\{-17 - 5n\}_{n=0}^{\infty}.$$ ∎

Geometric sequences

A **geometric sequence** is distinguished by the fact that the *ratio of any two consecutive terms is constant.* In other words, given any term, we obtain the next term by multiplying by a specific constant. The following are examples of geometric sequences:

$$1, 2, 4, 8, 16, \ldots \quad \text{(constant ratio} = 2\text{)}$$

$$3, \frac{3}{2}, \frac{3}{4}, \frac{3}{8}, \frac{3}{16}, \ldots \quad \text{(constant ratio} = \frac{1}{2}\text{)}$$

$$-7, \frac{7}{3}, -\frac{7}{9}, \frac{7}{27}, -\frac{7}{81}, \ldots \quad \text{(constant ratio} = -\frac{1}{3}\text{)}.$$

A geometric sequence can be spotted by examining the ratio of consecutive terms to see if it is constant. We can find a formula for any geometric

10.1 SEQUENCES

sequence provided that we know the value of the initial term and the constant ratio. Again, it is convenient to start with the index $n = 0$ so that a_0 is the initial term. If r is the constant ratio, we can write the geometric sequence as

$$\{a_0 r^n\}_{n=0}^{\infty}.$$

The formula $a_0 r^n$ generates

$$a_0,\ a_0 r,\ a_0 r^2,\ a_0 r^3,\ \ldots$$

for $n = 0, 1, 2, 3, \ldots$.

EXAMPLE 5 Find formulas for each of the geometric sequences described above.

Solution For the first sequence $1, 2, 4, 8, 16, \ldots$, we have $a_0 = 1$ and $r = 2$, so we can write

$$\{2^n\}_{n=0}^{\infty}.$$

For the second sequence $3, \frac{3}{2}, \frac{3}{4}, \frac{3}{8}, \frac{3}{16}, \ldots$ we have $a_0 = 3$ and $r = \frac{1}{2}$, so we can write

$$\left\{\frac{3}{2^n}\right\}_{n=0}^{\infty}.$$

For the third sequence $-7, \frac{7}{3}, -\frac{7}{9}, \frac{7}{27}, -\frac{7}{81}, \ldots$ we have $a_0 = -7$ and $r = -\frac{1}{3}$, so we can write

$$\left\{\frac{-7}{(-3)^n}\right\}_{n=0}^{\infty}.$$ ∎

EXAMPLE 6 Is the harmonic sequence arithmetic, geometric, or neither?

Solution The harmonic sequence is

$$1, \frac{1}{2}, \frac{1}{3}, \frac{1}{4}, \frac{1}{5}, \ldots$$

To check whether this is an arithmetic sequence, we note that $1/2 - 1 = -1/2$, but $1/3 - 1/2 = -1/6$, so the difference between consecutive terms is not constant, and the sequence cannot be arithmetic.

To check whether this is a geometric sequence, we note that $\frac{1/2}{1} = 1/2$, but $\frac{1/3}{1/2} = 2/3$, so the ratio of consecutive terms is not constant, and the sequence cannot be geometric.

We conclude that the harmonic sequence is neither arithmetic nor geometric. ∎

Convergence and divergence of sequences

A sequence of real numbers $\{a_n\}_{n=1}^{\infty}$ is said to *converge* if the terms in the sequence eventually "stabilize" toward some single limiting value. Otherwise, we say the sequence *diverges*. Two questions we'll explore are:

1. How can you tell whether a sequence converges or diverges?
2. If a sequence converges, how can you determine the limiting value?

The following formal definition states more precisely what we mean by convergence and divergence of a sequence.

Definition 1

> A sequence $\{a_n\}_{n=1}^{\infty}$ **converges to a limit** L, provided that for any given tolerance $\epsilon > 0$, there is a specific index N for which
> $$L - \epsilon < a_n < L + \epsilon$$
> whenever $n \geq N$. In this case, we write
> $$\lim_{n \to \infty} a_n = L.$$
> If a sequence does not converge, we say it **diverges**.

You can think of N as the position at which the "tail" of the sequence has only terms that are within ϵ of L (see Figure 10.2). No matter how small a positive tolerance ϵ we are given, we must be able to find a tail of the sequence with all the terms within that tolerance of L. Since we use decimal numbers so often to represent real numbers, you may find it convenient to think of the tolerance ϵ in terms of *decimal places of agreement*: given any number of decimal places, there must be a term in the sequence after

$$a_1, a_2, a_3, \ldots, \underbrace{a_N, a_{N+1}, a_{N+2}, \ldots}_{\text{All of these terms are within } \varepsilon \text{ of the limit } L.}$$

Figure 10.2 The tail of a sequence.

10.1 SEQUENCES

which all the terms are indistinguishable from the limit when rounded to that number of decimal places.

For example, consider the sequence

$$\{a_n\}_{n=1}^{\infty} = \{(1+\frac{1}{n})^n\}_{n=1}^{\infty}.$$

Let's examine the first few terms of this sequence:

$$a_1 = (1+1)^1 = 2$$

$$a_2 = (1+\frac{1}{2})^2 = 2.25$$

$$a_3 = (1+\frac{1}{3})^3 \approx 2.37037037037$$

$$a_4 = (1+\frac{1}{4})^4 = 2.44140625000$$

$$a_5 = (1+\frac{1}{5})^5 = 2.48832000000.$$

Now, let's move further down the sequence and examine a few more terms:

$$a_{10} = (1.1)^{10} \approx 2.59374246010$$

$$a_{100} = (1.01)^{100} \approx 2.70481382942$$

$$a_{1000} = (1.001)^{1000} \approx 2.71692393224$$

$$a_{10000} = (1.0001)^{10000} \approx 2.71814592683$$

$$a_{100000} = (1.00001)^{100000} \approx 2.71826823717$$

$$a_{1000000} = (1.000001)^{1000000} \approx 2.71828046932$$

$$a_{10000000} = (1.0000001)^{10000000} \approx 2.71828169254$$

$$\vdots$$

Note how the terms in the sequence appear to be stabilizing. Indeed, as n increases by a power of 10, all the subsequent terms agree to another decimal place. That is, all the terms following a_{100} agree to the first decimal place, all the terms following a_{1000} agree to two decimal places, and so on. This behavior suggests that the sequence converges to a single limiting value. In fact, it is known that

$$\lim_{n \to \infty} (1+\frac{1}{n})^n = e \approx 2.71828182846.$$

The index $n = N$ required to guarantee that all subsequent terms agree to a given number of decimal places gives us a measure of the *speed of convergence*. If we need only a relatively small increase in the index value N to guarantee more decimal places of agreement among the subsequent terms a_N, a_{N+1}, ..., then our sequence converges quickly. If a relatively large increase in index value N is required to guarantee another decimal place of agreement, then our sequence converges slowly.

Here's another sequence that converges to e:

$$b_1 = 1$$

$$b_2 = 1 + \frac{1}{1} = 2$$

$$b_3 = 1 + \frac{1}{1} + \frac{1}{1 \cdot 2} = 2.5$$

$$b_4 = 1 + \frac{1}{1} + \frac{1}{1 \cdot 2} + \frac{1}{1 \cdot 2 \cdot 3} \approx 2.66666666667$$

$$b_5 = 1 + \frac{1}{1} + \frac{1}{1 \cdot 2} + \frac{1}{1 \cdot 2 \cdot 3} + \frac{1}{1 \cdot 2 \cdot 3 \cdot 4} \approx 2.70833333334$$

$$b_5 = 1 + \frac{1}{1} + \frac{1}{1 \cdot 2} + \frac{1}{1 \cdot 2 \cdot 3} + \frac{1}{1 \cdot 2 \cdot 3 \cdot 4} + \frac{1}{1 \cdot 2 \cdot 3 \cdot 4 \cdot 5} \approx 2.71666666667$$

$$b_6 = 1 + \frac{1}{1} + \frac{1}{1 \cdot 2} + \frac{1}{1 \cdot 2 \cdot 3} + \frac{1}{1 \cdot 2 \cdot 3 \cdot 4} + \frac{1}{1 \cdot 2 \cdot 3 \cdot 4 \cdot 5} + \frac{1}{1 \cdot 2 \cdot 3 \cdot 4 \cdot 5 \cdot 6}$$
$$\approx 2.71805555556$$

$$b_7 = 1 + \frac{1}{1} + \frac{1}{1 \cdot 2} + \frac{1}{1 \cdot 2 \cdot 3} + \frac{1}{1 \cdot 2 \cdot 3 \cdot 4} + \frac{1}{1 \cdot 2 \cdot 3 \cdot 4 \cdot 5} + \frac{1}{1 \cdot 2 \cdot 3 \cdot 4 \cdot 5 \cdot 6}$$
$$+ \frac{1}{1 \cdot 2 \cdot 3 \cdot 4 \cdot 5 \cdot 6 \cdot 7} \approx 2.71825396826$$

$$b_8 = 1 + \frac{1}{1} + \frac{1}{1 \cdot 2} + \frac{1}{1 \cdot 2 \cdot 3} + \frac{1}{1 \cdot 2 \cdot 3 \cdot 4} + \frac{1}{1 \cdot 2 \cdot 3 \cdot 4 \cdot 5} + \frac{1}{1 \cdot 2 \cdot 3 \cdot 4 \cdot 5 \cdot 6}$$
$$+ \frac{1}{1 \cdot 2 \cdot 3 \cdot 4 \cdot 5 \cdot 6 \cdot 7} + \frac{1}{1 \cdot 2 \cdot 3 \cdot 4 \cdot 5 \cdot 6 \cdot 7 \cdot 8} \approx 2.71827876985$$

$$\vdots$$

The sequence $\{b_n\}_{n=1}^{\infty}$ converges much more quickly to e than the sequence $\{a_n\}_{n=1}^{\infty}$.

▶ ▶ ▶ **Speed of convergence can be misleading. If a sequence converges very slowly, the numerical evidence we obtain by ex-**

10.1 SEQUENCES

amining the first few terms may suggest that the sequence diverges. On the other hand, the first few terms of a divergent sequence might suggest that the sequence is converging.

To prove that a sequence $\{a_n\}_{n=1}^{\infty}$ has a limit L, we must demonstrate that for any given $\epsilon > 0$, we can find N such that

$$|a_n - L| < \epsilon \quad \text{whenever} \quad n \geq N.$$

EXAMPLE 7 Prove that the alternating harmonic sequence $\left\{\dfrac{(-1)^n}{n}\right\}_{n=1}^{\infty}$

$$-1, \frac{1}{2}, -\frac{1}{3}, \frac{1}{4}, -\frac{1}{5}, \ldots$$

converges to the limit $L = 0$.

Solution Given any positive tolerance ϵ, we can choose N so large that

$$\left|\frac{(-1)^n}{n} - 0\right| = \frac{1}{n} < \epsilon \quad \text{for all} \quad n \geq N.$$

(Just pick $N > \dfrac{1}{\epsilon}$ so that $\dfrac{1}{N} < \epsilon$.) ∎

Types of divergent behavior—unbounded, oscillatory, chaotic

Convergent behavior of a sequence is marked by the eventual stabilization of its terms to a single limiting value. If the terms in decimal form are flashed before our eyes in quick succession, we should see a "fixing" of more and more of the decimal places. Or, if enough terms are listed vertically in a long column with the decimal points aligned, we can often spot this stabilization (warning: *enough* could be a lot). Divergent behavior, on the other hand, can take on many different forms.

A **bounded** sequence is one whose terms fall within some bounded interval of real numbers, while an **unbounded** sequence contains *arbitrarily* large positive and/or negative terms. An unbounded sequence can never converge to a single finite limiting value.

The sequence of positive integers $\{n\}_{n=1}^{\infty}$

$$1, 2, 3, 4, 5, \ldots$$

is divergent, for the terms simply increase without bound. Another example of an unbounded sequence is

$$-2,\ 4,\ 6,\ -8,\ 10,\ 12,\ -14,\ \ldots.$$

(All the terms are even, and every third term is negative.) The geometric sequence $\{(1.01)^n\}_{n=1}^{\infty}$,

$$1.01,\ 1.0201,\ 1.030301,\ \ldots$$

is also an unbounded sequence, though its terms grow very slowly. (What's the thousandth term $a_{1000} = 1.01^{1000}$?)

However, even if a sequence is bounded, it need not converge. The terms of the sequence

$$0,\ 1,\ 0,\ 1,\ 0,\ 1,\ \ldots$$

oscillate forever between 0 and 1. The sequence is *bounded* (certainly all the terms lie within the interval $[0, 1]$), but it *diverges* because the terms do not tend toward a single limiting value.

This example shows that divergent behavior need not be wild, unbounded, or unpredictable. Sequences whose terms oscillate between two values or cycle periodically through some set of values are certainly "well-behaved," but they not convergent.

EXAMPLE 8 Find the 1000th term of the sequence $\{a_n\}_{n=1}^{\infty}$, where the terms a_n follow the pattern

$$1,\ 2,\ 3,\ 1,\ 2,\ 3,\ 1,\ 2,\ 3,\ \ldots.$$

Solution This sequence cycles through the values 1, 2, 3 over and over again. We can see that 3 appears every third term (in other words, $3 = a_3,\ a_6,\ a_9,\ \ldots$). So $a_{999} = 3$, and the 1000th term $a_{1000} = 1$. This sequence definitely does not converge. ∎

Of course, a sequence whose terms wildly fluctuate in a chaotic manner without ever stabilizing is divergent. We should point out that mathematicians and scientists alike are finding many examples of seemingly chaotic behavior that nevertheless can be described by fairly simple mathematical patterns and relations.

Monotonic sequences

Convergence and divergence are words used to describe whether or not a sequence has a limit. We can also describe how the terms of a sequence behave relative to one other.

10.1 SEQUENCES

Definition 2

> A **monotonically increasing** sequence $\{a_n\}_{n=1}^{\infty}$ has the property
> $$a_1 \leq a_2 \leq a_3 \leq \cdots \leq a_n \leq \cdots,$$
> and a **strictly monotonically increasing** sequence $\{a_n\}_{n=1}^{\infty}$ has the stronger property
> $$a_1 < a_2 < a_3 < \cdots < a_n < \cdots.$$
> Similarly, a **monotonically decreasing** sequence $\{a_n\}_{n=1}^{\infty}$ has the property
> $$a_1 \geq a_2 \geq a_3 \geq \cdots \geq a_n \geq \cdots,$$
> and a **strictly monotonically decreasing** sequence $\{a_n\}_{n=1}^{\infty}$ has the stronger property
> $$a_1 > a_2 > a_3 > \cdots > a_n > \cdots.$$
> Any sequence that is either monotonically increasing or decreasing may be referred to as **monotonic**.

EXAMPLE 9 Which of the following sequences are monotonic?

a) $\{\frac{1}{n}\}_{n=1}^{\infty}$ (b) $\{2^n\}_{n=0}^{\infty}$ (c) 0, 1, 0, 1, 0, 1, ...

Solution (a) The first few terms of the harmonic sequence $\{\frac{1}{n}\}_{n=1}^{\infty}$ are

$$1, \frac{1}{2}, \frac{1}{3}, \frac{1}{4}, \ldots, \frac{1}{n}, \ldots.$$

This sequence is monotonically decreasing. In fact, it is strictly monotonically decreasing, since $\frac{1}{n} > \frac{1}{n+1}$ for any $n = 1, 2, 3, \ldots$.

(b) The first few terms of the geometric sequence $\{2^n\}_{n=0}^{\infty}$ are

$$1, 2, 4, 8, 16, \ldots.$$

This sequence is monotonically increasing. In fact, it is strictly monotonically increasing, since $2^n < 2^{n+1}$ for any $n = 0, 1, 2, 3, \ldots$.

(c) The sequence

$$0, 1, 0, 1, 0, 1, \ldots$$

is neither monotonically increasing nor decreasing, since its terms oscillate between 0 and 1. ∎

We can conclude that a convergent sequence must be bounded. The reasoning here is that all the terms in a tail of the sequence must be close to a single value L. That accounts for all but the finitely many terms preceding the tail. This means we can find a bounded interval big enough to include the tail of the sequence, as well as all terms preceding the tail.

Even though a convergent sequence must be bounded, the converse is not true: there are certainly bounded sequences that don't converge (remember $0, 1, 0, 1, 0, 1, \ldots$). However, if a sequence is both bounded *and* monotonic, then it must converge. The reasoning makes use of a fundamental property of real numbers: given any bounded set of real numbers, there is a *smallest* closed interval containing all the numbers in the set. For example, the *smallest* closed interval containing all the numbers in the harmonic sequence $\{\frac{1}{n}\}_{n=1}^{\infty}$ is $[0, 1]$. Now, if the sequence is monotonic (as is the harmonic sequence), then the terms in the sequence will have to cluster at one end or the other of this closed interval. The endpoint is the limit of the sequence. (The harmonic sequence is monotonically decreasing, so its terms approach 0, the left endpoint of the interval.)

We summarize all these observations below.

Theorem 10.1

Hypothesis: $\{a_n\}_{n=1}^{\infty}$ is a sequence of real numbers.
Conclusion 1: If $\{a_n\}_{n=1}^{\infty}$ is convergent, then $\{a_n\}_{n=1}^{\infty}$ is bounded (but the converse need not hold).
Conclusion 2: If $\{a_n\}_{n=1}^{\infty}$ is bounded *and* monotonic, then $\{a_n\}_{n=1}^{\infty}$ is convergent.

Limit properties of sequences

If the *nth* term of a sequence is given by an explicit formula in terms of n,

$$a_n = f(n),$$

then we can investigate the behavior of the sequence both numerically and graphically.

Numerically, we can substitute the values $n = 1, 2, 3, \ldots$ into the formula to get a feel for the sequence. You can see convergent behavior numerically in the stabilization of the digits in the decimal form of each term. That is, if $a_1, a_2, a_3, \ldots$, are flashed before our eyes quickly in sequence (with the position of the decimal point fixed in place), then the result is like a movie where the digits become fixed from left to right. How long the movie must run for a convergent sequence to show stabilization of each digit is an indication of the speed of convergence of the sequence.

10.1 SEQUENCES

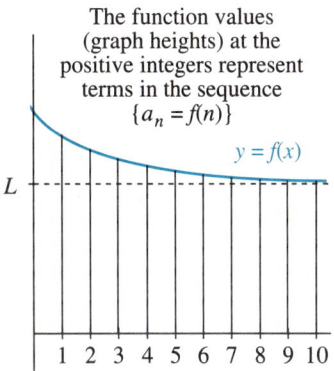

Figure 10.3 Investigating a sequence graphically.

Substituting large values of n may (or may not) give us a good approximation to the limit of a convergent sequence. Also, if you use a machine to compute the values a_n for large n, be particularly sensitive to round-off and cancellation errors that might occur.

Graphically, we can plot

$$y = f(x)$$

and examine the graph for large values of x. The y-coordinates of the points

$$(n, f(n))$$

that lie on this graph correspond to terms of the sequence. A limiting value L to the sequence will correspond to a horizontal asymptote $y = L$ to the graph of $y = f(x)$ (see Figure 10.3).

If you use a machine grapher to explore the behavior of a sequence, beware that it is subject to the same limitations as the machine's numerical computations. These numerical and graphical explorations can help give you a better feel for the behavior of the sequence, but ultimately neither can provide a definitive answer to the question: What is $\lim_{n\to\infty} a_n$?

If the formula describing a_n is composed or built up algebraically from familiar functions, then the same limit properties enjoyed by functions described in Chapter 2 are shared by sequences. More precisely, if $\lim_{n\to\infty} a_n = L_1$ and $\lim_{n\to\infty} b_n = L_2$, then all of the following are true:

$$\lim_{n\to\infty}(a_n + b_n) = L_1 + L_2 \qquad \lim_{n\to\infty}(a_n - b_n) = L_1 - L_2$$

$$\lim_{n\to\infty} a_n b_n = L_1 L_2 \qquad \lim_{n\to\infty} c a_n = c L_1 \quad (c \text{ a constant})$$

$$\lim_{n\to\infty} a_n/b_n = L_1/L_2 \qquad (\text{provided } b_n\text{'s and } L_2 \neq 0)$$

$$\lim_{n\to\infty} f(a_n) = f(L_1) \qquad (f \text{ a continuous function})$$

EXAMPLE 10 Consider the two sequences

$$\{a_n\}_{n=1}^{\infty} = \{(1+\frac{1}{n})^n\}_{n=1}^{\infty} \quad \text{and} \quad \{b_n\}_{n=1}^{\infty} = \{(2-\frac{1}{n})\}_{n=1}^{\infty}.$$

Find the limits of the following sequences:

(a) $\{a_n + b_n\}_{n=1}^{\infty}$ (b) $\{a_n - b_n\}_{n=1}^{\infty}$
(c) $\{a_n b_n\}_{n=1}^{\infty}$ (d) $\{a_n/b_n\}_{n=1}^{\infty}$
(e) $\{3b_n\}_{n=1}^{\infty}$ (f) $\{\ln(a_n)\}_{n=1}^{\infty}$.

Solution We noted earlier in this section that

$$\lim_{n\to\infty} a_n = \lim_{n\to\infty} (1+\frac{1}{n})^n = e.$$

Since $\frac{1}{n}$ approaches 0 as n grows larger,

$$\lim_{n\to\infty} b_n = \lim_{n\to\infty} (2-\frac{1}{n}) = 2.$$

Using the limit properties of sequences, we can compute the limits of each of the given sequences as follows:

(a) $\lim_{n\to\infty}(a_n+b_n) = e+2$ (b) $\lim_{n\to\infty}(a_n-b_n) = e-2$
(c) $\lim_{n\to\infty} a_n b_n = 2e$ (d) $\lim_{n\to\infty} a_n/b_n = e/2$
(e) $\lim_{n\to\infty} 3b_n = 3\cdot 2 = 6$ (f) $\lim_{n\to\infty} \ln(a_n) = \ln(e) = 1$.

EXERCISES for Section 10.1

For exercises 1–4: Give a formula for the nth term of the arithmetic sequence with given initial term and constant difference.

1. $a_0 = 3, d = -5$ **2.** $a_0 = -4, d = 0.5$
3. $a_0 = \pi, d = 0$ **4.** $a_0 = 0, d = \pi$

For exercises 5–8: Give a formula for the nth term of the geometric sequence with given initial term and constant ratio.

5. $a_0 = 3, r = -5$ **6.** $a_0 = -4, r = 0.5$
7. $a_0 = \pi, r = 2$ **8.** $a_0 = 2, r = \pi$

For exercises 9–12: Indicate whether each of the statements is true or false. If true, state why it is true. If false, give an example.

9. All bounded, monotonic sequences converge.

10. All non-monotonic sequences diverge.

10.1 SEQUENCES

11. All bounded sequences converge.

12. All unbounded sequences diverge.

For exercises 13–16: List the first 5 terms of each sequence $\{a_n\}_{n=1}^{\infty}$. For each, classify as bounded or not, monotonic or not, convergent or not.

13. $a_n = 2^{\cos n\pi}$

14. $a_n = \left(\frac{n-1}{n+1}\right)^n$

15. $-2, -3.1, -4.01, -5.001, \ldots$ (continues with same pattern)

16. the illustrated sequence continuing with the same pattern

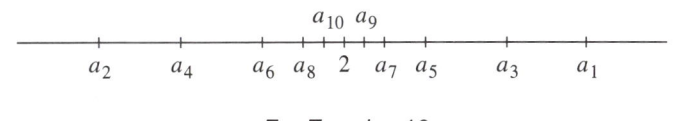

For Exercise 16

For exercises 17–20: Consider the sequence whose nth term is $a_n = 1 + (-1)^n\left(\frac{1}{n}\right)$. Note that $\lim_{n\to\infty} a_n = 1$.

17. Plot and label the first 7 terms of the sequence $\{a_n\}_{n=1}^{\infty}$.

18. Find the first term in the sequence that is within .1 less than the limit.

19. Find the first term in the sequence that is within .1 more than the limit.

20. Find the first 2 terms in the sequence that are within .1 of each other.

For exercises 21–25: Give three examples of sequences satisfying the given descriptions.

21. bounded, monotonic

22. bounded, non-monotonic, convergent

23. unbounded, monotonic

24. unbounded, non-monotonic

25. bounded, non-monotonic, divergent

---

26. Let $a_n = \left(1 + \frac{1}{n}\right)^n$. Find $a_1, a_2, a_3, a_{1000}, a_{2000}, a_{3000}$. Is the sequence $\{a_n\}$ bounded and/or monotonic? If so, what does the limit appear to be? Approximately, how close is a_{3000} to the limit?

27. The nth term of a sequence $\{a_n\}_{n=1}^{\infty}$ is given by $a_n = \frac{1}{n!}$. (Note: $n! = 1 \cdot 2 \cdot 3 \cdots \cdots n$.) Find the limit of the sequence, and then find the smallest value N such that all the terms after a_N in the sequence $\{a_n\}_{n=1}^{\infty}$ are within .001 of this limit.

28. The nth term of a sequence $\{a_n\}_{n=1}^{\infty}$ is given by $a_n = \dfrac{1}{2^n}$. Find the limit of the sequence, and then find the smallest value N such that all the terms after a_N in the sequence $\{a_n\}_{n=1}^{\infty}$ are within .001 of this limit.

29. Compare the rates of convergence of the sequences in the last two exercises.

30. The nth term of a sequence $\{a_n\}_{n=1}^{\infty}$ is given by $a_n = n\sin(\frac{\pi}{n})$. Find the limit of the sequence, and then find the smallest value N such that all the terms after a_N in the sequence $\{a_n\}_{n=1}^{\infty}$ are within .001 of this limit.

10.2 SERIES

We can roughly think of a series as the sum of the infinitely many terms in a sequence. Before embarking on our study of series, let's first take a look at an important property of real numbers.

The Archimedean property of real numbers

Suppose that we take a positive real number r and start adding it to itself over and over again. No matter how small r is ($r > 0$), this process will produce arbitrarily larger and larger sums. This fact is known as the **Archimedean property of real numbers**, and we can state it more precisely as follows: If r, M are any two positive real numbers, then there exists a positive integer N such that $Nr > M$. Note that there is absolutely no restriction on how small r can be (as long as it is positive) nor on how large M can be.

The quantity Nr represents a sum:

$$Nr = \sum_{n=1}^{N} r = r + r + r + \cdots r \qquad (N \text{ addends of } r).$$

The ancient Greek mathematicians did not have a language or notation that included anything like our symbol ∞ in their mathematics, but we can write the Archimedean property as follows:

$$\text{For any } r > 0, \qquad \lim_{N \to \infty} Nr = \infty.$$

Zeno's paradox

We might think of the Archimedean property as saying that a "sum" of *infinitely many* terms $r > 0$ is infinite. Is there any way that the "sum" of

10.2 SERIES

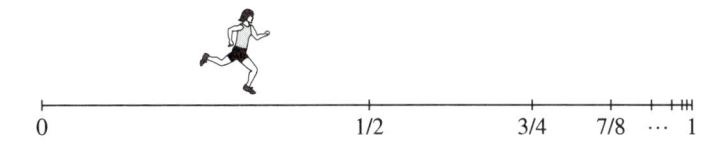

Figure 10.4 Zeno's racecourse.

infinitely many positive numbers could not be infinite? Many of the ancient Greek mathematicians did not think so, because of the Archimedean property. However, this belief was shaken by what is known as **Zeno's paradox**.

Zeno's paradox goes like this: Suppose that a person runs along a racecourse of length 1 mile. Before the race is completed, the person must first cover half the distance. Now, of the half-mile remaining, the person must cover half of it (or $1/4$ of a mile) to reach the $3/4$ mile mark. Of the final quarter-mile, the person must first cover half of it ($1/8$ mile) to reach the $7/8$ mile mark. This process occurs over and over: given any distance remaining, the person must first cover half of the distance before the race is completed (see Figure 10.4).

Each of these distances is positive, and the person must cover infinitely many of them. The Archimedean property might suggest that the total distance is infinite and that the person cannot possibly complete the race. It is clear, however, that people complete such races regularly. This paradox forces us to admit that it is reasonable to write something like

$$\frac{1}{2} + \frac{1}{4} + \frac{1}{8} + \cdots = 1.$$

The word *paradox* usually refers to a statement or situation contradicting itself or some firmly established truth. Zeno's paradox is not really a paradox, for it does *not* contradict the Archimedean property. Unlike the situation of adding the *same* positive number r to itself infinitely often, the racecourse example has us adding infinitely many *different* positive numbers, each one smaller (by the factor one-half) than the number preceding it. Evidently, in this different situation, it may be reasonable to assign a finite total to a sum of infinitely many positive terms.

What is a series?

The notion of a series is convenient for describing situations like Zeno's paradox or any other infinite process producing a sequence of running totals. In fact, using the terms in any sequence of real numbers $\{a_n\}_{n=1}^{\infty}$, we can build a new sequence of running totals. A **series** is the limit of this new sequence. The notation for a series uses summation notation for sums, and we write

$$\sum_{n=1}^{\infty} a_n$$

as shorthand for

$$a_1 + a_2 + a_3 + a_4 + \cdots + a_n + \cdots.$$

It is impossible to add infinitely many numbers together by actually summing them, so mathematically the series is the limit of a sequence of **partial sums** S_N:

$$S_1 = a_1$$
$$S_2 = a_1 + a_2$$
$$S_3 = a_1 + a_2 + a_3$$
$$S_4 = a_1 + a_2 + a_3 + a_4$$
$$\vdots$$
$$S_N = a_1 + a_2 + \cdots + a_N$$
$$\vdots$$

Using summation notation,

$$\sum_{n=1}^{\infty} a_n = \lim_{N \to \infty} \sum_{n=1}^{N} a_n = \lim_{N \to \infty} S_N.$$

EXAMPLE 11 Find the first four partial sums of the series $\sum_{n=1}^{\infty} \frac{1}{2^n}$.

Solution
$$S_1 = \frac{1}{2^1} = \frac{1}{2}$$
$$S_2 = \frac{1}{2^1} + \frac{1}{2^2} = \frac{3}{4}$$
$$S_3 = \frac{1}{2^1} + \frac{1}{2^2} + \frac{1}{2^3} = \frac{7}{8}$$
$$S_4 = \frac{1}{2^1} + \frac{1}{2^2} + \frac{1}{2^3} + \frac{1}{2^4} = \frac{15}{16}.$$

Can you come up with a formula that gives the Nth partial sum S_N? ■

10.2 SERIES

Definition 3 | A series is said to **converge** if the sequence of partial sums converges. Finding the **sum of a series** means finding the limit of the sequence of its partial sums. If a series does not converge, then we say the series **diverges** and it has no sum.

We can numerically investigate the convergence or divergence of a series in much the same way we investigate the limiting behavior of any sequence: by actually evaluating the partial sums S_N and examining their behavior as N grows larger without bound. Note that even if we have a formula for the nth term a_n of a series, it may be difficult to find a formula for the Nth partial sum S_N of the series $\sum_{n=1}^{\infty} a_n$.

Examples of series

A **telescoping series** gets its name from the behavior of its partial sums. The terms of a telescoping series can be written as a sum of differences of quantities, all of which cancel except for the first and last, collapsing (like an old spyglass telescope) to a single difference. Thus, we can find a formula for the Nth partial sum and more easily investigate its limiting behavior.

EXAMPLE 12 Does $\sum_{n=1}^{\infty}(\frac{1}{n} - \frac{1}{n+1})$ converge, and if so, what is its sum?

Solution The first few partial sums of the series are

$$S_1 = a_1 = \frac{1}{1} - \frac{1}{2} = \frac{1}{2}$$

$$S_2 = a_1 + a_2 = \left(1 - \frac{1}{2}\right) + \left(\frac{1}{2} - \frac{1}{3}\right) = 1 - \frac{1}{3} = \frac{2}{3}$$

$$S_3 = a_1 + a_2 + a_3 = \left(1 - \frac{1}{2}\right) + \left(\frac{1}{2} - \frac{1}{3}\right) + \left(\frac{1}{3} - \frac{1}{4}\right) = 1 - \frac{1}{4} = \frac{3}{4}$$

$$S_4 = a_1 + a_2 + a_3 + a_4 = \left(1 - \frac{1}{2}\right) + \left(\frac{1}{2} - \frac{1}{3}\right) + \left(\frac{1}{3} - \frac{1}{4}\right) + \left(\frac{1}{4} - \frac{1}{5}\right) = \frac{4}{5}.$$

Notice the cancellation involving the "internal terms." We can write

$$S_N = a_1 + a_2 + \cdots + a_N = 1 - \frac{1}{N+1}$$

and we can see that $\sum_{n=1}^{\infty} a_n = \lim_{N \to \infty}(1 - \frac{1}{N+1}) = 1$. ∎

Geometric series

The terms of a **geometric series** form a geometric sequence; each term is a constant ratio times the preceding term. For example, the series $\sum_{n=1}^{\infty} \frac{1}{2^n}$ of Zeno's paradox is a geometric series.

In general, a geometric series has the form

$$a + ar + ar^2 + ar^3 + ar^4 + \cdots ar^n + \cdots,$$

where a is the initial term and r is the constant ratio. If we use the power of r as the index, then we can start with $n = 0$ and write the series as

$$\sum_{n=0}^{\infty} ar^n.$$

Let's see how the sum of a geometric series depends on the ratio r. If $r = 1$, then

$$\sum_{n=0}^{\infty} ar^n = a + a + a + a + \cdots$$

will *diverge* for any *nonzero* a.

If $r = -1$, then

$$\sum_{n=0}^{\infty} ar^n = a - a + a - a + \cdots$$

also diverges because the partial sums oscillate between a and 0:

$$S_0 = a, \quad S_1 = 0, \quad S_2 = a, \quad S_3 = 0,$$

and so on.

Now let's look at the cases where $r \neq \pm 1$. The Nth partial sum is

$$S_N = a + ar + ar^2 + \cdots + ar^N.$$

Multiplying both sides by r, we obtain

$$rS_N = ar + ar^2 + ar^3 + \cdots + ar^N + ar^{N+1}.$$

Then, subtracting the two sides from the original equation for S_N, we have

$$S_N - rS_N = a + ar + ar^2 + \cdots + ar^N$$
$$- ar - ar^2 - \cdots - ar^N - ar^{N+1}$$
$$= a - ar^{N+1}.$$

10.2 SERIES

Since $S_N - rS_N = (1-r)S_N$, we can solve for S_N, provided that $r \neq 1$:

$$S_N = \frac{a - ar^{N+1}}{(1-r)} = \frac{a(1 - r^{N+1})}{(1-r)}.$$

Using this formula for S_N, we can determine the sum of the geometric series

$$\sum_{n=0}^{\infty} ar^n = \lim_{n \to \infty} \frac{a(1 - r^{N+1})}{1 - r} \quad (r \neq 1).$$

If $|r| > 1$ ($r > 1$ or $r < -1$), then r^{N+1} will become larger in magnitude and the limit will not exist.

If $|r| < 1$ ($-1 < r < 1$), then $\lim_{N \to \infty} r^{N+1} = 0$ and

$$\sum_{n=0}^{\infty} ar^n = \frac{a(1 - 0)}{1 - r} = \frac{a}{1 - r}.$$

Summarizing all the cases, we have the following theorem.

Theorem 10.2

If $a \neq 0$, then the geometric series $\sum_{n=0}^{\infty} ar^n$ *diverges* if $r \geq 1$ or $r \leq -1$ and *converges* if $-1 < r < 1$. If the geometric series $\sum_{n=0}^{\infty} ar^n$ converges, its sum is $\frac{a}{1-r}$.

EXAMPLE 13 Does the geometric series

$$7 + \frac{7}{5} + \frac{7}{25} + \frac{7}{125} + \cdots$$

converge or diverge? If the series converges, what is the sum?

Solution This series can also be written in summation notation as

$$\sum_{n=0}^{\infty} 7\left(\frac{1}{5}\right)^n.$$

The first term is $a = 7$, and the ratio is $r = 1/5$. Since $-1 < 1/5 < 1$, we know that the series must *converge*, and

$$\sum_{n=0}^{\infty} 7\left(\frac{1}{5}\right)^n = \frac{7}{1 - 1/5} = \frac{7}{4/5} = \frac{35}{4}.$$

EXAMPLE 14   Does the series

$$\sum_{n=0}^{\infty}(-3)^{1-n}$$

converge or diverge? If the series converges, what is the sum?

Solution   If we write the first few terms of the series, we may more easily recognize that it is a geometric series:

$$-3 + 1 - \frac{1}{3} + \frac{1}{9} - \frac{1}{27} + \frac{1}{81} + \cdots,$$

where the first term is $a = -3$, and the ratio is $r = -1/3$.

Since $-1 < -1/3 < 1$, we know that the series must *converge*, and

$$\sum_{n=0}^{\infty}(-3)^{1-n} = \frac{-3}{1-(-1/3)} = \frac{-3}{4/3} = \frac{-9}{4}.$$ ■

EXAMPLE 15   Does the series $.01 + .02 + .04 + .08 + \cdots$ converge or diverge? If the series converges, what is the sum?

Solution   This is a geometric series with first term $a = .01$ and ratio $r = 2$ (the terms double at each step). Since $r \geq 1$, the series *diverges*. ■

EXAMPLE 16   Does the series $\sum_{n=2}^{\infty} \frac{5}{4^n}$ converge or diverge? If the series converges, what is the sum?

Solution   Note that the starting index is $n = 2$. If we write the first few terms of the series, we have

$$\frac{5}{4^2} + \frac{5}{4^3} + \frac{5}{4^4} + \cdots,$$

so the first term is $a = 5/16$ and the ratio is $r = 1/4$.

Since $-1 < 1/4 < 1$, the series *converges*, and

$$\sum_{n=2}^{\infty}\frac{5}{4^n} = \frac{5/16}{1-1/4} = \frac{5}{12}.$$ ■

▶▶▶ **Be careful to distinguish between the limit of a** *sequence* $\{a_n\}_{n=1}^{\infty}$ **and the the associated** *series* $\sum_{n=1}^{\infty} a_n$. **It is quite possible that the sequence of individual** *terms* **converges, while the sequence of** *partial sums* **diverges.**

A simple example of this is the constant sequence

$$1, \ 1, \ 1, \ 1, \ \ldots,$$

10.2 SERIES

which certainly converges to 1, while the associated series

$$1 + 1 + 1 + 1 + 1 + 1 + \cdots$$

diverges (by the Archimedean property of real numbers).

Harmonic series

The **harmonic series** is the sum of the terms of the harmonic sequence:

$$\sum_{n=1}^{\infty} \frac{1}{n} = 1 + \frac{1}{2} + \frac{1}{3} + \frac{1}{4} + \cdots.$$

The *harmonic sequence converges to* 0, but the *harmonic series diverges*. One way to see this is to look at the partial sums S_1, S_2, S_4, S_8, S_{16}, $\cdots$. The first few of these partial sums are

$$S_1 = 1$$

$$S_2 = 1 + \frac{1}{2} = 1.5$$

$$S_4 = 1 + \frac{1}{2} + \frac{1}{3} + \frac{1}{4} > 2$$

$$S_8 = 1 + \frac{1}{2} + \frac{1}{3} + \frac{1}{4} + \frac{1}{5} + \frac{1}{6} + \frac{1}{7} + \frac{1}{8} > 2.5$$

$$S_{16} = 1 + \cdots + \frac{1}{16} > 3.$$

Each time we double the number of terms in the partial sum, we have added at least $1/2$ to the running total. This means that

$$S_{2^n} = 1 + \frac{1}{2} + \frac{1}{3} + \cdots + \frac{1}{2^n} \geq 1 + \frac{n}{2}.$$

The partial sums will grow larger without bound, since we can keep doubling the number of terms indefinitely. Hence, the harmonic series *diverges*.

The **alternating harmonic series** is the sum of the terms of the alternating harmonic sequence

$$\sum_{n=1}^{\infty} \frac{(-1)^{n+1}}{n} = 1 - \frac{1}{2} + \frac{1}{3} - \frac{1}{4} + \frac{1}{5} - \cdots.$$

It is unclear whether or not the alternating harmonic series converges or diverges from looking at the first few partial sums. Interestingly, it is known that the alternating harmonic series *converges* and has a sum $L = \ln(2)$.

Properties of series

Because a series is the limit of a sequence of partial sums, some of the limit properties of sequences also apply to properties of series.

If $\sum_{n=1}^{\infty} a_n = L_1$ and $\sum_{n=1}^{\infty} b_n = L_2$, then all of the following are true.

$$\sum_{n=1}^{\infty}(a_n + b_n) = L_1 + L_2$$

$$\sum_{n=1}^{\infty}(a_n - b_n) = L_1 - L_2$$

$$c\sum_{n=1}^{\infty} a_n = cL_1 \quad (c \text{ a constant}).$$

If $\sum_{n=1}^{\infty} a_n$ diverges and $\sum_{n=1}^{\infty} b_n$ converges, then

$$\sum_{n=1}^{\infty}(a_n + b_n) \quad \text{diverges}$$

$$\sum_{n=1}^{\infty}(a_n - b_n) \quad \text{diverges}$$

$$c\sum_{n=1}^{\infty} a_n \quad \text{diverges if } c \neq 0.$$

EXAMPLE 17 Assuming that the alternating harmonic series $\sum_{n=1}^{\infty} \frac{(-1)^{n+1}}{n}$ converges to the sum $\ln(2)$, find the sums of the following series:

(a) $\sum_{n=1}^{\infty}(\frac{(-1)^{n+1}}{n} + \frac{1}{2^n})$
(b) $\sum_{n=1}^{\infty}(\frac{(-1)^{n+1}}{n} - \frac{1}{2^n})$
(c) $\sum_{n=1}^{\infty}(\frac{1}{n} + \frac{1}{2^n})$
(d) $\sum_{n=1}^{\infty}(\frac{1}{n} - \frac{1}{2^n})$
(e) $\sum_{n=1}^{\infty} \frac{(-1)^{n+1} 5}{n}$
(f) $\sum_{n=1}^{\infty} \frac{5}{n}$.

Solution We know that the geometric series $\sum_{n=1}^{\infty} \frac{1}{2^n}$ converges to the sum 1, while the harmonic series $\sum_{n=1}^{\infty} \frac{1}{n}$ diverges. Using this fact, we can determine that:

(a) $\sum_{n=1}^{\infty}(\frac{(-1)^{n+1}}{n} + \frac{1}{2^n}) = \ln(2) + 1$.
(b) $\sum_{n=1}^{\infty}(\frac{(-1)^{n+1}}{n} - \frac{1}{2^n}) = \ln(2) - 1$.
(c) $\sum_{n=1}^{\infty}(\frac{1}{n} + \frac{1}{2^n})$ diverges.
(d) $\sum_{n=1}^{\infty}(\frac{1}{n} - \frac{1}{2^n})$ diverges.
(e) $\sum_{n=1}^{\infty}(\frac{(-1)^{n+1} 5}{n}) = 5\ln(2)$.
(f) $\sum_{n=1}^{\infty} \frac{5}{n}$ diverges.

10.2 SERIES

EXERCISES for Section 10.2

For exercises 1–16: Determine whether the given series converges or diverges. If the series converges, find the sum.

1. $\sum_{n=1}^{\infty}(-1/3)^n$
2. $\sum_{n=0}^{\infty}(1/3)^n$
3. $\sum_{n=1}^{\infty}e^{-n}$
4. $\sum_{n=0}^{\infty}\frac{3^n + 2^n}{4^n}$
5. $\sum_{n=1}^{\infty}(-1)^n$
6. $\sum_{n=0}^{\infty}(1/2)$
7. $\frac{1}{11} - \frac{10}{11^2} + \frac{100}{11^3} - \frac{1000}{11^4}\cdots$
8. $\sum_{n=1}^{\infty}n$
9. $.001 - .003 + .009 - .027\ldots$
10. $\sum_{n=1}^{\infty}\frac{2}{n}$
11. $\sum_{n=2}^{\infty}\frac{e^n}{3^{n-2}}$
12. $\sum_{n=1}^{\infty}\frac{(-1)^n}{2n}$
13. $\sum_{n=3}^{\infty}\left(\frac{1}{n-1} - \frac{1}{n+1}\right)$
14. $\sum_{n=1}^{\infty}\left(\frac{3}{2n-1} - \frac{3}{2n+1}\right)$
15. $\sum_{n=1}^{\infty}\left(\frac{1}{3^{n+1}} - \frac{1}{3^n}\right)$
16. $\sum_{n=0}^{\infty}\sin(\pi n)$

---

17. Find two different geometric series, both of which converge to 17/3.

18. If a geometric series converges to 1, then the sum of its first term a and the ratio r is 1. Why? Does the converse hold?

19. Find the first ten partial sums of $\sum_{n=0}^{\infty}1/n!$, where $n! = 1 \cdot 2 \cdot 3 \cdots n$ for $n \geq 1$ and $0! = 1$. Does the apparent limiting value look familiar?

20. Find S_{10}, S_{100}, and S_{1000} for the alternating harmonic series and compare these values to $\ln(2)$.

21. A ball is dropped from a height of six feet and begins bouncing up and down. The height of each bounce is 3/4 the height of the previous bounce. Find the total vertical distance traveled by the ball.

22. Let
$$c_n = 1 + \frac{1}{2} + \frac{1}{3} + \frac{1}{4} + \cdots + \frac{1}{n} - \ln(n).$$

Find c_{40}, c_{50}, and c_{60}. The sequence $\{c_n\}$ converges to a number known as **Euler's constant** $\gamma \approx 0.5772$.

Cantor's middle-third set, named for the mathematician Georg Cantor (1835-1918) is constructed as follows:

First, the open interval $(\frac{1}{3}, \frac{2}{3})$ is "erased" from the closed interval $[0, 1]$. This leaves two closed intervals, $[0, \frac{1}{3}]$ and $[\frac{2}{3}, 1]$.

Next, the middle third of each of these two remaining closed intervals is erased. That is, the open intervals $(\frac{1}{9}, \frac{2}{9})$ and $(\frac{7}{9}, \frac{8}{9})$ are erased, leaving four closed intervals, $[0, \frac{1}{9}]$, $[\frac{2}{9}, \frac{1}{3}]$, $[\frac{2}{3}, \frac{7}{9}]$, and $[\frac{8}{9}, 1]$.

We continue in the same way, erasing the open middle third of each remaining closed subinterval to obtain eight new closed subintervals. The illustration shows the original interval $[0, 1]$ and what remains after each of the first three stages. We continue this process indefinitely, at each stage erasing the open middle third of each remaining closed subinterval.

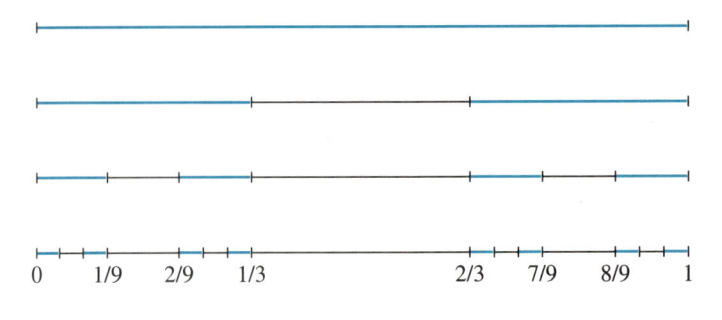

For Exercises 23–25

23. Find the total length of all the subintervals erased (using a geometric series).

24. Give an example of a point that is *never* erased.

25. How many points are left in the Cantor set?

10.3 CONVERGENCE AND DIVERGENCE OF SERIES

It may be difficult to find a formula that describes the Nth partial sum $S_N = \sum_{n=1}^{N} a_n$ of a series $\sum_{n=1}^{\infty} a_n$. With the use of computers or calculators, we can certainly calculate many of these partial sums quickly, and for large index values N we can get a feel for the limiting behavior of the series and approximate its sum if it converges.

However, note that any partial sum will have a finite value (perhaps large), so the convergence or divergence of a series is impossible to determine just by examining the values of some of these partial sums. For example, the harmonic series

$$\sum_{n=1}^{\infty} \frac{1}{n}$$

diverges, but very slowly. Indeed, the harmonic series will appear to converge if we calculate the partial sums using a machine with fixed precision.

If we could determine in advance that a given series $\sum_{n=1}^{\infty} a_n$ converges, then we might be able to approximate its sum closely with a partial sum $S_N = \sum_{n=1}^{N} a_n$ of suitably large index N. There are a variety of **convergence tests** for series. Sometimes a convergence test can also give

10.3 CONVERGENCE AND DIVERGENCE OF SERIES

us information regarding the number N of terms so that a partial sum achieves a given level of accuracy. In this section, we will discuss several of these convergence tests. For each of the tests we describe, it is important to pay special attention to:

1. the requirements that must be fulfilled so that the test can be applied,
2. the criteria for making a decision regarding convergence or divergence, and
3. any accuracy estimates for approximating the sum of a convergent series.

For each of the tests described in this section, we'll provide some discussion of how the test works and illustrate it with some examples.

Nth term test

Requirements:

The Nth **term test** can be applied to any series.

Criteria:

If $\lim_{n \to \infty} a_n \neq 0$, then $\sum_{n=1}^{\infty} a_n$ diverges.

If $\lim_{n \to \infty} a_n = 0$, then $\sum_{n=1}^{\infty} a_n$ *may* or *may not* converge.

Discussion:

For the partial sums to converge to a single limiting value, the individual terms a_n must approach 0. The Nth term test simply relies on this observation.

EXAMPLE 18 Use the Nth term test on each of the following series:

(a) $\sum_{n=1}^{\infty} n$ (b) $\sum_{n=1}^{\infty} (-1)^n$ (c) $\sum_{n=1}^{\infty} (1 - \frac{1}{n})$ (d) $\sum_{n=1}^{\infty} \frac{1}{n}$ (e) $\sum_{n=1}^{\infty} \frac{1}{2^n}$

Solution (a) $\sum_{n=1}^{\infty} n$ *diverges*, since $\lim_{n \to \infty} n = \infty \neq 0$.

(b) $\sum_{n=1}^{\infty} (-1)^n$ *diverges*, since $\lim_{n \to \infty} (-1)^n$ does not exist. (Note that $(-1)^n$ oscillates between 1 and -1, depending on whether n is even or odd.)

(c) $\sum_{n=1}^{\infty} (1 - \frac{1}{n})$ *diverges*, since $\lim_{n \to \infty} (1 - \frac{1}{n}) = 1 \neq 0$.

(d) The Nth term test provides no information regarding the convergence or divergence of the harmonic series

$$\sum_{n=1}^{\infty} \frac{1}{n},$$

since $\lim_{n\to\infty} \frac{1}{n} = 0$. (However, we do know that this series *diverges* from the previous section.)

(e) The Nth term test provides no information regarding the geometric series

$$\sum_{n=1}^{\infty} \frac{1}{2^n},$$

since $\lim_{n\to\infty} \frac{1}{2^n} = 0$. (However, we do know this series *converges* and has a sum equal to 1.) ∎

As parts (d) and (e) of this example show, $\lim_{n\to\infty} a_n = 0$ is a *necessary* but not *sufficient* condition for the series $\sum_{n=1}^{\infty} a_n$ to converge. The Nth term test can determine that a series diverges, but it does not give us conclusive information that a series converges.

Integral test

Requirements:

The **integral test** can be applied to any series whose terms $a_n = f(n)$, where f can be considered a *continuous, positive, decreasing* function of a real variable x.

Criteria:

If $\int_1^\infty f(x)\,dx$ converges, then $\sum_{n=1}^{\infty} a_n$ converges.

If $\int_1^\infty f(x)\,dx$ diverges, then $\sum_{n=1}^{\infty} a_n$ diverges.

Discussion:

The integral test allows us to decide the convergence or divergence of a series by comparing it to an improper integral. If all the terms a_n in

$$\sum_{n=1}^{\infty} a_n$$

are positive, we can represent them as the *areas* of rectangles under the curve $y = f(x)$ (see Figure 10.5). The base of each rectangle is 1, so the height of the nth rectangle must be a_n.

If $a_n = f(n)$, then the graph $y = f(x)$ passes through the upper right corner of each rectangle. If the function f is decreasing, then all of the rectangles starting with the second one fit *under* the graph. Hence, if the improper integral

$$\int_1^\infty f(x)\,dx = L,$$

10.3 CONVERGENCE AND DIVERGENCE OF SERIES

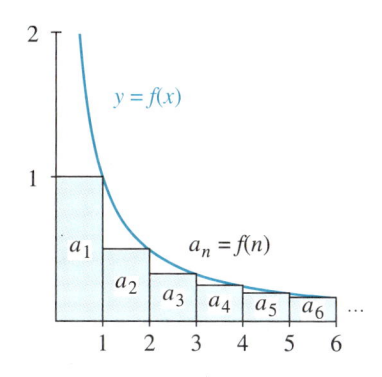

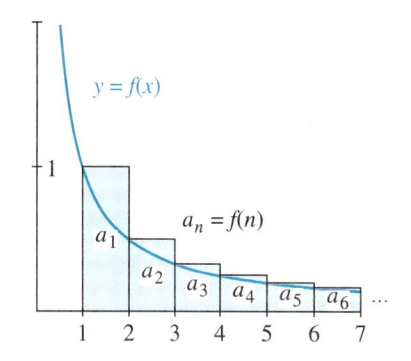

Figure 10.5 If $\int_1^\infty f(x)\,dx$ is finite, so is the total area $\sum_1^\infty a_n$.

Figure 10.6 If $\int_1^\infty f(x)\,dx$ is infinite, so is the total area $\sum_1^\infty a_n$.

then

$$\sum_{n=1}^\infty a_n \leq L + a_1.$$

Each term a_n makes a positive contribution to the total, so the partial sums S_N are increasing as N increases. Since the partial sums form a *monotonically increasing bounded sequence*, they must converge to a single limiting value.

Suppose that we considered the rectangles shifted one unit to the right, as in Figure 10.6. Now the graph $y = f(x)$ passes through all the upper left corners of the rectangles and

$$\int_1^\infty f(x)\,dx \leq \sum_{n=1}^\infty a_n.$$

Hence, if $\int_1^\infty f(x)\,dx$ *diverges*, then so must $\sum_{n=1}^\infty a_n$.

EXAMPLE 19 Use the integral test to test the convergence of the harmonic series $\sum_{n=1}^\infty \frac{1}{n}$.

Solution The integral test can be applied because $1/x$ is positive, continuous, and decreasing over the interval $[1, \infty)$:

$$\int_1^\infty \frac{1}{x}\,dx = \lim_{b\to\infty} \int_1^b \frac{1}{x}\,dx = \lim_{b\to\infty} \ln|b| = \infty.$$

This improper integral diverges, so the harmonic series $\sum_{n=1}^\infty \frac{1}{n}$ diverges. ∎

EXAMPLE 20 Use the integral test to test the convergence of the series $\sum_{n=1}^\infty \frac{1}{n^2}$.

Solution   The integral test can be applied because $1/x^2$ is positive, continuous, and decreasing over the interval $[1, \infty)$:

$$\int_1^\infty \frac{1}{x^2}\, dx = \lim_{b \to \infty} \int_1^b \frac{1}{x^2}\, dx = \lim_{b \to \infty} \frac{-1}{b} + 1 = 1.$$

This improper integral converges, so the series $\sum_{n=1}^\infty \frac{1}{n^2}$ converges. (Note: The sum of the *series* is not 1, however.) ∎

EXAMPLE 21   Why can't the integral test be applied to the alternating harmonic series?

Solution   The integral test cannot be applied to the alternating harmonic series

$$\sum_{n=1}^\infty \frac{(-1)^{n+1}}{n}$$

because the terms are not all positive. (Check the requirements for the integral test.) ∎

EXAMPLE 22   Why can't the integral test be applied to the series $\sum_{n=1}^\infty \sin^2(\pi n)$?

Solution   The integral test cannot be applied to the series

$$\sum_{n=1}^\infty \sin^2(\pi n)$$

because $\sin^2(\pi x)$ is not decreasing over $[1, \infty)$. (Check the requirements for the integral test.) ∎

Integral test estimate:

When the integral test indicates that a series $\sum_{n=1}^\infty a_n$ converges, we can get an estimate of how good an approximation the partial sum

$$S_N = \sum_{n=1}^N a_n = a_1 + a_2 + \cdots + a_N$$

is to the actual sum of the series. Let's use the notation R_N to indicate the remaining "tail" of the series,

$$R_N = \sum_{n=N+1}^\infty a_n = a_{N+1} + a_{N+2} + \cdots,$$

so that

$$S_N + R_N = \sum_{n=1}^\infty a_n.$$

10.3 CONVERGENCE AND DIVERGENCE OF SERIES

From our previous analysis, we can see that this remainder tail is "trapped" between two improper integrals

$$\int_{N+1}^{\infty} f(x)\,dx \leq R_N \leq \int_{N}^{\infty} f(x)\,dx.$$

If we use the lower bound $\int_{N+1}^{\infty} f(x)$ as an approximation for R_N, then the worst error possible is the difference between the lower bound and upper bound:

$$\int_{N}^{\infty} f(x)\,dx - \int_{N+1}^{\infty} f(x)\,dx = \int_{N}^{N+1} f(x)\,dx.$$

This all means that:

> If $\sum_{n=1}^{\infty} a_n$ converges by the integral test, then
> $$\sum_{n=1}^{\infty} a_n = \sum_{n=1}^{N} a_n + \int_{N+1}^{\infty} f(x)\,dx + \text{error},$$
> where $0 < \text{error} \leq \int_{N}^{N+1} f(x)\,dx$.

EXAMPLE 23 Estimate $\sum_{n=1}^{\infty} \frac{1}{n^2}$ to within .0001 of its sum.

Solution The approximation we need is

$$\sum_{n=1}^{\infty} \frac{1}{n^2} \approx \sum_{n=1}^{N} \frac{1}{n^2} + \int_{N+1}^{\infty} \frac{1}{x^2}\,dx.$$

We need to choose N large enough so that the

$$\text{error} \leq \int_{N}^{N+1} f(x)\,dx \leq .0001.$$

Now,

$$\int_{N}^{N+1} \frac{1}{x^2}\,dx = \left.\frac{-1}{x}\right]_{N}^{N+1} = \frac{-1}{N+1} + \frac{1}{N} = \frac{1}{N(N+1)},$$

and when $N = 100$, we have

$$\frac{1}{N(N+1)} = \frac{1}{(100)(101)} < .0001.$$

Thus, we can use $N = 100$ to make our estimate:

$$\sum_{n=1}^{\infty} \frac{1}{n^2} \approx \sum_{n=1}^{100} \frac{1}{n^2} + \int_{101}^{\infty} \frac{1}{x^2}\,dx.$$

We use a calculator or computer to compute

$$\sum_{n=1}^{100} \frac{1}{N^2} \approx 1.63498$$

to five decimal places, and

$$\int_{101}^{\infty} \frac{1}{x^2}\,dx = \lim_{b\to\infty} \frac{-1}{b} + \frac{1}{101} = \frac{1}{101} \approx .0099.$$

Thus,

$$\sum_{n=1}^{\infty} \frac{1}{n^2} \approx 1.64488,$$

accurate to within .0001. ∎

p-series

Let's use the tests we have developed on an important family of series of the form

$$\sum_{n=1}^{\infty} \frac{1}{n^p},$$

where p is a fixed power. In Section 7.7 on improper integrals, we noted that

$$\int_{1}^{\infty} \frac{1}{x^p}\,dx$$

converges for $p > 1$ and diverges for $p \leq 1$.

For $p > 0$, $1/x^p$ is positive and decreasing on $[1, \infty)$, and we can apply the integral test to "p-series" of the form

$$\sum_{n=1}^{\infty} \frac{1}{n^p}$$

and conclude that $\sum_{n=1}^{\infty} \frac{1}{n^p}$ converges for $p > 1$ and diverges for $0 < p \leq 1$.

For $p = 0$, $\lim_{n\to\infty} \frac{1}{n^p} = 1$, and for $p < 0$, $\lim_{n\to\infty} \frac{1}{n^p} = \infty$. In either case, $\sum_{n=1}^{\infty} \frac{1}{n^p}$ diverges by the Nth term test. Summarizing, we have:

10.3 CONVERGENCE AND DIVERGENCE OF SERIES

$$\sum_{n=1}^{\infty} \frac{1}{n^p} \text{ converges if } p > 1 \text{ and diverges if } p \leq 1.$$

EXAMPLE 24 Determine the convergence or divergence of the following p-series:
(a) $\sum_{n=1}^{\infty} \frac{1}{n^3}$
(b) $\sum_{n=1}^{\infty} \frac{1}{n^{3/2}}$
(c) $\sum_{n=1}^{\infty} \frac{1}{n^{1.01}}$

(d) $\sum_{n=1}^{\infty} \frac{1}{\sqrt{n}}$
(e) $\sum_{n=1}^{\infty} \frac{1}{\sqrt[3]{n}}$
(f) $\sum_{n=1}^{\infty} \frac{1}{n^{0.99}}$

Solution The first three series,

$$\text{(a)} \sum_{n=1}^{\infty} \frac{1}{n^3} \qquad \text{(b)} \sum_{n=1}^{\infty} \frac{1}{n^{3/2}} \qquad \text{(c)} \sum_{n=1}^{\infty} \frac{1}{n^{1.01}},$$

all *converge* ($p = 3$, $p = 3/2$, $p = 1.01$, respectively). The last three series,

$$\text{(d)} \sum_{n=1}^{\infty} \frac{1}{\sqrt{n}} \qquad \text{(e)} \sum_{n=1}^{\infty} \frac{1}{\sqrt[3]{n}} \qquad \text{(f)} \sum_{n=1}^{\infty} \frac{1}{n^{0.99}},$$

however, all *diverge* ($p = \frac{1}{2}$, $p = \frac{1}{3}$, $p = 0.99$, respectively). ■

Alternating series test

Requirements:

The terms in the series are alternately positive and negative. Such a series can be written

$$\sum_{n=1}^{\infty} (-1)^{n+1} c_n \qquad (c_n > 0)$$

if the first term is positive, or

$$\sum_{n=1}^{\infty} (-1)^n c_n \qquad (c_n > 0)$$

if the first term is negative.

Criteria:

If the magnitudes of the terms are strictly decreasing, so that

$$c_n > c_{n+1} > 0$$

for every n, and if

$$\lim_{n \to \infty} c_n = 0,$$

then the alternating series *converges*. If $\lim_{n\to\infty} c_n \neq 0$, then the series *diverges* by the *Nth* term test.

EXAMPLE 25 Does the alternating harmonic series $\sum_{n=1}^{\infty} \frac{(-1)^{n+1}}{n}$ converge or diverge?

Solution Here $c_n = \frac{1}{n}$. The alternating series test applies since the terms alternate signs and

$$\frac{1}{n} > \frac{1}{n+1} > 0$$

for every n.

Since $\lim_{n\to\infty} \frac{1}{n} = 0$, the alternating series test tells us that the alternating harmonic series *converges*. ∎

Discussion:

Let's compute the first few partial sums of the alternating harmonic series:

$$S_1 = 1$$
$$S_2 = 1 - \frac{1}{2} = \frac{1}{2}$$
$$S_3 = 1 - \frac{1}{2} + \frac{1}{3} = \frac{5}{6}$$
$$S_4 = 1 - \frac{1}{2} + \frac{1}{3} - \frac{1}{4} = \frac{7}{12}$$
$$S_5 = 1 - \frac{1}{2} + \frac{1}{3} - \frac{1}{4} + \frac{1}{5} = \frac{47}{60}$$
$$S_6 = 1 - \frac{1}{2} + \frac{1}{3} - \frac{1}{4} + \frac{1}{5} - \frac{1}{6} = \frac{37}{60}.$$

If we plot these on a number line, we see that the even partial sums $S_2, S_4, S_6, \ldots$ strictly increase, while the odd partial sums strictly decrease (see Figure 10.7). A similar configuration of partial sums will be the case for any alternating series satisfying the criteria of the alternating series test. From the starting point S_1 we take a sequence of smaller and smaller steps, reversing direction each time. At any point in our "trip," all future points will be between our last two consecutive points. Since the distance between our last two points, S_{n-1} and S_n, is simply the size of our last step, c_n, and c_n approaches 0, the partial sums must converge to a single point.

10.3 CONVERGENCE AND DIVERGENCE OF SERIES

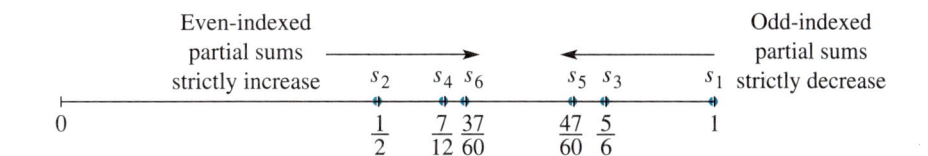

Figure 10.7 Partial sums of $\sum_{n=1}^{\infty} \frac{(-1)^{n+1}}{n}$.

Alternating series estimate:

This also means that we have a definite idea of how close we are to the limit L after any step, since L is between S_n and S_{n+1} for every n. If we use S_n as an approximation for L, then the remainder has magnitude

$$|R_n| < c_{n+1}.$$

We can tell whether S_n is on the right or left of the limit by the sign of the $(n+1)$st term.

EXAMPLE 26 Find the limit of the alternating harmonic series within .001.

Solution We choose $N = 1000$ so that

$$c_{N+1} = \frac{1}{N+1} = \frac{1}{1001} < .001.$$

Now we compute

$$\sum_{n=1}^{1000} \frac{(-1)^{n+1}}{n} \approx 0.69265$$

(to five decimal places). We mentioned that it is known that

$$\sum_{n=1}^{\infty} \frac{(-1)^{n+1}}{n} = \ln(2),$$

We note that $\ln(2) \approx 0.69315$ to five decimal places. ∎

EXERCISES for Section 10.3

For exercises 1–4: Apply the alternating series test to each of these series.
1. $\sum_{n=1}^{\infty} (-1)^n \ln(1 + \frac{1}{n})$
2. $\sum_{n=1}^{\infty} (-1)^n \frac{\ln(n)}{n}$
3. $\sum_{n=1}^{\infty} (-1)^n \frac{\arctan(n)}{n^9}$
4. $\sum_{n=1}^{\infty} \frac{(-1)^n}{2n^2 - 1}$

5. For each series in exercises 1-4 that converges, find its sum accurate to four decimal places.

For exercises 6–9: Apply the integral test to each of these series.

6. $\sum_{n=1}^{\infty} 3e^{-n}$ **7.** $\sum_{n=1}^{\infty} ne^{-n}$
8. $\sum_{n=1}^{\infty} \frac{\ln(n)}{n}$ **9.** $\sum_{n=1}^{\infty} \frac{1}{4n^2+9}$

10. For each series in exercises 6-9 that converges, find its sum accurate to four decimal places.

For exercises 11–12: Explain why the integral test does not apply to each of these series.

11. $\sum_{n=1}^{\infty} e^{-n} \sin n$ **12.** $\sum_{n=1}^{\infty} \frac{(-1)^n}{n}$

---

13. Given the series $\sum_{n=1}^{\infty} \frac{1}{n^6}$, find the least positive integer N such that the remainder tail $R_N = \sum_{n=N+1}^{\infty} \frac{1}{n^6}$ is less than 2×10^{-11}. Find S_N, and give an upper and lower bound for the sum of the series.

14. Given the series $\sum_{n=0}^{\infty} \frac{(-1)^n}{n!}$, find the least positive integer N such that $|R_N| < .000005$. The value S_N will approximate the series sum accurate to 5 decimal places. (Note: It is known that $\sum_{n=0}^{\infty} \frac{(-1)^n}{n!} = \frac{1}{e}$. Use this to check your result.)

15. Show that the series $\sum_{n=1}^{\infty} \ln(1 + \frac{1}{n})$ *diverges* using the integral test. Use integration by parts, letting $u = \ln(1 + \frac{1}{x})$ and $dv = dx$. (Note: When evaluating the integral, you will need to use L'Hôpital's rule, since the indeterminate form $0 \cdot \infty$ appears.)

For exercises 16–20: Explain why each of the given statements is *false* by providing an example that contradicts the statement.

16. If $\lim_{n \to \infty} a_n = 0$, then the series $\sum_{n=1}^{\infty} a_n$, $a_n > 0$ for all n, converges.

17. If the sequence $\{a_n\}_{n=1}^{\infty}$ converges, then the series $\sum_{n=1}^{\infty} a_n$ converges.

18. If $\sum_{n=1}^{\infty} a_n$ diverges (where $a_n > 0$ for all n), then the alternating series $\sum_{n=1}^{\infty} (-1)^n a_n$ also diverges.

19. If $\sum_{n=1}^{\infty} a_n$ and $\sum_{n=1}^{\infty} b_n$ both diverge, then $\sum_{n=1}^{\infty} (a_n + b_n)$ must also diverge.

20. If $\sum_{n=1}^{\infty} a_n$ converges, then $\sum_{n=1}^{\infty} (-1)^n a_n$ must also converge.

10.4 MORE TESTS OF CONVERGENCE—ABSOLUTE AND CONDITIONAL CONVERGENCE

Sometimes we can judge the convergence or divergence of one series by comparing it to another series whose behavior is known. The next two tests are of this type.

10.4 MORE TESTS OF CONVERGENCE—ABSOLUTE AND CONDITIONAL CONVERGENCE

Comparison test

Requirements:

The two series $\sum_{n=1}^{\infty} a_n$ and $\sum_{n=1}^{\infty} b_n$ must each have nonnegative terms. That is,

$$a_n \geq 0, \qquad b_n \geq 0 \qquad \text{for all } n.$$

Criteria:

If $a_n \leq b_n$ for each n, and $\sum_{n=1}^{\infty} b_n$ converges, then $\sum_{n=1}^{\infty} a_n$ converges.

If $b_n \leq a_n$ for each n, and $\sum_{n=1}^{\infty} b_n$ diverges, then $\sum_{n=1}^{\infty} a_n$ diverges.

Discussion:

The requirement that the terms a_n and b_n all be nonnegative guarantees that the partial sums of both series are *monotonically increasing*; hence, either series will converge *if and only if* its partial sums are *bounded*.

The way this test is used in practice is as follows: If the behavior of a given series

$$\sum_{n=1}^{\infty} a_n$$

is unknown and each term $a_n \geq 0$, then we search for (or devise) a comparison series $\sum_{n=1}^{\infty} b_n$ with $b_n \geq 0$.

If we suspect that $\sum_{n=1}^{\infty} a_n$ converges, then we search for a "larger" series $\sum_{n=1}^{\infty} b_n$ (in other words, $b_n \geq a_n$ for each n) that we know converges.

If we suspect that $\sum_{n=1}^{\infty} a_n$ diverges, then we search for a "smaller" series $\sum_{n=1}^{\infty} b_n$ (in other words, $b_n \leq a_n$ for each n) that we know diverges.

In either case, success in finding the series $\sum_{n=1}^{\infty} b_n$ with the property we desire means that we can use the comparison test to deduce the convergence or divergence of the original series $\sum_{n=1}^{\infty} a_n$.

What are good series $\sum_{n=1}^{\infty} b_n$ to look for? In general, any series whose behavior you know is a candidate. Geometric series and p series are good candidates, because they provide us with a large storehouse of both convergent and divergent series.

How do you choose a particular series $\sum_{n=1}^{\infty} b_n$? If you're looking for a "larger" series, try making some simplifying changes to the terms a_n that will result in larger terms (for example, decreasing denominators, increasing numerators, adding a positive amount, etc.). If you're looking for a "smaller" series, try making simplifying changes that will result in smaller terms (increasing denominators, decreasing numerators, subtracting a positive amount, etc.).

EXAMPLE 27 Does the series $\sum_{n=1}^{\infty} \frac{2^n}{3^n + n^4}$ converge or diverge?

Solution To get a feel for the series, let's write the first few terms in the sum:

$$\frac{2}{4} + \frac{4}{25} + \frac{8}{108} + \frac{16}{337} + \cdots.$$

These terms appear to be decreasing in size quite quickly, and the form of the series suggests that we try comparing it to a geometric series. Since we suspect that the series converges, we look for a "larger" series that we know converges.

If we remove n^4 from the denominator, we have

$$a_n = \frac{2^n}{3^n + n^4} \leq \frac{2^n}{3^n} = b_n$$

for each n. The geometric series $\sum_{n=1}^{\infty} (\frac{2}{3})^n$ converges ($-1 < r = 2/3 < 1$), so

$$\sum_{n=1}^{\infty} \frac{2^n}{3^n + n^4}$$

converges by the comparison test. ∎

EXAMPLE 28 Does the series $\sum_{n=1}^{\infty} \frac{1}{n + \sqrt{n}}$ converge or diverge?

Solution If we remove $\sqrt{n}$ from the denominator, we get a "larger" series,

$$\sum_{n=1}^{\infty} \frac{1}{n},$$

that diverges. Note, though, that this result is of no help to us. (Similarly, finding a "smaller" series that converges will tell us no information about the original series). If, however, we replace $\sqrt{n}$ by n, we obtain a *"smaller"* series,

$$\sum_{n=1}^{\infty} \frac{1}{n+n} = \sum_{n=1}^{\infty} \frac{1}{2n} = \frac{1}{2} \sum_{n=1}^{\infty} \frac{1}{n},$$

that we know diverges. We can conclude that $\sum_{n=1}^{\infty} \frac{1}{n+\sqrt{n}}$ *diverges* by the comparison test. ∎

EXAMPLE 29 Why can't the comparison test be used on the alternating harmonic series?

Solution The comparison test cannot be used on the alternating harmonic series

$$\sum_{n=1}^{\infty} \frac{(-1)^{n+1}}{n}$$

because some of its terms are negative. (Check the requirements of the comparison test.) ∎

Since any partial sum S_N is necessarily finite (adding finitely many real numbers always yields a finite result), it is always the remainder tail R_N that decides whether or not a series converges. This means that the requirements and criteria of the comparison test need only be satisfied by all the terms in a common remainder tail of two series being compared.

Consider the two series

$$\sum_{n=1}^{\infty} \frac{1}{2^n + 1} = \frac{1}{3} + \frac{1}{5} + \frac{1}{9} + \frac{1}{17} + \cdots$$

and

$$\sum_{n=1}^{\infty} \frac{1}{n^2 - 2} = -1 + \frac{1}{2} + \frac{1}{7} + \frac{1}{14} + \cdots.$$

Note that the first term of the second series is negative and smaller than the first term of the first series. However, starting with the second terms, the second series has consistently larger terms. This means that the comparison test can be used with these two series. That is, if the first diverges, so does the second. If the second converges, so does the first. (By the way, which is it?)

Limit comparison test

Requirements:

Both series $\sum_{n=1}^{\infty} a_n$ and $\sum_{n=1}^{\infty} b_n$ must have positive terms

$$a_n > 0, \quad b_n > 0 \quad \text{for all } n.$$

Criteria:

If $\lim_{n \to \infty} \frac{a_n}{b_n} = \infty$ and $\sum_{n=1}^{\infty} b_n$ diverges, then $\sum_{n=1}^{\infty} a_n$ diverges.

If $\lim_{n \to \infty} \frac{a_n}{b_n} = 0$ and $\sum_{n=1}^{\infty} b_n$ converges, then $\sum_{n=1}^{\infty} a_n$ converges.

If $\lim_{n \to \infty} \frac{a_n}{b_n} = L \neq 0$ (L a real number), then $\sum_{n=1}^{\infty} a_n$ and $\sum_{n=1}^{\infty} b_n$ behave the same. That is, either both series converge or both diverge.

Discussion:

Recall that a necessary (but not sufficient) condition for a series $\sum_{n=1}^{\infty} a_n$ to converge is that $\lim_{n \to \infty} a_n = 0$ (the Nth term test). Even if the terms do approach 0, they might not approach 0 "fast enough" for the series to converge. (The harmonic series is the best known example of this phenomenon.)

The limit comparison test looks at the ratio of the terms a_n/b_n from two series. If both $\lim_{n \to \infty} a_n = 0$ and $\lim_{n \to \infty} b_n = 0$, then

$$\lim_{n \to \infty} \frac{a_n}{b_n}$$

gives us a measure of the relative speeds that a_n and b_n approach 0.

If $\lim_{n \to \infty} \frac{a_n}{b_n} = \infty$, then a_n approaches 0 much "slower" than b_n.

If $\lim_{n \to \infty} \frac{a_n}{b_n} = 0$, then a_n approaches 0 much "faster" than b_n.

If $\lim_{n \to \infty} \frac{a_n}{b_n} = L \neq 0$, then a_n and b_n approach 0 at "comparable" speeds.

Like the comparison test, the limit comparison test is used to decide the divergence or convergence of a series $\sum_{n=1}^{\infty} a_n$ by choosing or devising a comparison sequence $\sum_{n=1}^{\infty} b_n$ whose behavior is known. If a_n can be described by a formula, then a good choice for b_n is to use only the dominant (fastest growing) terms of the formula. Constant coefficients can usually be changed to $+1$ or -1, depending on the sign.

EXAMPLE 30 Does $\sum_{n=1}^{\infty} \frac{6}{\sqrt{17n^3 + 5}}$ converge or diverge?

Solution The fastest growing term in the denominator under the radical sign is $17n^3$. We'll choose $b_n = \frac{1}{\sqrt{n^3}}$, for the purposes of comparison. Now,

$$\lim_{n \to \infty} \frac{a_n}{b_n} = \lim_{n \to \infty} \frac{\frac{6}{\sqrt{17n^3 + 5}}}{\frac{1}{\sqrt{n^3}}} = \lim_{n \to \infty} \frac{6\sqrt{n^3}}{\sqrt{17n^3 + 5}} = \lim_{n \to \infty} 6\sqrt{\frac{n^3}{17n^3 + 5}}.$$

Since $\lim_{n \to \infty} \frac{n^3}{17n^3 + 5} = \frac{1}{17}$, we have

$$\lim_{n \to \infty} \frac{a_n}{b_n} = \frac{6}{\sqrt{17}} \neq 0.$$

The limit comparison test tells us that both $\sum_{n=1}^{\infty} a_n$ and $\sum_{n=1}^{\infty} b_n$ have the same behavior. Since

$$\sum_{n=1}^{\infty} b_n = \sum_{n=1}^{\infty} \frac{1}{\sqrt{n^3}} = \sum_{n=1}^{\infty} \frac{1}{n^{3/2}}$$

converges, we conclude that $\sum_{n=1}^{\infty} \frac{6}{\sqrt{17n^3 + 5}}$ also *converges*. ∎

10.4 MORE TESTS OF CONVERGENCE—ABSOLUTE AND CONDITIONAL CONVERGENCE

EXAMPLE 31 Why can't the limit comparison test be applied to the alternating harmonic series?

Solution Again, the limit comparison test cannot be used on the alternating harmonic series

$$\sum_{n=1}^{\infty} \frac{(-1)^{n+1}}{n}$$

because some of the terms are negative (check the requirements of the limit comparison test). ∎

Absolute and conditional convergence

When a series has both positive and negative terms, such as an alternating series, the notions of *absolute convergence* and *conditional convergence* become relevant.

Definition 4

Suppose that a series $\sum_{n=1}^{\infty} a_n$ converges.
If the series $\sum |a_n|$ converges, we say that $\sum_{n=1}^{\infty} a_n$ is **absolutely convergent**.
If the series $\sum |a_n|$ diverges, we say that $\sum_{n=1}^{\infty} a_n$ is **conditionally convergent**.

EXAMPLE 32 Are the following series absolutely convergent, conditionally convergent, or divergent?

(a) $\sum_{n=1}^{\infty} \frac{(-1)^{n+1}}{n}$ (b) $\sum_{n=1}^{\infty} \left(-\frac{1}{2}\right)^n$

Solution (a) The alternating harmonic series is conditionally convergent because

$$\sum_{n=1}^{\infty} \frac{(-1)^{n+1}}{n}$$

converges, but

$$\sum_{n=1}^{\infty} \left| \frac{(-1)^{n+1}}{n} \right| = \sum_{n=1}^{\infty} \frac{1}{n}$$

diverges.

(b) The geometric series

$$\sum_{n=1}^{\infty}(-\frac{1}{2})^n$$

is absolutely convergent because

$$\sum_{n=1}^{\infty}\left|(-\frac{1}{2})^n\right| = \sum_{n=1}^{\infty}(\frac{1}{2})^n$$

converges. ∎

A useful result is the following theorem.

Theorem 10.3

Hypothesis: $\sum_{n=1}^{\infty}|a_n|$ converges.
Conclusion: $\sum_{n=1}^{\infty}a_n$ converges.

In other words, if we replace each term in a series by its absolute value and the resulting series converges, then the original series must converge. In fewer words, *an absolutely convergent series converges*.

Reasoning Suppose that we have a series $\sum_{n=1}^{\infty}a_n$ and we know that

$$\sum_{n=1}^{\infty}|a_n|$$

converges. If we let $b_n = a_n + |a_n|$, we'll either get $b_n = 0$ (when a_n is negative) or $b_n = 2|a_n|$ (when a_n is positive). Either way, we can say that

$$0 \leq b_n = a_n + |a_n| \leq 2|a_n|.$$

Using the comparison test, we can see that $\sum_{n=1}^{\infty}b_n$ converges because $\sum_{n=1}^{\infty}2|a_n|$ converges. Finally, that means $\sum_{n=1}^{\infty}a_n = \sum_{n=1}^{\infty}(b_n - |a_n|)$ must converge also. □

This theorem is useful because we have several tests requiring all our terms to be nonnegative (the integral test, the comparison test, and the limit comparison test). If we have a series $\sum_{n=1}^{\infty}a_n$ not satisfying this requirement, we can try replacing a_n by $|a_n|$. All the terms of $\sum_{n=1}^{\infty}|a_n|$ are nonnegative, so there are more tests that are available to use. If we find $\sum_{n=1}^{\infty}|a_n|$ converges, then this theorem tells us that $\sum_{n=1}^{\infty}a_n$ must also converge. However, if we find $\sum_{n=1}^{\infty}|a_n|$ diverges, we really don't have any more information about $\sum_{n=1}^{\infty}a_n$, because it could be conditionally convergent.

10.4 MORE TESTS OF CONVERGENCE—ABSOLUTE AND CONDITIONAL CONVERGENCE

There are two convergence tests for series that can be used to test for absolute convergence.

Ratio test

Requirements:

$\sum_{n=1}^{\infty} a_n$ can be any series.

Criteria:

If $\lim_{n \to \infty} \left| \frac{a_{n+1}}{a_n} \right| = L$, then

$\sum_{n=1}^{\infty} a_n$ is absolutely convergent when $L < 1$,

$\sum_{n=1}^{\infty} a_n$ is divergent when $L > 1$,

and the ratio test *fails to provide any information when* $L = 1$.

Discussion:

When $\lim_{n \to \infty} \left| \frac{a_{n+1}}{a_n} \right| = L < 1$, this tells us that the terms $|a_n|$ are decreasing fast enough for the series $\sum |a_n|$ to converge. (Essentially, for some remainder tail of the series $\sum |a_n|$, we can find a larger convergent geometric series by choosing a factor $L < r < 1$ and use the comparison test.) If $\lim_{n \to \infty} \left| \frac{a_{n+1}}{a_n} \right| = L > 1$, then the terms a_n are not approaching 0 and the series $\sum_{n=1}^{\infty} a_n$ must diverge.

The ratio test is often useful when the index n appears as an exponent in the formula for a_n (as it does for geometric series), or when the formula for a_n involves $n!$ (n factorial).

EXAMPLE 33 Use the ratio test to test for the absolute convergence of the series

$$\sum_{n=0}^{\infty} \frac{(-1)^n}{n!}.$$

Solution The first few terms in the sum are

$$1 + (-1) + \frac{1}{2} - \frac{1}{6} + \frac{1}{24} - \frac{1}{120} + \cdots.$$

This is an alternating series whose terms strictly decrease in magnitude. Since

$$\lim_{n \to \infty} \frac{1}{n!} = 0,$$

the series converges. To check for absolute convergence, we first form the ratio

$$\left|\frac{a_{n+1}}{a_n}\right| = \left|\frac{\frac{(-1)^{n+1}}{(n+1)!}}{\frac{(-1)^n}{n!}}\right| = \frac{n!}{(n+1)!} = \frac{1}{n+1}.$$

(If the last step in the above equality seems confusing to you, try evaluating $\frac{n!}{(n+1)!}$ for $n = 1, 2, 3, 4$ to see the pattern.)

Now, $\lim_{n\to\infty}\left|\frac{a_{n+1}}{a_n}\right| = \lim_{n\to\infty}\frac{1}{n+1} = 0 < 1$. So the ratio test tells us that the series $\sum_{n=0}^{\infty}\frac{(-1)^n}{n!}$ is absolutely convergent. ∎

EXAMPLE 34 Use the ratio test to test for the absolute convergence of the series $\sum_{n=1}^{\infty}\frac{2^n}{n^5}$.

Solution The first few terms in the sum are

$$2 + \frac{1}{8} + \frac{8}{243} + \frac{16}{1024} + \cdots$$

and at first glance, these terms appear to be approaching 0 rapidly. However, if we apply the ratio test, we find

$$\lim_{n\to\infty}\left|\frac{a_{n+1}}{a_n}\right| = \lim_{n\to\infty}\frac{\frac{2^{n+1}}{(n+1)^5}}{\frac{2^n}{n^5}} = \lim_{n\to\infty} 2\cdot\left(\frac{n}{n+1}\right)^5 = 2 > 1.$$

The ratio test tells us that this series diverges. ∎

EXAMPLE 35 Use the ratio test to test for the absolute convergence of the alternating harmonic series $\sum_{n=1}^{\infty}\frac{(-1)^{n+1}}{n}$.

Solution $\lim_{n\to\infty}\left|\frac{a_{n+1}}{a_n}\right| = \lim_{n\to\infty}\left|\frac{\frac{(-1)^{n+2}}{(n+1)}}{\frac{(-1)^{n+1}}{n}}\right| = \lim_{n\to\infty}\frac{n}{n+1} = 1$. The ratio test fails to provide us any information in this case (though we know that the alternating harmonic series conditionally converges). ∎

Root test

Requirements:

$\sum_{n=1}^{\infty} a_n$ can be any series.

10.4 MORE TESTS OF CONVERGENCE—ABSOLUTE AND CONDITIONAL CONVERGENCE

Criteria:

If $\lim_{n \to \infty} \sqrt[n]{|a_n|} = L$, then

$\sum_{n=1}^{\infty} a_n$ is absolutely convergent when $L < 1$,

$\sum_{n=1}^{\infty} a_n$ is divergent when $L > 1$,

and the root test *fails to provide any information* when $L = 1$.

Discussion:

The root test is more powerful than the ratio test in the following sense: whenever the ratio test provides information, so will the root test; however, there are instances where the ratio test fails but the root test does not. In practice, you will probably find the ratio test much easier to apply. (Essentially, if $L \neq 1$, then the root test follows from applying the limit comparison test to the series $\sum |a_n|$ and the geometric series $\sum r^n$, where r is chosen to be some number between 1 and L.)

EXAMPLE 36 Use the root test to test for the absolute convergence of the series

$$\sum_{n=1}^{\infty} \frac{(-1)^n}{n^n}.$$

Solution

$$\lim_{n \to \infty} \sqrt[n]{|a_n|} = \lim_{n \to \infty} \sqrt[n]{\frac{1}{n^n}} = \lim_{n \to \infty} \frac{1}{n} = 0 < 1.$$

The root test tells us that this series is absolutely convergent. ∎

Summary of Convergence Tests

Test	Form	Converges	Diverges
Nth term	$\sum_{n=1}^{\infty} a_n$	no information	$\lim_{n \to \infty} a_n \neq 0$
Integral test	$\sum_{n=1}^{\infty} a_n$ where $a_n = f(n)$	$\int_1^{\infty} f(x)\, dx$ converges	$\int_1^{\infty} f(x)\, dx$ diverges

(NOTE: f must be continuous, positive, and decreasing.)

Integral test estimate:

$$\sum_{n=1}^{\infty} a_n = \sum_{n=1}^{N} a_n + \int_{N+1}^{\infty} f(x)\, dx + \text{error}, \quad \text{where error} \leq \int_{N}^{N+1} f(x)\, dx.$$

Alternating	$\sum_{n=1}^{\infty} (-1)^{n+1} c_n$ $0 < c_{n+1} < c_n$	$\lim_{n \to \infty} c_n = 0$	$\lim_{n \to \infty} c_n \neq 0$

Alternating series estimate:

$$\sum_{n=1}^{\infty}(-1)^{n-1}c_n = \sum_{n=1}^{N}(-1)^{n-1}c_n \pm \text{error}, \quad \text{where the error} \leq c_{N+1}.$$

Summary of Convergence Tests

Test	Form	Converges	Diverges
Comparison	$\sum_{n=1}^{\infty} a_n$	$0 \leq a_n \leq b_n$, and $\sum_{n=1}^{\infty} b_n$ converges	$0 \leq b_n \leq a_n$, and $\sum_{n=1}^{\infty} b_n$ diverges
Limit comparison	$\sum_{n=1}^{\infty} a_n$ $(a_n, b_n > 0)$	$\lim_{n\to\infty} \frac{a_n}{b_n} = L \geq 0$ and $\sum_{n=1}^{\infty} b_n$ converges	$\lim_{n\to\infty} \frac{a_n}{b_n} = L > 0$ or ∞ and $\sum_{n=1}^{\infty} b_n$ diverges
Ratio	$\sum_{n=1}^{\infty} a_n$	$\lim_{n\to\infty} \left|\frac{a_{n+1}}{a_n}\right| < 1$	$\lim_{n\to\infty} \left|\frac{a_{n+1}}{a_n}\right| > 1$
Root	$\sum_{n=1}^{\infty} a_n$	$\lim_{n\to\infty} \sqrt[n]{|a_n|} < 1$	$\lim_{n\to\infty} \sqrt[n]{|a_n|} > 1$

Some Special Series

Series	Form	Converges	Diverges
Telescoping	$\sum_{n=1}^{\infty}(b_n - b_{n+1})$	$\lim_{n\to\infty} b_n = 0$	$\lim_{n\to\infty} b_n \neq 0$
Geometric	$\sum_{n=0}^{\infty} ar^n$	$|r| < 1$	$|r| \geq 1$
p-series	$\sum_{n=1}^{\infty} \frac{1}{n^p}$	$p > 1$	$p \leq 1$

EXERCISES for Section 10.4

For exercises 1–5: Determine whether the given series converges or diverges, using the indicated test.

1. $\sum_{n=0}^{\infty}(-1)^n \frac{e^n}{e^{2n}+1}$ (Use ratio test.)
2. $\sum_{n=1}^{\infty} \frac{(1-\pi)^n}{2^{n+1}}$ (Use root test.)
3. $\sum_{n=1}^{\infty} \frac{2^{3n}}{7^n}$ (Use root test.)
4. $\sum_{n=1}^{\infty} \frac{n! n^2}{(2n)!}$ (Use ratio test.)

10.4 MORE TESTS OF CONVERGENCE—ABSOLUTE AND CONDITIONAL CONVERGENCE

5. $\sum_{n=1}^{\infty} \frac{(-1)^{n+1} 2^{3n+2}}{n^n}$ (Use root test.)

6. Consider the series
$$\sum_{n=1}^{\infty} \frac{n}{2n^2 - 1}.$$
Does the ratio test indicate convergence, divergence, or no information?

7. Consider the series
$$\sum_{n=1}^{\infty} \frac{n}{2n^2 - 1}.$$
Does the root test indicate convergence, divergence, or no information?

8. Use the ratio test to show that the series $\sum_{n=0}^{\infty} \frac{(-1)^n}{(2n)!}$ absolutely converges. Find the smallest N such that S_N is accurate to 6 decimal places. Give the estimate for the sum of the series accurate to 6 decimal places.

9. Determine whether $\sum_{n=1}^{\infty} (-1)^{n+1} \frac{n^{2n}}{(3n^2+1)^n}$ absolutely converges, converges conditionally, or diverges.

10. Show that the ratio test is inconclusive when applied to the series
$$\sum_{n=1}^{\infty} (-1)^{n+1} \frac{\pi^{1/n}}{n}.$$
Then use any tests necessary to determine whether the series absolutely converges, conditionally converges, or diverges.

11. Suppose that I have a p-series $\sum_{n=1}^{\infty} \frac{1}{n^p}$ where p is a specific positive number. Show that the ratio test *always* fails to give any information about the convergence or divergence of this series.

12. Suppose that I have an alternating p-series $\sum_{n=1}^{\infty} \frac{(-1)^n}{n^p}$ where p is a specific positive number. For what values of p will this series converge?

13. Determine whether $\sum_{n=1}^{\infty} (-1)^{n+1} 3e^{-n}$ converges absolutely, converges conditionally, or diverges.

14. Find the sum of the series $\sum_{n=1}^{\infty} \frac{2^n}{3^{n-1}}$.

15. Decide whether $\sum_{n=1}^{\infty} \frac{1}{3+5^n}$ converges or diverges by using the comparison test with the geometric series $\sum_{n=1}^{\infty} \frac{1}{5^n}$.

16. Decide whether $\sum_{n=1}^{\infty} \frac{e^{1/n}}{n}$ converges or diverges by using the limit comparison test with $\sum_{n=1}^{\infty} \frac{1}{n}$.

For exercises 17–38: For each series, indicate whether it converges absolutely, converges conditionally, or diverges. Indicate what test or tests you use in each case.

17. $\sum_{n=1}^{\infty} \frac{n^2}{\sqrt{n^5}}$

18. $\sum_{n=1}^{\infty} \frac{1}{(-5)^n}$

19. $\sum_{n=1}^{\infty} \frac{1}{n+1000000}$

20. $\sum_{n=1}^{\infty} (-e)^{-n}$

21. $\sum_{n=1}^{\infty} \frac{\sin^2 \frac{1}{n}}{2^n}$

22. $\sum_{n=1}^{\infty} \frac{1}{2^{1/n}}$

23. $\sum_{n=1}^{\infty} \frac{n^n}{n!}$

24. $\sum_{n=1}^{\infty} (2-e)^n$

25. $\sum_{n=1}^{\infty} \frac{3^n}{n!}$

26. $\sum_{n=1}^{\infty} \frac{n \ln n}{n^2+1}$

27. $\sum_{n=1}^{\infty} \frac{1}{\sqrt{n}}$

28. $\sum_{n=1}^{\infty} \frac{(-1)^n}{\sqrt{n}}$

29. $\sum_{n=1}^{\infty} \frac{(-1)^n}{\sqrt[n]{10}}$

30. $\sum_{n=1}^{\infty} \frac{e^n}{2^n}$

31. $\sum_{n=1}^{\infty} \sin(n\pi/2)$

32. $\sum_{n=1}^{\infty} \frac{n^2}{n^3+n}$

33. $\sum_{n=2}^{\infty} \frac{(-1)^n}{\ln n}$

34. $\sum_{n=1}^{\infty} \left(\frac{n-1}{n}\right)^n$

35. $\sum_{n=1}^{\infty} \frac{3^n}{2^n+4^n}$

36. $\sum_{n=1}^{\infty} n \sin(1/n)$

37. $\sum_{n=1}^{\infty} \frac{(-5)^n}{n!}$

38. $\sum_{n=1}^{\infty} n e^{-n^2}$

10.5 POWER SERIES

A **power series** with real number coefficients has the form

$$\sum_{n=0}^{\infty} c_n x^n = c_0 + c_1 x + c_2 x^2 + c_3 x^3 + \cdots + c_n x^n + \cdots,$$

where each c_n is a real number and x is a variable. Roughly speaking, a power series is a polynomial with infinitely many terms. More precisely, it is the limit of a sequence of polynomials:

10.5 POWER SERIES

$$p_0(x) = c_0$$
$$p_1(x) = c_0 + c_1 x$$
$$p_2(x) = c_0 + c_1 x + c_2 x^2$$
$$p_3(x) = c_0 + c_1 x + c_2 x^2 + c_3 x^3$$
$$\vdots$$

where $p_n(x) = c_0 + c_1 x + c_2 x^2 + c_3 x^3 + \cdots + c_n x^n$ is a polynomial of degree n. These polynomials play the role of partial sums, and

$$\sum_{n=0}^{\infty} c_n x^n = \lim_{n \to \infty} p_n(x).$$

We'll see that power series may be manipulated much like polynomials. Polynomial functions are particularly easy to work with in calculus, and it is remarkable that so many non-polynomial functions (such as the trigonometric functions) can be represented as power series.

Sometimes it is convenient to have a particular real number a as a point of reference. In this case, we may wish to write a power series in terms of $(x - a)$ instead of x:

$$\sum_{n=0}^{\infty} c_n (x-a)^n = c_0 + c_1 (x-a) + c_2 (x-a)^2 + c_3 (x-a)^3 + \cdots + c_n (x-a)^n + \cdots$$

A power series written this way is said to be represented *about* $x = a$. A power series written in terms of powers of x is said to be represented about $x = 0$. For example,

$$\sum_{n=0}^{\infty} x^n = 1 + x + x^2 + x^3 + \cdots$$

is a power series represented about $x = 0$, and

$$\sum_{n=0}^{\infty} \frac{(x-1)^n}{n} = 1 + (x-1) + \frac{(x-1)^2}{2} + \frac{(x-1)^3}{3} + \cdots$$

is a power series represented about $x = 1$.

To evaluate a power series $\sum_{n=0}^{\infty} c_n x^n$ at a specific input value $x = x_0$, we substitute the value x_0 and obtain a "regular" series:

$$\sum_{n=0}^{\infty} c_n x_0^n$$

that may or may not converge to finite value. This raises the question: for exactly which input values x does the power series $\sum_{n=0}^{\infty} c_n x^n$ converge?

Radius and interval of convergence

If $\sum_{n=0}^{\infty} c_n x^n$ is a power series represented about $x = 0$, then there is always one value x that guarantees convergence—namely, $x = 0$. Note that substituting $x = 0$ makes all the terms 0 except, perhaps, the constant term c_0.

Similarly, if $\sum_{n=0}^{\infty} c_n (x-a)^n$ is a power series represented about $x = a$, then substituting $x = a$ guarantees convergence. To find the entire set of input values for which a power series converges, the ratio or root test may be used.

EXAMPLE 37 Find the set of input values x for which the power series $\sum_{n=0}^{\infty} \dfrac{x^n}{3^{2n}}$ converges.

Solution Using the ratio test, we find that

$$\lim_{n \to \infty} \left| \frac{x^{n+1}/3^{2(n+1)}}{x^n/3^{2n}} \right| = \lim_{n \to \infty} |x| \frac{3^{2n}}{3^{2n+2}} = \lim_{n \to \infty} \frac{|x|}{9} = \frac{|x|}{9}.$$

This limit value depends on x, and the ratio test tells us that the power series

$$\sum_{n=0}^{\infty} \frac{x^n}{3^{2n}} \text{ converges if } \frac{|x|}{9} < 1, \text{ and } \sum_{n=0}^{\infty} \frac{x^n}{3^{2n}} \text{ diverges if } \frac{|x|}{9} > 1.$$

We still need to determine the behavior for $\dfrac{|x|}{9} = 1$ or, equivalently, for $x = \pm 9$. We can substitute these two values directly back into the power series. For $x = 9$, we have

$$\sum_{n=0}^{\infty} \frac{9^n}{3^{2n}} = \sum_{n=0}^{\infty} 1,$$

which *diverges*. For $x = -9$, we have

$$\sum_{n=0}^{\infty} \frac{(-9)^n}{3^{2n}} = \sum_{n=0}^{\infty} (-1)^n,$$

which also *diverges*. We can conclude that the power series converges precisely when $\dfrac{|x|}{9} < 1$ or, equivalently, on the interval

$$\{x \ : \ -9 < x < 9\}.$$

10.5 POWER SERIES

We note that the same conclusion can be reached by using the root test. In this case, we examine

$$\lim_{n\to\infty} \sqrt[n]{\left|\frac{x^n}{3^{2n}}\right|} = \lim_{n\to\infty} \frac{|x|}{3^2} = \frac{|x|}{9}.$$

The criteria for the root test are the same as for the ratio test, so the analysis is exactly the same. ∎

The power series $\sum_{n=0}^{\infty} \frac{x^n}{3^{2n}}$ is represented about $x = 0$. Note that the set of values for which this power series converges is an *interval* $(-9, 9)$ *centered* at 0. This is no coincidence.

Theorem 10.4

Hypothesis: $\sum_{n=0}^{\infty} c_n(x-a)^n$ is a power series represented about $x = a$.
Conclusion: The set of values for which the power series converges is an *interval* centered at a.

This interval is called the **interval of convergence** for the power series. The radius R of this interval (in other words, the distance from the center a to either endpoint) is called the **radius of convergence**.

The interval of convergence may take any of the forms

$$[a-R, a+R], \quad [a-R, a+R), \quad (a-R, a+R], \quad (a-R, a+R),$$

depending on which endpoints, if any, are included.

Reasoning If we use the root test on any power series $\sum_{n=0}^{\infty} c_n(x-a)^n$, our criteria for convergence requires us to find the values x such that

$$\lim_{n\to\infty} \sqrt[n]{|c_n(x-a)^n|} = \lim_{n\to\infty} \sqrt[n]{|c_n|} \, |x-a| < 1.$$

If $\lim_{n\to\infty} \sqrt[n]{|c_n|} = \infty$, then the power series converges only for $x = a$. In this situation, $R = 0$ is the radius of convergence, and the interval of convergence is the single point $\{a\}$.

If $\lim_{n\to\infty} \sqrt[n]{|c_n|} = 0$, then the power series converges for *any* real number x. In this situation, we say $R = \infty$ is the radius of convergence, and the interval of convergence is the whole real line $(-\infty, \infty)$.

If $\lim_{n\to\infty} \sqrt[n]{|c_n|} = L \neq 0$, then the power series converges for those values x satisfying

$$L|x-a| < 1 \quad \text{or} \quad |x-a| < R,$$

where we have written $R = 1/L$. In addition, the power series may or may not converge at the two values $x = a - R$ or $x = a + R$. In any case, the set of values in this situation is a bounded interval centered at $x = a$. □

In general, we can use the root or ratio test to find an open interval of convergence $(a - R, a + R)$ and then check the endpoints by direct substitution into the power series.

EXAMPLE 38 Find the interval and radius of convergence for the power series

$$\sum_{n=0}^{\infty} \frac{3^n (x+2)^n}{n!}.$$

Solution Using the ratio test, we have

$$\lim_{n \to \infty} \frac{|3^{n+1}(x+2)^{n+1}/(n+1)!|}{|3^n(x+2)^n/n!|} = \lim_{n \to \infty} \frac{3|x+2|}{n+1} = 0$$

for any value x. Therefore, the interval of convergence is $(-\infty, \infty)$ and the radius of convergence is $R = \infty$. ■

EXAMPLE 39 Find the interval and radius of convergence for the power series

$$\sum_{n=0}^{\infty} \frac{3^n (x+2)^n}{n+1}.$$

Solution Using the ratio test, we have

$$\lim_{n \to \infty} \frac{|3^{n+1}(x+2)^{n+1}/(n+2)|}{|3^n(x+2)^n/(n+1)|} = \lim_{n \to \infty} \frac{3(n+1)|x+2|}{n+2} = 3|x+2|,$$

since $\lim_{n \to \infty} \frac{n+1}{n+2} = 1$. By the ratio test, the power series converges for any value x satisfying

$$3|x+2| < 1 \quad \text{or} \quad \frac{-7}{3} < x < \frac{-5}{3}.$$

We must check the behavior at the endpoints $x = -7/3$ and $x = -5/3$ by direct substitution into the power series $\sum_{n=0}^{\infty} \frac{3^n(x+2)^n}{n+1}$.

For $x = -7/3$, we have $\sum_{n=0}^{\infty} \frac{3^n(-1/3)^n}{n+1} = \sum_{n=0}^{\infty} \frac{(-1)^n}{n+1}$. This series satisfies the requirements of the alternating series test and *converges*. Therefore, we include $x = -7/3$ in the interval of convergence.

For $x = -5/3$, we have $\sum_{n=0}^{\infty} \frac{3^n(1/3)^n}{n+1} = \sum_{n=0}^{\infty} \frac{1}{n+1}$. This is just the harmonic series, which we know diverges. Therefore, we do *not* include $x = -5/3$ in the interval of convergence.

10.5 POWER SERIES

Hence, the interval of convergence is

$$[-7/3, -5/3),$$

with center at $a = -2$ and radius of convergence $R = 1/3$.

EXAMPLE 40 Find the interval and radius of convergence of $\sum_{n=0}^{\infty}(-1)^n n! x^n$.

Solution By the ratio test, we have

$$\lim_{n\to\infty} \frac{|(-1)^{n+1}(n+1)!x^{n+1}|}{(-1)^n n! x^n} = \lim_{n\to\infty} |x|(n+1),$$

which is infinite for any $x \neq 0$. Hence, the interval of convergence $\{0\}$ contains only one point $x = 0$, and the radius of convergence is $R = 0$.

Calculus and power series

If we use a power series to define a function f, so that

$$f(x) = \sum_{n=0}^{\infty} c_n (x-a)^n,$$

then the domain of f is the interval of convergence for the power series. A function defined by a power series in this way is continuous on the interval of convergence. In fact, it can be differentiated and antidifferentiated term-by-term just like a polynomial!

Theorem 10.5

Hypothesis: The function f is defined by a power series

$$f(x) = \sum_{n=0}^{\infty} c_n (x-a)^n.$$

Conclusion: The derivative

$$f'(x) = \sum_{n=0}^{\infty} n c_n (x-a)^{n-1}$$

and the antiderivative

$$\int f(x)\, dx = C + \sum_{n=0}^{\infty} \frac{c_n (x-a)^{n+1}}{n+1}$$

have the same radius of convergence as f.

EXAMPLE 41 Suppose that $f(x) = 1+x+x^2+x^3+\cdots = \sum_{n=0}^{\infty} x^n$. Express the derivative and antiderivative of f as power series, and find their radius of convergence.

Solution We have

$$f'(x) = 0 + 1 + 2x + 3x^2 + \cdots = \sum_{n=0}^{\infty} nx^{n-1},$$

and

$$\int f(x)\,dx = C + x + \frac{x^2}{2} + \frac{x^3}{3} + \frac{x^4}{4} + \cdots = C + \sum_{n=0}^{\infty} \frac{x^{n+1}}{n+1},$$

where C is an arbitrary constant.

The radius of convergence for $f(x) = \sum_{n=0}^{\infty} x^n$ is $R=1$. (It's a geometric series with ratio x.) By the theorem, the derivative f' and antiderivative $\int f(x)\,dx$ must have the same radius of convergence. (However, the *interval* of convergence for f and f' is $(-1,1)$, while the antiderivative also converges for $x=-1$.) ∎

▶▶▶ **Be careful! Two power series can have different intervals of convergence, even if the center a and radius of convergence R are the same. This is because the behavior at the *endpoints* $x = a - R$ and $x = a + R$ may be different for the two power series.**

Functions defined by power series

We have seen that we can define a function using a power series. The domain of the function is the interval of convergence for the power series. Many familiar functions may be represented by a suitable power series.

EXAMPLE 42 (a) Find the domain and a formula for the function f defined by

$$f(x) = \sum_{n=0}^{\infty} x^n.$$

(b) Find a power series representation for the derivative $f'(x)$ over the same domain.

(c) Find a power series representation and a formula for the antiderivative $F(x)$ satisfying $F(0) = 0$ over the same domain.

10.5 POWER SERIES

Solution (a) The power series for $f(x)$ can be thought of as a geometric series with ratio $r = x$. As such, it converges to $\dfrac{1}{1-x}$ for $-1 < x < 1$, and we can write

$$f(x) = \frac{1}{1-x} \quad \text{with domain } (-1, 1).$$

(b) To find a power series representation for

$$f'(x) = \frac{1}{(1-x)^2}$$

valid for $-1 < x < 1$, we can differentiate the power series representation for $f(x)$:

$$\frac{1}{(1-x)^2} = 1 + 2x + 3x^2 + \cdots = \sum_{n=0}^{\infty} nx^{n-1}.$$

(c) If we antidifferentiate f, we obtain the power series

$$\int f(x)\, dx = C + x + \frac{x^2}{2} + \frac{x^3}{3} + \frac{x^4}{4} + \cdots = C + \sum_{n=0}^{\infty} \frac{x^{n+1}}{n+1},$$

which must also converge for $-1 < x < 1$. Note that

$$F(x) = -\ln(1-x)$$

is an antiderivative for f satisfying the initial condition $F(0) = 0$ and whose formula is valid for $-1 < x < 1$. If we set $C = 0$, we have found a power series representation for $F(x)$:

$$-\ln(1-x) = x + \frac{x^2}{2} + \frac{x^3}{3} + \frac{x^4}{4} + \cdots = \sum_{n=0}^{\infty} \frac{x^{n+1}}{n+1}. \qquad \blacksquare$$

Since a function f defined by power series has a derivative f' that is also represented as a power series, we can continue differentiating to find power series representations for the higher-order derivatives f'', f''', $f^{(4)}$, and so on indefinitely. In other words, f is infinitely differentiable, and all its derivatives have the same radius of convergence.

Taylor and Maclaurin series

Suppose now that f is an infinitely differentiable function (as are almost all the functions we have dealt with), we can extend the sequence of Taylor polynomials indefinitely to give the **Taylor series** at $x = a$ for f,

$$\sum_{k=0}^{\infty} \frac{f^{(k)}(a)}{k!}(x-a)^k,$$

and, taking $a = 0$, the **Maclaurin series** for f,

$$\sum_{k=0}^{\infty} \frac{f^{(k)}(0)}{k!} x^k.$$

EXAMPLE 43   Compute the Maclaurin series for $f(x) = e^x$.

Solution   Since $f(x) = f'(x) = f''(x) = \cdots = e^x$ and $f(0) = 1$, we have

$$\sum_{k=0}^{\infty} \frac{f^{(k)}(0)}{k!} x^k = \sum_{k=0}^{\infty} \frac{1}{k!} x^k.$$

∎

We can ask two questions regarding this power series:

(1) For what values x does the Taylor (or Maclaurin) series converge?

(2) In particular, for what values x does the series converge to $f(x)$?

The partial sums of a Taylor series are Taylor polynomials, so we are really asking about the convergence of the sequence of Taylor polynomials.

If $p_n(x)$ is the nth-order Taylor polynomial for $f(x)$ about $x = a$, then the **remainder** is

$$R_n(x) = f(x) - p_n(x).$$

The convergence of the Taylor series to $f(x)$ is then equivalent to the convergence of the remainders $R_n(x)$ to 0. That is, the equality

$$f(x) = \sum_{k=0}^{\infty} \frac{f^{(k)}(a)}{k!} (x-a)^k$$

holds if and only if $\lim_{n \to \infty} R_n(x) = 0$. The appendix includes a discussion of this remainder term $R_n(x)$ and ways in which it can be expressed.

A Taylor series need not converge everywhere in the domain of the function. For example, Figure 10.8 shows the graphs of the $5th$- and $10th$-order Taylor polynomials at $x = 1$ of $f(x) = \sqrt{x}$. It appears from these graphs that the series is diverging from the function outside the region $(0, 2)$ and converging for some region inside this one. You can check that the interval of convergence for the Taylor series is $(0, 2)$ by using the ratio test.

A **point of validity** for a Taylor series is an input x for which the power series *converges to the function output* $f(x)$. For reference, following is a table of common Maclaurin series and their respective intervals of validity.

10.5 POWER SERIES

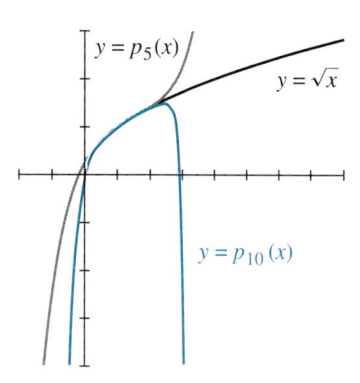

Figure 10.8 The $5th$- and $10th$-order Taylor polynomials at $x = 1$ of $\sqrt{x}$.

Maclaurin Series	Interval of Validity
$e^x = \sum_{k=0}^{\infty} \dfrac{x^k}{k!} = 1 + x + \dfrac{x^2}{2!} + \dfrac{x^3}{3!} + \dfrac{x^4}{4!} + \cdots$	$-\infty < x < +\infty$
$\sin x = \sum_{k=0}^{\infty} (-1)^k \dfrac{x^{2k+1}}{(2k+1)!} = x - \dfrac{x^3}{3!} + \dfrac{x^5}{5!} - \dfrac{x^7}{7!} + \cdots$	$-\infty < x < +\infty$
$\cos x = \sum_{k=0}^{\infty} (-1)^k \dfrac{x^{2k}}{(2k)!} = 1 - \dfrac{x^2}{2!} + \dfrac{x^4}{4!} - \dfrac{x^6}{6!} + \cdots$	$-\infty < x < +\infty$
$\ln(1+x) = \sum_{k=0}^{\infty} (-1)^k \dfrac{x^{k+1}}{k+1} = x - \dfrac{x^2}{2} + \dfrac{x^3}{3} - \dfrac{x^4}{4} + \cdots$	$-1 < x \leq 1$
$\arctan x = \sum_{k=0}^{\infty} (-1)^k \dfrac{x^{2k+1}}{2k+1} = x - \dfrac{x^3}{3} + \dfrac{x^5}{5} - \dfrac{x^7}{7} + \cdots$	$-1 \leq x \leq 1$
$\dfrac{1}{1-x} = \sum_{k=0}^{\infty} x^k = 1 + x + x^2 + x^3 + \cdots$	$-1 < x < 1$
$\sinh x = \sum_{k=0}^{\infty} \dfrac{x^{2k+1}}{(2k+1)!} = x + \dfrac{x^3}{3!} + \dfrac{x^5}{5!} + \dfrac{x^7}{7!} + \cdots$	$-\infty < x < +\infty$
$\cosh x = \sum_{k=0}^{\infty} \dfrac{x^{2k}}{(2k)!} = 1 + \dfrac{x^2}{2!} + \dfrac{x^4}{4!} + \dfrac{x^6}{6!} + \cdots$	$-\infty < x < +\infty$

EXERCISES for Section 10.5

1. Find the interval and radius of convergence for $\sum_{n=0}^{\infty} \frac{2^n x^n}{n!}$.

2. Find the interval and radius of convergence for $\sum_{n=0}^{\infty} \frac{(-1)^n x^n}{3^n n}$.

3. Find the interval and radius of convergence for $\sum_{n=0}^{\infty} \frac{(-1)^n (x-3)^n}{2^n}$.

Power series representations for new functions can be derived from known representations through substitution, differentiation, and antidifferentiation.

For exercises 4–7: Use the power series representation

$$f(x) = \frac{1}{1-x} = \sum_{n=0}^{\infty} x^n = 1 + x + x^2 + x^3 + \cdots \quad \text{for } -1 < x < 1$$

to answer these questions.

4. Find a power series representation for

$$g(x) = \frac{1}{1+x}$$

by substituting $-x$ for x. What is the interval of convergence for g?

5. Find a power series representation of $h(x) = \ln(1+x)$ by antidifferentiating the power series in exercise 4. (What's the initial condition?)

6. Use the ratio test to determine the interval of convergence for $h(x)$. Be sure to check the endpoints.

7. Let $x = 1$. Estimate $\ln(2)$ by adding the first 101 terms of the power series representation for $h(x)$. Use the alternating series remainder to determine the accuracy of the result. To how many decimal places does this agree with the value of $\ln(2)$?

For exercises 8–13: We can represent $f(x) = e^x$ by the power series

$$e^x = \sum_{n=0}^{\infty} \frac{x^n}{n!} = 1 + x + \frac{x^2}{2} + \frac{x^3}{6} + \frac{x^4}{24} + \cdots.$$

8. Show that the interval of convergence for this power series is $(-\infty, \infty)$.

9. Find a power series representation for e^{-x}.

10. Find a power series representation for $1 - e^{-x}$.

11. Find a power series representation for $\dfrac{1 - e^{-x}}{x}$.

12. Find a power series representation for the antiderivative of $\dfrac{1 - e^{-x}}{x}$.

13. Use your answer to exercise 12 to approximate $\int_0^{1/2} \frac{1-e^{-x}}{x} \, dx$ to within 1×10^{-10}.

10.6 ITERATION

For exercises 14–16: Here are power series representation of two trigonometric functions:

$$\sin(x) = x - \frac{x^3}{6} + \frac{x^5}{120} - \frac{x^7}{7!} + \cdots = \sum_{n=0}^{\infty} \frac{(-1)^n x^{2n+1}}{(2n+1)!}$$

$$\cos(x) = 1 - \frac{x^2}{2} + \frac{x^4}{24} - \frac{x^6}{6!} + \cdots = \sum_{n=0}^{\infty} \frac{(-1)^n x^{2n}}{(2n)!}$$

14. Show that the interval of convergence for both of these power series is $(-\infty, \infty)$.

15. Find a power series representation for

$$f(x) = \frac{x - \sin x}{x^3 \cos x}.$$

Graph the function.

16. Evaluate $\lim_{x \to 0} \frac{x - \sin x}{x^3 \cos x}$ using the result in exercise 15. Does the graph support your answer? Note that L'Hôpital's rule is appropriate here (but rather undesirable).

10.6 ITERATION

Iterative and recursive sequences

When we say that a sequence $\{a_n\}_{n=1}^{\infty}$ is described in **closed form,** we mean that we have an explicit formula for a_n written in terms of n. For example, the geometric sequence

$$a_0 = 1, \ a_1 = \frac{1}{2}, a_2 = \frac{1}{4}, \ a_3 = \frac{1}{8}, \ a_4 = \frac{1}{16}, \ \ldots$$

can be described by the formula

$$a_n = \frac{1}{2^n} \quad \text{for} \quad n = 0, 1, 2, \ldots.$$

In fact, we could write the sequence as

$$\{\frac{1}{2^n}\}_{n=0}^{\infty}.$$

Sometimes the terms in a sequence depend directly on one or more of the preceding terms in the sequence. When this is the case, it may be preferable or more convenient to describe the sequence using this relationship.

A sequence is said to be defined **recursively** or **inductively** if the first term is given (or the first few terms are given) and we have explicit instructions on how to obtain the subsequent terms using the values of preceding terms.

EXAMPLE 44 Describe the geometric sequence $1, \frac{1}{2}, \frac{1}{4}, \frac{1}{8}, \ldots$ recursively.

Solution
$$a_0 = 1$$
$$a_n = \frac{1}{2}a_{n-1} \quad \text{for} \quad n = 1, 2, 3, \ldots$$

A recursive description of a sequence allows us to "build" up the terms much like a stack of blocks. In this example, we are given

$$a_0 = 1.$$

The **recursion formula** $a_n = \frac{1}{2}a_{n-1}$ is used over and over again to obtain the subsequent terms:

$$a_1 = \frac{1}{2}a_{1-1} = \frac{1}{2}a_0 = \frac{1}{2} \cdot 1 = \frac{1}{2},$$

$$a_2 = \frac{1}{2}a_{2-1} = \frac{1}{2}a_1 = \frac{1}{2} \cdot \frac{1}{2} = \frac{1}{4},$$

$$a_3 = \frac{1}{2}a_{3-1} = \frac{1}{2}a_2 = \frac{1}{2} \cdot \frac{1}{4} = \frac{1}{8}.$$

$$\vdots$$

Any geometric sequence can be described recursively in a similar way. If a is the first term in the geometric sequence and r is the constant ratio, then the sequence

$$a, \ ar, \ ar^2, \ ar^3, \ \ldots$$

can be described recursively:

$$a_0 = a$$

$$a_n = ra_{n-1}.$$

Similarly, any arithmetic sequence can be described recursively.

10.6 ITERATION

EXAMPLE 45 Describe the arithmetic sequence 3, 7, 11, 15, 19, ... recursively.

Solution $a_0 = 3$ and $a_n = a_{n-1} + 4$ for $n = 1, 2, \ldots$.

In general, if a is the first term of an arithmetic sequence and d is the constant difference, then the sequence

$$a, \ a+d, \ a+2d, \ a+3d, \ \ldots$$

can be described recursively: $a_0 = a$ and $a_n = a_{n-1} + d$ for $n = 1, 2, \ldots$.

Perhaps the most famous of all sequences is the so called **Fibonacci sequence**. Fibonacci, also known as Leonardo di Pisa, was a thirteenth century Italian mathematician. The sequence named after him is usually described recursively as follows:

$$a_0 = 1$$
$$a_1 = 1$$
$$a_n = a_{n-1} + a_{n-2} \quad \text{for } n = 2, 3, 4, \ldots$$

In other words, the first two terms of the Fibonacci sequence are both 1, and each subsequent term is determined by adding the previous two terms. The Fibonacci sequence starts out

$$1, \ 1, \ 2, \ 3, \ 5, \ 8, \ 13, \ 21, \ 34, \ 55, \ 89, \ \ldots$$

The Fibonacci sequence appears quite often in number patterns arising in nature, including the population growth rates of rabbits, the genealogy of honey bees, and seed and leaf arrangements in vegetation, among others. The sequence has such a multitude of interesting properties, that an entire journal (*The Fibonacci Quarterly*) is devoted to it.

The recursive descriptions for arithmetic and geometric sequences are examples of *iteration*. An **iterative process** is one in which we start with an initial input x_0 and then successively apply a function f by using the output at each stage as the input to the next. For example, if

$$f(x) = x + 4$$

and our initial input (also called the initial *seed*) is $x_0 = 3$, then

$$x_1 = f(x_0) = f(3) = 7$$

$$x_2 = f(x_1) = f(7) = 11$$

$$x_3 = f(x_2) = f(11) = 15$$

$$x_4 = f(x_3) = f(15) = 19$$

$$\vdots$$

We can see that this function f, along with the initial input $x_0 = 3$, generates the arithmetic sequence 3, 7, 11, ...

▶▶▶ **Any function f and initial input x_0 can be used to define an iterative sequence, provided that the output at each step is in the domain of the function.**

EXAMPLE 46   Find the first 5 terms of the iterative sequence defined by

$$g(x) = \frac{x^2 + 2}{2x}$$

with initial input $x_0 = 1$.

Solution   With $x_0 = 1$, we have

$$x_1 = g(x_0) = g(1) = \frac{1^2 + 2}{2} = \frac{3}{2} = 1.5$$

$$x_2 = g(x_1) = g(1.5) = \frac{(1.5)^2 + 2}{3} \approx 1.41666666667$$

$$x_3 = g(x_2) \approx f(1.41666666667) \approx 1.41421568627$$

$$x_4 = g(x_3) \approx f(1.41421568627) \approx 1.41421356237$$

$$x_5 = g(x_4) \approx f(1.41421356237) \approx 1.41421356237.$$

The values listed for x_2, x_3, x_4, and x_5 are rounded to eleven decimal places, and from the result of x_5 we would anticipate that the subsequent terms x_6, x_7, and so on will stay "locked on" to the value 1.41421356237. ∎

Many approximation methods are essentially iterative processes—an initial guess for a solution is tested and then adjusted accordingly. This yields an infinite sequence of approximations, each better (one hopes) than the one preceding it. Both the bisection method and Newton's method are examples of iterative root-finding approximation procedures.

10.6 ITERATION

Fixed points

In the case of Newton's method, we developed an *iterating function* for a differentiable function f:

$$g(x) = x - \frac{f(x)}{f'(x)}.$$

Starting with an initial seed x_0, we compute a sequence of values

$$x_1 = g(x_0), \quad x_2 = g(x_1), \quad x_3 = g(x_2), \quad \ldots, \quad x_n = g(x_{n-1}), \quad \ldots.$$

In other words, we keep using the *output* at one stage as the *input* for the next. If this sequence converges to single limiting value x_r satisfying

$$x_r = g(x_r),$$

then x_r is a *root* of the function f.

In fact, the function g, where $g(x) = \dfrac{x^2 + 2}{2x}$, is the iterating function for

$$f(x) = x^2 - 2.$$

The value x_r is also called a **fixed point** of the function g, since feeding it in as an input to g results in the same fixed value as output.

The fixed points of a function f are simply the solutions to the equation

$$f(x) = x.$$

Graphically, the fixed points are the intersections of the graph of $y = f(x)$ and the diagonal line $y = x$. Figure 10.9 illustrates the fixed point of a linear function $f(x) = mx + b$ (with $m \neq 1$). If (x_r, x_r) is the point of intersection, then x_r is a fixed point of f. In this case, we can solve the equation

$$mx + b = x$$

to find that the fixed point

$$x = x_r = \frac{b}{1 - m}.$$

For a more general function f, the equation

$$f(x) = x$$

may be more difficult, or impossible, to solve using algebra.

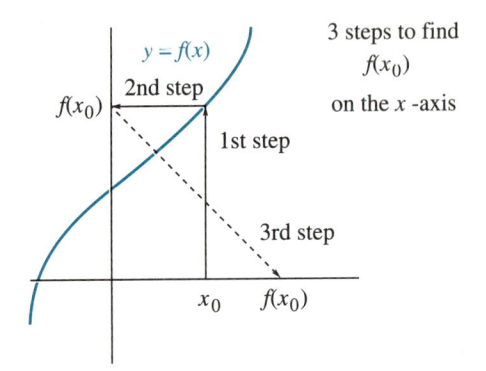

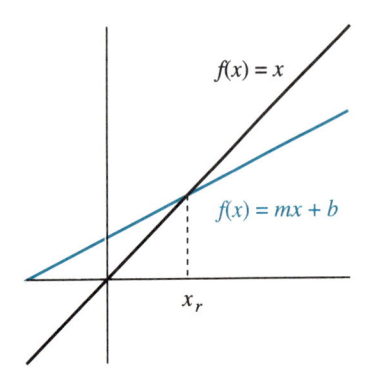

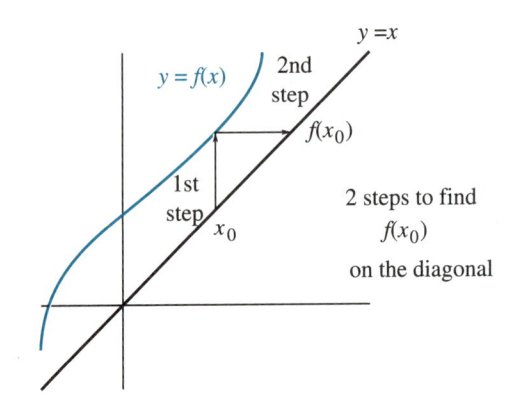

Figure 10.9 If $m \neq 1$, the graph of $y = mx + b$ must intersect the diagonal.

Figure 10.10 Graphing input-to-output relationship without and with use of the diagonal.

Iterating a function graphically

Normally, we think of an input to a function f as a point on the horizontal x-axis, and the output as a point on the vertical y-axis. We could graphically connect the input x_0 to the output $f(x_0)$ by drawing a vertical line from the x-axis to the graph of $y = f(x)$ and then a horizontal line from the graph to the y-axis. If we want to iterate the function, we then need to find the point representing the number $f(x_0)$ on the horizontal x-axis to use as input so that we can repeat the process again.

There is another simple but effective way of graphing the iteration process using the graph of the function $y = f(x)$ and the diagonal $y = x$. If we find the input x_0 on the diagonal $y = x$ (actually the point (x_0, x_0)), we can "bounce" vertically to the graph of $y = f(x)$ and then horizontally back to the diagonal to find $f(x_0)$ (actually the point $(f(x_0), f(x_0))$). This process is easy to iterate by just repeating it, using the new point. Figure 10.10 illustrates both techniques.

For example, using an initial seed $x_0 = .5$, we can graphically find 3 iterates of the linear function $f(x) = .8(x - 2) + 2$, as shown in the viewing

10.6 ITERATION

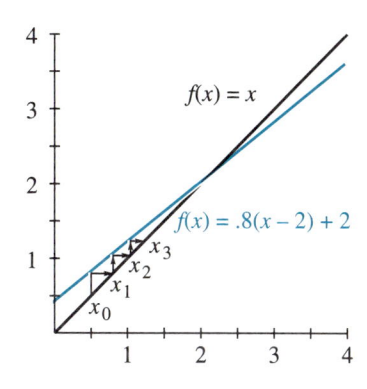

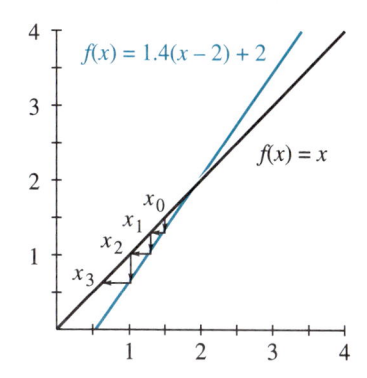

Figure 10.11 Graphing three iterates of $f(x) = .8(x-2) + 2$.

Figure 10.12 Graphing three iterates of $f(x) = 1.4(x-2) + 2$.

window $[0,4] \times [0,4]$. of Figure 10.11. This function has a fixed point $x_r = 2$, since $f(2) = 2$. Note that the iterates are "moving" *toward* the fixed point.

On the other hand, Figure 10.12 shows 3 iterates of the linear function $f(x) = 1.4(x-2) + 2$ in the viewing window $[0,4] \times [0,4]$, using an initial seed $x_0 = 1.5$. Note that this function also has a fixed point $x_r = 2$, but now the iterates are "moving" *away* from the fixed point.

Contractions and the fixed point theorem

A **contraction** is a function whose outputs are always closer together than the corresponding inputs. For example, a linear function

$$f(x) = mx + b$$

is a contraction, provided that $|m| < 1$. In the two previous examples, the first linear function had a slope of 0.8 and therefore is a contraction. The second linear function had a slope of 1.4 and therefore is *not* a contraction.

If we iterate a contraction by feeding its output back in as an input repeatedly, then the resulting sequence converges to a fixed point. Equipped with this graphing technique, we can look at four generic cases of iteration for linear functions: where the slope is between 0 and 1, where the slope is between -1 and 0, where the slope is greater than 1, and where the slope is less than -1. These cases are shown in Figure 10.13.

If the slope is positive, the iterates either "stairstep" toward or away from the fixed point, depending on whether the slope is less than or greater than 1, respectively. If the slope is negative, the iterates "spiral" toward or away from the fixed point, again as the slope is less than or greater than 1, respectively.

If we zoom in on the graph of a differentiable function, we expect to see a straight line. This suggests that we can use the information above

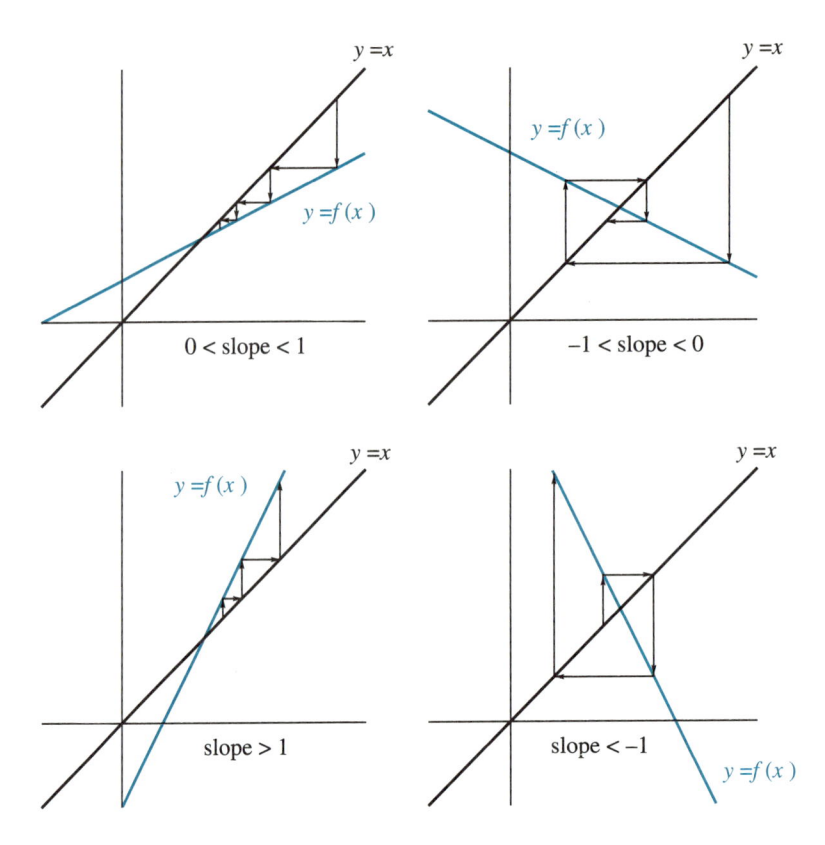

Figure 10.13 Four generic cases of iteration.

to find fixed points of non-linear functions, provided that we restrict our attention to a small enough neighborhood of a fixed point. Unfortunately, if we knew enough about the function to zoom in on the fixed point, we wouldn't need iteration to find the fixed point! Here's a theorem that gives us useful criteria for judging how close we need to be to a fixed point for this analysis to work.

Theorem 10.6

Fixed point theorem
Suppose that f is differentiable on $[a, b]$.
Hypothesis 1: $f(x) \in [a, b]$ for all $x \in [a, b]$.
Hypothesis 2: $|f'(x)| < 1$ for all $x \in [a, b]$.
Conclusion: There is exactly one point $a \leq x_r \leq b$ such that $f(x_r) = x_r$ and any iteration sequence starting with $x_0 \in [a, b]$ converges to x_r.

The next two examples discuss why each hypothesis is important.

EXAMPLE 47 If f is differentiable on $[a, b]$ but has *no* fixed points on $[a, b]$, which of the hypotheses must be violated?

10.6 ITERATION

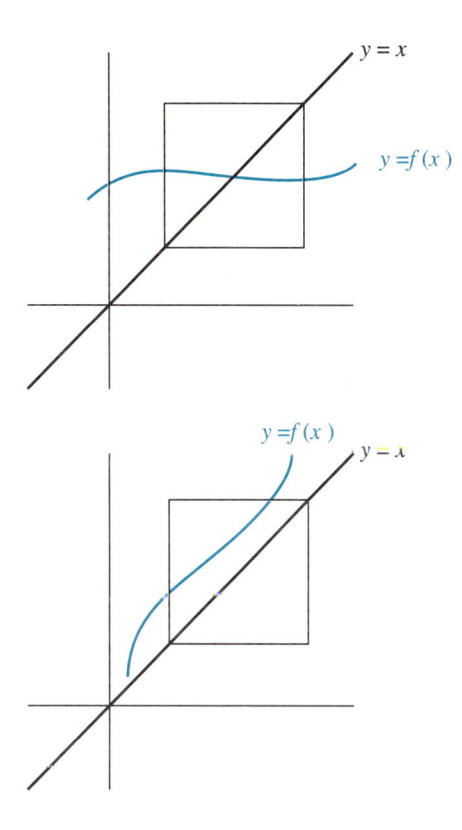

Figure 10.14 Geometric interpretation of hypothesis 1 and a violation of it.

Solution Geometrically, hypothesis 1 says that the graph of f over $[a,b]$ lies entirely within the square box, with corners (a,a), (b,b), (a,b), and (b,a) (see Figure 10.14). A differentiable function is continuous, and since the graph at the left endpoint must be on or above the diagonal, and the graph at the right endpoint must be on or below the diagonal, the intermediate value theorem tells us that somewhere in the interval the graph must cross the diagonal, yielding at least one fixed point. The only way the graph of $y = f(x)$ could connect the points $(a, f(a))$ and $(b, f(b))$ without crossing the diagonal would be for the graph to leave the box entirely, violating hypothesis 1. ∎

EXAMPLE 48 If f is differentiable on $[a,b]$ but has *two* (or *more*) fixed points on $[a,b]$, which of the hypotheses must be violated?

Solution If $f(x_1) = x_1$ and $f(x_2) = x_2$, then the mean value theorem tells us that there is some point $x = c$ such that

$$f'(c) = \frac{f(x_2) - f(x_1)}{x_2 - x_1} = \frac{x_2 - x_1}{x_2 - x_1} = 1.$$

But this violates hypothesis 2. ∎

Hypothesis 2 guarantees that our function is a contraction over the interval $[a, b]$. In fact, the value of the derivative over this interval gives us a measure of *how fast* we can expect the iteration to converge to the fixed point.

Criteria for the convergence of Newton's method

An especially important special case of the iteration method for finding roots is *Newton's method*, which we discussed in Chapter 4. To solve an equation $f(x) = 0$ by Newton's method, you use the iterating function

$$g(x) = x - \frac{f(x)}{f'(x)}.$$

To see why Newton's method is so effective, note that the derivative of g is

$$g'(x) = 1 - \frac{f'(x)f'(x) - f(x)f''(x)}{(f'(x))^2} = \frac{f(x)f''(x)}{(f'(x))^2}.$$

If x_r is a root of f, then $g'(x_r) = 0$, provided it is defined. This suggests that our iteration function will converge *very fast*, provided that we start with a seed x_0 sufficiently close to the root x_r. We can use the fixed-point theorem to derive criteria for convergence of Newton's method.

Suppose that $f'(x) \neq 0$ and $f''(x)$ is defined on $[a, b]$ (so that g is differentiable).

Criterion 1: $a < x - \dfrac{f(x)}{f'(x)} < b$ for all $x \in [a, b]$.

Criterion 2: $\left| \dfrac{f(x)f''(x)}{(f'(x))^2} \right| < 1$ for all $x \in [a, b]$.

Conclusion: For any $x_0 \in [a, b]$, Newton's method converges to a root of f.

The two criteria just correspond to the two hypotheses of the fixed-point theorem applied to the function $g(x) = x - \dfrac{f(x)}{f'(x)}$. Hypothesis 1 requires the graph of $g(x)$ to stay within the "box" $[a, b] \times [a, b]$. Hypothesis 2 requires that $|g'(x)|$ be less than 1 over the interval $[a, b]$.

EXAMPLE 49 Does the iteration function for Newton's method applied to the equation $x^3 - 19 = 0$ satisfy the criteria for convergence in the interval $[2, 3]$?

Solution We have

$$f(x) = x^3 - 19, \qquad f'(x) = 3x^2, \qquad f''(x) = 6x.$$

10.6 ITERATION

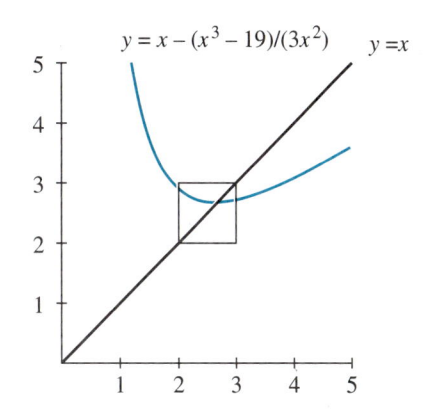

Figure 10.15 Graph of $y = g(x)$ satisfies the "box test" on $[2, 3]$.

We can see that $f'(x) \neq 0$ and $f''(x)$ is defined over the interval $[2,3]$, so the iteration function

$$g(x) = x - \frac{f(x)}{f'(x)} = x - \frac{x^3 - 19}{3x^2}$$

is differentiable. If we graph $y = g(x)$ and apply the "box test" to the interval $[2,3]$, we find that the first criterion holds (see Figure 10.15).

As for the second criterion, we have

$$\left| \frac{f(x)f''(x)}{(f'(x))^2} \right| = \left| \frac{6x^4 - 114x}{9x^4} \right| = \left| \frac{2}{3} - \frac{38}{3x^3} \right|,$$

which is less than 1 for $2 \leq x \leq 3$. Thus, both criteria are satisfied, and taking any initial seed $x_0 \in [2,3]$ will guarantee convergence. Starting with $x_0 = 3$, for example,

$$\left| \sqrt[3]{19} - x_0 \right| \approx 0.035302$$

$$\left| \sqrt[3]{19} - x_1 \right| \approx 0.00045893$$

$$\left| \sqrt[3]{19} - x_2 \right| \approx 0.00000007892$$

$$\left| \sqrt[3]{19} - x_3 \right| = 0 \quad \text{to more than 12 digits.}$$

EXERCISES for Section 10.6

For exercises 1–4: Describe the arithmetic sequence with given initial term a_0 and constant difference d with a recursive formula.

1. $a_0 = 3$, $d = -5$
2. $a_0 = -4$, $d = 0.5$
3. $a_0 = \pi$, $d = 0$
4. $a_0 = 0$, $d = \pi$

For exercises 5–8: Describe the geometric sequence with given initial term a_0 and constant ratio r with a recursive formula.

5. $a_0 = 3$, $r = -5$

6. $a_0 = -4$, $r = 0.5$

7. $a_0 = \pi$, $r = 2$

8. $a_0 = 2$, $r = \pi$

9. If $a_1 = \sqrt{3}$ and $a_n = \sqrt{3a_{n-1}}$ for $n \geq 2$, find the first five terms of the sequence (starting with $n = 1$).

10. If $a_1 = 1$ and $a_{n+1} = \frac{1}{1+a_n}$ for $n \geq 1$, find the first five terms of the sequence (starting with $n = 1$).

For exercises 11–14: The initial term and a function are given. Find the first five terms of the iterative sequence based on this information.

11. $x_0 = 0$, $f(x) = x - \cos(x)$

12. $x_0 = 1$, $f(x) = \dfrac{x+1}{2}$

13. $x_0 = 0.5$, $f(x) = x^2$

14. $x_0 = 10$, $f(x) = 1/x$

For exercises 15–20: The following sequence of numbers approaches the golden ratio $\dfrac{1+\sqrt{5}}{2}$:

$$\frac{1}{1}, \frac{2}{1}, \frac{3}{2}, \frac{5}{3}, \frac{8}{5}, \frac{13}{8}, \ldots$$

15. Describe the pattern you see in the numerators. Describe the pattern you see in the denominators.

16. What is the first term in the sequence that is within .001 less than the golden ratio?

17. What is the first term in the sequence that is within .001 more than the golden ratio?

18. What are the first two consecutive terms in the sequence that are within .000001 of each other?

19. Find the first 10 terms of the sequence whose nth term is given by

$$a_n = \frac{1}{\sqrt{5}}\left[\left(\frac{1+\sqrt{5}}{2}\right)^n - \left(\frac{1-\sqrt{5}}{2}\right)^n\right].$$

This is the closed-form formula for what sequence?

20. Use exercise 19 to find a closed-form formula for the sequence

$$\frac{1}{1}, \frac{2}{1}, \frac{3}{2}, \frac{5}{3}, \frac{8}{5}, \frac{13}{8}, \ldots .$$

For exercises 21–30: Verify that the corresponding Newton iteration function for each equation satisfies the criteria for convergence over the given

10.6 ITERATION

interval. If so, use the iteration function to find three successive approximations to the root, using the left endpoint as initial seed x_0.

21. $x^2 - 3.2 = 0;$ $[1, 2]$

22. $x^3 - 23 = 0;$ $[2, 3]$

23. $\sin(x) - 0.7 = 0;$ $[0.7, 0.8]$

24. $x^5 + 3.6x^4 - 2.51x^3 - 22.986x^2 - 28.24x - 10.4 = 0;$ $[2, 3]$

25. $x + \sqrt{x} + \sqrt{x^3} - 7 = 0;$ $[1.8, 2.6]$

26. $x \exp(x^2) - \sqrt{x^2 + 1} = 0;$ $[0.4, 1]$

27. $\ln(x^2 + \ln(x)) = 0;$ $[0.625, 1.35]$

28. $x^3 - \sqrt{x^2 + 1} + x - 3 = 0;$ $[1.1, 1.6]$

29. $3\cos(\cos(x)) - 2x = 0;$ $[1.1, 2.2]$

30. $\sqrt[4]{x} + x^3 - x - 0.5 = 0;$ $[0.2, 1.2]$

11

Fundamentals of Vectors

Many quantities that arise in applications simply cannot be described adequately by a single number. Take, for example, the following description of a traffic accident: "I was driving my car, going about 20 miles per hour, when suddenly this other car comes out of nowhere and hits me. The other car must have been going at least 30 miles per hour."

What do you think was the extent of the damage to the first car? Almost certainly you're thinking that you really need more information—namely, the *directions* the two cars were headed at the time of the collision. Did the second car come from the side, the rear, or head-on? And at what angle did it hit the first car? Certainly a rear-end collision at the speeds described would be much less damaging than a head-on collision.

In technical terms, we need to know the *velocities* of the two cars. In everyday language, *velocity* and *speed* are often used interchangeably. In mathematics and physics, however, there is a very important distinction to be made: *velocity* can be thought of as "speed with a direction." *Acceleration* and *force* are two other physical quantities whose descriptions require both a numerical *magnitude* and a *direction*. In mathematics, such a quantity is called a **vector quantity**, or simply a **vector**. Velocity, acceleration, and force are all examples of vector quantities. To distinguish them from those quantities whose descriptions require only a numerical magnitude, such as *mass*, *distance*, and *time*, the latter are sometimes called **scalar quantities**, or simply **scalars**.

In this chapter, we explain how vectors can be represented both geometrically and analytically (with the result that arithmetic operations with vectors have both geometric and analytic interpretations). Operations and functions involving vectors can sometimes be described conveniently by the use of rectangular arrays of numbers called *matrices*, and we discuss how matrices are combined arithmetically. Finally, we look at some applications of vectors to space geometry.

11.1 REPRESENTATIONS OF VECTORS

First, let's discuss notation for vectors. In this book, we denote vectors in boldface to distinguish them from scalars. Thus,

$$\mathbf{v} \quad \text{denotes a vector quantity,}$$

while

$$v \quad \text{denotes a scalar quantity.}$$

You may also see other notations for vectors using some kind of arrow over the letter, or perhaps underlining. Alternative notations for **v** are

$$\vec{v} \quad \text{or} \quad \overrightarrow{v} \quad \text{or} \quad \underline{v}.$$

You will quite likely want to use one of these notations or a similar one, since boldface is not practical for handwritten work.

Geometric representations of vectors—directed line segments

Geometrically, a vector can be represented by a *directed line segment* (a line segment with an arrowhead). The magnitude of the vector is indicated by the length of the segment. The direction of the vector is indicated by the direction of the arrow. In Figure 11.1, P denotes the **initial point** or "tail" of the vector **v**, and Q denotes the **terminal point** or "head" of **v**.

$$\mathbf{v} = \overrightarrow{PQ}$$

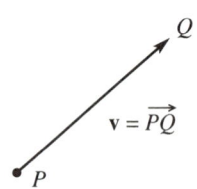

Figure 11.1 Illustration of a vector $\mathbf{v} = \overrightarrow{PQ}$.

If we refer to the vector by the initial and terminal points, we use the arrow notation:

$$\mathbf{v} = \overrightarrow{PQ}.$$

Note that $\overrightarrow{QP}$ is a *different* vector that points in the opposite direction of $\overrightarrow{PQ}$.

You can see that this geometric representation of a vector $\mathbf{v} = \overrightarrow{PQ}$ could possibly be confused with the common pictorial representation and notation for a *ray* in geometry. Since rays have infinite "length," the length of the arrow used to represent them is unimportant. In contrast, the length of the arrow used to represent a vector is crucially important. Context generally makes it clear whether arrows are being used to represent rays or vectors.

11.1 REPRESENTATIONS OF VECTORS

unequal
(same direction, but
different lengths)

unequal
(same length, but
different directions)

equal vectors
(same direction,
same length)

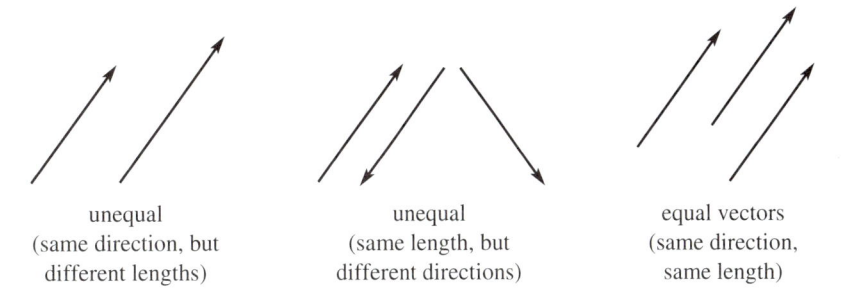

Figure 11.2 Equal vectors have both the same length and the same direction.

Two directed line segments having the same length and direction represent the same vector quantity. This means that any single vector has infinitely many representatives as directed line segments, all of which are equivalent in the sense of equal length and same direction. For this reason, a vector is sometimes referred to as an *equivalence class* of directed line segments. It is important to keep in mind that two arrows in different positions can represent the same vector as long as they have the same length and the same direction (see Figure 11.2).

One special vector is the vector of length 0. (Direction can be arbitrary on the zero vector.) We denote this vector as **0**, using boldface to distinguish it from the scalar 0. The initial and terminal points of the zero vector coincide. So, if P is any point, then

$$\mathbf{0} = \overrightarrow{PP}.$$

Analytic representation of vectors—components

Another very useful way to represent vectors, particularly for analysis, is by *components*. If our vectors all lie in a plane, we can set up a rectangular (Cartesian) coordinate system. The coordinate axes set up a reference system by which we can more easily compare two points' locations. The difference in the x-coordinates of the two points P and Q determines their horizontal distance apart, and the difference in the y-coordinates determines their vertical distance apart. We could specify how to travel from point P to point Q by stating these changes in coordinates.

This process leads us to the idea of specifying a vector by components. Suppose that we have a vector **v** represented by the directed line segment $\overrightarrow{PQ}$, where

$$P = (x_1, y_1) \quad \text{and} \quad Q = (x_2, y_2).$$

To determine the components of the vector **v** with respect to this coordinate system, we simply subtract the coordinates of its tail from the coordinates of its head. Hence, the x-component of **v** is $x_2 - x_1$ and the y-component of **v** is $y_2 - y_1$.

Now, $\overrightarrow{PQ}$ is only one representative of this vector. Imagine picking up this directed line segment and (without changing the direction) placing its initial point (tail) at the origin $(0,0)$. Now, we can read off the components of **v** directly as the coordinates of its terminal point (head), as shown in Figure 11.3.

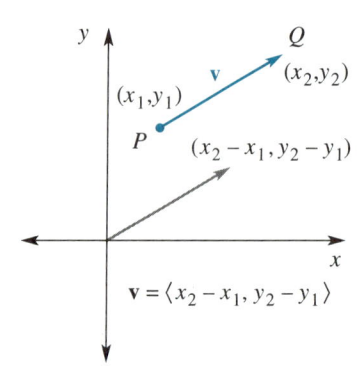

Figure 11.3 Determining the components of a vector **v**.

When we specify a vector by its components, we use the notation

$$\mathbf{v} = \langle v_1, v_2 \rangle$$

to distinguish between the *vector* $\mathbf{v} = \langle v_1, v_2 \rangle$ and the *point* (v_1, v_2).

In terms of components, $\mathbf{0} = \langle 0, 0 \rangle$.

Two vectors represented by components $\mathbf{v} = \langle v_1, v_2 \rangle$ and $\mathbf{w} = \langle w_1, w_2 \rangle$ are equal if and only if their components match exactly. In other words,

$$\mathbf{v} = \mathbf{w} \quad \text{if and only if} \quad v_1 = w_1 \text{ and } v_2 = w_2.$$

EXAMPLE 1 Find the components of $\mathbf{v} = \overrightarrow{PQ}$ if $P = (3, -4)$ and $Q = (-1.5, 2.3)$.

Solution $\mathbf{v} = \langle (-1.5) - 3, 2.3 - (-4) \rangle = \langle -4.5, 6.3 \rangle$.

EXAMPLE 2 Suppose that this same vector **v** is represented by the directed line segment $\overrightarrow{RS}$ with the initial point $R = (-1, -2.2)$. Find the coordinates of the terminal point S.

Solution Since $\mathbf{v} = \langle -4.5, 6.3 \rangle$, we must have

$$S = (-1 + (-4.5), -2.2 + 6.3) = (-5.5, 4.1).$$

11.1 REPRESENTATIONS OF VECTORS

Representing vectors in 3-dimensional space

There is no particular difficulty in representing vectors in 3-dimensional space until we run into the graphical difficulties of illustrating three dimensions on (2-dimensional) paper or blackboard. Geometrically, we can represent vectors as directed line segments in space, just as we do vectors lying in a single plane. To discuss such vectors analytically by components, we can use a rectangular coordinate system in 3-dimensional space.

A rectangular coordinate system for space has three mutually perpendicular coordinate axes, with x, y, and z used to represent the first, second, and third coordinates, respectively. Generally, such a system is illustrated on paper with the positive y-axis extending to the right, the positive z-axis extending straight up, and the positive x-axis drawn at an obtuse angle to both of the other axes to suggest that it extends straight *out* from the page. If you think of the origin as a corner point on the floor of a rectangular room, then these positive axes represent the two floor edges (x and y) and the wall corner (z) that all meet at that point. Figure 11.4 illustrates a 3-dimensional rectangular coordinate system.

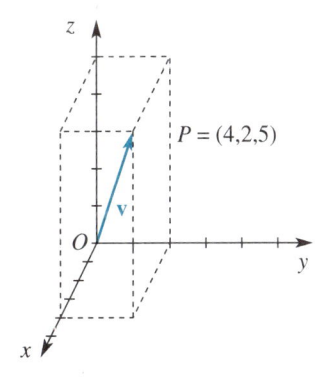

Figure 11.4 A 3-dimensional rectangular coordinate system.

Pictured are the positive x-, y-, and z-axes, along with the point $P = (4, 2, 5)$ and the origin $O = (0, 0, 0)$. The vector $\mathbf{v} = \overrightarrow{OP}$ in the picture has three components. Since its initial point is at O and its terminal point is at P, we have $\mathbf{v} = \langle 4, 2, 5 \rangle$. Note that the point $Q = (0, 1, 3)$ would appear to be at the same place in our picture as the point P. This points out both the difficulty in representing 3-dimensional space with a 2-dimensional picture as well as the need for care in labelling points in diagrams to avoid confusion.

Note that 3-dimensional space is naturally divided into eight octants (as opposed to four quadrants in 2-dimensional space). The octant where all coordinates are positive is the first octant (for example, P is in the first octant); the other seven octants have no standard numbering. For reference we note other areas of 3-dimensional space and their descriptions.

coordinate axes	description
x-axis	$\{(x,0,0) \ : \ x \in \mathbb{R}\}$
y-axis	$\{(0,y,0) \ : \ y \in \mathbb{R}\}$
z-axis	$\{(0,0,z) \ : \ z \in \mathbb{R}\}$

coordinate planes	description
xy-plane	$\{(x,y,0) \ : \ x,y \in \mathbb{R}\}$
xz-plane	$\{(x,0,z) \ : \ x,z \in \mathbb{R}\}$
yz-plane	$\{(0,y,z) \ : \ y,z \in \mathbb{R}\}$

Visually, the xy-plane is horizontal and perpendicular to the page. The xz-plane is vertical and perpendicular to the page, and the page itself lies in the yz-plane.

Representing vectors in n-dimensional space

There really is no restriction in talking about vectors with n components, even if $n > 3$. For example,

$$\mathbf{v} = \langle -1, 3, 5, -4 \rangle$$

is a 4-dimensional vector.

Certainly, a geometric picture of an n-dimensional vector becomes impossible if $n > 3$. Nevertheless, the idea of a directed line segment can be very useful in guiding our intuition, even in these higher dimensions. Our analytic view of vectors allows for any number of components. Indeed, in higher mathematics, vectors with *infinitely* many components may be studied.

One might ask what use there might be for such higher dimensional vectors. An example that is particularly important in physics is the notion of space-time. Space-time can be thought of as having four dimensions: the usual three dimensions of space and the fourth dimension of time.

A "point" in space-time describes a spatial location at a specific time. Additional components may be necessary if we want to specify additional "states" existing at that location and time. Some of the newer theories in physics suggest that as many as *eleven* dimensions may be needed to model all the forces at work in nature!

In some contexts, we may want to consider scalars as 1-dimensional vectors, that is, as vectors having only one component. The real number line can serve as the coordinate system ($a = \langle a \rangle$ in this notation), and we refer to this system as 1-space.

11.1 REPRESENTATIONS OF VECTORS

The Cartesian plane can be thought of as the Cartesian product of the set of real numbers with itself. If we write

$$\mathbb{R}^2 = \mathbb{R} \times \mathbb{R}$$

as a suggestive shorthand, we can generalize a concise notation for the various dimensional spaces of vectors with real number components:

$$\mathbb{R}^1 = 1\text{-space} = \{\langle x \rangle \; : \; x \in \mathbb{R}\}$$

$$\mathbb{R}^2 = 2\text{-space} = \{\langle x, y \rangle \; : \; x, y \in \mathbb{R}\}$$

$$\mathbb{R}^3 = 3\text{-space} = \{\langle x, y, z \rangle \; : \; x, y, z \in \mathbb{R}\}$$

$$\vdots$$

$$\mathbb{R}^n = n\text{-space} = \{\langle x_1, x_2, \ldots, x_n \rangle \; : \; x_1, x_2, \ldots, x_n \in \mathbb{R}\}.$$

EXERCISES for Section 11.1

For exercises 1-10: Suppose that in the picture below, each of the vectors p, q, u, and v has a length of three units, while n has a length of two units. Furthermore, suppose that the initial point of m is located at the point $(-1, 1)$ and the terminal point of m is located at the point $(-4, -1)$.

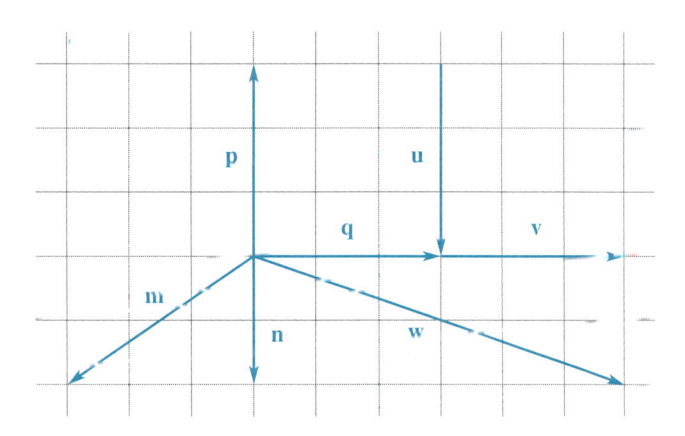

For Exercises 1-10

1. Find the coordinates of the initial points and terminal points of each of the vectors n, p, q, u, v, and w.

2. Find the components of each vector. For example, $p = \langle 0, 3 \rangle$.

3. Find the magnitudes (lengths) of the vectors m and w.

4. Which of the vectors are equal?

5. Which pairs of vectors have the same length but different directions?

6. Which pairs of vectors have the same direction but different lengths?

7. What is the area of the parallelogram having q and w as sides?

8. Which vectors have perpendicular directions?

9. Suppose that you started at the origin and proceeded to walk in the directions indicated by these vectors in alphabetical order. At what point would you end up?

10. Suppose that you started at the origin and proceeded to walk in the directions indicated by these vectors in reverse alphabetical order. At what point would you end up?

For exercises 11-15: Suppose that a vector in $\mathbb{R}^3$ has the given initial point P and terminal point Q. Write the vector in component notation $\langle a, b, c \rangle$.

11. $P = (1, 2, 3), \quad Q = (-2, -3, 4)$.

12. $P = (-5, 1, 0), \quad Q = (5, -1, 0)$.

13. $P = (2.5, -3.4, -4.1), \quad Q = (0, 0, 0)$.

14. $P = (\pi, \sqrt{2}, \ln 3), \quad Q = (\pi, \sqrt{2}, \ln 3)$.

15. $P = (-2, -3, 4), \quad Q = (1, 2, 3)$.

16. Suppose that $R = (-1, 2, 3)$. For each of exercises 11-15, find the point S such that $\overrightarrow{PQ} = \overrightarrow{RS}$.

17. Suppose that $U = (2.5, -3.4, -4.1)$. For each of exercises 11-15, find the point T such that $\overrightarrow{PQ} = \overrightarrow{TU}$.

Vectors in a plane are sometimes represented by using a distance (magnitude) and a compass direction. For example, wind velocity vectors may be indicated in this way. The convention is to assign 0° to due north and to measure positive angles *clockwise* from due north (in contrast to the counterclockwise convention for polar coordinates). Due east is therefore 90°, due south is 180°, and due west is 270° (or −90°).

For exercises 18-22: Express each of the vectors in component notation $\langle a, b \rangle$, if the positive y-axis is considered to point north.

18. 3 units at 135°

19. 4 units at −60°

20. 13.57 units at 213°

21. 0 units at 84°

22. 84 units at 0°

23. Find the compass direction of each of the vectors in the illustration for exercises 1-10.

11.2 ALGEBRA OF VECTORS

24. Here's a classic riddle: You start out and walk one mile due south, then one mile due east, then one mile due north, where you run into a bear standing at your starting point. What color is the bear?

25. Here's a real challenge. Find all the points on the surface of the earth for which the directions of the previous exercise would lead you to your starting point. (Hint: there is more than one point satisfying the conditions.)

11.2 ALGEBRA OF VECTORS

A vector can be defined by a magnitude and a direction by way of its geometric interpretation as a "directed line segment." Equivalently, we can define a vector by means of its real number components relative to a coordinate system, thus providing for an analytic interpretation. Vectors can be combined algebraically, but there are some important distinctions to be made between algebraic operations on vectors and algebraic operations on real numbers. In this section we discuss the most important operations that can be performed with vectors, as well as both their *geometric and analytic interpretations*.

Addition

If **v** and **w** are two vectors, then we obtain **v** + **w** geometrically by placing the tail of **w** on the head of **v** and noting **v** + **w** as the **resultant vector** with the same tail as **v** and same head as **w**. You can see in Figure 11.5 why vector addition is said to follow a "triangle law."

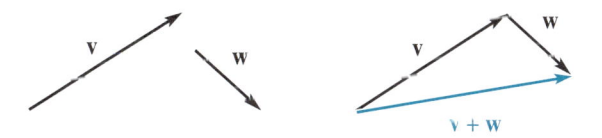

Figure 11.5 Addition of vectors.

This *resultant vector sum* also has a physical interpretation. If you imagine pushing an object the distance and direction indicated by **v**, and follow this by pushing the distance and direction indicated by **w**, the end position of the object is exactly the same as that obtained by the single push indicated by **v** + **w**.

Addition of vectors is simply performed *component-wise*. For example, in $\mathbb{R}^2$, if

$$\mathbf{v} = \langle v_1, v_2 \rangle \quad \text{and} \quad \mathbf{w} = \langle w_1, w_2 \rangle,$$

then

$$\mathbf{v} + \mathbf{w} = \langle v_1 + w_1, v_2 + w_2 \rangle.$$

Additive inverses

If **v** is a vector, then geometrically −**v** is the vector with the same length but opposite direction (see Figure 11.6).

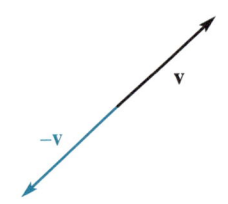

Figure 11.6 Additive inverse of a vector.

Analytically, the additive inverse of a vector is obtained by taking the additive inverse of each component. For example, in $\mathbb{R}^3$,

$$\text{if} \quad \mathbf{v} = \langle v_1, v_2, v_3 \rangle, \quad \text{then} \quad -\mathbf{v} = \langle -v_1, -v_2, -v_3 \rangle.$$

The additive inverse has the property that for any vector **v**, we have

$$\mathbf{v} + (-\mathbf{v}) = \mathbf{0}.$$

Note that $-\mathbf{0} = \mathbf{0}$.

Subtraction

One way of thinking of subtraction of real numbers is as "adding the opposite." This is also one way to think of subtraction of vectors. If **v** and **w** are two vectors, then

$$\mathbf{v} - \mathbf{w} = \mathbf{v} + (-\mathbf{w}).$$

Figure 11.7 illustrates the geometric interpretation of vector subtraction.

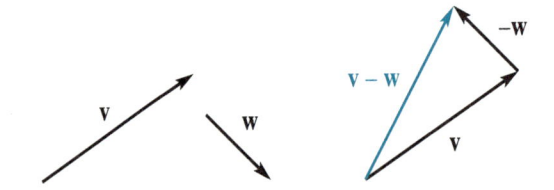

Figure 11.7 Subtraction of vectors—"adding the opposite."

An equivalent method for determining the difference of two vectors **v** − **w** geometrically is to place the tail of **w** on the tail of **v**; then **v** − **w**

11.2 ALGEBRA OF VECTORS

has its tail at the head of w and its head on the head of v. This method corresponds to the "missing addend" approach to subtraction. That is, we can think of v − w as the missing vector to add to w to obtain v. Figure 11.8 illustrates this approach.

Analytically, subtraction is performed component-wise. For example, in $\mathbb{R}^2$, if

$$\mathbf{v} = \langle v_1, v_2 \rangle \quad \text{and} \quad \mathbf{w} = \langle w_1, w_2 \rangle,$$

then

$$\mathbf{v} - \mathbf{w} = \langle v_1 - w_1, v_2 - w_2 \rangle.$$

Scalar multiplication and division

Vectors can be multiplied by scalars. Suppose that v is a vector and a is a scalar (in other words, that a is simply a real number). Geometrically, $a\mathbf{v}$ is a vector whose length is $|a|$ times the original length of v. As for the direction:

if $a > 0$, then $a\mathbf{v}$ has the same direction as v;

if $a < 0$, then $a\mathbf{v}$ has the opposite direction as v;

if $a = 0$, then $a\mathbf{v} = \mathbf{0}$.

Note that this means $(-1)\mathbf{v} = -\mathbf{v}$. Figure 11.9 illustrates several scalar multiples of a particular vector v.

Analytically, scalar multiplication is accomplished by distributing the scalar over each component. For example, in $\mathbb{R}^3$,

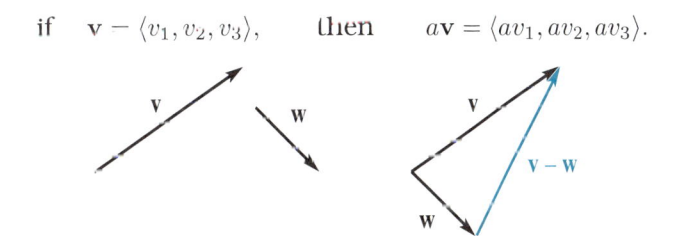

$$\text{if} \quad \mathbf{v} = \langle v_1, v_2, v_3 \rangle, \quad \text{then} \quad a\mathbf{v} = \langle av_1, av_2, av_3 \rangle.$$

Figure 11.8 Subtraction of vectors—"missing addend" approach.

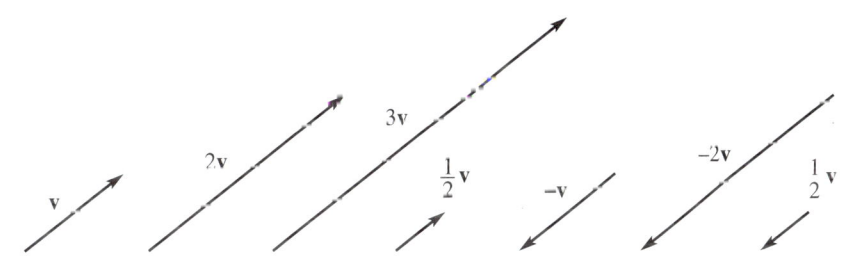

Figure 11.9 Scalar multiples of a vector.

EXAMPLE 3 Suppose that $\mathbf{v} = \langle 2, 3 \rangle$ and $\mathbf{w} = \langle -1, 2 \rangle$. Represent $\mathbf{v}$ and $\mathbf{w}$ geometrically. Find $2\mathbf{v}$, $-\mathbf{w}$, $\mathbf{v} + \mathbf{w}$ and $\mathbf{v} - \mathbf{w}$ analytically by components. Illustrate each result geometrically.

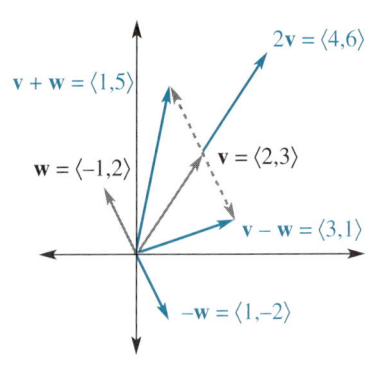

Figure 11.10 Algebraic operations on vectors.

Solution   Figure 11.10 illustrates the solution both analytically and geometrically. ∎

To divide a vector $\mathbf{v}$ by a scalar $a \neq 0$ is equivalent to multiplying by the scalar $1/a$.

EXAMPLE 4 In $\mathbb{R}^5$, suppose that $\mathbf{v} = \langle 5, 2, -1, 0, -5 \rangle$ and $\mathbf{w} = \langle 0, 1, 2, -7, 4 \rangle$. Find

$$\mathbf{v} + \mathbf{w}, \quad \mathbf{v} - \mathbf{w}, \quad 2\mathbf{v}, \quad \text{and} \quad \frac{\mathbf{w}}{-5}.$$

Solution   We have

$$\mathbf{v} + \mathbf{w} = \langle 5, 3, 1, -7, -1 \rangle$$
$$\mathbf{v} - \mathbf{w} = \langle 5, 1, -3, 7, -9 \rangle$$
$$2\mathbf{v} = \langle 10, 4, -2, 0, -10 \rangle$$
$$\frac{\mathbf{w}}{-5} = -\frac{1}{5}\langle 0, 1, 2, -7, 4 \rangle$$
$$= \langle 0, -0.2, -0.4, 1.4, -0.8 \rangle. \quad ∎$$

Algebraic properties of vectors

We can see that there are two ways of thinking of a vector: (1) geometrically, using a directed line segment of specific length and direction, or (2) analytically, using components. Which is better? The particular situation under investigation has a lot to do with whether we use a geometric representation or an analytic representation of a vector (or both). You should strive to feel comfortable with both representations, because each has its advantages.

11.2 ALGEBRA OF VECTORS

As an illustration, we will list some of the algebraic properties of vectors and show how these properties can be understood in either the geometric sense or the analytic sense.

Theorem 11.1

For any vectors **u**, **v**, and **w** (all of the same dimension) and for any scalars a and b, the following properties hold:

1. $\mathbf{u} + \mathbf{v} = \mathbf{v} + \mathbf{u}$ commutative law for addition
2. $(\mathbf{u} + \mathbf{v}) + \mathbf{w} = \mathbf{u} + (\mathbf{v} + \mathbf{w})$ associative law for addition
3. $\mathbf{u} + \mathbf{0} = \mathbf{0} + \mathbf{u} = \mathbf{u}$ additive identity law for **0**
4. $\mathbf{u} + (-\mathbf{u}) = \mathbf{0}$ additive inverse law
5. $a(b\mathbf{u}) = (ab)\mathbf{u}$ associative law for scalars
6. $a(\mathbf{u} + \mathbf{v}) = a\mathbf{u} + a\mathbf{v}$ first distributive law for scalars
7. $(a + b)\mathbf{u} = a\mathbf{u} + b\mathbf{u}$ second distributive law for scalars
8. $1\mathbf{u} = \mathbf{u}$ scalar multiplication identity law for 1
9. $0\mathbf{u} = \mathbf{0}$ scalar multiplication property of 0
10. $-1\mathbf{u} = -\mathbf{u}$ scalar multiplication property of -1

Reasoning These properties can be illustrated or verified using either geometric or analytic reasoning. We will discuss the first two here for 2-dimensional vectors. In the exercises, you are asked to verify properties 3 through 10 for yourself.

1. (using components) We will verify the commutative law analytically for 2-dimensional vectors. Similar reasoning can be used for vectors of higher dimensions. Let $\mathbf{u} = \langle u_1, u_2 \rangle$ and $\mathbf{v} = \langle v_1, v_2 \rangle$ be any two vectors in $\mathbb{R}^2$. Then,

$$\mathbf{u} + \mathbf{v} = \langle u_1, u_2 \rangle + \langle v_1, v_2 \rangle$$
$$= \langle u_1 + v_1, u_2 + v_2 \rangle$$
$$= \langle v_1 + u_1, v_2 + u_2 \rangle$$
$$= \langle v_1, v_2 \rangle + \langle u_1, u_2 \rangle$$
$$= \mathbf{v} + \mathbf{u}.$$

(Note that we are using the commutativity of *real number* addition in the third line.)

1. (geometric illustration) If we represent **u** and **v** with directed line segments, then $\mathbf{u} + \mathbf{v}$ and $\mathbf{v} + \mathbf{u}$ represent the same (directed) diagonal of a parallelogram. Figure 11.11 illustrates this **parallelogram law** of vector addition.

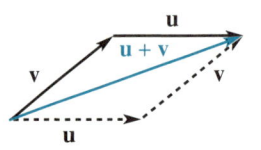

Figure 11.11 Parallelogram law for vector addition.

2. (using components)   Let $\mathbf{u} = \langle u_1, u_2 \rangle$, $\mathbf{v} = \langle v_1, v_2 \rangle$, and $\mathbf{w} = \langle w_1, w_2 \rangle$. Then,

$$(\mathbf{u} + \mathbf{v} + \mathbf{w}) = (\langle u_1, u_2 \rangle + \langle v_1, v_2 \rangle) + \langle w_1, w_2 \rangle$$
$$= \langle u_1 + v_1, u_2 + v_2 \rangle + \langle w_1, w_2 \rangle$$
$$= \langle (u_1 + v_1) + w_1, (u_2 + v_2) + w_2 \rangle$$
$$= \langle u_1 + (v_1 + w_1), u_2 + (v_2 + w_2) \rangle$$
$$= \langle u_1, u_2 \rangle + \langle v_1 + w_1, v_2 + w_2 \rangle$$
$$= \mathbf{u} + (\mathbf{v} + \mathbf{w}).$$

(Note that we used the associativity of real numbers in the fourth line.)

2. (geometric illustration)   If we represent all three vectors with directed line segments, the associative law can be seen by noting that we obtain the same resultant vector for the sum, regardless of whether we compute $\mathbf{u} + \mathbf{v}$ or $\mathbf{v} + \mathbf{w}$ first (see Figure 11.12). Since it doesn't matter whether we write $(\mathbf{u} + \mathbf{v}) + \mathbf{w}$ or $\mathbf{u} + (\mathbf{v} + \mathbf{w})$, we can write $\mathbf{u} + \mathbf{v} + \mathbf{w}$ without ambiguity. □

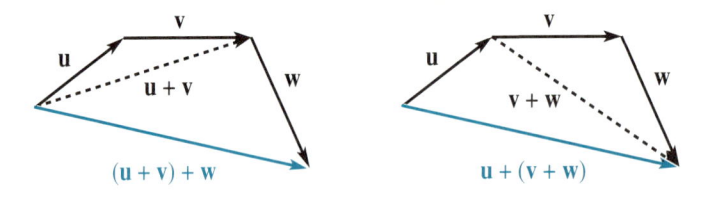

Figure 11.12 Associative law for vector addition.

EXERCISES for Section 11.2

For exercises 1-6: Suppose that, in the picture below, each of the vectors **p**, **q**, **u**, and **v** has a length of three units, while **n** has a length of two units. Furthermore, suppose that the initial point of **m** is located at the point $(-1, 1)$ and the terminal point of **m** is located at the point $(-4, -1)$.

11.2 ALGEBRA OF VECTORS

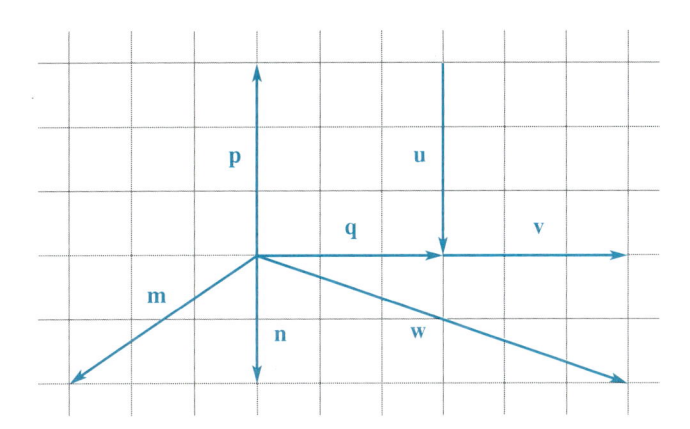

For Exercises 1-6

1. What vector represents $\mathbf{p} + \mathbf{q} + \mathbf{u}$?
2. What vector represents $\mathbf{n} + \mathbf{q} + \mathbf{v}$?
3. What vector represents $\mathbf{q} - \mathbf{v}$?
4. What vector represents $\mathbf{n} - \mathbf{q}$?
5. What vector represents $\mathbf{w} - \mathbf{n}$?
6. What vector represents $\mathbf{m} - \mathbf{n} + \mathbf{p} + \mathbf{q} + \mathbf{u} - 2\mathbf{v} + \mathbf{w}$?

For exercises 7-8: Suppose that six vectors of equal length are arranged to form a regular hexagon as shown below, with two of the consecutive sides labelled $\mathbf{v}$ and $\mathbf{w}$.

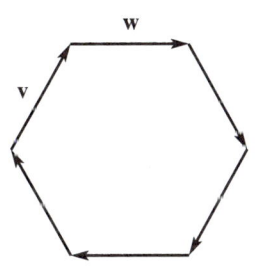

For Exercises 7-8

7. Express each of the other four vector sides in terms of $\mathbf{v}$ and $\mathbf{w}$.
8. What is the vector sum of all six sides?

For exercises 9-16: Suppose that $\mathbf{v} = \langle -4, -3 \rangle$ and $\mathbf{w} = \langle 2.72, 3.14 \rangle$. Find the indicated vectors.

9. $2\mathbf{v}$
10. $-3\mathbf{w}$
11. $\mathbf{v} + \mathbf{w}$
12. $\mathbf{v} - \mathbf{w}$
13. $\mathbf{w} - \mathbf{v}$
14. $2\mathbf{v} - 3\mathbf{w}$

15. $\mathbf{w}/5 - 4\mathbf{v}$ **16.** $\mathbf{v} + (0.5)\langle 1, -1\rangle$

For exercises 17-24: Suppose that $\mathbf{v} = \langle 2, -3, 1\rangle$ and $\mathbf{w} = \langle \frac{1}{2}, 1, -7\rangle$. Find the indicated vector.

17. $2\mathbf{v}$ **18.** $-3\mathbf{w}$

19. $\mathbf{v} + \mathbf{w}$ **20.** $\mathbf{v} - \mathbf{w}$

21. $\mathbf{w} - \mathbf{v}$ **22.** $2\mathbf{v} - 3\mathbf{w}$

23. $\mathbf{w}/5 - 4\mathbf{v}$ **24.** $\mathbf{v} + (0.5)\langle 1, 0, -1\rangle$

---

25. Verify properties 3 to 10 in Theorem 11.1 both geometrically and analytically for vectors in $\mathbb{R}^2$.

11.3 DOT PRODUCTS OF VECTORS

While we have discussed the product of a scalar and a vector, we haven't yet mentioned the product of two vectors. The **dot product** is one type of product that proves to be useful and is sometimes referred to as an *inner product* or a *scalar product*. As we did for the other arithmetic operations, we will give both a geometric and an analytic explanation of the dot product. Here's the definition in terms of rectangular components.

Definition 1

> The **dot product** of two vectors $\mathbf{v}$ and $\mathbf{w}$ (of the same dimension) is the sum of the component-wise products, and it is denoted by $\mathbf{v} \cdot \mathbf{w}$.

In $\mathbb{R}^2$: $\langle v_1, v_2\rangle \cdot \langle w_1, w_2\rangle = v_1 w_1 + v_2 w_2$

In $\mathbb{R}^3$: $\langle v_1, v_2, v_3\rangle \cdot \langle w_1, w_2, w_3\rangle = v_1 w_1 + v_2 w_2 + v_3 w_3$

In $\mathbb{R}^n$: $\langle v_1, v_2, \ldots v_n\rangle \cdot \langle w_1, w_2, \ldots w_n\rangle = v_1 w_1 + v_2 w_2 + \cdots + v_n w_n$

(In $\mathbb{R}^1$, the dot product is just the usual product of two numbers.)

11.3 DOT PRODUCTS OF VECTORS

▶▶▶ **The term *scalar product* makes sense since the result of a dot product is always a scalar, *not* a vector.**

EXAMPLE 5 Calculate the following dot products of vectors:

$$\langle 2,3 \rangle \cdot \langle 5,-1 \rangle \qquad \langle 1,2,3 \rangle \cdot \langle -5,3,-7 \rangle \qquad \langle -1,2,4 \rangle \cdot \langle 10,3,1 \rangle$$

$$\langle 0,0,0 \rangle \cdot \langle 1000, 10000, 1000000 \rangle \qquad \langle 1,2,3,4,5 \rangle \cdot \langle 5,4,3,2,1 \rangle$$

Solution We have

$$\langle 2,3 \rangle \cdot \langle 5,-1 \rangle = 2 \cdot 5 + 3(-1) = 10 - 3 = 7$$

$$\langle 1,2,3 \rangle \cdot \langle -5,3,-7 \rangle = -5 + 6 - 21 = -20$$

$$\langle -1,2,4 \rangle \cdot \langle 10,3,1 \rangle = -10 + 6 + 4 = 0$$

$$\langle 0,0,0 \rangle \cdot \langle 1000, 10000, 1000000 \rangle = 0$$

$$\langle 1,2,3,4,5 \rangle \cdot \langle 5,4,3,2,1 \rangle = 5 + 8 + 9 + 8 + 5 = 35 \qquad \blacksquare$$

Some algebraic properties of the dot product are easy to verify using the definition.

Theorem 11.2

For any vectors $\mathbf{u}$, $\mathbf{v}$, and $\mathbf{w}$ (all of the same dimension) and for any scalar a, the following properties hold:

$$\mathbf{u} \cdot \mathbf{v} = \mathbf{v} \cdot \mathbf{u}$$

$$\mathbf{u} \cdot (\mathbf{v} + \mathbf{w}) = \mathbf{u} \cdot \mathbf{v} + \mathbf{u} \cdot \mathbf{w}$$

$$a(\mathbf{u} \cdot \mathbf{v}) = (a\mathbf{u}) \cdot \mathbf{v} = \mathbf{u} \cdot (a\mathbf{v}).$$

Reasoning Let's verify these properties for vectors in $\mathbb{R}^2$. (The verifications for higher dimensions are very similar.) Suppose that $\mathbf{u} = \langle u_1, u_2 \rangle$, $\mathbf{v} = \langle v_1, v_2 \rangle$, and $\mathbf{w} = \langle w_1, w_2 \rangle$. The first property tells us that the dot product is *commutative*. Note that

$$\mathbf{u} \cdot \mathbf{v} = u_1 v_1 + u_2 v_2 = v_1 u_1 + v_2 u_2 = \mathbf{v} \cdot \mathbf{u}.$$

The second property tells us that the dot product *distributes over vector sums*:

$$\mathbf{u} \cdot (\mathbf{v} + \mathbf{w}) = \langle u_1, u_2 \rangle \cdot \langle v_1 + w_1, v_2 + w_2 \rangle$$
$$= u_1(v_1 + w_1) + u_2(v_2 + w_2)$$
$$= (u_1 v_1 + u_1 w_1) + (u_2 v_2 + u_2 w_2)$$
$$= (u_1 v_1 + u_2 v_2) + (u_1 w_1 + u_2 w_2)$$
$$= \langle u_1, u_2 \rangle \cdot \langle v_1, v_2 \rangle + \langle u_1 u_2 \rangle \cdot \langle w_1, w_2 \rangle$$
$$= \mathbf{u} \cdot \mathbf{v} + \mathbf{u} \cdot \mathbf{w}.$$

The last property describes how scalar multiplication and dot products behave together. Just compute each of the three quantities indicated, and you'll see that all three have the same value:

$$a\,(\mathbf{u} \cdot \mathbf{v}) = (a\mathbf{u}) \cdot \mathbf{v} = \mathbf{u} \cdot (a\mathbf{v}) = au_1 v_1 + au_2 v_2. \qquad \square$$

The dot product is an extremely useful computational tool whose worth will become more evident once we provide a geometric description.

Length of a vector—norm

The **norm** of a vector in two or three dimensions is simply its *length*. We denote the norm of a vector $\mathbf{v}$ as $||\mathbf{v}||$. If such a vector's components are given, the norm of the vector $||\mathbf{v}||$ can be computed using the Pythagorean theorem, as illustrated in Figure 11.13.

While the geometric notion of length might not seem to make sense for higher dimensional vectors, we can still generalize the Pythagorean theorem to provide an analytic definition of norm in $\mathbb{R}^n$:

If $\mathbf{v} = \langle v_1, v_2, \ldots, v_n \rangle$, then $||\mathbf{v}|| = \sqrt{v_1^2 + v_2^2 + \cdots + v_n^2}$.

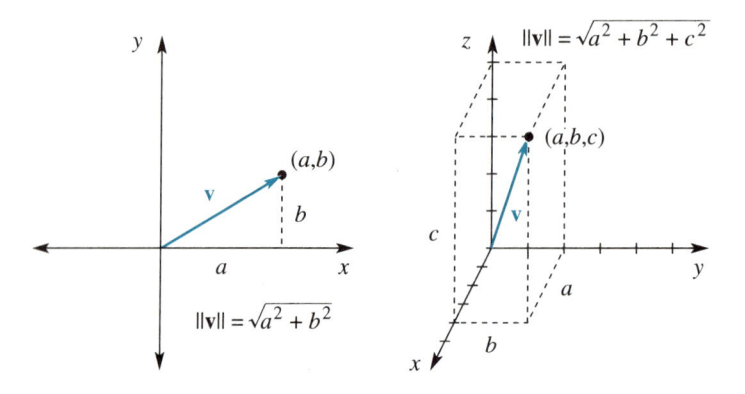

Figure 11.13 The norm of a vector is its length.

11.3 DOT PRODUCTS OF VECTORS

In the special case of $\mathbb{R}^1$, the norm of a 1-dimensional vector is simply its absolute value: $||\langle a \rangle|| = |a|$.

EXAMPLE 6 Calculate the norms of each of the following vectors:

$$\langle 2, 3 \rangle \qquad \langle -1, 2, 4 \rangle \qquad \langle 0, 0, 0 \rangle \qquad \langle 1, -0.2, 3.4, 4, -1.8 \rangle$$

Solution We have
$$||\langle 2, 3 \rangle|| = \sqrt{2^2 + 3^2} = \sqrt{4 + 9} = \sqrt{13} \approx 3.61$$
$$||\langle -1, 2, 4 \rangle|| = \sqrt{(-1)^2 + 2^2 + 4^2} = \sqrt{21} \approx 4.58$$
$$||\langle 0, 0, 0 \rangle|| = \sqrt{0^2 + 0^2 + 0^2} = 0$$
$$||\langle 1, -0.2, 3.4, 4, -1.8 \rangle|| = \sqrt{1^2 + (-0.2)^2 + (3.4)^2 + 4^2 + (-1.8)^2} \approx 5.6427 \quad \blacksquare$$

▶▶▶ **It is important to realize that $||\mathbf{v}||$ is always a (nonnegative) scalar.**

How does scalar multiplication affect the norm of a vector? Look back at Figure 11.9 and note that the *sign* of the scalar determines the direction (a negative scalar "flips" the vector in the opposite direction), but the absolute value of the scalar determines the new magnitude (by "scaling" the length). We can verify this analytically.

If $\mathbf{v} = \langle v_1, v_2, \ldots, v_n \rangle$ is a vector and a is a scalar, then

$$a\mathbf{v} = \langle av_1, av_2, \ldots, av_n \rangle$$

and

$$||a\mathbf{v}|| = \sqrt{a^2 v_1^2 + a^2 v_2^2 + \cdots + a^2 v_n^2} = \sqrt{a^2}\sqrt{v_1^2 + v_2^2 + \cdots + v_n^2} = |a|\,||\mathbf{v}||.$$

▶▶▶ **Geometrically, $a\mathbf{v}$ points in the same or opposite direction as $\mathbf{v}$, depending on whether a is positive or negative. The length of $a\mathbf{v}$ is $|a|$ times the length of the original vector $\mathbf{v}$.**

Unit vectors and the standard basis vectors

A **unit vector** is a vector of norm 1.

For example, in $\mathbb{R}^2$, $\mathbf{v} = \langle \frac{3}{5}, \frac{4}{5} \rangle$ is a unit vector, since

$$||\mathbf{v}|| = \sqrt{\frac{9}{25} + \frac{16}{25}} = 1.$$

In $\mathbb{R}^3$, $\mathbf{w} = \langle -\frac{2}{3}, -\frac{1}{3}, \frac{2}{3} \rangle$ is a unit vector, since

$$||\mathbf{w}|| = \sqrt{\frac{4}{9} + \frac{1}{9} + \frac{4}{9}} = 1.$$

Some special unit vectors in $\mathbb{R}^2$ are:

$$\mathbf{i} = \langle 1, 0 \rangle \quad \text{and} \quad \mathbf{j} = \langle 0, 1 \rangle.$$

These are called the standard basis vectors for $\mathbb{R}^2$. Note that any vector $\langle a, b \rangle$ can be written as a **linear combination** (meaning a sum of scalar multiples) of i and j:

$$\langle a, b \rangle = a\mathbf{i} + b\mathbf{j}.$$

To see this, just carry out the indicated scalar multiplication and vector addition:

$$a\mathbf{i} + b\mathbf{j} = a\langle 1, 0 \rangle + b\langle 0, 1 \rangle = \langle a, 0 \rangle + \langle 0, b \rangle = \langle a, b \rangle.$$

The standard basis vectors for $\mathbb{R}^3$ are the three vectors

$$\mathbf{i} = \langle 1, 0, 0 \rangle, \quad \mathbf{j} = \langle 0, 1, 0 \rangle, \quad \mathbf{k} = \langle 0, 0, 1 \rangle,$$

and $\langle a, b, c \rangle = a\mathbf{i} + b\mathbf{j} + c\mathbf{k}$ for any vector in $\mathbb{R}^3$.

EXAMPLE 7 Write $\langle -2, 3.2, -9 \rangle$ in terms of the standard basis vectors.

Solution $\langle -2, 3.2, -9 \rangle = -2\mathbf{i} + 3.2\mathbf{j} - 9\mathbf{k}.$ ∎

Unit vectors are often used when only a particular *direction* needs to be indicated for some purpose. Since the magnitude of the vector we use in this case is immaterial, we often agree to choose a unit vector.

Given a nonzero vector **v**, we can always find a unit vector having the same direction by simply dividing **v** by its own length (a positive scalar). In other words, $\mathbf{v}/\|\mathbf{v}\|$ is a unit vector having the same direction as **v**, provided that $\mathbf{v} \neq \mathbf{0}$.

EXAMPLE 8 Find a unit vector having the same direction as $\mathbf{v} = \langle -2, 3, -6 \rangle$.

Solution $\dfrac{\mathbf{v}}{\|\mathbf{v}\|} = \dfrac{\langle -2, 3, -6 \rangle}{\sqrt{(-2)^2 + 3^2 + (-6)^2}} = \dfrac{\langle -2, 3, -6 \rangle}{\sqrt{49}} = \dfrac{1}{7}\langle -2, 3, -6 \rangle = \left\langle -\dfrac{2}{7}, \dfrac{3}{7}, -\dfrac{6}{7} \right\rangle.$

Since this vector is a positive scalar multiple ($\frac{1}{7}$) of our original vector, we know that it points in the same direction. We can also check that it has unit length:

$$\left\| \left\langle -\frac{2}{7}, \frac{3}{7}, -\frac{6}{7} \right\rangle \right\| = \sqrt{\left(-\frac{2}{7}\right)^2 + \left(\frac{3}{7}\right)^2 + \left(-\frac{6}{7}\right)^2} = \sqrt{\frac{4}{49} + \frac{9}{49} + \frac{36}{49}} = \sqrt{\frac{49}{49}} = \sqrt{1} = 1.$$

Hence, the vector $\left\langle -\frac{2}{7}, \frac{3}{7}, -\frac{6}{7} \right\rangle$ satisfies both requirements. ∎

11.3 DOT PRODUCTS OF VECTORS

Since
$$||\mathbf{v}|| = \sqrt{v_1^2 + v_2^2 + \cdots + v_n^2}$$

and
$$\mathbf{v} \cdot \mathbf{v} = v_1^2 + v_2^2 + \cdots + v_n^2,$$

we can write
$$||\mathbf{v}|| = \sqrt{\mathbf{v} \cdot \mathbf{v}},$$

or
$$\mathbf{v} \cdot \mathbf{v} = ||\mathbf{v}||^2.$$

▶▶▶ **In other words, the dot product of any vector with itself is the square of the norm.**

Geometric description of dot product

We are now in a position to give a geometric description of the dot product of two vectors. In $\mathbb{R}^2$ or $\mathbb{R}^3$, the dot product of vectors **u** and **v** is

$$\mathbf{u} \cdot \mathbf{v} = ||\mathbf{u}||\, ||\mathbf{v}|| \cos\theta,$$

where θ is the smallest positive angle between **u** and **v** (so $0 \leq \theta \leq \pi$), provided that both **u** and **v** are nonzero. Figure 11.14 illustrates the angle θ in several instances. (If $\mathbf{u} = 0$ or $\mathbf{v} = 0$, then $\mathbf{u} \cdot \mathbf{v} = 0$.)

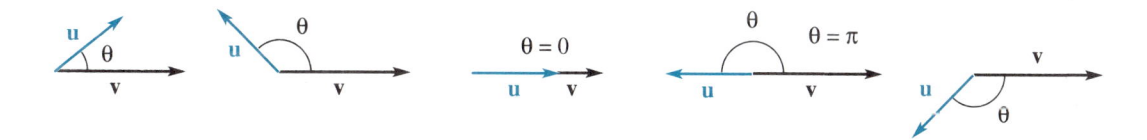

Figure 11.14 Smallest positive angle θ between two vectors.

This description of the dot product looks quite different from our original definition. Let's compute the dot product using both the definition and the geometric description in a specific example.

EXAMPLE 9 Find the dot product $\mathbf{v} \cdot \mathbf{w}$ where $\mathbf{v} = \langle 2, 2 \rangle$ and $\mathbf{w} = \langle 0, 3 \rangle$, using both the analytic definition and the geometric description.

Solution Figure 11.15 shows the two vectors. In this case we can see that the angle between the vectors is $45° = \pi/4$.

First, we compute the lengths of the two vectors:

$$||\mathbf{v}|| = \sqrt{2^2 + 2^2} = \sqrt{8}$$

$$||\mathbf{w}|| = \sqrt{0^2 + 3^2} = 3$$

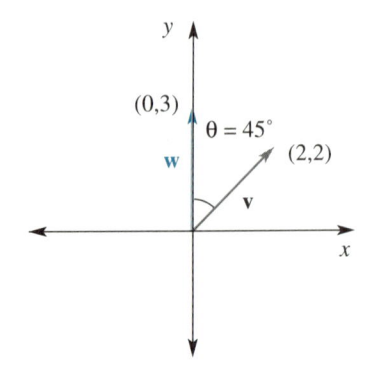

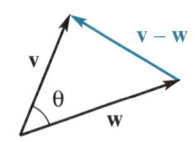

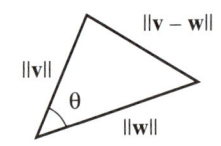

Figure 11.15 Vectors $\mathbf{v} = \langle 2, 2 \rangle$ and $\mathbf{w} = \langle 0, 3 \rangle$.

Figure 11.16 Two vectors and their difference.

Figure 11.17 Triangle formed by the three vectors.

Now, using the geometric description of dot product, we have

$$\mathbf{v} \cdot \mathbf{w} = ||\mathbf{v}||\ ||\mathbf{w}|| \cos 45° = (\sqrt{8})(3)(\sqrt{2}/2) = 6.$$

Using the analytic definition of dot product, we have

$$\mathbf{v} \cdot \mathbf{w} = \langle 2, 2 \rangle \cdot \langle 0, 3 \rangle = 2(0) + 2(3) = 6.$$

Hence, we obtain the same result using either method. ∎

Let's see why the geometric description always provides the same result as the analytic definition.

Reasoning   Suppose that $\mathbf{v}$ and $\mathbf{w}$ are any two *nonzero* vectors.

Figure 11.16 illustrates two such vectors, along with their difference $\mathbf{v} - \mathbf{w}$.

Now, let's examine the side lengths of the triangle formed by the three vectors $\mathbf{v}$, $\mathbf{w}$, and $\mathbf{v} - \mathbf{w}$ (see Figure 11.17).

We can calculate $||\mathbf{v} - \mathbf{w}||^2$ in two different ways. First, we can use the law of cosines to write

$$||\mathbf{v} - \mathbf{w}||^2 = ||\mathbf{v}||^2 + ||\mathbf{w}||^2 - 2||\mathbf{v}||||\mathbf{w}|| \cos \theta.$$

We can also calculate $||\mathbf{v} - \mathbf{w}||^2$ using the dot product:

$$||\mathbf{v} - \mathbf{w}||^2 = (\mathbf{v} - \mathbf{w}) \cdot (\mathbf{v} - \mathbf{w})$$
$$= (\mathbf{v} - \mathbf{w}) \cdot \mathbf{v} - (\mathbf{v} - \mathbf{w}) \cdot \mathbf{w}$$
$$= \mathbf{v} \cdot \mathbf{v} - \mathbf{v} \cdot \mathbf{w} - \mathbf{w} \cdot \mathbf{v} + \mathbf{w} \cdot \mathbf{w}$$
$$= ||\mathbf{v}||^2 + ||\mathbf{w}||^2 - 2\mathbf{v} \cdot \mathbf{w}.$$

11.3 DOT PRODUCTS OF VECTORS

When we equate these two ways of calculating $||\mathbf{v} - \mathbf{w}||^2$, we obtain

$$||\mathbf{v}||^2 + ||\mathbf{w}||^2 - 2\mathbf{v} \cdot \mathbf{w} = ||\mathbf{v}||^2 + ||\mathbf{w}||^2 - 2||\mathbf{v}||\,||\mathbf{w}||\cos\theta.$$

Note that we must have

$$\mathbf{v} \cdot \mathbf{w} = ||\mathbf{v}||\,||\mathbf{w}||\cos\theta.$$

Computing the angle between two vectors—orthogonal vectors

We now have a means of computing the angle between two nonzero vectors. Using

$$\mathbf{v} \cdot \mathbf{w} = ||\mathbf{v}||\,||\mathbf{w}||\cos\theta,$$

we have

$$\cos\theta = \frac{\mathbf{v} \cdot \mathbf{w}}{||\mathbf{v}||\,||\mathbf{w}||}.$$

Solving for θ gives us

$$\theta = \arccos\left(\frac{\mathbf{v} \cdot \mathbf{w}}{||\mathbf{v}||\,||\mathbf{w}||}\right).$$

EXAMPLE 10 Under what conditions on $\mathbf{v}$ and $\mathbf{w}$ is the dot product $\mathbf{v} \cdot \mathbf{w} = 0$?

Solution Certainly, if either $\mathbf{v} = 0$ or $\mathbf{w} = 0$, then $\mathbf{v} \cdot \mathbf{w} = 0$. If both $\mathbf{v}$ and $\mathbf{w}$ are nonzero, then their lengths are nonzero, and we can conclude that $\cos\theta = 0$, since

$$0 = \mathbf{v} \cdot \mathbf{w} = ||\mathbf{v}||\,||\mathbf{w}||\cos\theta.$$

But $\cos\theta = 0$ means that $\theta = 90°$ or $\pi/2$ radians. ■

▶▶▶ **The dot product of two nonzero vectors is *zero* precisely when the vectors are *perpendicular* to each other.**

Definition 2 Two vectors $\mathbf{v}$ and $\mathbf{w}$ are said to be **orthogonal** if and only if $\mathbf{v} \cdot \mathbf{w} = 0$.

▶▶▶ **In other words, we say that two vectors are orthogonal if and only if they are perpendicular to each other, or at least one of them is the zero vector.**

Since $\cos\theta < 0$ for $\pi/2 < \theta \leq \pi$, and $\cos\theta > 0$ for $0 \leq \theta < \pi/2$, we can see how the sign of the dot product can be used to determine the type of angle θ between two nonzero vectors **v** and **w**:

θ is acute ($< 90°$)	if and only if	$\mathbf{v} \cdot \mathbf{w} > 0$
θ is obtuse ($> 90°$)	if and only if	$\mathbf{v} \cdot \mathbf{w} < 0$
$\theta = \pi/2 (= 90°)$	if and only if	$\mathbf{v} \cdot \mathbf{w} = 0$

EXAMPLE 11 Given the vectors $\mathbf{u} = \mathbf{i} + \mathbf{j}$, $\mathbf{v} = 2\mathbf{j} - 3\mathbf{k}$, and $\mathbf{w} = 5\mathbf{k}$ in $\mathbb{R}^3$, find the angle between each pair of vectors. Which (if any) of the vectors are orthogonal to each other?

Solution First we compute the norm of each vector and the dot products of each pair of vectors:

$$\|\mathbf{u}\| = \sqrt{2} \qquad \|\mathbf{v}\| = \sqrt{13} \qquad \|\mathbf{w}\| = 5$$

$$\mathbf{u} \cdot \mathbf{v} = 2 \qquad \mathbf{v} \cdot \mathbf{w} = -15 \qquad \mathbf{u} \cdot \mathbf{w} = 0$$

This tells us immediately that **u** and **w** are orthogonal to each other. (Think about the position of these vectors in space. **u** can be represented geometrically as lying in the xy-plane, while **w** can be represented as lying along the z-axis.) Hence, the angle between **u** and **w** is $\pi/2$ radians, or $90°$.

To find the angles between the other two pairs of vectors, we compute

$$\frac{\mathbf{u} \cdot \mathbf{v}}{\|\mathbf{u}\|\,\|\mathbf{v}\|} = \frac{2}{\sqrt{26}} \approx 0.39223 \quad \text{and} \quad \frac{\mathbf{v} \cdot \mathbf{w}}{\|\mathbf{v}\|\,\|\mathbf{w}\|} = \frac{-15}{5\sqrt{13}} \approx -0.83205.$$

The angle between **u** and **v** is approximately

$$\arccos(0.39223) \approx 1.17 \text{ radians, or } 66.9°,$$

while the angle between **v** and **w** is approximately

$$\arccos(-0.83205) \approx 2.55 \text{ radians, or } 146.3°. \qquad \blacksquare$$

Component of a vector relative to a direction

We have seen that one way to specify a vector is in terms of its standard components (relative to the standard basis vectors). For example,

$$\mathbf{v} = \langle a, b, c \rangle = a\mathbf{i} + b\mathbf{j} + c\mathbf{k}$$

has standard components $a\mathbf{i}$, $b\mathbf{j}$, and $c\mathbf{k}$. We can also refer to these three vectors as the **vector components** of **v** in the directions **i**, **j**, and **k**, respectively. The scalars a, b, and c are called the **scalar components** of **v** in

11.3 DOT PRODUCTS OF VECTORS

the directions **i**, **j**, and **k**, respectively. Sometimes it is useful to describe a vector relative to a direction other than those given by the standard basis vectors. Specifically, suppose a direction is specified by some given *unit* vector **u**. Then for any other vector **v**, we can find its **vector component** in the direction **u**.

We can visualize this component geometrically by placing the tail of **v** on the tail of **u** and "projecting" **v** down on the line determined by **u** (see Figure 11.18). (The line segment formed in this way is called an *orthogonal projection* because of the right triangle formed in this picture.) This vector is called the **vector component of v in direction u**.

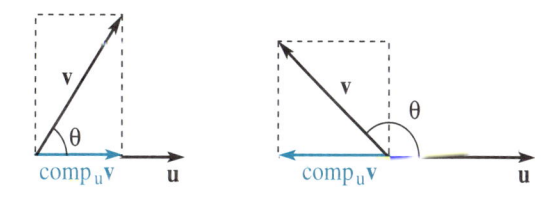

Figure 11.18 Vector component of **v** in direction **u**.

In Figure 11.18, we have indicated this vector component as $\text{comp}_\mathbf{u} \mathbf{v}$. Note that the vector component may point in either the same or opposite direction as **u**. If **v** and **u** are orthogonal, then the vector component of **v** in direction **u** is simply the zero vector **0**. In any case, we always have

$$\text{comp}_\mathbf{u} \mathbf{v} = a\mathbf{u}$$

for some scalar a. This scalar a is called the **scalar component of v in direction u**. The scalar component is negative if the vector component points in the opposite direction of **u**, and is positive if the vector component points in the same direction as **u**. Since **u** is a unit vector,

$$|a| = \|\text{comp}_\mathbf{u} \mathbf{v}\|.$$

Now we'll look at a very convenient way of computing vector and scalar components. Note that the angle θ in Figure 11.18 is the angle between **v** and **u**. From trigonometry, we can see that

$$\cos\theta = \frac{a}{\|\mathbf{v}\|}.$$

However, we know from the geometric description of the dot product that

$$\cos\theta = \frac{\mathbf{v} \cdot \mathbf{u}}{\|\mathbf{v}\| \|\mathbf{u}\|}.$$

Equating these two expressions for $\cos\theta$, and using the fact that $\|\mathbf{u}\| = 1$ (remember, **u** is a *unit* vector), we can find a formula for the scalar component a:

> The scalar component of **v** in the direction of unit vector **u** is
> $$a = \|\mathbf{v}\| \cos\theta = \|\mathbf{v}\|\frac{\mathbf{v}\cdot\mathbf{u}}{\|\mathbf{v}\|} = \mathbf{v}\cdot\mathbf{u}.$$

That's neat! To find the scalar component of **v** in the direction of unit vector **u**, we need only take the dot product of the two vectors. To find the vector component, we simply multiply this scalar by the unit vector **u**:

> The vector component of **v** in the direction of unit vector **u** is
> $$\mathrm{comp}_{\mathbf{u}}\mathbf{v} = a\mathbf{u} = (\mathbf{v}\cdot\mathbf{u})\mathbf{u}.$$

EXAMPLE 12 Find the scalar and vector components of $\mathbf{v} = \langle 1, -2, 3\rangle$ in the direction of the unit vector $\mathbf{u} = \langle -\frac{2}{3}, \frac{1}{3}, \frac{2}{3}\rangle$.

Solution First, note that **u** really is a unit vector, since

$$\|\mathbf{u}\| = \sqrt{\left(-\frac{2}{3}\right)^2 + \left(\frac{1}{3}\right)^2 + \left(\frac{2}{3}\right)^2} = \sqrt{\frac{4}{9} + \frac{1}{9} + \frac{4}{9}} = 1.$$

The scalar component of **v** in the direction **u** is

$$\mathbf{v}\cdot\mathbf{u} = \langle 1, -2, 3\rangle \cdot \left\langle -\frac{2}{3}, \frac{1}{3}, \frac{2}{3}\right\rangle = -\frac{2}{3} - \frac{2}{3} + \frac{6}{3} = \frac{2}{3}.$$

The vector component is

$$(\mathbf{v}\cdot\mathbf{u})\mathbf{u} = \frac{2}{3}\left\langle -\frac{2}{3}, \frac{1}{3}, \frac{2}{3}\right\rangle = \left\langle -\frac{4}{9}, \frac{2}{9}, \frac{4}{9}\right\rangle. \blacksquare$$

If the direction is specified by a vector **w** that does not have unit length, then we can simply use $\mathbf{u} = \mathbf{w}/\|\mathbf{w}\|$ for the purposes of determining scalar and vector components in that direction.

EXAMPLE 13 Find the scalar and vector components of $\mathbf{v} = \langle 3.5, -2.4\rangle$ in the direction of the vector $\mathbf{w} = \langle -3, 4\rangle$.

Solution First we find a unit vector **u** in the same direction as **w**:

$$\mathbf{u} = \frac{\mathbf{w}}{\|\mathbf{w}\|} = \frac{\langle -3, 4\rangle}{\sqrt{(-3)^2 + 4^2}} = \frac{\langle -3, 4\rangle}{5} = \langle -0.6, 0.8\rangle.$$

11.3 DOT PRODUCTS OF VECTORS

Now, the scalar component of **v** in the direction **w** is

$$\mathbf{v} \cdot \mathbf{u} = \langle 3.5, -2.4 \rangle \cdot \langle -0.6, 0.8 \rangle = (3.5)(-0.6) + (-2.4)(0.8) = -4.02,$$

and the vector component is

$$(\mathbf{v} \cdot \mathbf{u})\mathbf{u} = (-4.02)\langle -0.6, 0.8 \rangle = \langle 2.412, -3.216 \rangle.$$ ■

Once we have found the vector component of **v** in the direction of unit **u**, we can find another vector component in the direction orthogonal or perpendicular to **u** (and lying in the same plane as both **u** and **v**). These two vector components sum up to the original vector **v**, so that

$$\text{component of } \mathbf{v} \text{ orthogonal to } \mathbf{u} = \mathbf{v} - \text{comp}_\mathbf{u} \mathbf{v}.$$

The illustrations in Figure 11.19 indicate this orthogonal vector component.

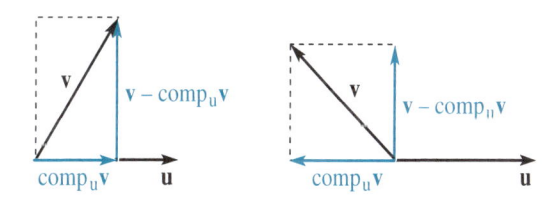

Figure 11.19 $\mathbf{v} - \text{comp}_\mathbf{u}\mathbf{v}$ is the orthogonal vector component of **v**.

EXERCISES for Section 11.3

For exercises 1-6: Suppose that in the picture below, each of the vectors **p**, **q**, **u**, and **v** has a length of three units, while **n** has a length of two units. Furthermore, suppose that the initial point of **m** is located at the point $(-1, 1)$ and the terminal point of **m** is located at the point $(-4, -1)$.

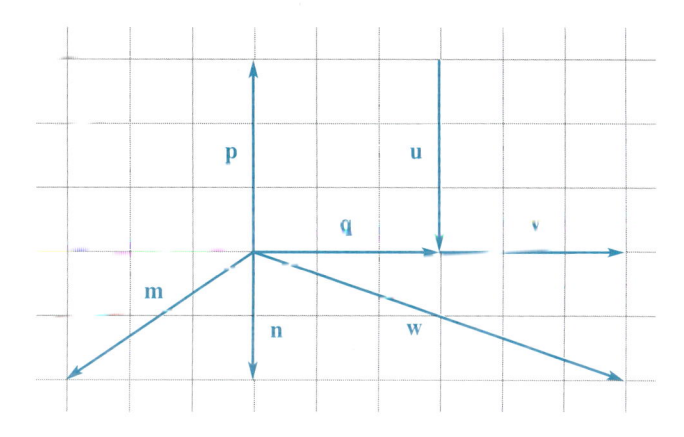

For Exercises 1-6

1. Find **p** · **n**.
2. Find **p** · **q**.
3. Find **u** · **p**.
4. Find **w** · **n**.

5. Find the vector component of **m** in the direction of **u**.

6. Which pairs of vectors in the picture have a positive dot product? negative dot product? zero dot product?

For exercises 7-8: Suppose that six vectors of unit length are arranged to form a regular hexagon as shown below, with two of the consecutive sides labelled **v** and **w**.

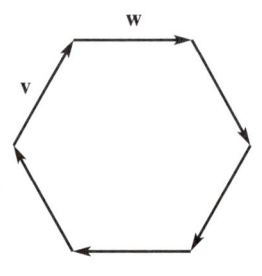

For Exercises 7-8

7. Find **v** · **w**.

8. Find the dot product of **v** with each of the other four sides.

For exercises 9-22: Suppose that $\mathbf{v} = \langle -4, -3 \rangle$ and $\mathbf{w} = \langle 2.72, 3.14 \rangle$. Find the indicated scalars or vectors.

9. **v** · **w**
10. $\mathbf{w} \cdot (0.5)(\mathbf{i} - \mathbf{j})$
11. $\|\mathbf{v}\|$
12. $\|\mathbf{w}\|$
13. $\|\mathbf{v} + \mathbf{w}\|$
14. $\|\mathbf{v}\| + \|\mathbf{w}\|$
15. $\|\mathbf{v} - \mathbf{w}\|$
16. $\|\mathbf{v}\| - \|\mathbf{w}\|$

17. a unit vector having the same direction as **v**

18. a unit vector orthogonal to **w**

19. the angle θ between **v** and **w**

20. the scalar component of **v** in the direction **w**

21. the vector component of **v** in the direction **w**

22. the area of the triangle determined by the origin and the terminal points of **v** and **w** (if each has its initial point at the origin)

11.3 DOT PRODUCTS OF VECTORS

For exercises 23-36: Suppose that $\mathbf{v} = 2\mathbf{i} - 3\mathbf{j} + \mathbf{k}$ and $\mathbf{w} = \dfrac{\mathbf{i}}{2} + \mathbf{j} - 7\mathbf{k}$. Find the indicated scalars or vectors.

23. $\mathbf{v} \cdot \mathbf{w}$

24. $\mathbf{w} \cdot (0.5)(\mathbf{i} - \mathbf{k})$

25. $\|\mathbf{v}\|$

26. $\|\mathbf{w}\|$

27. $\|\mathbf{v} + \mathbf{w}\|$

28. $\|\mathbf{v}\| + \|\mathbf{w}\|$

29. $\|\mathbf{v} - \mathbf{w}\|$

30. $\|\mathbf{v}\| - \|\mathbf{w}\|$

31. a unit vector having the same direction as $\mathbf{v}$

32. a unit vector orthogonal to both $\mathbf{v}$ and $\mathbf{w}$

33. the angle θ between $\mathbf{v}$ and $\mathbf{w}$

34. the scalar component of $\mathbf{v}$ in the direction $\mathbf{w}$

35. the vector component of $\mathbf{v}$ in the direction $\mathbf{w}$

36. the area of the triangle determined by the origin and the terminal points of $\mathbf{v}$ and $\mathbf{w}$ (if each has its initial point at the origin)

---

37. If $\mathbf{v} \cdot \mathbf{w} = 0$, must $\mathbf{v} = 0$ or $\mathbf{w} = 0$?

38. If $\|\mathbf{v} - \mathbf{w}\| = 0$, must $\mathbf{v} = \mathbf{w}$?

39. Show that if $\mathbf{u}$ is a unit vector, $\mathbf{v} - \text{comp}_\mathbf{u}\mathbf{v}$ is orthogonal to $\mathbf{u}$ for all other vectors $\mathbf{v}$.

40. Show that the scalar components of the vector $\langle a, b, c \rangle$ in the directions of the positive coordinate axes are simply the standard components a, b, and c.

41. Suppose that $\mathbf{v} = \langle x, y \rangle$. If $\mathbf{v}$ has its tail at the origin, then its head is at the point (x, y). Express the *polar coordinates* (r, θ) of this point in terms of $\mathbf{v}$. (Note: r is the distance to the origin, and θ is the angle that the line segment connecting (x, y) to the origin makes with the positive x-axis.)

The **direction cosines** of a vector $\mathbf{u}$ are the cosines of the angles $\mathbf{u}$ makes with the positive coordinate axes. Since the standard basis vectors are unit vectors that lie in these directions, we can use them to compute the direction cosines as follows (where either $\mathbf{u} = \langle u_1, u_2 \rangle$ in $\mathbb{R}^2$ or $\mathbf{u} = \langle u_1, u_2, u_3 \rangle$ in $\mathbb{R}^3$):

$$\cos \alpha = \frac{\mathbf{u} \cdot \mathbf{i}}{\|\mathbf{u}\| \, \|\mathbf{i}\|} = \frac{u_1}{\|\mathbf{u}\|},$$

$$\cos \beta = \frac{\mathbf{u} \cdot \mathbf{j}}{\|\mathbf{u}\| \, \|\mathbf{j}\|} = \frac{u_2}{\|\mathbf{u}\|},$$

$$\cos \gamma = \frac{\mathbf{u} \cdot \mathbf{k}}{\|\mathbf{u}\| \, \|\mathbf{k}\|} = \frac{u_3}{\|\mathbf{u}\|}.$$

The **direction angles** are α, β, and γ, respectively.

For exercises 42-48: Use the formulas for the direction cosines given above to answer exercises.

42. Find the direction angles α and β for the vectors **v** and **w** given in the instructions for exercises 9-22.

43. Compute $\cos^2 \alpha + \cos^2 \beta$ for the vectors **v** and **w** given in the instructions for exercises 9-22.

44. Verify that the result of the previous exercise is true for any nonzero vector **u** in $\mathbb{R}^2$.

45. Find the direction angles α, β, and γ for the vectors **v** and **w** given in the instructions for exercises 23-36.

46. Compute $\cos^2 \alpha + \cos^2 \beta + \cos^2 \gamma$ for the vectors **v** and **w** given in the instructions for exercises 23-36.

47. Verify that the result of the previous exercise is true for any nonzero vector **u** in $\mathbb{R}^3$.

48. Is it possible for a nonzero vector in $\mathbb{R}^2$ or $\mathbb{R}^3$ to be perpendicular to *all* the coordinate axes at once? If so, give an example. If not, explain why it is not possible.

49. Show that if $\mathbf{v} \neq \mathbf{0}$, then $\dfrac{\mathbf{v}}{\|\mathbf{v}\|}$ is always a unit vector.

50. Is it possible for $\mathbf{v} - \mathbf{w}$ to have a greater length than either **v** or **w**? Justify your answer (either give an example or show why it is not possible).

11.4 MATRICES

A **matrix** is a rectangular array of numbers. You may have used matrices before as useful computational tools for dealing with linear equations. Matrices are also extremely useful for working with vectors. In this section we discuss some basic material on matrices necessary for our purposes.

If a matrix has n rows and m columns, we call it an $n \times m$ matrix ("n by m" matrix). We call the element in the ith row and j^{th} column the ij-entry of the matrix. We will be concerned with *real* matrices, whose entries are all real numbers.

For example, here are three matrices:

$$A = \begin{bmatrix} 2 & 1 & -3 \\ -2 & 4 & 1 \end{bmatrix}, \quad B = \begin{bmatrix} -3 & 2 & 7 \\ 1 & 4 & -2 \end{bmatrix}, \quad \text{and} \quad C = \begin{bmatrix} 1 & 2 & 3 \\ -2 & -1 & 0 \\ 0 & 4 & -3 \end{bmatrix}.$$

A and B are 2×3 matrices; C is a 3×3 matrix.

If a matrix has the same number of rows as columns, we call it a **square matrix**. In the example above, C is a 3×3 square matrix.

Two matrices are considered equal if and only if they are exactly the same size and match entry-by-entry throughout.

If a matrix has 0's in all its entries, then we call it a **zero matrix**. For example, $O = \begin{bmatrix} 0 & 0 \\ 0 & 0 \\ 0 & 0 \end{bmatrix}$ is a 3×2 zero matrix.

If a *square* matrix has 1's along the *main diagonal* (from upper left to lower right) and 0's for all its other entries, we call it an **identity matrix**. For example, $I = \begin{bmatrix} 1 & 0 & 0 \\ 0 & 1 & 0 \\ 0 & 0 & 1 \end{bmatrix}$ is a 3×3 identity matrix.

We'll reserve the following notation for zero and identity matrices:

I_n represents an $n \times n$ identity matrix.

O_n represents an $n \times n$ zero matrix.

EXAMPLE 14 $I_3 = \begin{bmatrix} 1 & 0 & 0 \\ 0 & 1 & 0 \\ 0 & 0 & 1 \end{bmatrix}$ and $O_2 = \begin{bmatrix} 0 & 0 \\ 0 & 0 \end{bmatrix}.$ ∎

Algebra of matrices

We can perform arithmetic operations on matrices in a manner very similar to the arithmetic operations we perform on vectors. We summarize these operations below, and illustrate each with some examples.

Scalar multiplication: We multiply a matrix by a scalar simply by multiplying each entry by the scalar.

EXAMPLE 15 Consider the matrices

$$A = \begin{bmatrix} 2 & 1 & -3 \\ -2 & 4 & 1 \end{bmatrix} \quad \text{and} \quad B = \begin{bmatrix} -3 & 2 & 7 \\ 1 & 4 & -2 \end{bmatrix}.$$

Find $2A$ and $(-1)B$.

Solution We have

$$2A = 2 \begin{bmatrix} 2 & 1 & -3 \\ -2 & 4 & 1 \end{bmatrix} = \begin{bmatrix} 4 & 2 & -6 \\ -4 & 8 & 2 \end{bmatrix}$$

$$(-1)B = (-1) \begin{bmatrix} -3 & 2 & 7 \\ 1 & 4 & -2 \end{bmatrix} = \begin{bmatrix} 3 & -2 & -7 \\ -1 & -4 & 2 \end{bmatrix}.$$ ∎

Matrix addition and subtraction: Two matrices can be added or subtracted only if they are the same size (same number of rows, same number of columns). We add and subtract matrices "entry-wise."

EXAMPLE 16 Consider the matrices

$$A = \begin{bmatrix} 2 & 1 & -3 \\ -2 & 4 & 1 \end{bmatrix}, \quad B = \begin{bmatrix} -3 & 2 & 7 \\ 1 & 4 & -2 \end{bmatrix}, \quad \text{and} \quad C = \begin{bmatrix} 1 & 2 & 3 \\ -2 & -1 & 0 \\ 0 & 4 & -3 \end{bmatrix}.$$

Which pairs of these matrices can be added?

Solution The matrices A and C *cannot* be added because they are not the same size. Similarly, B and C *cannot* be added because they are not the same size. We can, however, add A and B:

$$A + B = \begin{bmatrix} 2 & 1 & -3 \\ -2 & 4 & 1 \end{bmatrix} + \begin{bmatrix} -3 & 2 & 7 \\ 1 & 4 & -2 \end{bmatrix} = \begin{bmatrix} -1 & 3 & 4 \\ -1 & 8 & -1 \end{bmatrix}.$$ ∎

EXAMPLE 17 If $B = \begin{bmatrix} -3 & 2 & 7 \\ 1 & 4 & -2 \end{bmatrix}$, what is $B + (-1)B$?

Solution We have

$$B + (-1)B = \begin{bmatrix} 0 & 0 & 0 \\ 0 & 0 & 0 \end{bmatrix},$$

a zero matrix. ∎

11.4 MATRICES

Note that this holds for any matrix M:

$$M + (-1)M = O$$

where O is a zero matrix with the same number of rows and columns as M. For this reason, we write

$$-M = (-1)M$$

and call $-M$ the *additive inverse* of M.

Matrix multiplication: Two matrices can be multiplied if and only if *the number of columns of the first matrix is the same as the number of rows of the second matrix*. To get the ij entry of the matrix product, we take the *dot product* of the ith *row* of the first matrix and the jth *column* of the second (as if they were vectors).

EXAMPLE 18 Consider the matrices

$$A = \begin{bmatrix} 2 & 1 & -3 \\ -2 & 4 & 1 \end{bmatrix}, \quad B = \begin{bmatrix} -3 & 2 & 7 \\ 1 & 4 & -2 \end{bmatrix}, \quad \text{and} \quad C = \begin{bmatrix} 1 & 2 & 3 \\ -2 & -1 & 0 \\ 0 & 4 & -3 \end{bmatrix}.$$

Find the matrix products AB and AC.

Solution The matrix product AB does *not* make sense. (A has 3 columns, B has only 2 rows.) The matrix product AC, however, does make sense. (A has 3 columns and C has 3 rows.)

The first entry of AC (first row, first column) is the dot product of the first row of A and the first column of C:

$$\langle 2, 1, -3 \rangle \cdot \langle 1, \ 2, 0 \rangle = 2(1) + 1(-2) + -3(0) = 0.$$

The middle entry of the first row is the dot product of the first row of A and the second column of C:

$$\langle 2, 1, -3 \rangle \cdot \langle 2, -1, 4 \rangle = 2(2) + 1(-1) + (-3)(4) = -9.$$

Continuing in this manner, we can complete the computation of the matrix product:

$$AC = \begin{bmatrix} 2 & 1 & -3 \\ -2 & 4 & 1 \end{bmatrix} \begin{bmatrix} 1 & 2 & 3 \\ -2 & -1 & 0 \\ 0 & 4 & -3 \end{bmatrix} = \begin{bmatrix} 0 & -9 & 15 \\ -10 & -4 & -9 \end{bmatrix}. \quad \blacksquare$$

▶▶▶ **If we multiply an $(n \times m)$ matrix times an $(m \times p)$ matrix, the result is an $(n \times p)$ matrix.**

Properties of matrix algebra

Here we summarize some of the important properties of matrix algebra. Below, S, T, and R are matrices and a and b are scalars.

> *Matrix addition is associative:*
> $$S + (T + R) = (S + T) + R.$$
>
> *Matrix addition is commutative:*
> $$S + T = T + S.$$
>
> *Scalar multiplication is distributive over matrix addition:*
> $$a(S + T) = aS + aT.$$
>
> In all three of the properties above, we assume that S, T, and R are the same size (same number of rows in each, same number of columns in each).

> *Scalar multiplication is associative with matrix multiplication:*
> $$a(ST) = (aS)T = S(aT).$$
>
> *Matrix multiplication is associative:*
> $$S(TR) = (ST)R.$$
>
> *Matrix multiplication is distributive over matrix addition:*
> $$S(T + R) = ST + SR \quad \text{and} \quad (S + T)R = SR + TR.$$
>
> In all three of these properties, we assume that the products make sense (i.e., that the numbers of rows and columns in each matrix are appropriate).

Notice that we have said nothing about matrix multiplication being *commutative*. Is it? (Try finding the products BC and CB using the matrices B and C from the previous examples.)

> If S is a square $n \times n$ matrix, then
> $$S + O_n = O_n + S = S \quad \text{and} \quad SI_n = I_nS = S,$$
> so the zero and identity matrices behave in a similar way to the real numbers 0 and 1, respectively.

11.4 MATRICES

Vectors as matrices

When multiplying two matrices together, we essentially treat the rows and columns as if they were vectors and use the dot product to compute the entries of the product matrix. In general, a vector can be considered either as a matrix with a single row or a matrix with a single column.

For example, suppose $\mathbf{v} = \langle 2, 1, -5 \rangle = 2\mathbf{i} + \mathbf{j} - 5\mathbf{k}$. If it is convenient, we could consider $\mathbf{v}$ as a matrix with one row, such as

$$\mathbf{v} = \begin{bmatrix} 2 & 1 & -5 \end{bmatrix},$$

or as a matrix with one column, such as

$$\mathbf{v} = \begin{bmatrix} 2 \\ 1 \\ -5 \end{bmatrix}.$$

To distinguish these two ways of writing the vector, we say in the first case, that $\mathbf{v}$ is a **row vector**, and in the second case, that $\mathbf{v}$ is a **column vector**. A 1-dimensional vector $\langle a \rangle$ can be considered a 1×1 matrix $[a]$.

When it is written as a row or column vector, we can multiply the vector by a matrix (provided that the matrix has the appropriate size).

EXAMPLE 19 Using $A = \begin{bmatrix} 2 & 1 & -3 \\ -2 & 4 & 1 \end{bmatrix}$ and $\mathbf{v} = \langle 2, 1, -5 \rangle$ as a column vector, find $A\mathbf{v}$.

Solution

$$A\mathbf{v} = \begin{bmatrix} 2 & 1 & -3 \\ -2 & 4 & 1 \end{bmatrix} \begin{bmatrix} 2 \\ 1 \\ -5 \end{bmatrix} = \begin{bmatrix} 2(2) + 1(1) + (-3)(-5) \\ -2(2) + 4(1) + 1(-5) \end{bmatrix} = \begin{bmatrix} 20 \\ -5 \end{bmatrix}.$$

Note that since A is a 2×3 matrix, and $\mathbf{v}$ is a 3×1 matrix (3 rows and 1 column), the product $A\mathbf{v}$ is a 2×1 matrix. We could interpret this result as a 2-dimensional column vector. ∎

EXAMPLE 20 Using $A = \begin{bmatrix} 2 & 1 & -3 \\ -2 & 4 & 1 \end{bmatrix}$ and $\mathbf{w} = \langle -1, 4 \rangle$ as a row vector, find $\mathbf{w}A$.

Solution

$$\mathbf{w}A = \begin{bmatrix} -1 & 4 \end{bmatrix} \begin{bmatrix} 2 & 1 & -3 \\ -2 & 4 & 1 \end{bmatrix}$$

$$= \begin{bmatrix} (-1)(2) + 4(-2) & (-1)(1) + 4(4) & (-1)(-3) + 4(1) \end{bmatrix}$$

$$= \begin{bmatrix} -10 & 15 & 7 \end{bmatrix}.$$

The result can be considered a 3-dimensional row vector. ∎

EXAMPLE 21   Using **v** and **w** as in the previous two examples, compute **v**A and A**w**.

Solution   Neither product makes sense, regardless of whether we write the vectors in row or column format. (The sizes are inappropriate for matrix products). ∎

Determinants

The *determinant* of a square matrix is a single number. If A is a square matrix, we denote its determinant by

$$\det A \text{ or } |A|.$$

The determinant of a 1×1 matrix is the value of its single entry:

$$\det \, [a] = a.$$

The determinant of a 2×2 matrix is the difference of the products of the diagonal entries:

$$\det \begin{bmatrix} a & b \\ c & d \end{bmatrix} = \begin{vmatrix} a & b \\ c & d \end{vmatrix} = ad - bc.$$

EXAMPLE 22   Find the determinant of $\begin{bmatrix} -2 & 3 \\ -1 & -4 \end{bmatrix}$.

Solution

$$\det \begin{bmatrix} -2 & 3 \\ -1 & -4 \end{bmatrix} = \begin{vmatrix} -2 & 3 \\ -1 & -4 \end{vmatrix} = (-2)(-4) - 3(-1) = 11. \quad \blacksquare$$

The determinant of a 3×3 matrix has a more complicated formula:

$$\det \begin{bmatrix} a_1 & a_2 & a_3 \\ b_1 & b_2 & b_3 \\ c_1 & c_2 & c_3 \end{bmatrix} = \begin{vmatrix} a_1 & a_2 & a_3 \\ b_1 & b_2 & b_3 \\ c_1 & c_2 & c_3 \end{vmatrix}$$

$$= a_1(b_2c_3 - b_3c_2) - a_2(b_1c_3 - b_3c_1) + a_3(b_1c_2 - b_2c_1).$$

11.4 MATRICES

EXAMPLE 23 Find the determinant of $C = \begin{bmatrix} 1 & 2 & 3 \\ -2 & -1 & 0 \\ 0 & 4 & -3 \end{bmatrix}$.

Solution

$$\det C = \begin{vmatrix} 1 & 2 & 3 \\ -2 & -1 & 0 \\ 0 & 4 & -3 \end{vmatrix}$$

$$= 1(-1(-3) - 0(4)) - 2(-2(-3) - 0(0)) + 3(-2(4) - (-1)(0))$$

$$= -33.$$

■

There is a handy method of remembering this determinant formula for 3×3 matrices called **expansion by minors** or **cofactors**. Here's how it works: If we pick any entry in the 3×3 matrix and cross out all the entries in the same row and column, we are left with a 2×2 matrix. The determinant of this smaller matrix is called the **minor** of that entry.

For example, the minor of b_2 in $\begin{bmatrix} a_1 & a_2 & a_3 \\ b_1 & b_2 & b_3 \\ c_1 & c_2 & c_3 \end{bmatrix}$ is $a_1c_3 - a_3c_1$.

Now, we multiply the minor by $+1$ or -1 to obtain the **cofactor** of the entry. To determine which sign to use, just take the row number i and the column number j of the entry and compute $(-1)^{i+j}$. For example, the cofactor of b_2 in $\begin{bmatrix} a_1 & a_2 & a_3 \\ b_1 & b_2 & b_3 \\ c_1 & c_2 & c_3 \end{bmatrix}$ is $(-1)^{2+2}(a_1c_3 - a_3c_1) = a_1c_3 - a_3c_1$, since b_2 is in the second row and second column.

Some people refer to a "checkerboard" of "+" and "−" signs to determine the choice $+1$ or -1. If you make a 3×3 matrix with a "+" in the upper left and alternate signs in both directions, you obtain the pattern

$$\begin{bmatrix} + & - & + \\ - & + & - \\ + & - & + \end{bmatrix}.$$

Note that the sign corresponding to b_2 (second row and second column) tells us that $+1$ is the correct choice.

Finally, the determinant of the 3×3 matrix is computed by:

1. choosing any single row or any single column,
2. multiplying each entry in that row or column by its cofactor, and
3. adding the results.

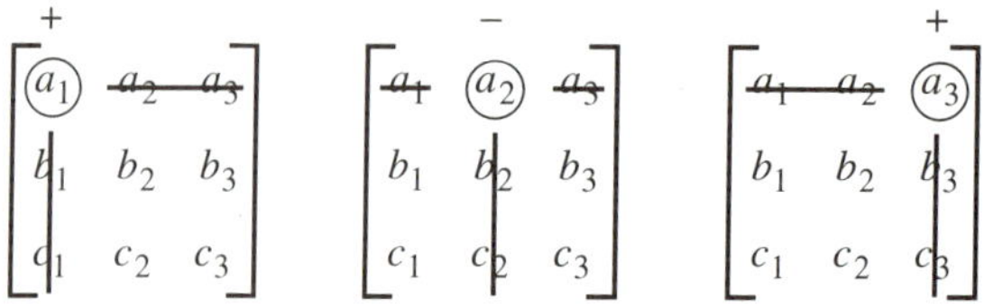

$$\begin{bmatrix} \boxed{a_1} & \cancel{a_2} & \cancel{a_3} \\ \cancel{b_1} & b_2 & b_3 \\ \cancel{c_1} & c_2 & c_3 \end{bmatrix}^{+} \quad \begin{bmatrix} \cancel{a_1} & \boxed{a_2} & \cancel{a_3} \\ b_1 & \cancel{b_2} & b_3 \\ c_1 & \cancel{c_2} & c_3 \end{bmatrix}^{-} \quad \begin{bmatrix} \cancel{a_1} & \cancel{a_2} & \boxed{a_3} \\ b_1 & b_2 & \cancel{b_3} \\ c_1 & c_2 & \cancel{c_3} \end{bmatrix}^{+}$$

$$a_1(b_2 c_3 - b_3 c_2) \;-\; a_2(b_1 c_3 - b_3 c_1) \;+\; a_3(b_1 c_2 - b_2 c_1)$$

Figure 11.20 Calculating the determinant using expansion by minors.

If you examine the formula for the determinant of the 3×3 matrix above, you'll see that it is written using the first row and expansion by minors (see Figure 11.20).

For 4×4, 5×5, 6×6, and higher dimensions, the same expansion by minors technique is used. Note that for a 4×4 matrix, the minors are determinants of 3×3 matrices, each of which may be calculated using expansion by minors! Hence, you can see that the computation of determinants quickly gets more complicated and tedious as the size of the matrix increases. Fortunately, certain calculators and computer software are equipped to calculate determinants of matrices.

▶▶▶ **Beware, however, that the numerical limitations of calculators and computers can be sorely tested by the calculation of determinants. This is because the calculation requires finding several differences of products whose values have been rounded off. This process leads to cancellation errors much like what we can experience when computing difference quotients with a machine.**

EXERCISES for Section 11.4

For exercises 1-15: Consider the following matrices:

$$S = \begin{bmatrix} 2 & -1 \\ -3 & 4 \\ 0 & 1 \end{bmatrix}, \quad T = \begin{bmatrix} 3 & -2 \\ 7 & 0 \\ 1 & -3 \end{bmatrix}, \quad R = \begin{bmatrix} 1 & 4 \\ 2 & 2 \\ -6 & -8 \end{bmatrix}$$

$$A = \begin{bmatrix} 2 & 1 & -3 \\ -2 & 4 & 1 \end{bmatrix}, \quad B = \begin{bmatrix} -3 & 2 & 7 \\ 1 & 4 & -2 \end{bmatrix}, \quad \text{and} \quad C = \begin{bmatrix} 1 & 2 & 3 \\ -2 & -1 & 0 \\ 0 & 4 & -3 \end{bmatrix}.$$

1. Find $U = S + T$.

2. Find $V = U + R$.

3. Find $X = T + R$.

4. Find $Y = S + X$.

5. Compare matrix V and matrix Y. What matrix algebra property does this comparison illustrate?

6. Illustrate that matrix addition is commutative by verifying that $S + T = T + S$.

11.4 MATRICES

7. Find $W = -2S - 2T$.

8. Find $Z = -2U$ (where matrix U is computed in exercise 1).

9. Compare matrix W and matrix Z. What matrix algebra property does this comparison illustrate?

10. Find $D = AS$ and $E = SA$. Is matrix multiplication commutative?

11. Find $F = AT$.

12. Find ET and SF. What matrix algebra property does this illustrate?

13. Find $I_3 C$ and $C I_3$.

14. Find $\det(D)$.

15. Find $\det(E)$.

16. Find $\det(I_2)$.

17. Find $\det(I_3)$.

18. Find $\det(I_4)$. What is $\det(I_n)$ for any n?

19. What is $\det(O_n)$ for any n?

20. Let $M = \begin{bmatrix} a_1 & b_1 \\ c_1 & d_1 \end{bmatrix}$ and $N = \begin{bmatrix} a_2 & b_2 \\ c_2 & d_2 \end{bmatrix}$. Find $|M|$, $|N|$, MN, and $|MN|$, and verify that $\det(MN) = \det(M)\det(N)$.

For exercises 21-30: Let

$$P = \begin{bmatrix} 1 & 2 & 3 \\ -2 & 4 & -1 \\ ? & ? & ? \end{bmatrix},$$

where you will consider different possibilities for the entries in the third row.

21. If the third row is replaced by 1 1 1, what is $\det(P)$?

22. If the third row is replaced by 2 2 2, what is $\det(P)$?

23. If the third row is replaced by $-3 \ -3 \ -3$, what is $\det(P)$?

24. If the third row is replaced by 0 0 0, what is $\det(P)$?

25. If the third row is replaced by $a \ a \ a$, what is $\det(P)$?

26. If the third row is replaced by 1 1 1 but the first two rows are interchanged, what is $\det(P)$?

27. If the third row is the same as the first row, what is $\det(P)$?

28. If the third row is the same as the second row, what is $\det(P)$?

29. Replace the third row by the difference of twice the first row and three times the second row. What is $\det(P)$?

30. Replace the third row by the sum of a times the first row and b times the second row. What is $\det(P)$?

For exercises 31-36: In these exercises you will verify several properties of matrices for the 3×3 case. Consider a general matrix of the form

$$\begin{bmatrix} a_1 & a_2 & a_3 \\ b_1 & b_2 & b_3 \\ c_1 & c_2 & c_3 \end{bmatrix}.$$

31. Show that multiplying any row of the matrix by a scalar results in the determinant being multiplied by that same scalar.

32. Show that replacing any row of the matrix by 0 0 0 results in a matrix whose determinant is 0.

33. Show that interchanging two rows of the matrix results in a matrix with a determinant of opposite sign but same absolute value as the original matrix.

34. Show that replacing one of the rows with another row (so that two of the rows are exactly the same) results in a matrix whose determinant is 0.

35. Show that replacing one of the rows with a scalar multiple of another results in a matrix whose determinant is 0.

36. Show that replacing one of the rows with a sum of scalar multiples of the remaining two rows (called a *linear combination*) results in a matrix whose determinant is 0.

11.5 GEOMETRIC APPLICATIONS OF VECTORS IN $\mathbb{R}^3$

In this section we examine how vectors can be used to solve geometric problems in space.

We have already seen that a geometric interpretation can be given to the dot product of two vectors. In $\mathbb{R}^3$, if

$$\mathbf{v} = \langle v_1, v_2, v_3 \rangle \quad \text{and} \quad \mathbf{w} = \langle w_1, w_2, w_3 \rangle,$$

then

$$\mathbf{v} \cdot \mathbf{w} = v_1 w_1 + v_2 w_2 + v_3 w_3.$$

11.5 GEOMETRIC APPLICATIONS OF VECTORS IN $\mathbb{R}^3$

If **v** and **w** are nonzero, then

$$\mathbf{v} \cdot \mathbf{w} = ||\mathbf{v}||\,||\mathbf{w}||\cos\theta,$$

where θ is the smallest angle between **v** and **w**. This allows us to use the dot product to find the angle between two vectors. In particular, we can use the dot product to test the orthogonality of two vectors. (Recall that two vectors are orthogonal if and only if their dot product is zero.)

The *dot product* is defined for vectors in $\mathbb{R}^n$ for any dimension n. Now let's introduce some special products used just for vectors in $\mathbb{R}^3$.

Cross product of vectors in $\mathbb{R}^3$

The **cross product** is a product defined *only* for vectors in $\mathbb{R}^3$. Unlike the dot product, which produces a scalar result, the cross product produces a *vector* result. The notation and analytic formula for the cross product of two 3-dimensional vectors **v** and **w** is

$$\mathbf{v}\times\mathbf{w} = (v_2 w_3 - v_3 w_2)\mathbf{i} - (v_1 w_3 - v_3 w_1)\mathbf{j} + (v_1 w_2 - v_2 w_1)\mathbf{k}$$

$$= \langle v_2 w_3 - v_3 w_2,\ v_3 w_1 - v_1 w_3,\ v_1 w_2 - v_2 w_1 \rangle.$$

This somewhat cumbersome formula is most easily remembered in the form of a "determinant." If we create a 3 by 3 matrix having the standard basis vectors **i**, **j**, and **k** in the first row, the components of **v** in the second row, and the components of **w** in the third row, then we can compute $\mathbf{v}\times\mathbf{w}$ as if it were the determinant:

$$\mathbf{v}\times\mathbf{w} = \begin{vmatrix} \mathbf{i} & \mathbf{j} & \mathbf{k} \\ v_1 & v_2 & v_3 \\ w_1 & w_2 & w_3 \end{vmatrix}.$$

EXAMPLE 24 Find $\mathbf{v}\times\mathbf{w}$ if $\mathbf{v} = \langle 1, 2, 3 \rangle$ and $\mathbf{w} = \langle -3, 1, 5 \rangle$.

Solution

$$\mathbf{v}\times\mathbf{w} = \langle 1, 2, 3 \rangle \times \langle -3, 1, 5 \rangle$$

$$= \begin{vmatrix} \mathbf{i} & \mathbf{j} & \mathbf{k} \\ 1 & 2 & 3 \\ -3 & 1 & 5 \end{vmatrix}$$

$$= (2(5) - 3(1))\mathbf{i} - (1(5) - 3(-3))\mathbf{j} + (1(1) - 2(-3))\mathbf{k}$$

$$= 7\mathbf{i} - 14\mathbf{j} + 7\mathbf{k} = 7\langle 1, -2, 1 \rangle.$$

Theorem 11.3

For any vectors u, v, and w in $\mathbb{R}^3$, and for any scalar a, the following properties hold:

$$\mathbf{v} \times \mathbf{w} = -(\mathbf{w} \times \mathbf{v})$$

$$(a\mathbf{v}) \times \mathbf{w} = \mathbf{v} \times (a\mathbf{w}) = a(\mathbf{v} \times \mathbf{w})$$

$$\mathbf{u} \times (\mathbf{v} + \mathbf{w}) = \mathbf{u} \times \mathbf{v} + \mathbf{u} \times \mathbf{w}$$

$$\mathbf{v} \times \mathbf{v} = \mathbf{0}.$$

Reasoning   If we let $\mathbf{u} = \langle u_1, u_2, u_3 \rangle$, $\mathbf{v} = \langle v_1, v_2, v_3 \rangle$, and $\mathbf{w} = \langle w_1, w_2, w_3 \rangle$, then all of these properties can be verified by simply computing the indicated vector quantities on both sides of the "=" sign and checking that the same result is obtained. □

Some of these properties deserve some special mention. Note that the first property tells us that the cross product is definitely *not* commutative. Indeed, the cross product is **anti-commutative**, meaning that a reversal of the order of the factors changes the sign of the result. The third property tells us that the cross product does distribute over vector sums. The last property is actually a consequence of the first ($\mathbf{u} \times \mathbf{u} = -(\mathbf{u} \times \mathbf{u})$ implies that $\mathbf{u} \times \mathbf{u}$ must be the zero vector). Together with the second property, this means

$$(a\mathbf{u}) \times \mathbf{u} = \mathbf{u} \times (a\mathbf{u}) = a(\mathbf{u} \times \mathbf{u}) = \mathbf{0}.$$

In other words, the cross product of a vector with any scalar multiple of itself is the zero vector.

Just as the dot product can be used to check that two vectors are *perpendicular*, the cross product can be used to check that two vectors in $\mathbb{R}^3$ are *parallel*.

▶ ▶ ▶   **The cross product $\mathbf{v} \times \mathbf{w} = \mathbf{0}$ if either v or w is the zero vector, or if v and w have the same or opposite directions.**

Triple scalar product

Given three vectors u, v, and w in $\mathbb{R}^3$, we can combine the dot and cross products to obtain what is known as the **triple scalar product** $\mathbf{u} \cdot (\mathbf{v} \times \mathbf{w})$. Computationally, you can verify that the triple scalar product is simply the *determinant* of the 3×3 matrix obtained by using the components of u, v,

11.5 GEOMETRIC APPLICATIONS OF VECTORS IN $\mathbb{R}^3$

and w as the rows of the matrix:

$$\mathbf{u} \cdot (\mathbf{v} \times \mathbf{w}) = \begin{vmatrix} u_1 & u_2 & u_3 \\ v_1 & v_2 & v_3 \\ w_1 & w_2 & w_3 \end{vmatrix}.$$

Note that the final result of this computation is a *scalar*.

Definition 3

> Vectors **u**, **v**, and **w** in $\mathbb{R}^3$ are said to be **linearly dependent** if there are scalars a, b, and c not all equal to 0 such that
>
> $$a\mathbf{u} + b\mathbf{v} + c\mathbf{w} = \mathbf{0}.$$
>
> If the *only* choice of scalars satisfying this equation are
>
> $$a = b = c = 0,$$
>
> then we say the vectors are **linearly independent**.

Equivalently, if any one of the vectors can be written as a linear combination of the other two (so that all three vectors lie in the same plane), then the three vectors are linearly dependent. From properties of the determinant (see Section 11.4, exercises 31-36), the triple scalar product $\mathbf{u} \cdot (\mathbf{v} \times \mathbf{w})$ can be used to determine whether the vectors **u**, **v**, and **w** all lie in the same plane.

> Vectors **u**, **v**, and **w** in $\mathbb{R}^3$ are linearly dependent if and only if
>
> $$\mathbf{u} \cdot (\mathbf{v} \times \mathbf{w}) = \mathbf{0}.$$

EXAMPLE 25 Determine whether the three vectors $\mathbf{u} = \langle -1, 5, 11 \rangle$, $\mathbf{v} = \langle 1, 2, 3 \rangle$, and $\mathbf{w} = \langle -3, 1, 5 \rangle$ are linearly independent or linearly dependent.

Solution We need only check whether or not the triple scalar product $\mathbf{u} \cdot (\mathbf{v} \times \mathbf{w})$ is nonzero. In Example 24, we found that $\mathbf{v} \times \mathbf{w} = 7\langle 1, -2, 1 \rangle$. Hence

$$\mathbf{u} \cdot (\mathbf{v} \times \mathbf{w}) = \langle -1, 5, 11 \rangle \cdot 7\langle 1, -2, 1 \rangle = 7(-1 - 10 + 11) = 0,$$

and we can conclude that the three vectors are linearly dependent. ∎

Geometric interpretations

Just as the dot product has a nice geometric interpretation, so does the cross product. To see this for ourselves, let's write out the calculation of the *squared norm* of the cross product for two general vectors $\mathbf{v} = \langle v_1, v_2, v_3 \rangle$ and $\mathbf{w} = \langle w_1, w_2, w_3 \rangle$:

$$||\mathbf{v} \times \mathbf{w}||^2 = (v_2 w_3 - v_3 w_2)^2 + (v_3 w_1 - v_1 w_3)^2 + (v_1 w_2 - v_2 w_1)^2.$$

If you write out the following calculation, you will find that it is equivalent:

$$||\mathbf{v} \times \mathbf{w}||^2 = (v_1^2 + v_2^2 + v_3^2)(w_1^2 + w_2^2 + w_3^2) - (v_1 w_1 + v_2 w_2 + v_3 w_3)^2$$
$$= ||\mathbf{v}||^2 ||\mathbf{w}||^2 - (\mathbf{v} \cdot \mathbf{w})^2.$$

For two nonzero vectors $\mathbf{v}$ and $\mathbf{w}$, we know that $\mathbf{v} \cdot \mathbf{w} = ||\mathbf{v}|| \; ||\mathbf{w}|| \; \cos \theta$, where θ is the smallest angle between the two vectors. Substituting into this last expression for $||\mathbf{v} \times \mathbf{w}||^2$, we have

$$||\mathbf{v} \times \mathbf{w}||^2 = ||\mathbf{v}||^2 ||\mathbf{w}||^2 - ||\mathbf{v}||^2 ||\mathbf{w}||^2 \cos^2 \theta$$
$$= ||\mathbf{v}||^2 ||\mathbf{w}||^2 (1 - \cos^2 \theta)$$
$$= ||\mathbf{v}||^2 ||\mathbf{w}||^2 (\sin^2 \theta).$$

The magnitude (length) of $\mathbf{v} \times \mathbf{w}$ is

$$||\mathbf{v} \times \mathbf{w}|| = ||\mathbf{v}|| \; ||\mathbf{w}|| \; |\sin \theta|.$$

Note that $\sin \theta = 0$ when $\theta = 0$ or $\theta = 180° = \pi$ radians. This is consistent with our earlier observation that the cross product of parallel vectors is the zero vector. However, now we can give a geometric interpretation to the magnitude of the cross product.

Using trigonometry, Figure 11.21 shows that we can interpret this magnitude as the *area of the parallelogram* determined by $\mathbf{v}$ and $\mathbf{w}$.

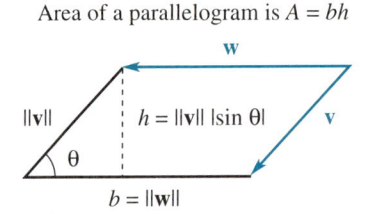

Figure 11.21 The area of the parallelogram is $||\mathbf{v} \times \mathbf{w}|| = ||\mathbf{v}|| \; ||\mathbf{w}|| |\sin \theta|$.

The cross product $\mathbf{v} \times \mathbf{w}$ is orthogonal to both $\mathbf{v}$ and $\mathbf{w}$. (The triple scalar products $\mathbf{v} \cdot (\mathbf{v} \times \mathbf{w})$ and $\mathbf{w} \cdot (\mathbf{v} \times \mathbf{w})$ are both 0 since they each represent determinants of matrices with two equal rows.)

If $\mathbf{v}$ and $\mathbf{w}$ are not parallel, there are exactly two possible directions (both perpendicular to the plane containing both $\mathbf{v}$ and $\mathbf{w}$). The correct

11.5 GEOMETRIC APPLICATIONS OF VECTORS IN $\mathbb{R}^3$

choice is determined by the **right-hand rule**: If you hold your right hand so that your fingers point in the direction of v and can curl toward w, then your extended thumb will point in the direction of v × w. Figure 11.22 illustrates the right-hand rule.

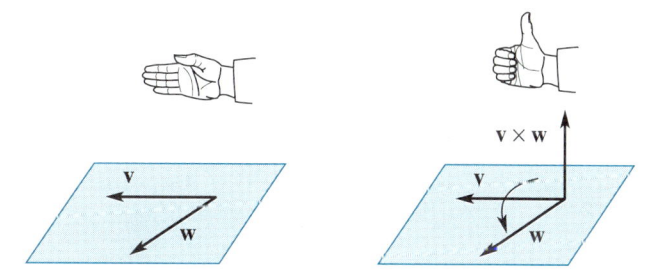

Figure 11.22 The right-hand rule determines the direction of the cross product.

You can verify that using the right-hand rule for w × v forces your thumb to point in the opposite direction.

EXAMPLE 26 Compute all the possible dot and cross products of the standard basis vectors i, j, and k.

Solution

$$\mathbf{i} \cdot \mathbf{j} = \mathbf{j} \cdot \mathbf{i} = 0 \qquad \mathbf{i} \cdot \mathbf{i} = 1$$
$$\mathbf{i} \cdot \mathbf{k} = \mathbf{k} \cdot \mathbf{i} = 0 \qquad \mathbf{j} \cdot \mathbf{j} = 1$$
$$\mathbf{j} \cdot \mathbf{k} = \mathbf{k} \cdot \mathbf{j} = 0 \qquad \mathbf{k} \cdot \mathbf{k} = 1$$
$$\mathbf{i} \times \mathbf{j} = \mathbf{k} \qquad \mathbf{j} \times \mathbf{k} = \mathbf{i} \qquad \mathbf{k} \times \mathbf{i} = \mathbf{j}$$
$$\mathbf{i} \times \mathbf{i} = \mathbf{j} \times \mathbf{j} = \mathbf{k} \times \mathbf{k} = 0$$
$$\mathbf{j} \times \mathbf{i} = -\mathbf{k} \qquad \mathbf{k} \times \mathbf{j} = -\mathbf{i} \qquad \mathbf{i} \times \mathbf{k} = -\mathbf{j}$$

∎

The diagram in Figure 11.23 can be useful for remembering cross products of the standard basis vectors: The cross product of any two standard basis vectors is either +1 (if the order goes clockwise) or −1 (if the order goes counterclockwise) times the remaining basis vector.

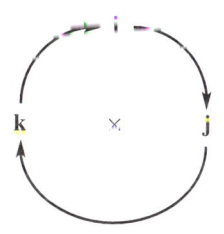

Figure 11.23 Cross products diagram for standard basis vectors.

EXAMPLE 27 Find the area of the parallelogram determined by $\mathbf{u} = \mathbf{j} + 2\mathbf{k}$ and $\mathbf{v} = \mathbf{i} - 2\mathbf{j}$.

Solution $\mathbf{u} \times \mathbf{v} = \begin{vmatrix} \mathbf{i} & \mathbf{j} & \mathbf{k} \\ 0 & 1 & 2 \\ 1 & -2 & 0 \end{vmatrix} = 4\mathbf{i} + 2\mathbf{j} - \mathbf{k}$. The area of the parallelogram is

$$\|\mathbf{u} \times \mathbf{v}\| = \sqrt{4^2 + 2^2 + (-1)^2} = \sqrt{21} \approx 4.58.$$

EXAMPLE 28 Find all unit vectors that are perpendicular to both $\mathbf{u} = \langle 1, 2, 3 \rangle$ and $\mathbf{v} = \langle -2, 0, -4 \rangle$.

Solution We can find a vector perpendicular to both $\mathbf{u}$ and $\mathbf{v}$ by means of their cross product:

$$\mathbf{u} \times \mathbf{v} = \begin{vmatrix} \mathbf{i} & \mathbf{j} & \mathbf{k} \\ 1 & 2 & 3 \\ -2 & 0 & -4 \end{vmatrix} = -8\mathbf{i} - 2\mathbf{j} + 4\mathbf{k}.$$

Now, we can scale this vector to unit length. Since

$$\|\mathbf{u} \times \mathbf{v}\| = \sqrt{64 + 4 + 16} = \sqrt{84} = 2\sqrt{21},$$

$\langle \frac{-8}{2\sqrt{21}}, \frac{-2}{2\sqrt{21}}, \frac{4}{2\sqrt{21}} \rangle = \langle -\frac{4}{\sqrt{21}}, -\frac{1}{\sqrt{21}}, \frac{2}{\sqrt{21}} \rangle$ is a unit vector.

We check that this vector is orthogonal to both $\mathbf{u}$ and $\mathbf{v}$ by means of the dot product:

$$\langle 1, 2, 3 \rangle \cdot \left\langle -\frac{4}{\sqrt{21}}, -\frac{1}{\sqrt{21}}, \frac{2}{\sqrt{21}} \right\rangle = -\frac{4}{\sqrt{21}} - \frac{2}{\sqrt{21}} + \frac{6}{\sqrt{21}} = 0,$$

and

$$\langle -2, 0, -4 \rangle \cdot \left\langle -\frac{4}{\sqrt{21}}, -\frac{1}{\sqrt{21}}, \frac{2}{\sqrt{21}} \right\rangle = \frac{8}{\sqrt{21}} + 0 - \frac{8}{\sqrt{21}} = 0.$$

The additive inverse of this vector is $\langle \frac{4}{\sqrt{21}}, \frac{1}{\sqrt{21}}, -\frac{2}{\sqrt{21}} \rangle$, and this is the only other unit vector perpendicular to both $\mathbf{u}$ and $\mathbf{v}$.

The right-hand rule for cross products of vectors is dependent on our choice of a *right-handed coordinate system* for $\mathbb{R}^3$. A different labelling for the coordinate axes can result in a left-handed coordinate system, in the sense that we must use a left-hand rule for cross products. Figure 11.24 illustrates a left-handed coordinate system.

Note that the right-hand rule applied to this coordinate system would result in $\mathbf{i} \times \mathbf{j} = -\mathbf{k}$, while the left-hand rule furnishes the usual result $\mathbf{i} \times \mathbf{j} = \mathbf{k}$.

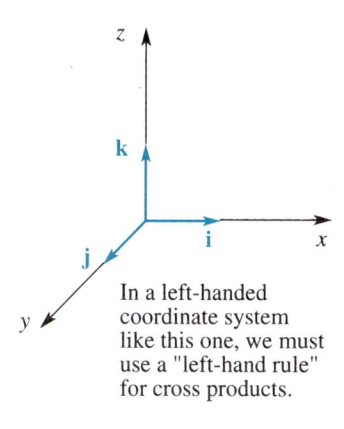

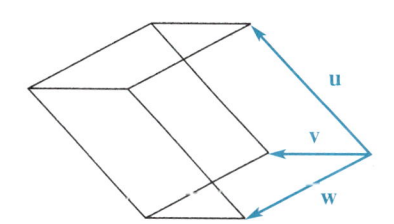

11.5 GEOMETRIC APPLICATIONS OF VECTORS IN $\mathbb{R}^3$

Figure 11.24 A left-handed coordinate system.

In a left-handed coordinate system like this one, we must use a "left-hand rule" for cross products.

Figure 11.25 The volume of the "box" is $|\mathbf{u} \cdot (\mathbf{v} \times \mathbf{w})|$.

Geometrically, the absolute value of the triple scalar product gives us the volume of the parallelepiped (slanted box) determined by the three vectors **u**, **v**, and **w** (see Figure 11.25).

The triple product is 0 exactly when the three vectors all lie in the same plane (resulting in a "flat" box).

EXERCISES for Section 11.5

For exercises 1-15: Refer to the following vectors:

$$\mathbf{a} = 3\mathbf{i} + \mathbf{j} - \mathbf{k}$$
$$\mathbf{b} = 2\mathbf{i} - \mathbf{j} + \mathbf{k}$$
$$\mathbf{c} = \mathbf{i} + 2\mathbf{j} + 3\mathbf{k}$$

1. Find $\mathbf{b} \times \mathbf{a}$.
2. Find $\mathbf{c} \times \mathbf{a}$.
3. Find $(\mathbf{b} - \mathbf{a}) \times (\mathbf{c} - \mathbf{a})$.
4. Find a unit vector perpendicular to both **a** and **c**.
5. Find a unit vector perpendicular to both $-3\mathbf{i}$ and **a**.
6. Find the area of the parallelogram determined by the two vectors $\langle 1, 4, 7 \rangle$ and $\langle -2, 1, -6 \rangle$.
7. Find the area of the *triangle* determined by **a** and **c**.
8. Find the angle between **a** and $\mathbf{b} \times \mathbf{c}$.
9. Find the area of the parallelogram determined by **b** and **c**.
10. Find the volume of the parallelepiped determined by all three vectors.

11. Show that these three vectors do not lie in the same plane. (Hint: consider the triple scalar product.)

12. Find the volume of the parallelepiped determined by the three vectors $\langle 1, 4, 7 \rangle$, $\langle -2, 1, -6 \rangle$ and $\langle 3, 0, -5 \rangle$.

13. Find a general equation for the plane determined by the three points $(1, 4, 7)$, $(-2, 1, -6)$, and $(3, 0, -5)$.

14. Determine the angle between **a** and **c** using the dot product.

15. Determine the sine of the angle between **b** and **c** using the cross product. What are the two possible angles (between 0 and π) that have this sine value? Now determine which is the actual angle between **b** and **c**.

For exercises 16-26: Answer "True" or "False."

16. For any two vectors **a**, **b** in $\mathbb{R}^3$, $\mathbf{a} \cdot \mathbf{b} = \mathbf{b} \cdot \mathbf{a}$.

17. For any two vectors **a**, **b** in $\mathbb{R}^3$, $\mathbf{a} \times \mathbf{b} = \mathbf{b} \times \mathbf{a}$.

18. For any vector **a** in $\mathbb{R}^3$, $\mathbf{a} \times \mathbf{a} = \mathbf{0}$.

19. If $\mathbf{a} \cdot \mathbf{b} = 0$, then either $\mathbf{a} = \mathbf{0}$ or $\mathbf{b} = \mathbf{0}$.

20. For any two vectors **a**, **b** in $\mathbb{R}^3$, $\mathbf{a} \times \mathbf{b} = ||\mathbf{a}||\,||\mathbf{b}|| \sin \theta$.

21. If $\mathbf{a} \times \mathbf{b} = \mathbf{0}$, then either $\mathbf{a} = \mathbf{0}$ or $\mathbf{b} = \mathbf{0}$.

22. For any three vectors in $\mathbb{R}^3$, **a**, **b**, and **c**, we have $\mathbf{a} \cdot (\mathbf{b} \times \mathbf{c}) = (\mathbf{a} \cdot \mathbf{b}) \times (\mathbf{a} \cdot \mathbf{c})$.

23. $\mathbf{i} \cdot (\mathbf{j} \times \mathbf{k}) = 1$.

24. For any two vectors in $\mathbb{R}^3$, **a** and **b**, $|\mathbf{a} \cdot \mathbf{b}| \leq ||\mathbf{a}||\,||\mathbf{b}||$.

25. If **a** and **b** are orthogonal, then $\mathbf{a} \times \mathbf{b} = \mathbf{0}$.

26. If **a** and **b** lie in the same plane, then $\mathbf{a} \cdot \mathbf{b} = 0$.

11.6 EQUATIONS OF LINES AND PLANES IN $\mathbb{R}^3$

Parametric equations of lines

You are already familiar with the **slope-intercept equation**

$$y = mx + b$$

and the **general equation** for a line in the Cartesian plane:

$$ax + by + c = 0.$$

A line can also be represented by parametric equations expressing both x and y in terms of a parameter t.

11.6 EQUATIONS OF LINES AND PLANES IN $\mathbb{R}^3$

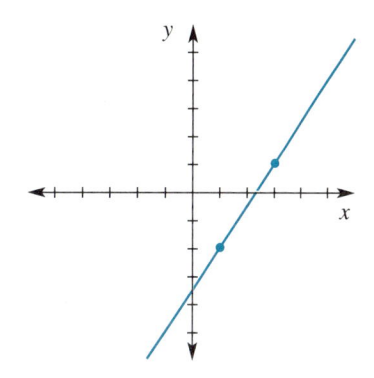

Figure 11.26 Graph of line represented by $x = 2t + 1$ and $y = 3t - 2$.

EXAMPLE 29 Graph and find both the slope-intercept and the general equations for the line represented by the parametric equations

$$x = 2t + 1,$$
$$y = 3t - 2.$$

Solution We can find two points on this line very quickly by substituting $t = 0$ and $t = 1$. The corresponding points are $(1, -2)$ and $(3, 1)$. Figure 11.26 shows the graph of the line.

We can compute the slope of the line as $m = \frac{1-(-2)}{3-1} = \frac{3}{2}$ and see that the y-intercept is $b = -\frac{7}{2}$. The slope-intercept form is

$$y = \frac{3}{2}x - \frac{7}{2},$$

and the general equation is

$$3x - 2y - 7 = 0.$$

You know that a line is determined by its slope and one point on the line. In terms of vectors, we can think of a line as being determined by a *direction vector* **v** and an initial *position vector* **u**. The points of the line are determined by the terminal points (heads) of all vectors of the form

$$\mathbf{u} + t\mathbf{v},$$

where t is any real number. This is called the **vector parametric form** of the line.

EXAMPLE 30 Find a position vector **u** and a direction vector **v** for the line of the previous example, and express it in vector parametric form.

Solution   The initial position vector can be obtained by substituting $t = 0$:

$$\mathbf{u} = \langle 1, -2 \rangle,$$

and a direction vector is given by the coefficients of t in the equations for x and y:

$$\mathbf{v} = \langle 2, 3 \rangle.$$

The vector parametric form of the line is

$$\mathbf{u} + t\mathbf{v} = \langle 1, -2 \rangle + t \langle 2, 3 \rangle,$$

where t is any real number. Note that the components of the vector parametric form are simply the original parametric equations:

$$\langle x(t), y(t) \rangle = \langle 1 + 2t, -2 + 3t \rangle. \blacksquare$$

EXAMPLE 31   Find a vector parametric form for the line passing through the point $(-3, 5)$ and with direction vector $\langle 2, -1 \rangle$.

Solution $\langle x(t), y(t) \rangle = \langle -3 + 2t, 5 - t \rangle.$ $\blacksquare$

EXAMPLE 32   Find a vector parametric form for the line passing through the points $P = (2, 2)$ and $Q = (-2, 3)$.

Solution   We can determine a direction vector by taking

$$\overrightarrow{PQ} = \langle -2 - 2,\ 3 - 2 \rangle = \langle -4, 1 \rangle.$$

Either point can serve as the initial position vector. If we use $P = (2, 2)$, we have

$$\langle x(t), y(t) \rangle = \langle 2 - 4t, 2 + t \rangle.$$

We can check that $\langle x(1), y(1) \rangle = \langle -2, 3 \rangle$ corresponds to point Q. $\blacksquare$

Vector parametric equations for lines in 3-dimensional space can be found in exactly the same way. All we need is an initial position vector and a direction vector.

11.6 EQUATIONS OF LINES AND PLANES IN $\mathbb{R}^3$

EXAMPLE 33 Find a vector parametric equation for the line that passes through the point $(-1, -3, 0)$ and with direction vector $\langle 5, -1, 2 \rangle$.

Solution $\langle x(t), y(t), z(t) \rangle = \langle -1 + 5t, -3 - t, 2t \rangle$. ∎

EXAMPLE 34 Find a vector parametric form for the line passing through the two points $P = (1, 1, 1)$ and $Q = (-1, -2, -3)$.

Solution Using $\overrightarrow{PQ} = \langle -2, -3, -4 \rangle$ as the direction vector and $\langle 1, 1, 1 \rangle$ as the initial position vector, we have

$$\langle x(t), y(t), z(t) \rangle = \langle 1 - 2t, 1 - 3t, 1 - 4t \rangle.$$

Note that $\langle x(1), y(1), z(1) \rangle = \langle -1, -2, -3 \rangle$ corresponds to Q. ∎

Equations for a plane

A line in space is 1-dimensional in the sense that all three of its components can be expressed in terms of a single parameter t. A plane in space is 2-dimensional, so we will need two parameters, s and t, to adequately describe it.

Just as a line is determined by an initial position vector and a direction vector, a plane is determined by an initial position vector **u** and *two* nonparallel direction vectors **v** and **w** that lie in the plane. The vector parametric form of the plane is

$$\mathbf{u} + s\mathbf{v} + t\mathbf{w}.$$

As the values of s and t range over the set of real numbers, this form specifies the points in the plane (see Figure 11.27).

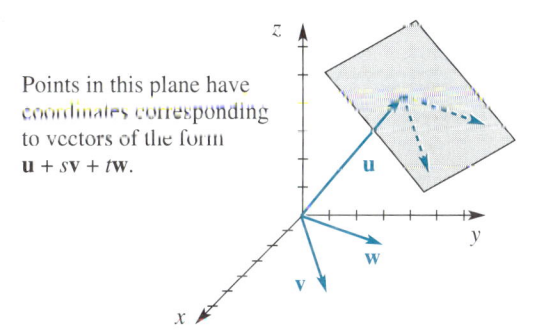

Points in this plane have coordinates corresponding to vectors of the form
u + s**v** + t**w**.

Figure 11.27 Parametrizing a plane.

EXAMPLE 35 Find the vector parametric form for the plane containing the point $(1, 2, 3)$ and with direction vectors $\mathbf{v} = \mathbf{i} + \mathbf{j}$ and $\mathbf{w} = \mathbf{j} - \mathbf{k}$.

Solution $\mathbf{u} = \langle 1, 2, 3 \rangle$ is the initial position vector, $\mathbf{v} = \langle 1, 1, 0 \rangle$ and $\mathbf{w} = \langle 0, 1, -1 \rangle$ are the direction vectors. We can write the vector parametric form of the plane as

$$\langle x(s,t), y(s,t), z(s,t) \rangle = \langle 1 + s, 2 + s + t, 3 - t \rangle.$$

We can locate three noncollinear points in the plane by substituting $(0, 0)$, $(1, 0)$, and $(0, 1)$ for (s, t) in the vector parametric form:

$$\langle x(0,0), y(0,0), z(0,0) \rangle = \langle 1, 2, 3 \rangle$$

$$\langle x(1,0), y(1,0), z(1,0) \rangle = \langle 2, 3, 3 \rangle$$

$$\langle x(0,1), y(0,1), z(0,1) \rangle = \langle 1, 3, 2 \rangle.$$

∎

A plane in space also has a **general equation** of the form

$$ax + by + cz + d = 0.$$

The points (x, y, z) in the plane are precisely those that satisfy this equation.

EXAMPLE 36 $x + 2y - 3z - 6 = 0$ is the general equation of a plane. Three points that satisfy the equation are $(6, 0, 0)$, $(0, 3, 0)$, and $(0, 0, -2)$. ∎

Remarkably, the vector $\langle a, b, c \rangle$ must be orthogonal to every vector that lies in this plane! (In the previous example, this means that the vector $\langle 1, 2, -3 \rangle$ is orthogonal to any vector lying in the plane.) The general reasoning goes like this: If we pick any two points in the plane $ax + by + cz + d = 0$, say $P = (x_1, y_1, z_1)$ and $Q = (x_2, y_2, z_2)$, then we know that

$$ax_1 + by_1 + cz_1 + d = 0$$

$$ax_2 + by_2 + cz_2 + d = 0.$$

If we subtract the first equation from the second, we obtain

$$a(x_2 - x_1) + b(y_2 - y_1) + c(z_2 - z_1) = 0.$$

Note that the left-hand side of this equation is the dot product of $\langle a, b, c \rangle$ and $\overrightarrow{PQ} = \langle (x_2 - x_1), (y_2 - y_1), (z_2 - z_1) \rangle$. Since this dot product is zero, the two vectors must be orthogonal.

The vector $\langle a, b, c \rangle$ is said to be a **normal vector** to the plane. This observation gives a way of finding the general equation of a plane, provided

11.6 EQUATIONS OF LINES AND PLANES IN $\mathbb{R}^3$

that we can determine a normal vector and a point in the plane (see Figure 11.28).

Equation of the plane containing (x_0, y_0, z_0) and perpendicular to the vector $\langle a,b,c \rangle$ is
$a(x - x_0) + b(y - y_0) + c(z - z_0) = 0.$

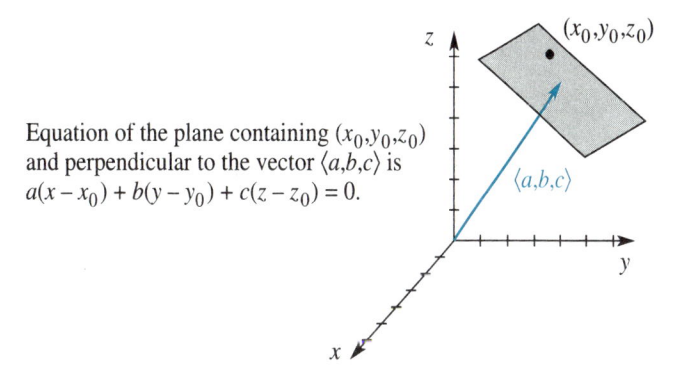

Figure 11.28 A normal vector determines a plane.

EXAMPLE 37 Find the general equation of the plane containing the point $(-2, 3, -5)$ and with normal vector $\langle 1, -3, 2 \rangle$.

Solution From the normal vector, we can see that the general equation must be of the form

$$x - 3y + 2z + d = 0.$$

We can determine d by substituting the point and solving

$$(-2) - 3(3) + 2(-5) + d = 0.$$

This tells us that $d = 21$, and the general equation of the plane is

$$x - 3y + 2z + 21 = 0.$$

EXERCISES for Section 11.6

For exercises 1-3: Refer to the following vectors:

$$\mathbf{a} = 3\mathbf{i} + \mathbf{j} - \mathbf{k}$$
$$\mathbf{b} = 2\mathbf{i} - \mathbf{j} + \mathbf{k}$$
$$\mathbf{c} = \mathbf{i} + 2\mathbf{j} + 3\mathbf{k}$$

1. If **b** and **c** both have their initial points at the origin, find the vector parametric form of the line passing through their terminal points.

2. If **a**, **b**, and **c** all have their initial points at the origin, find both the vector parametric form and the general equation of the plane passing through their terminal points.

3. Find the vector parametric form of the line containing the origin and with direction vector a.

4. Find a vector parametric form for the line determined by the two points $(1, 4, 7)$ and $(-2, 1, -6)$.

5. Find a vector parametric form for the plane determined by the three points $(1, 4, 7)$, $(-2, 1, -6)$, and $(3, 0, -5)$.

6. Find a general equation for the plane determined by the three points $(1, 4, 7)$, $(-2, 1, -6)$, and $(3, 0, -5)$.

7. Find a vector parametric form for the line containing the point $(1, 3, 5)$ and having direction vector $\langle -2, 1, 3 \rangle$.

8. Find a vector parametric form for the line passing through the points $(1, 2, -3)$ and $(0, 4, 2)$.

9. Find a vector parametric form for the line passing through $(1, -5, 7)$ and having direction vector $\langle 1, 0, -1 \rangle$.

10. Find a parametric form for the line passing through the point $(1, 2, 3)$ and parallel to the line having vector parametric form

$$\langle x(t), y(t), z(t) \rangle = \langle 1 + 2t, 1 - 3t, 2 + t \rangle.$$

11. Find a parametric form for the line passing through the origin and parallel to the line having vector parametric form

$$\langle x(t), y(t), z(t) \rangle = \langle 3 - t, 2t, 4 + t \rangle.$$

12. Find a vector parametric form for the line in the xy-plane that passes through the origin and is perpendicular to the line

$$\langle x(t), y(t), z(t) \rangle = \langle 1 + t, 2 - t, 0 \rangle.$$

13. Find the point of intersection of the two lines

$$\langle x(t), y(t), z(t) \rangle = \langle 1 + t, -2t, 1 - t \rangle \text{ and } \langle x(s), y(s), z(s) \rangle = \langle 2 - s, 1 - s, s \rangle.$$

14. Find a vector parametric form for the line that passes through the point of intersection of the two lines $\langle x(t), y(t), z(t) \rangle = \langle 1 + 3t, 3 - t, -1 + 2t \rangle$ and $\langle x(s), y(s), z(s) \rangle = \langle 3 - s, 1 + s, 5 - 3s \rangle$ and is perpendicular to both lines.

15. Find a vector parametric form for the plane determined by the three points $(1, 1, 1), (2, 4, 6)$, and $(3, 4, 5)$.

16. Find a general equation for the plane determined by the three points $(1, 1, 1), (2, 4, 6)$, and $(3, 4, 5)$.

11.6 EQUATIONS OF LINES AND PLANES IN $\mathbb{R}^3$

17. Find a general equation for the plane that passes through the point $(3, -1, -2)$ and is perpendicular to the line

$$\langle x(t), y(t), z(t) \rangle = \langle 1 - t, 2 + 3t, 5 - t \rangle.$$

(A line is perpendicular to a plane if it is perpendicular to every line in that plane.)

18. Find a general equation for the plane containing the two lines

$$\langle x(t), y(t), z(t) \rangle = \langle 1 - 2t, 2 + t, 1 + 6t \rangle$$

and

$$\langle x(t), y(t), z(t) \rangle = \langle 1 - 2t, 1 - t, 3 + 10t \rangle.$$

19. Does the line with parametric form

$$\langle x(t), y(t), z(t) \rangle = \langle 1 + t, 2 - t, 5 - \tfrac{9}{2}t \rangle$$

intersect the plane $3x - 6y + z = 1$? If so, where? If not, why not? Does the line

$$\langle x(t), y(t), z(t) \rangle = \langle 3t, -6t, t \rangle$$

intersect this plane?

20. Find a vector parametric form for the line that passes through $(2, 1, 3)$ and is perpendicular to the plane $3x - 6y + z = 1$.

21. Find a vector parametric form of the plane containing the point $(1, 3, 4)$ and the line

$$\langle x(t), y(t), z(t) \rangle = \langle 4 - t, 3 + 5t, 2t \rangle.$$

22. Find a general equation of the plane containing the point $(1, 3, 4)$ and the line

$$\langle x(t), y(t), z(t) \rangle = \langle 4 - t, 3 + 5t, 2t \rangle.$$

23. Find a vector parametric form of the plane containing the two lines

$$\langle x(t), y(t), z(t) \rangle = \langle 1 + t, 2 + 3t, 4 - t \rangle \text{ and } \langle x(t), y(t), z(t) \rangle = \langle 3 - t, 1 - 3t, 2 + t \rangle.$$

24. Find a general equation of the plane containing the two lines

$$\langle x(t), y(t), z(t) \rangle = \langle 1 + t, 2 + 3t, 4 - t \rangle$$

and

$$\langle x(t), y(t), z(t)\rangle = \langle 3-t, 1-3t, 2+t\rangle.$$

25. Find a vector parametric form for the line of intersection of the two planes $x + 3y - z = 1$ and $x + 2y + z = 3$.

26. Find the general equation of the plane that passes through $(-1, 1, -2)$ and is perpendicular to both the plane $x + 3y - z = 1$ and $x + 2y + z = 3$.

27. Under what conditions on the given points, lines, and planes is it true that

(a) two lines determine a unique plane
(b) a line and a point determine a unique plane
(c) two planes intersect in a unique line
(d) a given line and plane are perpendicular
(e) a given line and plane intersect

How could you tell that the given points, lines, and planes satisfy the conditions from looking at their equations?

12

Calculus of Curves

In this chapter, we discuss **vector-valued functions** that take a *single real number as an input* and produce a *vector for an output*. We often denote the independent variable as t (to suggest *time*) and the output vector as $\mathbf{r}(t)$. Note the use of boldface to distinguish between scalar and vector quantities: t represents a *scalar* input, while $\mathbf{r}$ is a function producing a *vector* output.

You can think of these vector-valued functions as describing some vector quantity that changes with time. We've already used calculus to study real-valued functions; now we'll extend the use of derivatives and integrals to study these new vector-valued functions.

12.1 POSITION FUNCTIONS AND GRAPHICAL REPRESENTATIONS

If $\mathbf{r}$ is a function that takes a real number as input from a domain D and its outputs are 2-dimensional vectors, we write

$$\mathbf{r}: D \longrightarrow \mathbb{R}^2.$$

The assignment process defining this function has two components corresponding to the two **coordinate functions** x and y, each of which is a real-valued function:

$$\mathbf{r}(t) = \langle x(t), y(t) \rangle = x(t)\mathbf{i} + y(t)\mathbf{j}.$$

Here, the single independent variable is t, while $x(t)$ and $y(t)$ represent the two real number components of $\mathbf{r}(t)$.

Similarly, if the outputs of a function $\mathbf{r}$ are 3-dimensional, we write

$$\mathbf{r}: D \longrightarrow \mathbb{R}^3,$$

and the assignment process defining $\mathbf{r}$ has three components:

$$\mathbf{r}(t) = \langle x(t), y(t), z(t) \rangle = x(t)\mathbf{i} + y(t)\mathbf{j} + z(t)\mathbf{k}.$$

In general, if the function r has real numbers for inputs and n-dimensional vectors for outputs, we can write

$$\mathbf{r} : D \longrightarrow \mathbb{R}^n$$

and
$$\mathbf{r}(t) = \langle x_1(t), x_2(t), \ldots, x_n(t) \rangle.$$

If the outputs of a vector-valued function are 2- or 3-dimensional, then we can think of them as designating positions in a plane or in space, respectively. Here's how: if we represent a vector with its initial point (tail) at the origin, then the vector's components are exactly described by the coordinates of its terminal point (head). As the input t takes on values in the domain, the terminal points determined by $\mathbf{r}(t)$ trace out a curve in the plane or space (see Figure 12.1). This curve gives us a way to visualize the behavior of the function. Indeed, much of the language we use to describe these functions is intimately tied to this visual idea of a curve.

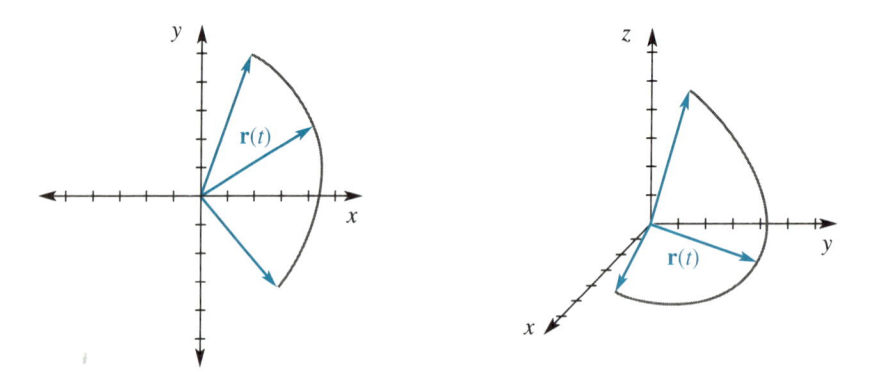

Figure 12.1 The terminal point of $\mathbf{r}(t)$ traces out a curve.

If t is thought of as representing *time*, then we can think of the curve as the *path* of a moving object. This physical interpretation is also the source of much of the terminology we'll use.

EXAMPLE 1 Find a parametrization for the line in the plane passing through the point $(1, -2)$ and having direction $\langle 2, 3 \rangle$.

Solution   We saw in the last chapter that this line can be parametrized by the equations

$$x = 2t + 1$$
$$y = 3t - 2.$$

12.1 POSITION FUNCTIONS AND GRAPHICAL REPRESENTATIONS

Thus, we can represent this line as the curve defined by the position function $\mathbf{r}: \mathbb{R} \longrightarrow \mathbb{R}^2$, where

$$\mathbf{r}(t) = \langle 2t+1,\ 3t-2 \rangle = (2t+1)\mathbf{i} + (3t-2)\mathbf{j}.$$

Figure 12.2 shows the *image curve* of **r**, with the points corresponding to $\mathbf{r}(0)$ and $\mathbf{r}(1)$ indicated.

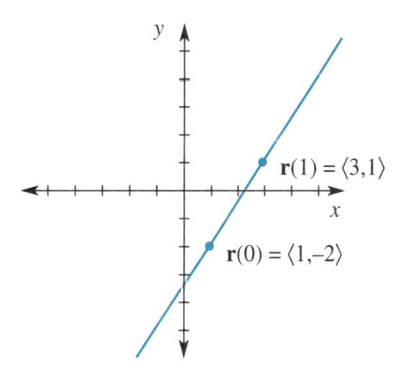

Figure 12.2 Line parametrized by $\mathbf{r}(t) = \langle 2t+1,\ 3t-2 \rangle$.

We say that the position function **r** **parametrizes** this line. ∎

One curve, many parametrizations

A single curve can have many parametrizations, and the orientations of these parametrizations can be different. For example, the circle of radius 1 centered at the origin has parametric equations

$$x = \cos t$$

$$y = \sin t$$

for $0 \leq t \leq 2\pi$. The corresponding position function is $\mathbf{r}: [0, 2\pi] \longrightarrow \mathbb{R}^2$ (see Figure 12.3), where

$$\mathbf{r}(t) = \langle \cos t,\ \sin t \rangle = (\cos t)\mathbf{i} + (\sin t)\mathbf{j}.$$

The same circle is traced out by the function $\mathbf{u}: [0, 2\pi] \longrightarrow \mathbb{R}^2$, where

$$\mathbf{u}(t) = \langle \sin t,\ \cos t \rangle = (\sin t)\mathbf{i} + (\cos t)\mathbf{j},$$

and also by the function $\mathbf{w}: [0, \pi] \longrightarrow \mathbb{R}^2$, where

$$\mathbf{w}(t) = \langle \cos 2t,\ \sin 2t \rangle = (\cos 2t)\mathbf{i} + (\sin 2t)\mathbf{j}.$$

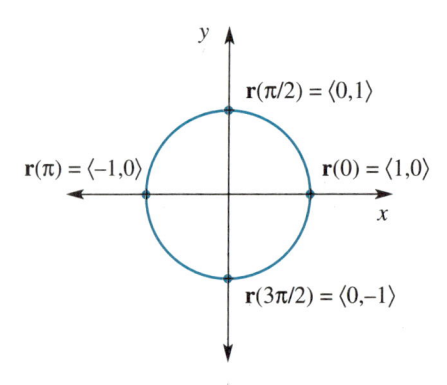

Figure 12.3 Circle parametrized by $\mathbf{r}(t) = (\cos t)\mathbf{i} + (\sin t)\mathbf{j}$.

▶▶▶ **Note that a single curve can be parametrized in many different ways. That is, *different* position functions can have the *same* image curve.**

Parametrized curves have orientation

Curves parametrized by position functions have an *orientation* given by "increasing t". The orientation of the parametrized line in Example 1 is from lower left to upper right as t increases.

The orientations of the three parametrizations of the unit circle given above are different: r and w trace the circle in a counterclockwise direction (starting at (1,0)) while u traces the circle clockwise as t increases (0,1). If we parametrized the unit circle with the position function

$$\mathbf{v}(t) = \langle \cos(-t), \sin(-t) \rangle$$

for $0 \leq t \leq 2\pi$, is the orientation clockwise or counterclockwise?

Space curves

Curves in space can also be parametrized by position functions. For example, the space curve with parametric equations

$$x = 1 - t$$
$$y = t^2 + 1$$
$$z = 2t^3/3 + 1$$

for $-1 \leq t \leq 1$ has position function

$$\mathbf{r} : [-1, 1] \longrightarrow \mathbb{R}^3,$$

12.1 POSITION FUNCTIONS AND GRAPHICAL REPRESENTATIONS

where

$$\mathbf{r}(t) = \langle 1-t,\ t^2+1,\ \frac{2t^3}{3}+1 \rangle = (1-t)\mathbf{i} + (t^2+1)\mathbf{j} + (\frac{2t^3}{3}+1)\mathbf{k}.$$

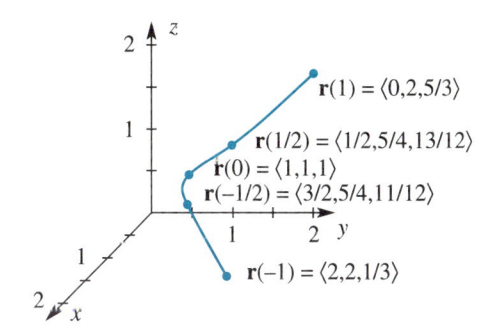

Figure 12.4 The space curve defined by $\mathbf{r}(t)a = (1-t)\mathbf{i} + (t^2+1)\mathbf{j} + (\frac{2t^3}{3}+1)\mathbf{k}$ for $t \in [-1, 1]$.

Figure 12.4 shows the curve with the points corresponding to $\mathbf{r}(-1)$, $\mathbf{r}(-1/2)$, $\mathbf{r}(0)$, $\mathbf{r}(1/2)$, and $\mathbf{r}(1)$.

Here's another example that we'll use for an illustration. The space curve with parametric equations

$$x = \cos t$$
$$y = \sin t$$
$$z = t$$

for $0 \le t \le 2\pi$ has position function

$$\mathbf{r} : [0, 2\pi] \longrightarrow \mathbb{R}^3,$$

where $\mathbf{r}(t) = \langle \cos t, \sin t, t \rangle = (\cos t)\mathbf{i} + (\sin t)\mathbf{j} + (t)\mathbf{k}$.

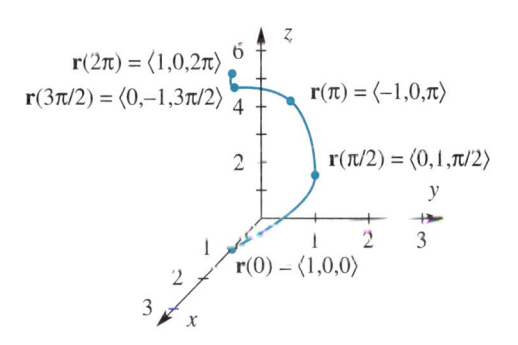

Figure 12.5 The space curve defined by $\mathbf{r}(t) = (\cos t)\mathbf{i} + (\sin t)\mathbf{j} + (t)\mathbf{k}$ for $t \in [0, 2\pi]$.

Figure 12.5 shows this spiral or *helix* with the points corresponding to $\mathbf{r}(0)$, $\mathbf{r}(\pi/2)$, $\mathbf{r}(\pi)$, $\mathbf{r}(3\pi/2)$, and $\mathbf{r}(2\pi)$.

Figures 12.4 and 12.5 illustrate the importance of labeling points in a 2-dimensional representation of a curve in 3-dimensional space. Without them, it is quite impossible to visualize exactly where these curves lie in space or how the curve is traced with respect to t from the pictures alone. The orientations of these parametrized space curves are upwards as t increases.

Using machine graphics to visualize space curves

Even if you have a machine grapher that will plot a parametrized space curve quickly on a computer or calculator screen, this "flat" representation can leave much to be desired.

There is a way that you can use your graphing calculator or computer software to obtain additional information about the position of a curve in space by means of *multiple views*. In many design specifications for objects, three **principal views** are given. These usually are taken to be the *front view*, the *right-side view*, and the *top view*. In terms of our Cartesian coordinate system for space, we can think of these as corresponding to views obtained by looking toward the origin to see the xz-plane, the yz-plane, and the xy-plane, respectively, as shown in Figure 12.6.

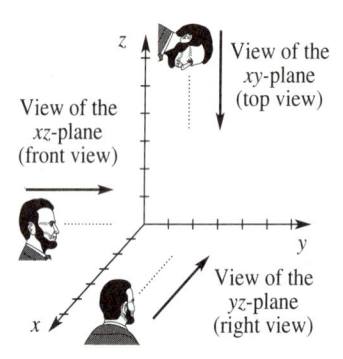

Figure 12.6 Three views of space.

You can obtain these three views easily, provided that you have a 2-dimensional parametric plotter, as found on many graphing calculators and graphing software. Just plot each possible pair out of the three parametric equations that define the coordinate functions. More specifically, if the space curve has position function

$$\mathbf{r}(t) = \langle f(t),\ g(t),\ h(t) \rangle,$$

1. plot $f(t)$ horizontal and $g(t)$ vertical for a "top" view of the xy-plane;
2. plot $f(t)$ horizontal and $h(t)$ vertical for a "front" view of the xz-plane;
3. plot $g(t)$ horizontal and $h(t)$ vertical for a "side" view of the yz-plane.

12.1 POSITION FUNCTIONS AND GRAPHICAL REPRESENTATIONS

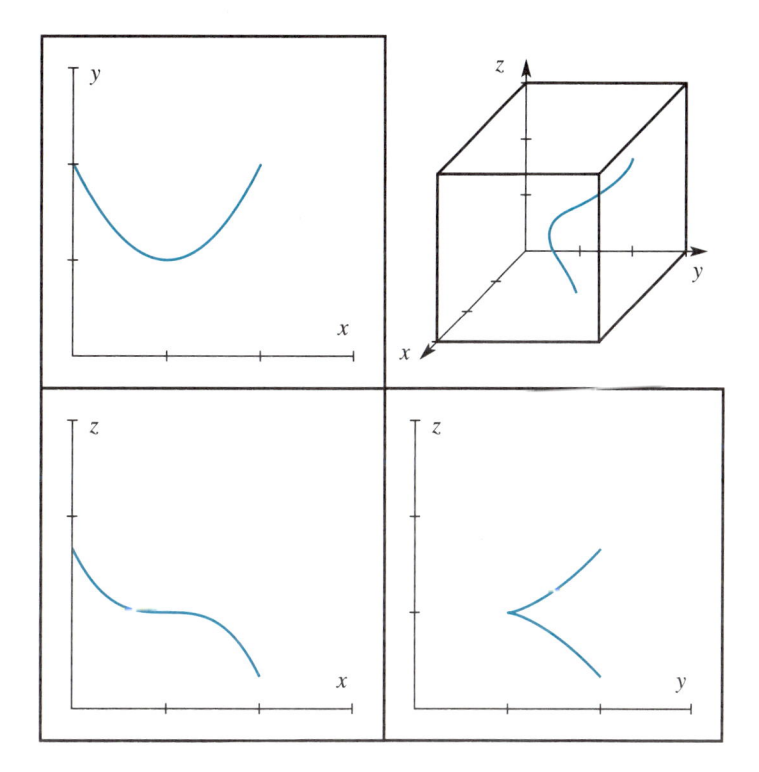

Figure 12.7 Three views of the curve defined by $\mathbf{r}(t) = (1-t)\mathbf{i} + (t^2+1)\mathbf{j} + (\frac{2t^3}{3}+1)\mathbf{k}$ for $t \in [-1, 1]$.

EXAMPLE 2 Use a 2-dimensional parametric plotter to obtain three views of the space curve

$$\mathbf{r}(t) = (1-t)\mathbf{i} + (t^2+1)\mathbf{j} + (\frac{2t^3}{3}+1)\mathbf{k} \quad \text{for } t \in [-1, 1].$$

Solution Figure 12.7 illustrates the desired views of our first space curve example.

Note the relationship between the three views as shown in Figure 12.7. You can think of each view as the picture seen through a clear window placed parallel with each of the three coordinate planes:

The xy-view (top) corresponds to the parametric plot of

$$x = 1 - t \quad \text{and} \quad y = t^2 + 1.$$

The xz-view (front) corresponds to the parametric plot of

$$x = 1 - t \quad \text{and} \quad z = 2t^3/3 + 1.$$

The yz-view (side) corresponds to the parametric plot of

$$y = t^2 + 1 \quad \text{and} \quad z = 2t^3/3 + 1.$$ ■

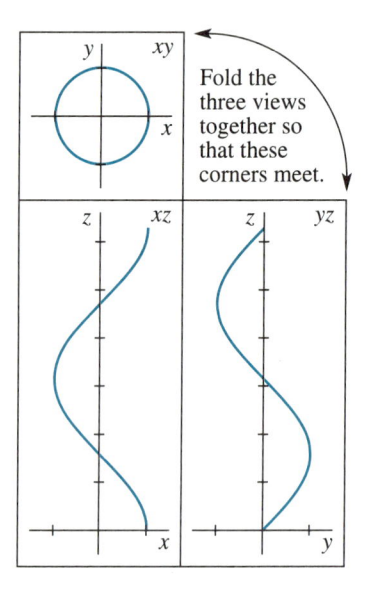

Figure 12.8 The space curve defined by $\mathbf{r}(t) = (\cos t)\mathbf{i} + (\sin t)\mathbf{j} + (t)\mathbf{k}$ for $t \in [0, 2\pi]$.

EXAMPLE 3 Use a 2-dimensional parametric plotter to obtain three views of the space curve defined by

$$\mathbf{r}(t) = (\cos t)\mathbf{i} + (\sin t)\mathbf{j} + (t)\mathbf{k} \quad \text{for } t \in [0, 2\pi].$$

Solution Figure 12.8 shows the three views of the helix. If you cut out the three views exactly as shown and fold them so that the corners indicated in Figure 12.8 align, you have a 3-dimensional model showing all three views (see Figure 12.9, on facing page).

Since machine graphics programs generally plot parametric curves dynamically as t increases, we are able to witness the orientation as we watch the plot progress.

EXERCISES for Section 12.1

For exercises 1-8: For each pair of parametric equations, graph the image curve of the indicated position function for the interval $0 \leq t \leq 2\pi$.

1. $\mathbf{r}(t) = 2\cos(t)\,\mathbf{i} + 3\sin(t)\,\mathbf{j}$.
2. $\mathbf{r}(t) = 2\cos(t)\,\mathbf{i} - 3\sin(t)\,\mathbf{j}$.
3. $\mathbf{r}(t) = -2\cos(t)\,\mathbf{i} + 3\sin(t)\,\mathbf{j}$.
4. $\mathbf{r}(t) = -2\cos(t)\,\mathbf{i} - 3\sin(t)\,\mathbf{j}$.
5. $\mathbf{r}(t) = 3\cos(t)\,\mathbf{i} + 2\sin(t)\,\mathbf{j}$.
6. $\mathbf{r}(t) = 3\cos(t)\,\mathbf{i} - 2\sin(t)\,\mathbf{j}$.
7. $\mathbf{r}(t) = -3\cos(t)\,\mathbf{i} + 2\sin(t)\,\mathbf{j}$.
8. $\mathbf{r}(t) = -3\cos(t)\,\mathbf{i} - 2\sin(t)\,\mathbf{j}$.

9. Describe the orientation (clockwise or counterclockwise) of each of the curves described by the position functions in exercises 1-8. What type of conic section is traced out by each of these position functions?

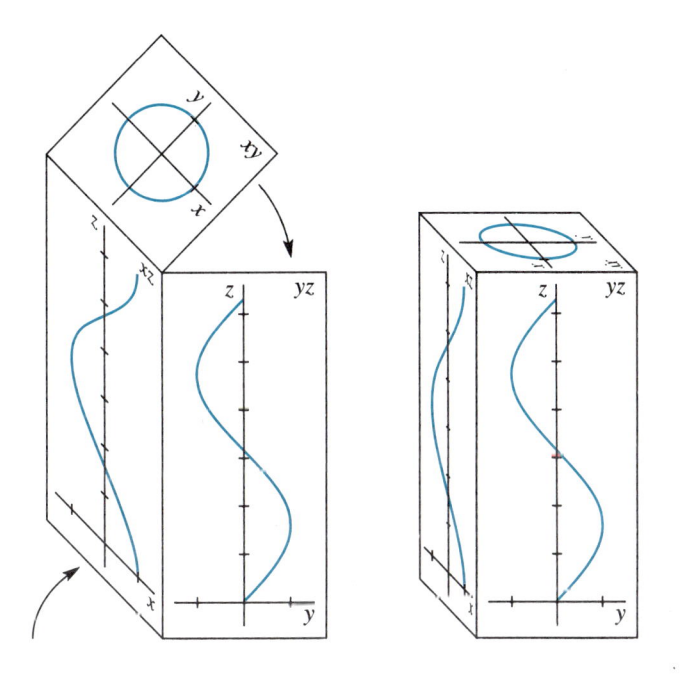

12.1 POSITION FUNCTIONS AND GRAPHICAL REPRESENTATIONS

Figure 12.9 Folding three views to make a model of the helix.

10. Describe as completely as possible, including orientation, the curve traced out by

$$\mathbf{r}(t) = a\cos(t)\mathbf{i} + b\sin(t)\mathbf{j}$$

for $0 \leq t \leq 2\pi$. Be specific regarding the effects of a and b on the shape and orientation of the curve.

For exercises 11-18: For each pair of parametric equations, graph the curve for the interval $-2 \leq t \leq 2$.

11. $\mathbf{r}(t) = 2\cosh(t)\,\mathbf{i} + 3\sinh(t)\,\mathbf{j}$.
12. $\mathbf{r}(t) = 2\cosh(t)\,\mathbf{i} - 3\sinh(t)\,\mathbf{j}$.
13. $\mathbf{r}(t) = -2\cosh(t)\,\mathbf{i} + 3\sinh(t)\,\mathbf{j}$.
14. $\mathbf{r}(t) = -2\cosh(t)\,\mathbf{i} - 3\sinh(t)\,\mathbf{j}$.
15. $\mathbf{r}(t) = 3\cosh(t)\,\mathbf{i} + 2\sinh(t)\,\mathbf{j}$.
16. $\mathbf{r}(t) = 3\cosh(t)\,\mathbf{i} - 2\sinh(t)\,\mathbf{j}$.
17. $\mathbf{r}(t) = -3\cosh(t)\,\mathbf{i} + 2\sinh(t)\,\mathbf{j}$.
18. $\mathbf{r}(t) = -3\cosh(t)\,\mathbf{i} - 2\sinh(t)\,\mathbf{j}$.

19. Describe the orientation of each of the curves described by the position functions in exercises 11-18. What type of conic section is traced out by each of these position functions?

20. Describe as completely as possible, including orientation, the curve traced out by

$$\mathbf{r}(t) = a\cosh(t) + b\sinh(t)$$

for $-2 \leq t \leq 2$. Be specific regarding the effects of a and b on the shape and orientation of the curve.

For exercises 21-30: Plot each of the three principal views of the space curve described by the indicated position function for $-3 \leq t \leq 3$. Then use your three views to sketch the curve on a 3-dimensional Cartesian coordinate system.

21. $\mathbf{r}(t) = (2t+3)\,\mathbf{i} + (3t+2)\,\mathbf{j} + (5-t)\,\mathbf{k}$.

22. $\mathbf{r}(t) = (\sqrt{9-t^2})\,\mathbf{i} + 2\mathbf{j} - t\,\mathbf{k}$.

23. $\mathbf{r}(t) = 2\mathbf{i} + (\sqrt{9-t^2})\,\mathbf{j} + t\,\mathbf{k}$.

24. $\mathbf{r}(t) = t\,\mathbf{i} + 2\mathbf{j} - (\sqrt{9-t^2})\,\mathbf{k}$.

25. $\mathbf{r}(t) = t\cos(t)\,\mathbf{i} + t\sin(t)\,\mathbf{j} - e^t\,\mathbf{k}$.

26. $\mathbf{r}(t) = t\,\mathbf{i} + (t^2/2)\,\mathbf{j} + (t^3/3)\,\mathbf{k}$.

27. $\mathbf{r}(t) = t\,\mathbf{i} + (1-\cos t)\,\mathbf{j} + (\sin t)\,\mathbf{k}$.

28. $\mathbf{r}(t) = \left(\frac{1}{1+t^2}\right)\mathbf{i} + (\arctan t)\,\mathbf{j} + t^2\,\mathbf{k}$.

29. $\mathbf{r}(t) = 3\cos(2t)\,\mathbf{i} + 3\sin(2t)\,\mathbf{j} + 4t\,\mathbf{k}$.

30. $\mathbf{r}(t) = 2\cos(3t)\,\mathbf{i} + 2\sin(3t)\,\mathbf{j} + (t/3)\,\mathbf{k}$.

12.2 DIFFERENTIAL CALCULUS OF CURVES

The derivatives of a position function r have some natural geometric and physical interpretations when we imagine the *image curve* of r as describing the path of a moving object.

Limits of position functions

Derivatives are defined in terms of limits of difference quotients, so first we need to discuss what it means to find the limit of a vector-valued function.

Definition 1

The **limit** of the position vector $\mathbf{r}(t)$ as t approaches a is
$$\lim_{t \to a} \mathbf{r}(t) = \mathbf{L},$$
provided that
$$\lim_{t \to a} ||\mathbf{r}(t) - \mathbf{L}|| = 0.$$

12.2 DIFFERENTIAL CALCULUS OF CURVES

If the difference between the vectors $\mathbf{r}(t)$ and $\mathbf{L}$ vanishes as t approaches a, then $\mathbf{L}$ is the limit vector. If no vector satisfies this requirement, we say that the limit *does not exist*.

Note that $\lim_{t \to a} \mathbf{r}(t) = \mathbf{L}$ requires that each component of $\mathbf{r}(t)$ approach the corresponding component of $\mathbf{L}$ as t approaches a. In other words, we can evaluate limits "component-wise."

EXAMPLE 4 Find $\lim_{t \to 0} \mathbf{r}(t)$, if $\mathbf{r}(t) = \langle \frac{\sin t}{t}, t^2, 5-t \rangle$ for $t \neq 0$, and $\mathbf{r}(0) = 5\mathbf{k}$.

Solution Note that the actual value $\mathbf{r}(0) = 5\mathbf{k}$ is ignored with regard to finding the limit of $\mathbf{r}(t)$ as t approaches 0. Since

$$\mathbf{r}(t) = \left\langle \frac{\sin t}{t}, t^2, 5-t \right\rangle \qquad \text{for } t \neq 0,$$

we have

$$\lim_{t \to 0} \mathbf{r}(t) = \left\langle \lim_{t \to 0} \frac{\sin t}{t}, \lim_{t \to 0} t^2, \lim_{t \to 0} 5-t \right\rangle = \langle 1, 0, 5 \rangle. \qquad \blacksquare$$

Continuity of position functions

Continuity of position functions is defined using exactly the same criteria as for real-valued functions.

Definition 2

> The position vector $\mathbf{r}$ is **continuous at** $t = a$ provided that
>
> $$\lim_{t \to a} \mathbf{r}(t) = \mathbf{r}(a).$$
>
> If $\mathbf{r}$ is continuous at every value in its domain, then we say that $\mathbf{r}$ is a **continuous** position function.

EXAMPLE 5 Is the position function $\mathbf{r}$ of the previous example continuous at $t = 0$? For what values t is $\mathbf{r}$ continuous?

Solution The position function $\mathbf{r}$ of the previous example is *not continuous at* $t = 0$, because

$$\lim_{t \to 0} \mathbf{r}(t) = \langle 1, 0, 5 \rangle \neq \langle 0, 0, 5 \rangle = \mathbf{r}(0).$$

The function $\mathbf{r}$ is continuous at every other real value t. For instance, at $t = \pi$ we have

$$\lim_{t \to \pi} \mathbf{r}(t) = \langle \lim_{t \to \pi} \frac{\sin t}{t}, \lim_{t \to \pi} t^2, \lim_{t \to \pi} 5-t \rangle = \langle 0, \pi^2, 5-\pi \rangle = \mathbf{r}(\pi). \qquad \blacksquare$$

Derivatives of vector position functions

The derivative of a position function is defined as the limit of a difference quotient.

Definition 3

> The position function **r** is said to be **differentiable** at t if and only if the limit
> $$\lim_{\Delta t \to 0} \frac{\mathbf{r}(t + \Delta t) - \mathbf{r}(t)}{\Delta t}$$
> exists. This value is called the **derivative** of **r** at t, and it is denoted
> $$\mathbf{r}'(t) \quad \text{or} \quad \frac{d\mathbf{r}}{dt}.$$
> If the limit doesn't exist, we say that **r** is *not differentiable at t*. A curve is called **smooth** if the function $\mathbf{r}'$ is continuous and $\mathbf{r}'(t)$ is never 0.

Let's make a few comments about this definition. Suppose that we have a position function **r** with vector values $\mathbf{r}(t) = \langle x(t), y(t), z(t) \rangle$, and we examine the difference quotient
$$\frac{\mathbf{r}(t + \Delta t) - \mathbf{r}(t)}{\Delta t}.$$

First, note that the numerator of the difference quotient is a difference of two *vectors* $\mathbf{r}(t + \Delta t)$ and $\mathbf{r}(t)$, while the denominator is a scalar Δt. Since subtraction is performed component-wise and scalar division is distributed to each of the components, this difference quotient can be written
$$\left\langle \frac{x(t + \Delta t) - x(t)}{\Delta t}, \frac{y(t + \Delta t) - y(t)}{\Delta t}, \frac{z(t + \Delta t) - z(t)}{\Delta t} \right\rangle.$$

Given the fact that limits of vector functions are also calculated component-wise, we can see that the derivative
$$\mathbf{r}'(a) = \lim_{\Delta t \to 0} \left\langle \frac{x(t + \Delta t) - x(t)}{\Delta t}, \frac{y(t + \Delta t) - y(t)}{\Delta t}, \frac{z(t + \Delta t) - z(t)}{\Delta t} \right\rangle$$
$$= \langle x'(t), y'(t), z'(t) \rangle.$$

Hence, to find the derivative $\mathbf{r}'(t)$, we differentiate each component.

EXAMPLE 6 Find $\mathbf{r}'(t)$ if $\mathbf{r}(t) = \langle 2t + 1, 3t - 2 \rangle = (2t + 1)\mathbf{i} + (3t - 2)\mathbf{j}$.

Solution $\frac{d\mathbf{r}}{dt} = \mathbf{r}'(t) = \langle 2, 3 \rangle$ for all real values of t. Note that the derivative gives us the direction vector for the line parametrized by this position function. ■

12.2 DIFFERENTIAL CALCULUS OF CURVES

EXAMPLE 7 Find $\mathbf{r}'(t)$ if $\mathbf{r}(t) = \langle \cos t, \sin t \rangle = (\cos t)\mathbf{i} + (\sin t)\mathbf{j}$, for $0 \leq t \leq 2\pi$. Evaluate $\mathbf{r}'(\pi/2)$ and $\mathbf{r}'(\pi)$.

Solution $\dfrac{d\mathbf{r}}{dt} = \mathbf{r}'(t) = \langle -\sin t, \cos t \rangle = (-\sin t)\mathbf{i} + (\cos t)\mathbf{j}$ for $0 < t < 2\pi$.

Note that we do not assign a derivative value at the two endpoints, since the domain of the function does not allow us to approach $t = 0$ through values $t < 0$ nor to approach $t = 2\pi$ through values $t > 2\pi$. We have $\mathbf{r}'(\pi/2) = \langle -\sin \pi/2, \cos \pi/2 \rangle = \langle -1, 0 \rangle$ and $\mathbf{r}'(\pi) = \langle -\sin \pi, \cos \pi \rangle = \langle 0, -1 \rangle$. ■

EXAMPLE 8 Suppose that $\mathbf{r} : [0, 2\pi] \longrightarrow \mathbb{R}^3$,

where $\mathbf{r}(t) = \langle \cos t, \sin t, t \rangle = (\cos t)\mathbf{i} + (\sin t)\mathbf{j} + t\,\mathbf{k}$. Find $\dfrac{d\mathbf{r}}{dt} = \mathbf{r}'(t)$.

Solution The derivative is $\dfrac{d\mathbf{r}}{dt} = \mathbf{r}'(t) = \langle -\sin t, \cos t, 1 \rangle$ (for $0 < t < 2\pi$). ■

Graphical and physical interpretations of the derivative

If we imagine the image curve of $\mathbf{r}$ as representing the path of an object, then the vector represented by the difference

$$\mathbf{r}(t + \Delta t) - \mathbf{r}(t)$$

tells us the *displacement* (how far and in what direction the object moved) over the time interval of length Δt (see Figure 12.10).

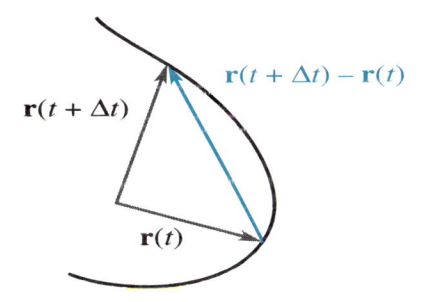

Figure 12.10 $\mathbf{r}(t + \Delta t) - \mathbf{r}(t)$ is a displacement vector.

If the object does not change direction suddenly, then for very small positive values of Δt, this vector also gives a close approximation to the actual path traveled by the object, and the length of this vector approximates the actual distance traveled by the object over this tiny time interval. If we divide by the scalar Δt to obtain the difference quotient

$$\frac{\mathbf{r}(t + \Delta t) - \mathbf{r}(t)}{\Delta t},$$

this new vector has the same direction, but it now has a magnitude that represents the *average speed* of the object (distance covered divided by elapsed time). Since the difference quotient vector gives us both the direction traveled and the average speed, we call it the *average velocity vector* over the time interval $[t, t + \Delta t]$. Note that if Δt is negative, then the numerator of the difference quotient points in the opposite direction, but the denominator sets us right again. In other words, the difference quotient gives us the average velocity for both positive and negative values Δt.

As Δt approaches 0, the limiting value of the difference quotient represents the *instantaneous velocity vector*. Its direction gives us the *instantaneous direction of movement*, and its magnitude gives us the *instantaneous speed*.

$$\mathbf{r}'(t) = \frac{d\mathbf{r}}{dt} = \text{instantaneous velocity at } t,$$

$$||\mathbf{r}'(t)|| = ||\frac{d\mathbf{r}}{dt}|| = \text{instantaneous speed at } t.$$

The vector value of the derivative points in the direction of the parametrized curve. For example, consider again the position function $\mathbf{r}'(t)$ if $\mathbf{r}(t) = \langle \cos t, \sin t \rangle = (\cos t)\mathbf{i} + (\sin t)\mathbf{j}$, for $0 \leq t \leq 2\pi$. If you imagine an object constrained to travel in the circular path defined by this position function (like a swinging ball tied to the end of a rope), then the derivative vector gives us the direction the object would travel if freed from the constraint at that point (when the rope breaks). One way to illustrate this is to place the derivative vector so that its initial point coincides with the point corresponding to the value of the position vector (see Figure 12.11).

We write **v** for the **velocity** vector function $\mathbf{r}'$.

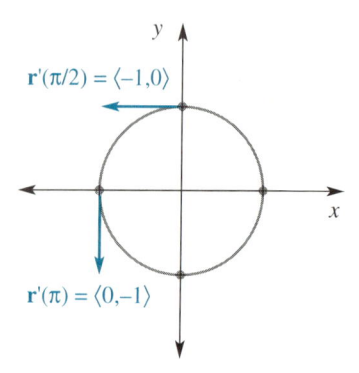

Figure 12.11 The derivative vector points in the direction of the curve.

12.2 DIFFERENTIAL CALCULUS OF CURVES

EXAMPLE 9 Suppose that $\mathbf{r}: [-1, 1] \longrightarrow \mathbb{R}^3$, where

$$\mathbf{r}(t) = \langle 1-t,\ t^2+1,\ \frac{2t^3}{3}+1 \rangle = (1-t)\mathbf{i} + (t^2+1)\mathbf{j} + (\frac{2t^3}{3}+1)\mathbf{k}.$$

Find the velocity and speed when $t = 1/2$.

Solution The velocity is

$$\frac{d\mathbf{r}}{dt} = \mathbf{r}'(t) = \mathbf{v}(t) = \langle -1, 2t, 2t^2 \rangle = -\mathbf{i} + 2t\mathbf{j} + 2t^2\mathbf{k}$$

for $-1 < t < 1$. The velocity at $t = 1/2$ is

$$\mathbf{r}'(1/2) = \mathbf{v}(1/2) = \langle -1, 1, 1/2 \rangle$$

and the speed is

$$\|\mathbf{v}(1/2)\| = \sqrt{(-1)^2 + 1^2 + (1/2)^2} = \sqrt{9/4} = 3/2.$$

Figure 12.12 shows the image curve and the velocity vector $\mathbf{v}(1/2)$. ∎

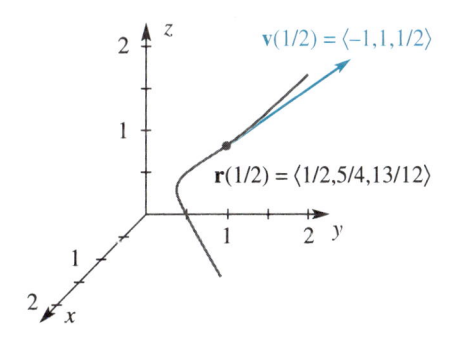

Figure 12.12 A velocity vector.

Tangent lines and unit tangent vectors

The velocity vector $\mathbf{v}(t) = \mathbf{r}'(t)$ gives us the tangent direction to the image curve at the point corresponding to $\mathbf{r}(t)$. Using this point and the velocity vector as a direction vector, we can parametrize the tangent line to the image curve at this point.

EXAMPLE 10 Find a parametrization for the tangent line to the image curve of

$$\mathbf{r}(t) = (1-t)\mathbf{i} + (t^2+1)\mathbf{j} + (\frac{2t^3}{3}+1)\mathbf{k}$$

at $t = 1/2$.

Solution   The velocity vector is

$$\mathbf{v}(t) = \mathbf{r}'(t) = -\mathbf{i} + (2t)\,\mathbf{j} + (2t^2)\,\mathbf{k}.$$

Hence, at $t = 1/2$, we have a direction vector $\mathbf{v}(1/2) = \langle -1, 1, 1/2 \rangle$. The point of tangency corresponds to $\mathbf{r}(1/2) = \langle 1/2, 5/4, 13/12 \rangle$, so the tangent line is parametrized by $\mathbf{r}(\tfrac{1}{2}) + t\mathbf{v}(\tfrac{1}{2}) = \langle \tfrac{1}{2} - t, \tfrac{5}{4} + t, \tfrac{13}{12} + t/2 \rangle$. Figure 12.13 illustrates this tangent line. ∎

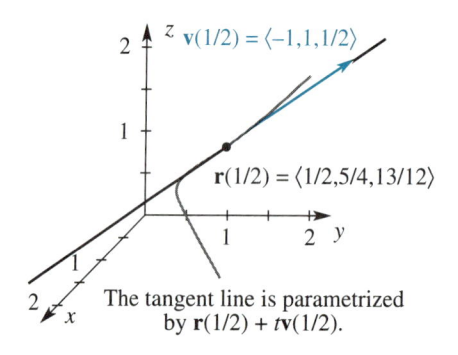

Figure 12.13 Tangent line to image curve of $\mathbf{r}(t)$ at $t = 1/2$.

If we are interested only in measuring the *direction* of the image curve at any point, we can simply scale the velocity vector $\mathbf{r}'(t)$ to unit length by dividing by the speed $\|\mathbf{r}'(t)\|$. This result is called the **unit tangent vector**, and it is denoted

$$\mathbf{T}(t) = \frac{\mathbf{v}(t)}{\|\mathbf{v}(t)\|} = \frac{\mathbf{r}'(t)}{\|\mathbf{r}'(t)\|}.$$

EXAMPLE 11   Compute the unit tangent vector $\mathbf{T}(1/2)$ for the position vector function $\mathbf{r}$ of the previous example.

Solution   We have $\mathbf{v}(1/2) = \langle -1, 1, 1/2 \rangle$. Therefore,

$$T(1/2) = \frac{\langle -1, 1, 1/2 \rangle}{\|\langle -1, 1, 1/2 \rangle\|} = \frac{\langle -1, 1, 1/2 \rangle}{\sqrt{(-1)^2 + 1^2 + (1/2)^2}} = \frac{\langle -1, 1, 1/2 \rangle}{\sqrt{9/4}} = \frac{2}{3}\langle -1, 1, 1/2 \rangle.$$

Figure 12.14 illustrates the unit tangent vector. ∎

Higher order derivatives—acceleration

Of course, we can take higher order derivatives of position vector functions by simply computing the corresponding higher order derivatives of their

12.2 DIFFERENTIAL CALCULUS OF CURVES

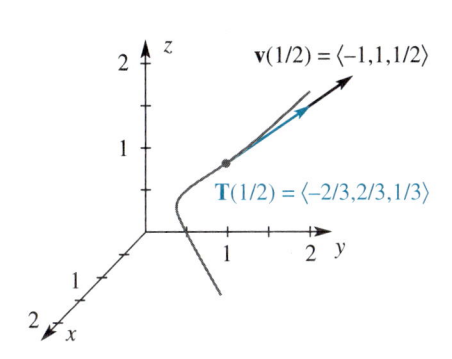

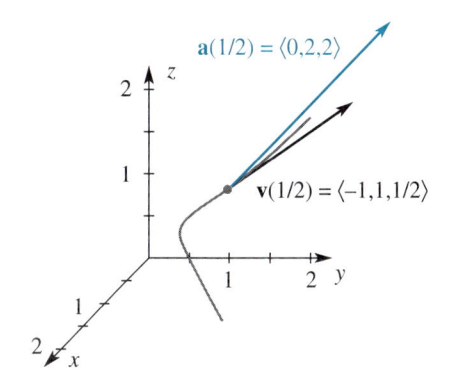

Figure 12.14 A unit tangent vector.

Figure 12.15 An acceleration vector.

component functions. The second derivative of a position vector function is called the **acceleration** vector function, and it is denoted

$$\mathbf{a}(t) = \mathbf{v}'(t) = \mathbf{r}''(t).$$

EXAMPLE 12 Compute the acceleration vector function $\mathbf{a}(t)$ for the position vector function $\mathbf{r}$ of the previous example. Use it to find the acceleration vector $\mathbf{a}(1/2)$.

Solution We have $\mathbf{v}(t) = \mathbf{r}'(t) = -\mathbf{i} + 2t\,\mathbf{j} + 2t^2\,\mathbf{k}$. Therefore,

$$\mathbf{a}(t) = \mathbf{v}'(t) = \mathbf{r}''(t) = 2\,\mathbf{j} + 4t\,\mathbf{k}.$$

The acceleration vector at $t = 1/2$ is

$$\mathbf{a}(1/2) = 2\,\mathbf{j} + 4(1/2)\,\mathbf{k} = \langle 0, 2, 2 \rangle.$$

The acceleration vector is illustrated in Figure 12.15.

EXERCISES for Section 12.2

For exercises 1-5: Consider the position vector function for a curve in $\mathbb{R}^2$ with

$$\mathbf{r}(t) = \langle 3t^2, \sqrt{t} \rangle \qquad \text{for } 1 \le t \le 3.$$

1. Find $\mathbf{v}(t)$ and $\mathbf{v}(2)$.
2. Find $\mathbf{a}(t)$ and $\mathbf{a}(2)$.
3. Find $\mathbf{T}(t)$ and $\mathbf{T}(2)$.
4. Find the speed at $t = 2$.
5. Find a parametrization of the tangent line to the curve at $\mathbf{r}(2)$.

For exercises 6-10: Suppose that a particle moves on the path

$$\mathbf{r} : [0, 2] \longrightarrow \mathbb{R}^3$$

with $\quad\quad\quad\quad \mathbf{r}(t) = \langle \sin 3t,\ \cos 3t,\ 2t^{3/2} \rangle.$

6. Find $\mathbf{v}(t)$ and $\mathbf{v}(1)$.
7. Find $\mathbf{a}(t)$ and $\mathbf{a}(1)$.
8. Find $\mathbf{T}(t)$ and $\mathbf{T}(1)$.
9. Find the speed at $t = 1$.
10. Find a parametrization of the tangent line to the curve at $\mathbf{r}(1)$.

For exercises 11-15: Consider the path of a moving object described by:

$$\mathbf{r} : [0, 2] \longrightarrow \mathbb{R}^3$$

with $\quad\quad\quad\quad \mathbf{r}(t) = \langle \dfrac{t^2}{2},\ \dfrac{4}{3}t^{3/2},\ 2t \rangle.$

11. Find $\mathbf{v}(t)$ and $\mathbf{v}(1)$.
12. Find $\mathbf{a}(t)$ and $\mathbf{a}(1)$.
13. Find $\mathbf{T}(t)$ and $\mathbf{T}(1)$.
14. Find the speed at $t = 1$.
15. Find a parametrization of the tangent line to the curve at $\mathbf{r}(1)$.

For exercises 16-25: Consider the path of a moving object described by:

$$\mathbf{r} : [\tfrac{1}{2}, e] \longrightarrow \mathbb{R}^3$$

with $\quad\quad\quad\quad \mathbf{r}(t) = \langle \sin \pi t,\ \ln t,\ e^t \rangle.$

16. Find $\mathbf{v}(t)$ and $\mathbf{v}(1)$.
17. Find $\mathbf{a}(t)$ and $\mathbf{a}(1)$.
18. Find $\mathbf{T}(t)$ and $\mathbf{T}(1)$.
19. Find the speed at $t = 1$.

20. At what time t during the time interval $[\tfrac{1}{2}, e]$ is the object *farthest* from the origin?

21. At what time t during the time interval $[\tfrac{1}{2}, e]$ does the object achieve *maximum speed*?

22. At what time t during the time interval $[\tfrac{1}{2}, e]$ is the length of the acceleration vector at a maximum?

23. At what times does the object touch one of the coordinate axes?

24. During what times t is the object in the first octant?

25. Find a parametrization of the tangent line to the curve at $\mathbf{r}(1)$.

For exercises 26-34: Consider the path of a moving object described by:

$$\mathbf{r} : [0, 6] \longrightarrow \mathbb{R}^3,$$

where $\quad\quad\quad\quad \mathbf{r}(t) = \langle 2t - \cos(\dfrac{\pi}{2}t),\ \sin(\dfrac{\pi}{2}t),\ t^2 + 16 \rangle.$

12.2 DIFFERENTIAL CALCULUS OF CURVES

26. Find $\mathbf{v}(t)$ and $\mathbf{v}(3)$.

27. Find $\mathbf{a}(t)$ and $\mathbf{a}(3)$.

28. Find the speed at $t = 3$.

29. At what time t during the interval $[3, 4]$ is the object farthest from the origin?

30. At what time t during the time interval $[3, 4]$ does the object achieve maximum speed?

31. At what time t during the time interval $[3, 4]$ is the length of the acceleration vector at a maximum?

32. At what time(s) in $[3, 4]$ does the object touch one of the coordinate axes?

33. At what time(s) in $[3, 4]$ does the object cross one of the coordinate planes?

34. Find a parametrization of the tangent line to this curve at $\mathbf{r}(3.5)$.

For exercises 35-43: Consider the path of a moving object described by the following position function:

$$\mathbf{r} : [0, 6] \longrightarrow \mathbb{R}^3,$$

where
$$\mathbf{r}(t) = \langle 6\cos(\tfrac{\pi}{2}t),\ \sin(\tfrac{\pi}{2}t),\ t \rangle.$$

35. Find $\mathbf{v}(t)$ and $\mathbf{v}(1)$.

36. Find $\mathbf{a}(t)$ and $\mathbf{a}(1)$.

37. Find the speed at $t = 2$.

38. At what time t during the time interval $[1, 3]$ is the object farthest from the origin?

39. At what time t during the time interval $[1, 3]$ does the object achieve maximum speed? Minimum speed?

40. At what time t during the time interval $[1, 3]$ is the length of the acceleration vector at a maximum?

41. At what time(s) in $[1, 3]$ does the object touch one of the coordinate axes?

42. At what time(s) in $[1, 3]$ does the object touch one of the coordinate planes?

43. Find a parametrization of the tangent line to this curve at $\mathbf{r}(2)$.

12.3 PROPERTIES OF DERIVATIVES

Derivatives of position vector functions satisfy many properties similar to derivative properties of real-valued functions. For example, we know that the derivative of a constant function g such that

$$g(x) = c$$

for all x has derivative $g'(x) = 0$ for all x. Similarly, the constant position function with

$$\mathbf{g}(t) = \mathbf{c}$$

for all t (where $\mathbf{c} = \langle c_1, c_2, \ldots, c_n \rangle$ is a constant vector) has derivative

$$\mathbf{g}'(t) = \mathbf{0},$$

the zero position function. This follows from the observation that the derivative of each of the constant components is 0.

EXAMPLE 13 Find $\mathbf{r}'(t)$ if $\mathbf{r}(t) = \langle 1, -2, 3 \rangle$ for all t.

Solution Since $\mathbf{r}(t)$ is a constant vector, $\mathbf{r}'(t) = \langle 0, 0, 0 \rangle$. ∎

We can also "factor out" constant (scalar) factors. That is, if a is a scalar and $\mathbf{r}$ is a differentiable vector function, then

$$\frac{d}{dt}(a\mathbf{r}(t)) = a\mathbf{r}'(t).$$

EXAMPLE 14 Find $\mathbf{r}'(t)$ if $\mathbf{r}(t) = \langle 5\ln t, 5e^t \rangle$ for all $t > 0$.

Solution Since $\mathbf{r}(t) = \langle 5\ln t, 5e^t \rangle$, we have

$$\mathbf{r}'(t) = \left\langle \frac{5}{t}, 5e^t \right\rangle = 5\left\langle \frac{1}{t}, e^t \right\rangle.$$ ∎

If $\mathbf{r}_1$ and $\mathbf{r}_2$ are any two differentiable position functions, then we have

$$\frac{d}{dt}[\mathbf{r}_1 + \mathbf{r}_2] = \frac{d\mathbf{r}_1}{dt} + \frac{d\mathbf{r}_2}{dt}.$$

This is verified by simply carrying out the computation indicated on both sides of the equation and noting that they yield the same result.

Taken together, we say that the scalar multiplication property and this addition property of vector-valued functions establish the *linearity properties* of the derivative.

12.3 PROPERTIES OF DERIVATIVES

If a and b are any scalars, then
$$\frac{d}{dt}[a\mathbf{r}_1 + b\mathbf{r}_2] = a\frac{d\mathbf{r}_1}{dt} + b\frac{d\mathbf{r}_2}{dt}.$$

Product rules

The product rule for real-valued functions can be written
$$\frac{d}{dt}(fg) = \frac{df}{dt}g + f\frac{dg}{dt}.$$

There are different product rules for vector-valued functions corresponding to the different types of multiplication we can perform (scalar, dot, and cross). It is quite fortunate (and somewhat amazing) that the product rule in each case has essentially the same form as the familiar product rule for real-valued functions.

First, suppose that we have a position vector function $\mathbf{r}$ and a *scalar-valued function* f of a real variable t.

The *scalar product rule* states
$$\frac{d}{dt}f(t)\mathbf{r}(t) = f'(t)\mathbf{r}(t) + f(t)\mathbf{r}'(t).$$

EXAMPLE 15 Suppose that $\mathbf{r}(t) = \langle \cos t, \sin t, t \rangle$ and $f(t) = t^2$ for all t. Find the derivative with respect to t of $f(t)\mathbf{r}(t)$.

Solution We have
$$f(t)\mathbf{r}(t) = \langle t^2 \cos t, t^2 \sin t, t^3 \rangle,$$

and the derivative of this vector function is
$$\frac{d}{dt}f(t)\mathbf{r}(t) = \langle 2t \cos t - t^2 \sin t, 2t \sin t + t^2 \cos t, 3t^2 \rangle.$$

Note that $f'(t) = 2t$ and $\mathbf{r}'(t) = \langle -\sin t, \cos t, 1 \rangle$.

If we apply the product rule, we obtain exactly the same result:
$$f'(t)\mathbf{r}(t) + f(t)\mathbf{r}'(t) = \langle 2t \cos t, 2t \sin t, 2t^2 \rangle + \langle t^2(-\sin t), t^2(\cos t), t^2 \rangle$$
$$= \langle 2t \cos t - t^2 \sin t, 2t \sin t + t^2 \cos t, 3t^2 \rangle.$$

We can verify the *product rule for scalar multiplication* by carrying out this same computation in general. If we have any 3-dimensional differentiable position function $\mathbf{r}(t) = \langle x(t), y(t), z(t) \rangle$ and any differentiable scalar-valued function $f(t)$, then

$$\frac{d}{dt} f(t) \mathbf{r}(t) = \frac{d}{dt} \langle f(t) x(t), f(t) y(t), f(t) z(t) \rangle$$

$$= \langle \frac{df}{dt} x(t) + f(t) \frac{dx}{dt}, \frac{df}{dt} y(t) + f(t) \frac{dy}{dt}, \frac{df}{dt} z(t) + f(t) \frac{dz}{dt} \rangle$$

$$= \frac{df}{dt} \langle x(t), y(t), z(t) \rangle + f(t) \langle \frac{dx}{dt}, \frac{dy}{dt}, \frac{dz}{dt} \rangle$$

$$= \frac{df}{dt} \mathbf{r}(t) + f(t) \frac{d\mathbf{r}}{dt}.$$

The verification for any other dimension is entirely similar.

As for derivatives of the dot product of two vector functions, assume that $\mathbf{r}_1$ and $\mathbf{r}_2$ are any differentiable position vector functions of t.

> The *dot product rule* states
> $$\frac{d}{dt}(\mathbf{r}_1 \cdot \mathbf{r}_2) = \frac{d\mathbf{r}_1}{dt} \cdot \mathbf{r}_2 + \mathbf{r}_1 \cdot \frac{d\mathbf{r}_2}{dt}.$$

Again, we can verify this property by simply carrying out the computation indicated on both sides of the equation and noting that it yields the same result in both cases. If

$$\mathbf{r}_1(t) = \langle x_1(t), y_1(t), z_1(t) \rangle \quad \text{and} \quad \mathbf{r}_2(t) = \langle x_2(t), y_2(t), z_2(t) \rangle$$

are any two differentiable 3-dimensional position vector functions, then

$$\mathbf{r}_1(t) \cdot \mathbf{r}_2(t) = x_1(t) x_2(t) + y_1(t) y_2(t) + z_1(t) z_2(t).$$

Note that this is a *scalar* function. Applying the usual differentiation rules gives us

$$\frac{d}{dt}(\mathbf{r}_1 \cdot \mathbf{r}_2) = (\frac{dx_1}{dt} x_2 + x_1 \frac{dx_2}{dt}) + (\frac{dy_1}{dt} y_2 + y_1 \frac{dy_2}{dt}) + (\frac{dz_1}{dt} z_2 + z_1 \frac{dz_2}{dt})$$

$$= (\frac{dx_1}{dt} x_2 + \frac{dy_1}{dt} y_2 + \frac{dz_1}{dt} z_2) + (x_1 \frac{dx_2}{dt} + y_1 \frac{dy_2}{dt} + z_1 \frac{dz_2}{dt})$$

$$= \frac{d\mathbf{r}_1}{dt} \cdot \mathbf{r}_2 + \mathbf{r}_1 \cdot \frac{d\mathbf{r}_2}{dt}.$$

The verification for other dimensions is similar.

For vector functions having outputs in $\mathbb{R}^3$, we also have the *cross product rule*.

12.3 PROPERTIES OF DERIVATIVES

> The *cross product rule* states
> $$\frac{d}{dt}(\mathbf{r}_1 \times \mathbf{r}_2) = \frac{d\mathbf{r}_1}{dt} \times \mathbf{r}_2 + \mathbf{r}_1 \times \frac{d\mathbf{r}_2}{dt}.$$

This statement can also be verified by direct computation of both sides of this equation.

EXAMPLE 16 Suppose that

$$\mathbf{r}_1(t) = \langle \arctan(t), \ln(t), e^t \rangle \quad \text{and} \quad \mathbf{r}_2(t) = \langle t, t^2, t^3 \rangle$$

for $0 \leq t \leq 1$. Find $\mathbf{r}_1(t) \cdot \mathbf{r}_2(t)$ and $\mathbf{r}_1(t) \times \mathbf{r}_2(t)$. Then, compute

$$\frac{d}{dt}(\mathbf{r}_1(t) \cdot \mathbf{r}_2(t)) \quad \text{and} \quad \frac{d}{dt}(\mathbf{r}_1(t) \times \mathbf{r}_2(t)),$$

both directly and by use of the dot and cross product rules.

Solution The dot product of the two functions is

$$\mathbf{r}_1(t) \cdot \mathbf{r}_2(t) = t \arctan t + t^2 \ln t + t^3 e^t,$$

while the cross product can be computed using the "determinant" method:

$$\mathbf{r}_1(t) \times \mathbf{r}_2(t) = \begin{vmatrix} \mathbf{i} & \mathbf{j} & \mathbf{k} \\ \arctan(t) & \ln t & e^t \\ t & t^2 & t^3 \end{vmatrix}$$

$$= (t^3 \ln t - t^2 e^t)\,\mathbf{i} - (t^3 \arctan(t) - t e^t)\,\mathbf{j} + (t^2 \arctan(t) - t \ln t)\,\mathbf{k}.$$

The derivatives of the two given vector functions are

$$\mathbf{r}_1'(t) = \langle \frac{1}{1+t^2}, \frac{1}{t}, e^t \rangle \quad \text{and} \quad \mathbf{r}_2'(t) = \langle 1, 2t, 3t^2 \rangle.$$

Direct computation of the derivative of the dot product gives us

$$\frac{d}{dt}(\mathbf{r}_1(t) \cdot \mathbf{r}_2(t)) = \frac{d}{dt}(t \arctan t + t^2 \ln t + t^3 e^t)$$

$$= t\frac{1}{1+t^2} + \arctan t + t^2 \frac{1}{t} + 2t \ln t + t^3 e^t + 3t^2 e^t$$

$$= \frac{t}{1+t^2} + t + t^3 e^t + \arctan t + 2t \ln t + 3t^2 e^t.$$

However, the dot product rule gives us

$$\frac{d\mathbf{r}_1}{dt} \cdot \mathbf{r}_2 + \mathbf{r}_1 \cdot \frac{d\mathbf{r}_2}{dt}$$

$$= \langle \frac{1}{1+t^2}, \frac{1}{t}, e^t \rangle \cdot \langle t, t^2, t^3 \rangle + \langle \arctan(t), \ln(t), e^t \rangle \cdot \langle 1, 2t, 3t^2 \rangle$$

$$= (\frac{t}{1+t^2} + t + t^3 e^t) + (\arctan t + 2t \ln t + 3t^2 e^t).$$

We note that the result is the same in either case.

Direct computation of the derivative of the cross product gives us

$$\frac{d}{dt}(\mathbf{r}_1(t) \times \mathbf{r}_2(t))$$

$$= \frac{d}{dt}\left((t^3 \ln t - t^2 e^t)\mathbf{i} - (t^3 \arctan(t) - te^t)\mathbf{j} + (t^2 \arctan(t) - t \ln t)\mathbf{k}\right)$$

$$= (3t^2 \ln t + t^3 \frac{1}{t} - 2te^t - t^2 e^t)\mathbf{i}$$

$$- (3t^2 \arctan(t) + t^3 \frac{1}{1+t^2} - e^t - te^t)\mathbf{j}$$

$$+ (2t \arctan(t) + t^2 \frac{1}{1+t^2} - \ln t - 1)\mathbf{k}.$$

But, the cross product rule yields

$$\frac{d\mathbf{r}_1}{dt} \times \mathbf{r}_2 + \mathbf{r}_1 \times \frac{d\mathbf{r}_2}{dt}$$

$$= \begin{vmatrix} \mathbf{i} & \mathbf{j} & \mathbf{k} \\ \frac{1}{1+t^2} & \frac{1}{t} & e^t \\ t & t^2 & t^3 \end{vmatrix} + \begin{vmatrix} \mathbf{i} & \mathbf{j} & \mathbf{k} \\ \arctan(t) & \ln t & e^t \\ 1 & 2t & 3t^2 \end{vmatrix}$$

$$= [(t^3 \frac{1}{t} - t^2 e^t)\mathbf{i} - (t^3 \frac{1}{1+t^2} - te^t)\mathbf{j} + (t^2 \frac{1}{1+t^2} - 1)\mathbf{k}]$$

$$+ [(3t^2 \ln t - 2te^t)\mathbf{i} - (3t^2 \arctan(t) - e^t)\mathbf{j} + (2t \arctan(t) - \ln t)\mathbf{k}].$$

You can check that this matches the result of the direct computation. ■

▶▶▶ **The order of the functions is important in the cross product rule because the cross product is *not* commutative.**

Chain rule

For a composition of real-valued functions $g \circ f$, the chain rule provides us with the derivative formula:

$$(g \circ f)'(t) = g'(f(t))f'(t).$$

12.3 PROPERTIES OF DERIVATIVES

Using the Leibniz notation, and writing $u = g(w)$ and $w = f(t)$, we have the equivalent formulation

$$\frac{du}{dt} = \frac{du}{dw}\frac{dw}{dt}.$$

Now, suppose that $\mathbf{u} = \mathbf{r}(w)$ is a *vector position function* of w where, in turn, $w = f(t)$ is a scalar-valued function of t. We can then ask what the velocity vector of $\mathbf{u}$ is with respect to t. The answer is again supplied by the *chain rule*:

$$(\mathbf{r} \circ f)'(t) = \mathbf{r}'(f(t))f'(t),$$

or, using Leibniz notation:

$$\frac{d\mathbf{u}}{dt} = \frac{d\mathbf{u}}{dw}\frac{dw}{dt}.$$

This chain rule is verified by direct computation and use of the chain rule for real-valued functions. For a 3-dimensional vector position function $\mathbf{u} = \mathbf{r}(w) = \langle x(w), y(w), z(w) \rangle$, we have

$$\frac{d\mathbf{u}}{dt} = \frac{d}{dt}\left(\langle x(f(t)), y(f(t)), z(f(t)) \rangle\right)$$
$$= \langle x'(f(t))f'(t), y'(f(t))f'(t), z'(f(t))f'(t) \rangle$$
$$= f'(t)\langle x'(f(t)), y'(f(t)), z'(f(t)) \rangle$$
$$= f'(t)\mathbf{r}'(f(t)).$$

EXAMPLE 17 Suppose that $\mathbf{u} = \mathbf{r}(w) = \langle \cos w, \sin w, w \rangle$ and $w = f(t) = t^2$. Find $\dfrac{d\mathbf{u}}{dt}$, both directly and by using the chain rule.

Solution We have $\mathbf{u} = \langle \cos(t^2), \sin(t^2), t^2 \rangle$, so

$$\frac{d\mathbf{u}}{dt} = \langle -2t\sin(t^2), 2t\cos(t^2), 2t \rangle.$$

Using the chain rule, we have

$$\frac{d\mathbf{u}}{dt} = \frac{d\mathbf{u}}{dw}\frac{dw}{dt} = \langle -\sin w, \cos w, 1 \rangle (2t).$$

Substituting t^2 for w and multiplying each component by the scalar $2t$ gives us the same result as the direct computation. ■

EXERCISES for Section 12.3

For exercises 1-10: Consider the vector function for a curve in $\mathbb{R}^2$:

$$\mathbf{r}(t) = \langle 3t^2, \sqrt{t}\rangle \quad \text{for } 0 \le t \le 1$$

and the scalar function $f(t) = 1 - t^2$ for $0 \le t \le 1$.

1. Find $\mathbf{r}'(t)$ and $\mathbf{r}'(1/2)$.

2. Find $\mathbf{r}''(t)$ and $\mathbf{r}''(1/2)$.

3. Find the angle between the velocity and acceleration vectors when $t = 1/2$.

4. Is there any time $t \in [0, 1]$ at which the velocity and acceleration vectors are orthogonal?

5. Find $\mathbf{T}(t)$ and $\mathbf{T}(1/2)$.

6. Let $u(t) = \|\mathbf{r}(t)\|$ and find $\dfrac{du}{dt}$.

7. Find the speed $\|\mathbf{r}'(t)\|$.

8. True or false: The length of the derivative is the same as the derivative of the length. (Compare your answers with those to the previous two exercises.)

9. Find $\dfrac{d(f\mathbf{r})}{dt}$ and verify the scalar product rule for the given functions.

10. Find $\dfrac{d(\mathbf{r} \circ f)}{dt}$ and verify the chain rule for the given functions.

For exercises 11-20: Consider the two paths

$$\mathbf{r}_1 : [0, 2] \to \mathbb{R}^3$$

$$\mathbf{r}_1(t) = \langle \sin 3t,\ \cos 3t,\ 2t^{3/2}\rangle.$$

$$\mathbf{r}_2 : [0, 2] \to \mathbb{R}^3$$

$$r_2(t) = \left\langle \frac{t^2}{2},\ \frac{4}{3}t^{3/2},\ 2t \right\rangle.$$

11. Find $\mathbf{v}_1(t)$ and $\mathbf{v}_2(t)$.

12. Find $\mathbf{a}_1(t)$ and $\mathbf{a}_2(t)$.

13. Find $\mathbf{T}_1(t)$ and $\mathbf{T}_2(t)$.

14. Find the velocity of the object whose path is described by $-2\mathbf{r}_1 + 3\mathbf{r}_2$.

15. Find the acceleration of the object whose position at time t is given by $\mathbf{r}_1(t^2)$.

12.4 INTEGRATION AND ARC LENGTH

16. Find the unit tangent vector of the object whose position at time t is given by $t^2 \mathbf{r}_2(t)$.

17. Find $\mathbf{r}_1 \cdot \mathbf{r}_2$, then compute $\dfrac{d(\mathbf{r}_1 \cdot \mathbf{r}_2)}{dt}$ both directly and by the dot product rule.

18. Find $\mathbf{r}_1 \times \mathbf{r}_2$, then compute $\dfrac{d(\mathbf{r}_1 \times \mathbf{r}_2)}{dt}$ both directly and by the cross product rule.

19. Find $\dfrac{d}{dt}(\|\mathbf{r}_1\|)$.

20. Find $\dfrac{d}{dt}(\|\mathbf{r}_2\|)$.

---

21. Verify the cross product rule in general for two vector functions $\mathbf{r}_1(t) = \langle x_1(t), y_1(t), z_1(t)\rangle$ and $\mathbf{r}_2(t) = \langle x_2(t), y_2(t), z_2(t)\rangle$ by computing the derivative of $\mathbf{r}_1 \times \mathbf{r}_2$ both directly and by using the cross product rule.

22. Verify that if a vector position function has constant length (in other words, $\|\mathbf{r}(t)\| = c$ for some scalar c), then $\mathbf{r}'(t)$ is orthogonal to $\mathbf{r}(t)$. (Hint: write $c = \|\mathbf{r}(t)\| = \sqrt{\mathbf{r}(t) \cdot \mathbf{r}(t)}$ and differentiate.)

23. Find an example of a vector position function $\mathbf{r}$ such that $\|\mathbf{r}(t)\|$ is constant, but $\mathbf{r}'(t) \neq \mathbf{0}$ (that is, $\mathbf{r}(t)$ is not constant).

24. Find a formula for the derivative of the triple scalar product $\mathbf{r}_1 \cdot (\mathbf{r}_2 \times \mathbf{r}_3)$, where $\mathbf{r}_1$, $\mathbf{r}_2$, and $\mathbf{r}_3$ are 3-dimensional differentiable vector functions.

25. Let $\mathbf{T}(t) = \mathbf{r}'(t)/\|\mathbf{r}'(t)\|$ be the unit tangent vector function for a 3-dimensional differentiable vector function $\mathbf{r}$. Verify that

$$\frac{d\mathbf{T}}{dt} \times \mathbf{T}(t) = \frac{\mathbf{r}''(t)}{\|\mathbf{r}'(t)\|} \times \mathbf{T}(t).$$

12.4 INTEGRATION AND ARC LENGTH

Indefinite integrals

Since the derivative of a position vector function is computed component-wise, to find an antiderivative we simply need to find an antiderivative of each component. In other words, indefinite integrals of position vector functions are also computed component-wise.

In finding the antiderivative of a position vector function, it is important to realize that each component will involve its own arbitrary constant.

EXAMPLE 18   Find the general antiderivative of the vector function $\mathbf{v}(t) = \langle t, t^2, t^3 \rangle$.

Solution   We write
$$\int \langle t, t^2, t^3 \rangle \, dt = \langle \frac{t^2}{2} + C_1, \frac{t^3}{3} + C_2, \frac{t^4}{4} + C_3 \rangle$$
$$= \langle \frac{t^2}{2}, \frac{t^3}{3}, \frac{t^4}{4} \rangle + \mathbf{C},$$

where $\mathbf{C} = \langle C_1, C_2, C_3 \rangle$ is an arbitrary *constant vector*. Note that the constants C_1, C_2, C_3 may or may not have different values in specifying any particular antiderivative. ■

Definite integrals

Both limits and sums of vector functions are determined component-wise. Since a definite integral is the limit of a Riemann sum, it is natural that definite integrals of vector functions are also computed component-wise. We make this procedure more precise with the following definitions.

Definition 4

A **Riemann sum** for a position vector function $\mathbf{r}$ over the interval $a \leq t \leq b$ is obtained by partitioning the interval $[a, b]$ into n subintervals of equal length $\Delta t = \dfrac{b - a}{n}$, choosing one value t_i ($i = 1, 2, \ldots, n$) from each subinterval, and calculating
$$\sum_{i=1}^{n} \mathbf{r}(t_i) \, \Delta t.$$

This Riemann sum will be a vector itself. If, as the partitions are taken finer and finer, all the possible Riemann sums tend to a single limiting vector, this limit is defined as the definite integral of $\mathbf{r}(t)$ over $[a, b]$.

Definition 5

$$\int_a^b \mathbf{r}(t) \, dt = \lim_{n \to \infty} \sum_{i=1}^{n} \mathbf{r}(t_i) \, \Delta t$$

provided that this limit exists.

EXAMPLE 19   Find $\int_1^2 \langle t, t^2, t^3 \rangle \, dt$.

Solution   We have

12.4 INTEGRATION AND ARC LENGTH

$$\int_1^2 \langle t, t^2, t^3 \rangle \, dt = \langle \frac{t^2}{2}, \frac{t^3}{3}, \frac{t^4}{4} \rangle \Big|_1^2$$

$$= \langle 2 - \frac{1}{2}, \frac{8}{3} - \frac{1}{3}, 4 - \frac{1}{4} \rangle$$

$$= \langle \frac{3}{2}, \frac{7}{3}, \frac{15}{4} \rangle.$$

If the vector function we are integrating has a physical interpretation as a velocity vector function of time t, then the definite integral represents a *net vector change in position*. For example, if $\mathbf{v}(t) = \langle t, t^2, t^3 \rangle$ represents velocity of an object at time t, then

$$\int_1^2 \langle t, t^2, t^3 \rangle \, dt = \langle \frac{3}{2}, \frac{7}{3}, \frac{15}{4} \rangle$$

tells us that this object had a net change in position between $t = 1$ and $t = 2$ of $\frac{3}{2}$ in the x-direction, $\frac{7}{3}$ in the y-direction, and $\frac{15}{4}$ in the z-direction. The actual path traveled between the two times is determined by the antiderivative

$$\mathbf{r}(t) = \langle \frac{t^2}{2}, \frac{t^3}{3}, \frac{t^4}{4} \rangle + \langle C_1, C_2, C_3 \rangle.$$

Note that some initial condition is necessary to determine the actual values $\mathbf{r}(1)$ and $\mathbf{r}(2)$. We don't know the value of the arbitrary constant vector. What we do know is that

$$\mathbf{r}(2) - \mathbf{r}(1) = \langle \frac{3}{2}, \frac{7}{3}, \frac{15}{4} \rangle,$$

regardless of the particular starting and ending points of the path in question.

Arc length

We have just discussed how a velocity vector function can be integrated to find the net change in position (also a vector). Unless the moving object always travels in the *same* direction along a straight line, this net change in position will not give us a true indication of the actual distance traveled. For example, consider an object moving in a circle of radius 3, whose path is parametrized by

$$\mathbf{r}(t) = \langle 3\cos t, \ 3\sin t \rangle$$

for $0 \leq t \leq 2\pi$. The starting point and ending point are

$$\mathbf{r}(0) = \langle 3\cos 0, 3\sin 0 \rangle = \langle 3, 0 \rangle = \langle 3\cos 2\pi, 3\sin 2\pi \rangle = \mathbf{r}(2\pi).$$

Hence, the net change in position is

$$\mathbf{r}(2\pi) - \mathbf{r}(0) = \langle 0, 0 \rangle = \mathbf{0},$$

the zero vector. (After all, the object has traveled in a complete circle.) Yet, the actual distance traveled by the object is 6π units, since the path is a circle of radius 3.

Let's think for a moment about a similar situation we encountered for real-valued functions. For an object traveling in one dimension (a straight line), integrating the *velocity* between time $t = a$ and $t = b$ gives us the *net distance* traveled during that time interval. If, instead, we integrate the *speed* (the absolute value of the velocity) between time $t = a$ and $t = b$, we obtain the *total distance* traveled.

To measure the actual total distance traveled by an object moving in a plane or in space with position function $\mathbf{r}$, the same approach may be valid. If the curve traced out by $\mathbf{r}(t)$ is *smooth* over the interval $[a, b]$, meaning that the velocity vector $\mathbf{v}(t) = \mathbf{r}'(t)$ is *continuous* and *nonzero* for all $t \in [a, b]$, then the object does not change direction suddenly, and we know the object's speed (the norm or magnitude $\|\mathbf{v}(t)\| = \|\mathbf{r}'(t)\|$) at each instant t. In this case, we can integrate the *scalar-valued* function $\|\mathbf{v}(t)\|$ between time $t = a$ and $t = b$ to find the *total distance* traveled:

$$\text{total distance traveled} = \int_a^b \|\mathbf{v}(t)\| \, dt = \int_a^b \|\mathbf{r}'(t)\| \, dt.$$

In terms of the image curve of a position vector function, we are saying that the **arc length** of the curve from $\mathbf{r}(a)$ to $\mathbf{r}(b)$ is given by $\int_a^b \|\mathbf{r}'(t)\| \, dt$. Note that $\|\mathbf{r}'\|$ is integrable on $[a, b]$ since we have already assumed that $\mathbf{r}'$ is continuous. We can justify this arc length formula for the path of a position function $\mathbf{r}$ in a similar way to our justification of the arc length formula for the graph of a function $y = f(x)$.

Consider the path $\mathbf{r}(t)$ traces out from $t = a$ to $t = b$, as shown in Figure 12.16.

If we subdivide the time interval $[a, b]$ into n subintervals of length Δt, then the path of the object between time t and time $t + \Delta t$ can be approximated by a line segment of length

$$\|\mathbf{r}(t + \Delta t) - \mathbf{r}(t)\|,$$

which we could also write as

$$\frac{\|\mathbf{r}(t + \Delta t) - \mathbf{r}(t)\| \Delta t}{\Delta t},$$

12.4 INTEGRATION AND ARC LENGTH

We can use the velocity vector $\mathbf{v}(t)$ to approximate the path traced by $\mathbf{r}(t)$ with a polygonal path.

Each line segment's length can be approximated by $||\mathbf{v}(t)||\Delta t$.

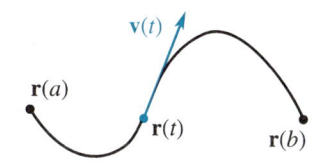

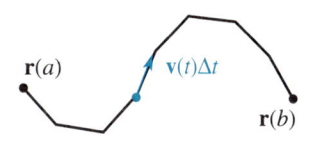

Figure 12.16 Approximating an image curve with a polygonal path.

by multiplying the numerator and denominator by Δt. For Δt very small, this is closely approximated by

$$||\mathbf{v}(t)||\Delta t.$$

The total length of the curve from $\mathbf{r}(a)$ to $\mathbf{r}(b)$ is found by summing up the lengths of the n line segments

$$\sum_{i=1}^{n} ||\mathbf{r}'(t_i)||\,\Delta t = \sum_{i=1}^{n} ||\mathbf{v}(t_i)||\,\Delta t,$$

where t_i is the time at the beginning of the ith time interval. As Δt approaches 0, the approximation becomes better and better, with the limit providing the formula:

$$\text{arc length} \;=\; \int_a^b ||\mathbf{r}'(t)||\,dt \;=\; \int_a^b ||\mathbf{v}(t)||\,dt.$$

EXAMPLE 20 Use the arc length formula to find the distance traveled by an object whose position is given by

$$\mathbf{r}(t) = \langle 3\cos t, 3\sin t \rangle$$

for $0 \leq t \leq 2\pi$.

Solution Since $\mathbf{r}'(t) = \langle -3\sin t, 3\cos t \rangle$, we have

$$||\mathbf{r}'(t)|| = \sqrt{9\sin^2 t + 9\cos^2 t} = \sqrt{9(\sin^2 t + \cos^2 t)} = \sqrt{9} = 3,$$

and

$$\int_0^{2\pi} 3\,dt = 3t \Big]_{t=0}^{t=2\pi} = 6\pi.$$

This is precisely the circumference of a circle of radius 3.

EXAMPLE 21   Find the arc length of the path traced out by an object whose position is given by

$$\mathbf{r}(t) = \langle 1 - t, t^2 + 1, 2t^3/3 + 1 \rangle$$

between $t = 0$ and $t = 1/2$.

Solution   The speed of the object is given by

$$\begin{aligned}||\mathbf{r}'(t)|| &= ||\langle -1, 2t, 2t^2 \rangle|| \\ &= \sqrt{(-1)^2 + (2t)^2 + (2t^2)^2} \\ &= \sqrt{1 + 4t^2 + 4t^4} \\ &= \sqrt{(1 + 2t^2)^2} \\ &= |1 + 2t^2| = 1 + 2t^2.\end{aligned}$$

(The absolute value signs can be removed, since $1 + 2t^2$ is positive for all t.) The arc length between $t = 0$ and $t = 1/2$ is given by

$$\int_0^{1/2} ||\mathbf{r}'(t)|| = \int_0^{1/2} (1 + 2t^2)\, dt = \left(t + \frac{2t^3}{3} \right) \Bigg]_{t=0}^{t=1/2} = \frac{7}{12}.$$ ∎

Arc length function—parametrization by arc length

We can define an **arc length function**

$$s(t) = \int_a^t ||\mathbf{r}'(u)||\, du$$

by integrating the speed from a fixed value $u = a$ and allowing the upper limit $u = t$ to vary. The arc length function output $s(t)$ tells us the arc length of the image curve of $\mathbf{r}$ from $\mathbf{r}(a)$ to $\mathbf{r}(t)$.

Now, the fundamental theorems of calculus tell us that

$$\frac{ds}{dt} = ||\mathbf{r}'(t)||.$$

This makes perfect sense: the rate of change of the distance traveled s with respect to time t is simply the speed! One important consequence of this fact is simply that, with respect to arc length, $\mathbf{r}$ has *constant unit speed*. In other words,

$$\left\| \frac{d\mathbf{r}}{ds} \right\| = 1.$$

To see why, note that by the chain rule we have

$$\mathbf{r}'(t) = \frac{d\mathbf{r}}{dt} = \frac{d\mathbf{r}}{ds}\frac{ds}{dt} = \frac{d\mathbf{r}}{ds}||\mathbf{r}'(t)||.$$

12.4 INTEGRATION AND ARC LENGTH

Now, if we take the norm of both sides of the equation, we have

$$||\mathbf{r}'(t)|| = \left|\left|\frac{d\mathbf{r}}{ds}\right|\right| \, ||\mathbf{r}'(t)||.$$

Dividing both sides by $||\mathbf{r}'(t)||$ gives us $||d\mathbf{r}/ds|| = 1$.

EXAMPLE 22 Find the arc length function for the path traced out by an object whose position is given by

$$\mathbf{r}(t) = \langle 1 - t, t^2 + 1, \frac{2t^3}{3} + 1 \rangle$$

using $t = 0$ as the starting time.

Solution Since $||\mathbf{r}'(t)|| = 1 + 2t^2$ (as found in the last example), we have

$$s(t) = \int_0^t (1 + 2u^2) \, du = \left(u + \frac{2u^3}{3}\right)\Big]_{u=0}^{u=t} = t + \frac{2t^3}{3}.$$

You can check that $\dfrac{ds}{dt} = 1 + 2t^2 = ||\mathbf{r}'(t)||$. ■

The arc length s is particularly useful as a "time-independent" parameter for describing a smooth curve. That is, we can use the arc length s as our independent variable in parametrizing a curve by fixing a reference point on the curve and then describing other positions in terms of the distance traveled from the reference point.

Given a parametrization $\mathbf{r}(t)$ of a smooth curve in terms of t, we can *reparametrize by arc length s* by choosing an initial time $t = a$ and:

1. finding the resulting arc length function s in terms of t;
2. solving for t in terms of s; and
3. expressing the position function $\mathbf{r}$ in terms of s by substitution.

Here is an example illustrating the steps in reparametrizing a curve in terms of arc length.

EXAMPLE 23 Parametrize by arc length the curve traced out by an object whose position is given by

$$\mathbf{r}(t) = \langle \cos t, \sin t, t \rangle$$

for $0 \le t \le 2\pi$.

Solution   This curve is our helix example from earlier discussion. The first step is to find the arc length function. The velocity vector is given by

$$\mathbf{v}(t) = \mathbf{r}'(t) = \langle -\sin t, \cos t, 1 \rangle$$

and the speed is

$$\|\mathbf{v}(t)\| = \|\mathbf{r}'(t)\| = \sqrt{(-\sin t)^2 + (\cos t)^2 + 1^2} = \sqrt{2}.$$

The arc length function is

$$s(t) = \int_0^t \sqrt{2}\, du = \sqrt{2}u \Big]_{u=0}^{u=t} = \sqrt{2}t.$$

Solving for t in terms of s, we have

$$t = s/\sqrt{2},$$

and the parametrization by arc length is

$$\langle \cos(s/\sqrt{2}),\ \sin(s/\sqrt{2}),\ s/\sqrt{2} \rangle.$$

The domain of this new parametrization is $[0, 2\sqrt{2}\pi]$, since $s = \sqrt{2}t$ and $0 \le t \le 2\pi$. ∎

A few comments are in order. Because the curve is assumed to be *smooth*, the speed is always positive ($\mathbf{r}'(t) \ne \mathbf{0}$, so $\|\mathbf{r}'(t)\| > 0$) and the distance traveled is strictly increasing. This means that each different value t results in a different value s, and so t is determined uniquely by s. Theoretically, this allows us to solve for t in terms of s (although that might be quite difficult algebraically.)

You can think of parametrizing a curve by arc length as "standardizing" our description of the curve, in the sense that the speed with respect to arc length s is guaranteed to be the constant 1 unit. Over the interval $a \le s \le b$, the distance traveled will be simply $b-a$, the length of the interval (see Figure 12.17).

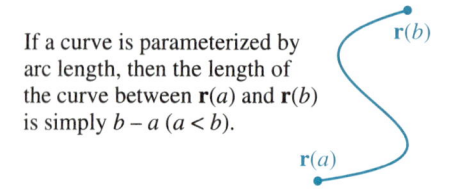

If a curve is parameterized by arc length, then the length of the curve between $\mathbf{r}(a)$ and $\mathbf{r}(b)$ is simply $b-a$ ($a < b$).

Figure 12.17 A curve parametrized by arc length.

12.4 INTEGRATION AND ARC LENGTH

EXAMPLE 24 In the previous example, the parametrization of the helix relative to arc length is

$$\langle \cos(s/\sqrt{2}), \sin(s/\sqrt{2}), s/\sqrt{2} \rangle.$$

Verify that the speed with respect to arc length is 1 unit.

Solution The velocity with respect to arc length is

$$\left\langle -\frac{\sin(s/\sqrt{2})}{\sqrt{2}}, \frac{\cos(s/\sqrt{2})}{\sqrt{2}}, \frac{1}{\sqrt{2}} \right\rangle,$$

and, hence, the speed with respect to arc length is

$$\sqrt{\left(-\frac{\sin(s/\sqrt{2})}{\sqrt{2}}\right)^2 + \left(\frac{\cos(s/\sqrt{2})}{\sqrt{2}}\right)^2 + \left(\frac{1}{\sqrt{2}}\right)^2}$$

$$= \sqrt{\frac{\sin^2(s/\sqrt{2}) + \cos^2(s/\sqrt{2})}{2} + \frac{1}{2}} = \sqrt{1} = 1. \quad \blacksquare$$

The fact that the speed is always one unit for a curve parametrized by arc length tells us that the velocity vector with respect to arc length is the same as the unit tangent vector. In other words, if $\mathbf{r}(s)$ is the position in terms of arc length s, then

$$\frac{d\mathbf{r}}{ds} = \mathbf{T}.$$

EXERCISES for Section 12.4

For exercises 1-8: Consider an object moving in $\mathbb{R}^2$ with velocity

$$\mathbf{v}(t) = \langle 3t^2, \sqrt{t} \rangle \quad \text{for } 1 \leq t \leq 4.$$

1. Find the acceleration $\mathbf{a}(t)$ of the object.

2. Find $\int_2^3 \mathbf{v}(t)\,dt$.

3. Find the net distance traveled by the object between $t = 1$ and $t = 4$.

4. Find the total distance traveled by the object between $t = 1$ and $t = 4$.

5. If the object's location is $(-2, 3)$ at $t = 1$, find its position at $t = 4$.

6. If the object's location is $(-2, 3)$ at $t = 1$, find its position function $\mathbf{r}(t) = \langle x(t), y(t) \rangle$ for $1 \leq t \leq 4$.

7. Find the arc length function for $\mathbf{r}(t)$.

8. How far does the object travel between $t = 2$ and $t = 3$?

For exercises 9-14: Suppose that a particle moves on the path

$$\mathbf{r} : [0, 2] \to \mathbb{R}^3$$

$$\mathbf{r}(t) = \langle \sin 3t, \cos 3t, 2t^{3/2} \rangle.$$

9. Find $\int \mathbf{r}(t)\, dt$.

10. Find $\int_0^2 \mathbf{r}(t)\, dt$.

11. Find the net distance traveled by the particle.

12. Find the total distance traveled by the particle.

13. Find the arc length function s for $\mathbf{r}$.

14. Reparametrize by arc length the path traced by $\mathbf{r}$.

For exercises 15-18: Consider the path of a moving object whose acceleration is described by

$$\mathbf{a} : [0, 2] \to \mathbb{R}^3$$

$$\mathbf{a}(t) = \langle t^2, t^3, 2t \rangle.$$

15. Find the velocity $\mathbf{v}(t)$ if the initial velocity is $\mathbf{v}(0) = \langle 1, 2, 3 \rangle$. What is the terminal velocity?

16. Find the position $\mathbf{r}(t)$ if $\mathbf{r}(0) = \langle -1, 0, -2 \rangle$. What is the terminal position?

17. Find the net distance traveled by the object.

18. Find the total distance traveled by the object.

19. Consider the path of a moving object described by the following position function:

$$\mathbf{r} : [0, 2\pi] \to \mathbb{R}^3$$

with
$$\mathbf{r}(t) = \langle 3 \sin t, 3 \cos t, 4t \rangle.$$

Find the total distance traveled by the object and reparametrize its path by arc length.

20. Two of the ways to parametrize the unit circle in $\mathbb{R}^2$ starting at the point $(1, 0)$ are

$$\mathbf{r}_1(t) = \langle \cos(t/2), \sin(t/2) \rangle \quad \text{for} \quad 0 \leq t \leq 4\pi,$$

and
$$\mathbf{r}_2(t) = \langle \cos(2t), \sin(2t) \rangle \quad \text{for} \quad 0 \leq t \leq \pi.$$

The image curves are identical. The major difference between these two ways of parametrizing the unit circle is that $\mathbf{r}_2$ traces out the circle *four times as fast*. (Note that the time interval $[0, \pi]$ is one-fourth as long as the time interval $[0, 4\pi]$ for $\mathbf{r}_1$.) Show that starting with either $\mathbf{r}_1$ or $\mathbf{r}_2$, reparametrizing by arc length yields identical results.

12.5 NORMAL VECTORS TO A CURVE

At each point along a smooth curve described by the differentiable position function $\mathbf{r}$, we have a tangent direction provided by the velocity vector $\mathbf{v}(t) = \mathbf{r}'(t)$. As we have discussed, if we scale this vector to have unit length, then we call the resulting vector the **unit tangent vector**

$$\mathbf{T}(t) = \frac{\mathbf{v}(t)}{\|\mathbf{v}(t)\|}.$$

We can use the direction of this vector to parametrize the tangent line to the curve at that point.

A vector *perpendicular* to the tangent vector is called a **normal** vector. If a normal vector has length 1 unit, we call it a **unit normal vector**. For a smooth curve in the plane $\mathbb{R}^2$, there are exactly two unit normal vectors at each point (having opposite directions). For a smooth curve in space $\mathbb{R}^3$, there are *infinitely* many unit normal vectors at each point, all lying in the plane normal to the tangent vector (see Figure 12.18).

It will be convenient to make a specific choice of unit normal at each point of the curve, which we will call the **principal unit normal** and denote by $\mathbf{N}$.

Two unit normals for a planar curve.

Infinitely many unit normals for a space curve.

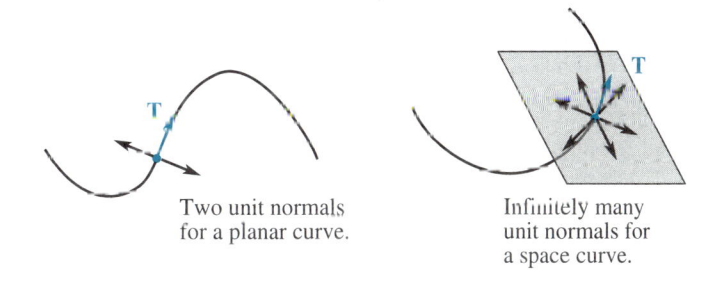

Figure 12.18 Unit normal vectors to a plane curve and a space curve.

> We agree to choose as the principal unit normal the vector
> $$\mathbf{N} = \frac{d\mathbf{T}/dt}{||d\mathbf{T}/dt||},$$
> where $d\mathbf{T}/dt$ is the derivative of the unit tangent vector $\mathbf{T}$ with respect to t.

We can see that this vector $\mathbf{N}$ has unit length (note that we are dividing $d\mathbf{T}/dt$ by its length), but it is not immediately clear why this vector is necessarily orthogonal to $\mathbf{T}$. To see why, note that

$$\mathbf{T} \cdot \mathbf{T} = 1,$$

since $\mathbf{T}$ is a *unit vector*. If we differentiate both sides with respect to t using the dot product rule, we see that

$$\frac{d\mathbf{T}}{dt} \cdot \mathbf{T} + \mathbf{T} \cdot \frac{d\mathbf{T}}{dt} = 0.$$

Using the fact that the dot product is commutative, we can rewrite this expression as

$$2\frac{d\mathbf{T}}{dt} \cdot \mathbf{T} = 0.$$

If we divide both sides by 2, we can see that $\frac{d\mathbf{T}}{dt} \cdot \mathbf{T} = 0$, so that $\frac{d\mathbf{T}}{dt}$ is orthogonal to $\mathbf{T}$. Hence, we indeed have

$$\mathbf{N} \cdot \mathbf{T} = \frac{d\mathbf{T}/dt}{||d\mathbf{T}/dt||} \cdot \mathbf{T} = 0.$$

In the plane, the principal unit normal is always on the same side of the tangent line as the curve itself. If $\mathbf{T} = \langle u_1, u_2 \rangle$, then we must have $\mathbf{N} = \langle -u_2, u_1 \rangle$ or $\mathbf{N} = \langle u_2, -u_1 \rangle$, depending on whether we must choose the unit normal 90° counterclockwise or clockwise from the unit tangent vector.

EXAMPLE 25   Find the principal unit normal $\mathbf{N}$ to the curve described by the position function

$$\mathbf{r}(t) = \langle 3\cos t, 3\sin t \rangle$$

at $t = 3\pi/4$.

Solution   The velocity vector is

$$\mathbf{v}(t) = \langle -3\sin t, 3\cos t \rangle,$$

12.5 NORMAL VECTORS TO A CURVE

with speed $||\mathbf{v}(t)|| = \sqrt{9\sin^2 t + 9\cos^2 t} = \sqrt{9} = 3$. So, the unit tangent vector is

$$\mathbf{T}(t) = \langle -\sin t, \cos t \rangle,$$

and at $t = 3\pi/4$ we have $\mathbf{T} = \langle -\sqrt{2}/2, -\sqrt{2}/2 \rangle$.

Since $\mathbf{r}$ is a plane curve, there are only two choices for $\mathbf{N}$:

$$\langle -\sqrt{2}/2, \sqrt{2}/2 \rangle \quad \text{or} \quad \langle \sqrt{2}/2, -\sqrt{2}/2 \rangle.$$

An inspection of the circle makes it clear that the correct choice is

$$\mathbf{N} = \langle \sqrt{2}/2, -\sqrt{2}/2 \rangle.$$

However, we can check this by computing

$$\frac{d\mathbf{T}}{dt} = \langle -\cos t, -\sin t \rangle.$$

At $t = 3\pi/4$, this vector is $\langle \sqrt{2}/2, -\sqrt{2}/2 \rangle$, and it is the principal unit normal vector. (It already has unit length.) ∎

EXAMPLE 26 Find the principal unit normal $\mathbf{N}$ to the curve described by the position function

$$\mathbf{r}(t) = \langle 1 - t, t^2 + 1, 2t^3/3 + 1 \rangle$$

at $t = 1/2$.

Solution The velocity vector is $\mathbf{r}'(t) = \langle -1, 2t, 2t^2 \rangle$, and the unit tangent vector is given by

$$\mathbf{T}(t) = \frac{\mathbf{r}'(t)}{||\mathbf{r}'(t)||} = \frac{\langle -1, 2t, 2t^2 \rangle}{1 + 2t^2} = \left\langle -\frac{1}{1+2t^2}, \frac{2t}{1+2t^2}, \frac{2t^2}{1+2t^2} \right\rangle.$$

Now we take the derivative of the unit tangent vector function with respect to t:

$$\frac{d\mathbf{T}}{dt} = \left\langle \frac{4t}{(1+2t^2)^2}, \frac{2 - 4t^2}{(1+2t^2)^2}, \frac{4t}{(1+2t^2)^2} \right\rangle.$$

At $t = 1/2$, we have

$$\left. \frac{d\mathbf{T}}{dt} \right|_{t=1/2} = \left\langle \frac{8}{9}, \frac{4}{9}, \frac{8}{9} \right\rangle.$$

The length of this vector is

$$\sqrt{\frac{64}{81} + \frac{16}{81} + \frac{64}{81}} = \sqrt{\frac{144}{81}} = \frac{12}{9} = \frac{4}{3}.$$

Dividing by this scalar gives the principal unit normal

$$\mathbf{N}(1/2) = \left\langle \frac{2}{3}, \frac{1}{3}, \frac{2}{3} \right\rangle.$$

We note that $\mathbf{T}(1/2) = \langle -\frac{2}{3}, \frac{2}{3}, \frac{1}{3} \rangle$, and we verify that

$$\mathbf{N}(1/2) \cdot \mathbf{T}(1/2) = \left\langle \frac{2}{3}, \frac{1}{3}, \frac{2}{3} \right\rangle \cdot \left\langle -\frac{2}{3}, \frac{2}{3}, \frac{1}{3} \right\rangle = -\frac{4}{9} + \frac{2}{9} + \frac{2}{9} = 0,$$

showing that $\mathbf{N}(1/2)$ is, indeed, orthogonal to $\mathbf{T}(1/2)$. (You can also check that $\mathbf{N}(1/2)$ has length 1 unit.) ∎

EXAMPLE 27 Find the principal unit normal $\mathbf{N}$ to the curve described by the position function

$$\mathbf{r}(t) = \langle \cos t, \sin t, t \rangle$$

at $t = \pi$.

Solution The velocity vector is $\mathbf{r}'(t) = \langle -\sin t, \cos t, 1 \rangle$, and the unit tangent vector is given by

$$\mathbf{T}(t) = \frac{\mathbf{r}'(t)}{||\mathbf{r}'(t)||} = \frac{\langle -\sin t, \cos t, 1 \rangle}{\sqrt{2}}.$$

Now we take the derivative of the unit tangent vector function with respect to t:

$$\frac{d\mathbf{T}}{dt} = \left\langle -\frac{\cos t}{\sqrt{2}}, -\frac{\sin t}{\sqrt{2}}, 0 \right\rangle,$$

and that gives us

$$\left. \frac{d\mathbf{T}}{dt} \right|_{t=\pi} = \left\langle -\frac{\cos \pi}{\sqrt{2}}, -\frac{\sin \pi}{\sqrt{2}}, 0 \right\rangle = \left\langle \frac{1}{\sqrt{2}}, 0, 0 \right\rangle.$$

A unit vector in the same direction is

$$\mathbf{N}(\pi) = \langle 1, 0, 0 \rangle = \mathbf{i}.$$

Since $\mathbf{T}(\pi) = \langle 0, -1/\sqrt{2}, 1/\sqrt{2} \rangle$, we can check that

$$\mathbf{N}(\pi) \cdot \mathbf{T}(\pi) = \langle 1, 0, 0 \rangle \cdot \langle 0, -1/\sqrt{2}, 1/\sqrt{2} \rangle = 0 + 0 + 0 = 0,$$

showing that $\mathbf{N}(\pi)$ is orthogonal to $\mathbf{T}(\pi)$. ∎

12.5 NORMAL VECTORS TO A CURVE

The osculating plane

The plane that is determined by **N** and **T** at a particular point on a smooth curve is called the **osculating plane** to the curve at that point. In a very real sense, this plane "best fits" the curve at this point.

Let's clarify this statement a bit. Consider a specific time t_0 and the point corresponding to the position $\mathbf{r}(t_0)$ on the smooth curve described by **r**. Now, suppose that we choose three *distinct* points in time t_1, t_2, t_3, and the three points on the curve corresponding to $\mathbf{r}(t_1)$, $\mathbf{r}(t_2)$, and $\mathbf{r}(t_3)$. Unless these three points are collinear, they determine a plane. As all three time values approach t_0, you can think of the osculating plane as representing the "limiting plane" determined by these three points.

EXAMPLE 28 Find the general equation of the osculating plane to the curve described by the position function

$$\mathbf{r}(t) = \langle 1 - t, t^2 + 1, 2t^3/3 + 1 \rangle$$

at $t = 1/2$.

Solution We've already computed the unit tangent vector and the principal unit normal vector to the curve at $t = 1/2$. A vector giving the normal direction of the plane determined by the two vectors $\mathbf{N}(1/2)$ and $\mathbf{T}(1/2)$ is

$$\mathbf{N}(1/2) \times \mathbf{T}(1/2) = \begin{vmatrix} \mathbf{i} & \mathbf{j} & \mathbf{k} \\ \frac{2}{3} & \frac{1}{3} & \frac{2}{3} \\ -\frac{2}{3} & \frac{1}{3} & \frac{2}{3} \end{vmatrix} = \left\langle -\frac{1}{3}, -\frac{2}{3}, \frac{2}{3} \right\rangle,$$

and the plane contains the point corresponding to $\mathbf{r}(1/2) = \langle 1/2, 5/4, 13/12 \rangle$.

The general equation of the osculating plane is of the form

$$ax + by + cz + d = 0.$$

The vector $\langle a, b, c \rangle$ can be taken to be $\langle -\frac{1}{3}, -\frac{2}{3}, \frac{2}{3} \rangle$, so we have

$$-\frac{1}{3}x - \frac{2}{3}y + \frac{2}{3}z + d = 0.$$

To find d, we substitute $(1/2, 5/4, 13/12)$ for (x, y, z) and solve for d:

$$d = \left(\frac{1}{3}\right)\left(\frac{1}{2}\right) + \left(\frac{2}{3}\right)\left(\frac{5}{4}\right) - \left(\frac{2}{3}\right)\left(\frac{13}{12}\right) = \frac{5}{18}.$$

Substituting $d = 5/18$, we have

$$-\frac{1}{3}x - \frac{2}{3}y + \frac{2}{3}z + \frac{5}{18} = 0$$

as an equation of the osculating plane. ■

Another property of the osculating plane is that the acceleration vector **a** must lie in this plane. In other words, we must have scalars p and q such that

$$\mathbf{a} = p\mathbf{T} + q\mathbf{N}.$$

That is, the acceleration vector must be a linear combination of the unit tangent vector and the principal unit normal vector. You will be asked to verify this in general in the exercises, but here is an example that illustrates the statement.

EXAMPLE 29 Verify that the acceleration vector is a linear combination of the unit tangent vector and the principal unit normal vector to the curve described by

$$\mathbf{r}(t) = \langle 1-t, t^2+1, 2t^3/3 + 1 \rangle$$

at $t = 1/2$.

Solution The acceleration vector $\mathbf{a}(t) = \langle 0, 2, 4t \rangle$, so

$$\mathbf{a}(1/2) = \langle 0, 2, 2 \rangle.$$

The question calls for us to verify that there are scalars p and q such that

$$\langle 0, 2, 2 \rangle = p \left\langle -\frac{2}{3}, \frac{2}{3}, \frac{1}{3} \right\rangle + q \left\langle \frac{2}{3}, \frac{1}{3}, \frac{2}{3} \right\rangle = \left\langle \frac{2q-2p}{3}, \frac{2p+q}{3}, \frac{p+2q}{3} \right\rangle.$$

We note that this is, indeed, the case if we have $p = q = 2$. ∎

Torsion

The *torsion* of a space curve at a particular point is a measure of the "twist" in the curve at that instant. The idea is to determine how fast the osculating plane changes its normal direction with respect to arc length.

To find the torsion, we first need to define the **unit binormal vector**

$$\mathbf{B}(t) = \mathbf{N}(t) \times \mathbf{T}(t).$$

Note that the unit binormal vector is mutually orthogonal to both unit tangent vector and the principal unit normal vector because it is the cross product of the two vectors. (You can verify that this vector also has unit length.)

The unit binormal vector **B** determines the osculating plane (it is perpendicular to the plane determined by **N** and **T**). The unit binormal **B** certainly does not change length, so the derivative with respect to arc length

$$\frac{d\mathbf{B}}{ds}$$

12.5 NORMAL VECTORS TO A CURVE

is strictly a measure of its change in direction.

In the exercises, you will verify that

$$\frac{d\mathbf{B}}{ds} = a\mathbf{N}$$

for some scalar a.

> The **torsion** is $\tau = -a$.

The absolute value of the torsion $|\tau|$ is $||d\mathbf{B}/ds||$.

EXERCISES for Section 12.5

For exercises 1-5: Consider the vector function for a curve in $\mathbb{R}^2$

$$\mathbf{r}(t) = \langle 3t^2, \sqrt{t} \rangle \quad \text{for } 1 \leq t \leq 3.$$

1. Find $\mathbf{T}(t)$.
2. Find $\mathbf{T}(3/2)$, $\mathbf{T}(2)$, and $\mathbf{T}(5/2)$.
3. Find $\mathbf{N}(t)$.
4. Find $\mathbf{N}(3/2)$, $\mathbf{N}(2)$, and $\mathbf{N}(5/2)$.
5. Draw a sketch of the curve carefully on graph paper and indicate each of the vectors you found in exercises 2 and 4.

For exercises 6-10: Suppose that $\mathbf{T}$ is the unit tangent vector, $\mathbf{N}$ is the principal unit normal vector, and $\mathbf{B}$ is the unit binormal vector ($\mathbf{B} = \mathbf{T} \times \mathbf{N}$) at a point on a space curve.

6. Explain why $\mathbf{B} \cdot \mathbf{T} = 0$ and $\mathbf{B} \cdot \mathbf{N} = 0$.

7. Verify that $\mathbf{B} \cdot \mathbf{B} = 1$.

8. Differentiate both sides of the equation $\mathbf{B} \cdot \mathbf{T} = 0$ with respect to s, and use the fact that $d\mathbf{T}/ds$ has the same direction as $\mathbf{N}$ to explain why $d\mathbf{B}/ds$ must be orthogonal to $\mathbf{T}$.

9. Differentiate $\mathbf{B} \cdot \mathbf{B} = 1$ with respect to s and use that result to explain why $d\mathbf{B}/ds$ must be orthogonal to $\mathbf{B}$.

10. Since $d\mathbf{B}/ds$ is orthogonal to both $\mathbf{T}$ and $\mathbf{B}$, then it must have the same or opposite direction as $\mathbf{N}$. Explain why. (This means that there is a scalar a such that

$$d\mathbf{B}/ds = a\mathbf{N}.$$

The **torsion** is $\tau = -a$.)

For exercises 11-15: Suppose that a particle moves on the path

$$\mathbf{r} : [0, 2] \to \mathbb{R}^3$$

$$\mathbf{r}(t) = \langle \sin 3t, \cos 3t, 2t^{3/2} \rangle.$$

11. Find $\mathbf{T}(t)$ and $\mathbf{T}(1)$.
12. Find $\mathbf{N}(t)$ and $\mathbf{N}(1)$.
13. Find the equation of the osculating plane at $t = 1$.
14. Find the unit binormal vector $\mathbf{B}(1)$.
15. Find the torsion τ at $t = 1$.

For exercises 16-20: Consider the path of a moving object described by the position function $\mathbf{r} : [0, 2] \longrightarrow \mathbb{R}^3$ where

$$\mathbf{r}(t) = \left\langle \frac{t^2}{2}, \frac{4}{3} t^{3/2}, 2t \right\rangle.$$

16. Find $\mathbf{T}(1)$.
17. Find $\mathbf{N}(1)$.
18. Find the equation of the osculating plane at $t = 1$.
19. Find the unit binormal vector $\mathbf{B}(1)$.
20. Find the torsion τ at $t = 1$.

For exercises 21-25: Consider the path of a moving object described by the position function $\mathbf{r} : [\frac{1}{2}, e] \longrightarrow \mathbb{R}^3$ where

$$\mathbf{r}(t) = \langle t^2, \ln t, e^t \rangle.$$

21. Find $\mathbf{T}(1)$.
22. Find $\mathbf{N}(1)$.
23. Find the equation of the osculating plane at $t = 1$.
24. Find the unit binormal vector $\mathbf{B}(1)$.
25. Find the torsion τ at $t = 1$.

12.6 CURVATURE

In this section, we take a look at how we can use calculus to describe more completely certain characteristics of the path traced out by a vector position function, including "how curved" the path is.

12.6 CURVATURE

Measuring curvature for curves in a plane

We can think of *curvature* as a measure of how "fast" the direction of a curve changes. Of course, the speed of change in direction depends in part on simply how quickly a curve is traversed, and that depends entirely on the particular parametrization we choose for the curve. Certainly, we want any measure of curvature to depend intrinsically on the shape of the curve, and not on some arbitrary choice of the infinitely many ways that a curve can be parametrized.

A natural and objective choice to make is to insist on parametrization by arc length for purposes of measuring curvature. The unit tangent vector **T** at any point tells us the direction of the curve at that point, so we define the **curvature** κ at the point to be

$$\kappa = \|d\mathbf{T}/ds\|$$

which is the length of the derivative of the unit tangent vector with respect to arc length. The *vector* $d\mathbf{T}/ds$ is sometimes called the **curvature vector**, but it is its *scalar* length that gives us the measure of curvature.

Exactly what does this value κ tell us? If we were to find a circle that "best fits" the curve at a particular point on a curve, then the curvature κ at that point represents the *reciprocal of the radius* of this circle (see Figure 12.19).

Circle of "best fit"

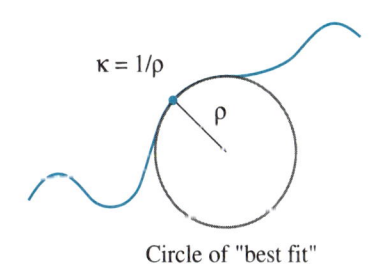

Figure 12.19 The relationship between curvature and radius of curvature.

This idea makes sense. If a circle has a small radius, its curvature is great (it turns in a tight space). If a circle has a large radius, its curvature is small (a person can walk in a large circle while believing that the path is straight). The radius ρ of this circle of best fit is called the **radius of curvature** and, as we have stated, it has an inverse relationship to the curvature κ:

$$\kappa = 1/\rho \quad \text{and} \quad \rho = 1/\kappa.$$

If the curve is a straight line (or the part of the curve including the point in question is straight), then $\kappa = 0$ (zero curvature) and we agree to say the radius of curvature is infinite and write $\rho = \infty$.

It may be difficult to parametrize a curve relative to arc length; fortunately, however, there are several alternative, but equivalent, formulas for measuring curvature. For example, by the chain rule, we have

$$\frac{d\mathbf{T}}{dt} = \frac{d\mathbf{T}}{ds}\frac{ds}{dt}.$$

We can solve this equation for $d\mathbf{T}/ds$ and, noting that

$$\frac{ds}{dt} = ||\mathbf{v}(t)||,$$

we have

$$\frac{d\mathbf{T}}{ds} = \frac{d\mathbf{T}/dt}{||\mathbf{v}(t)||}.$$

We can see that this tells us that *the curvature vector points in the same direction as the principal unit normal* (recall that $\mathbf{N} = (d\mathbf{T}/dt)/||d\mathbf{T}/dt||$). We now have another way to express curvature:

$$\kappa = ||d\mathbf{T}/ds|| = \frac{||d\mathbf{T}/dt||}{||\mathbf{v}(t)||}.$$

EXAMPLE 30   Compute the curvature κ and the radius of curvature ρ for the curve defined by

$$\mathbf{r}(t) = \langle 1 - t, t^2 + 1, 2t^3/3 + 1\rangle$$

at $t = 1/2$.

Solution   As we have computed before for this position vector function, the derivative of the unit tangent vector with respect to t is

$$\frac{d\mathbf{T}}{dt} = \left\langle \frac{4t}{(1+2t^2)^2}, \frac{2-4t^2}{(1+2t^2)^2}, \frac{4t}{(1+2t^2)^2} \right\rangle,$$

from which we have

$$\left.\frac{d\mathbf{T}}{dt}\right|_{t=1/2} = \langle \frac{8}{9}, \frac{4}{9}, \frac{8}{9}\rangle.$$

The length of this vector is

$$||d\mathbf{T}/dt|| = 4/3.$$

12.6 CURVATURE

The speed is $||\mathbf{v}(t)|| = 1 + 2t^2$, from which we have

$$||\mathbf{v}(1/2)|| = 3/2.$$

We can now evaluate the curvature at time $t = \frac{1}{2}$,

$$\kappa(1/2) = \frac{4/3}{3/2} = \frac{8}{9},$$

and the radius of curvature at time $t = \frac{1}{2}$,

$$\rho(1/2) = \frac{1}{\kappa} = \frac{9}{8}.$$

EXAMPLE 31 Compute the curvature $\kappa(t)$ and the radius of curvature $\rho(t)$ for the curve defined by

$$\mathbf{r}(t) = \langle \cos t, \sin t, t \rangle$$

for any value t.

Solution As we have computed before, the derivative of the unit tangent vector with respect to t is

$$\frac{d\mathbf{T}}{dt} = \left\langle -\frac{\cos t}{\sqrt{2}}, -\frac{\sin t}{\sqrt{2}}, 0 \right\rangle,$$

whose length is

$$||d\mathbf{T}/dt|| = \sqrt{\frac{\cos^2 t}{2} + \frac{\sin^2 t}{2} + 0} = 1/\sqrt{2}$$

for any value t. The speed is $||\mathbf{v}(t)|| = \sqrt{2}$ for any value t, so we have

$$\kappa = \frac{1/\sqrt{2}}{\sqrt{2}} = \frac{1}{2} \quad \text{and} \quad \rho = 2$$

for any value t. This helix is a curve with *constant curvature* κ.

Relation between curvature and $d^2\mathbf{r}/ds^2$

If we have a parametrization by arc length, the position vector $\mathbf{r}(s)$ has constant unit speed (in other words, $||d\mathbf{r}/ds|| = 1$). That means that any change in the velocity vector $\mathbf{r}'(s)$ is purely *directional*. The acceleration vector $\mathbf{r}''(s)$ measures the rate of change in the velocity vector, so it makes sense that the magnitude of the acceleration vector $||\mathbf{r}''(s)||$ would be related in some way to the curvature κ. Indeed, we have

$$\kappa = \left\|\frac{d^2\mathbf{r}}{ds^2}\right\| = \|\mathbf{r}''(s)\|.$$

To see this, note that

$$\frac{d^2\mathbf{r}}{ds^2} = \frac{d}{ds}\left(\frac{d\mathbf{r}}{ds}\right) = \frac{d}{ds}\left(\frac{d\mathbf{r}/dt}{ds/dt}\right) = \frac{d}{ds}\left(\frac{\mathbf{v}(t)}{\|\mathbf{v}(t)\|}\right) = \frac{d\mathbf{T}}{ds},$$

which is precisely the curvature vector.

Restated, the curvature κ can be considered as the length of the second derivative of position with respect to arc length $d^2\mathbf{r}/ds^2$.

Curvature of plane curves—an alternative formula

Another formula for κ if $\mathbf{r}(t) = \langle x(t), y(t)\rangle$ is simply

$$\kappa = \frac{|x'y'' - y'x''|}{((x')^2 + (y')^2)^{3/2}}.$$

This formula can be verified by first noting that

$$\mathbf{T} = \frac{\mathbf{v}}{\|\mathbf{v}\|} = \left\langle \frac{x'}{\sqrt{(x')^2 + (y')^2}}, \frac{y'}{\sqrt{(x')^2 + (y')^2}} \right\rangle$$

and then computing

$$\frac{d\mathbf{T}}{ds} = \frac{d\mathbf{T}/dt}{\|\mathbf{v}(t)\|} = \frac{d\mathbf{T}/dt}{\sqrt{((x')^2 + (y')^2)}}.$$

The length of the resulting vector is precisely the expression above.

Curvature of space curves—an alternative formula

For space curves, we have an equivalent formulation of arc length that in practice is usually more directly computable (in the sense that we can avoid the intermediary step of finding the unit tangent vector function). This formula is

$$\kappa = \frac{\|\mathbf{r}''(t) \times \mathbf{r}'(t)\|}{\|\mathbf{r}'(t)\|^3} = \frac{\|\mathbf{a}(t) \times \mathbf{v}(t)\|}{\|\mathbf{v}(t)\|^3}.$$

12.6 CURVATURE

So, for the special case of a curve in $\mathbb{R}^3$, the curvature κ is the length of the cross product of acceleration and velocity, divided by the cube of the speed. We postpone verifying this formula to a set of exercises at the end of this section. However, let's check it out on our previous examples.

EXAMPLE 32 Compute the curvature κ for the curve defined by

$$\mathbf{r}(t) = \langle 1 - t, t^2 + 1, 2t^3/3 + 1 \rangle$$

at $t = 1/2$.

Solution The velocity vector is $\mathbf{v}(t) = \langle -1, 2t, 2t^2 \rangle$, from which we can compute

$$\mathbf{v}(1/2) = \langle -1, 1, 1/2 \rangle \quad \text{and} \quad \|\mathbf{v}(1/2)\| = \sqrt{9/4} = 3/2.$$

The acceleration vector is $\mathbf{a}(t) = \langle 0, 2, 4t \rangle$, from which we can compute

$$\mathbf{a}(1/2) = \langle 0, 2, 2 \rangle.$$

The cross product of the acceleration and velocity vectors is

$$\mathbf{a}(1/2) \times \mathbf{v}(1/2) = \begin{vmatrix} \mathbf{i} & \mathbf{j} & \mathbf{k} \\ 0 & 2 & 2 \\ -1 & 1 & 1/2 \end{vmatrix} = \langle -1, -2, 2 \rangle,$$

so our alternative formula for curvature yields

$$\kappa = \frac{\|\mathbf{a}(1/2) \times \mathbf{v}(1/2)\|}{\|\mathbf{v}(1/2)\|^3} = \frac{\sqrt{(-1)^2 + (-2)^2 + 2^2}}{(3/2)^3} = \frac{\sqrt{9}}{27/8} = \frac{8}{9},$$

matching our previous computation. ∎

EXAMPLE 33 Compute the curvature $\kappa(t)$ for the curve defined by

$$\mathbf{r}(t) = \langle \cos t, \sin t, t \rangle$$

for any value t.

Solution The velocity and acceleration vectors are

$$\mathbf{v}(t) = \langle -\sin t, \cos t, 1 \rangle \quad \text{and} \quad \mathbf{a}(t) = \langle -\cos t, -\sin t, 0 \rangle.$$

The speed at any time t is

$$\|\mathbf{v}(t)\| = \sqrt{(-\sin t)^2 + (\cos t)^2 + 1} = \sqrt{2},$$

and

$$\mathbf{a}(t) \times \mathbf{v}(t) = \begin{vmatrix} \mathbf{i} & \mathbf{j} & \mathbf{k} \\ -\cos t & -\sin t & 0 \\ -\sin t & \cos t & 1 \end{vmatrix} = \langle -\sin t, \cos t, -1 \rangle.$$

Hence, at any time t we have curvature

$$\kappa = \frac{||\mathbf{a}(t) \times \mathbf{v}(t)||}{||\mathbf{v}(t)||^3} = \frac{\sqrt{(-\sin t)^2 + (\cos t)^2 + 1}}{(\sqrt{2})^3} = \frac{\sqrt{2}}{2\sqrt{2}} = \frac{1}{2},$$

the same constant curvature we computed before. ∎

EXERCISES for Section 12.6

1. Find the curvature κ and the radius of curvature ρ at $t = 2$ for the curve

$$\mathbf{r}(t) = \langle 3t^2, \sqrt{t} \rangle \qquad \text{for } 1 \leq t \leq 3.$$

2. Suppose that a particle moves on the path

$$\mathbf{r} : [0, 2] \to \mathbb{R}^3$$

$$\mathbf{r}(t) = \langle \sin 3t, \cos 3t, 2t^{3/2} \rangle.$$

Find the curvature κ and the radius of curvature ρ at $t = 1$.

3. Consider the path of a moving object described by the position function $\mathbf{r} : [0, 2] \longrightarrow \mathbb{R}^3$ where

$$\mathbf{r}(t) = \left\langle \frac{t^2}{2}, \frac{4}{3}t^{3/2}, 2t \right\rangle.$$

Find the curvature κ and the radius of curvature ρ at $t = 1$.

4. Consider the path of a moving object described by the position function $\mathbf{r} : [\frac{1}{2}, e] \longrightarrow \mathbb{R}^3$ where

$$\mathbf{r}(t) = \langle t^2, \ln t, e^t \rangle.$$

Find the curvature κ and the radius of curvature ρ at $t = 1$.

5. Verify the curvature formula for plane curves

$$\kappa(t) = \frac{|x'y'' - y'x''|}{((x')^2 + (y')^2)^{3/2}}$$

by considering a position function $\mathbf{r}(t) = \langle x(t), y(t) \rangle$ and computing $\kappa = \dfrac{d\mathbf{T}/dt}{||\mathbf{r}'(t)||}$ directly.

12.6 CURVATURE

For exercises 6-13: Let **r** be a smooth 3-dimensional position function and let

$$v(t) = ||\mathbf{v}(t)|| = ||\mathbf{r}'(t)||.$$

6. Show that $\mathbf{v}(t) = v(t)\mathbf{T}(t)$.

7. Differentiate both sides of $\mathbf{v}(t) = v(t)\mathbf{T}(t)$ with respect to t to verify that

$$\mathbf{a}(t) = \frac{dv}{dt}\mathbf{T}(t) + v(t)\frac{d\mathbf{T}}{dt}.$$

8. If $\kappa(t)$ is the curvature at time t, show that $d\mathbf{T}/dt = \kappa(t)v(t)\mathbf{N}(t)$. (Hint: substitute the definitions for $\kappa(t)$, $v(t)$ and $\mathbf{N}(t)$.)

9. Use the results of exercises 7 and 8 to verify that

$$\mathbf{a}(t) = \frac{dv}{dt}\mathbf{T}(t) + \kappa(t)v^2(t)\frac{d\mathbf{T}}{dt}.$$

10. Substitute the result of exercise 9 for $\mathbf{a}(t)$ and substitute the result of exercise 6 for $\mathbf{v}(t)$ and compute

$$\mathbf{a}(t) \times \mathbf{v}(t) = \kappa(t)v^3(t)\mathbf{N}(t) \times \mathbf{T}(t).$$

11. Use exercise 10 to verify that

$$||\mathbf{a}(t) \times \mathbf{v}(t)|| = \kappa(t)v(t)^3.$$

Dividing both sides by $v(t)^3$ gives us the curvature formula for space curves.

12. Use exercise 10 to verify that

$$\frac{\mathbf{a} \times \mathbf{v}}{v^3} = \kappa \mathbf{B},$$

where **B** is the unit binormal vector.

13. Suppose that $||\mathbf{a}(t)|| = a$ and $||\mathbf{v}(t)|| = b$ for all t, where a and b are constants. Show that the curvature κ is also constant, and express it in terms of a and b.

The graph of a function over an interval $[a, b]$ is a curve in the plane. If $y = f(x)$, then we can parametrize the function's graph with the position vector function

$$\mathbf{r}(t) = \langle t, \ f(t) \rangle,$$

where $a \leq t \leq b$.

The **curvature of the graph** of $y = f(x)$ at the point $(x_0, f(x_0))$ is sometimes defined as

$$\kappa = \frac{|f''(x_0)|}{(1 + [f'(x_0)]^2)^{3/2}}.$$

As usual, the *radius of curvature* is $\rho = 1/\kappa$ for $\kappa \neq 0$ and $\rho = \infty$ for $\kappa = 0$.

14. Show that the definition of κ given above matches that for a plane curve given in this section, namely

$$\kappa = \frac{|x'y'' - y'x''|}{((x')^2 + (y')^2)^{3/2}}.$$

15. Calculate the curvature κ and the radius of curvature ρ at the point $(1, -1)$ on the graph of $y = f(x) = x^3 - 2x$.

16. Find the curvature and radius of curvature at the point $(3, 4)$ on a circle of radius 5 centered at the origin by considering the top "half" of the circle as the graph of the function

$$f(x) = \sqrt{25 - x^2}$$

over the interval $[-5, 5]$.

13

Fundamentals of Multivariable Functions

So far, we have focused virtually all our attention on *functions of a single variable*—those functions having a single real number input as the independent variable. Functions of the form

$$y = f(x)$$

have a single real number as input and a single real number as output. We have seen how calculus can be used to measure and investigate the behavior of these functions. Differentiation and integration have geometrical interpretations in terms of slopes and areas, as well as physical interpretations in terms of rates of change and net effect.

In the last chapter, we studied single-variable functions having *vectors*

$$\mathbf{r}(t)$$

as outputs. Thinking of the independent variable t as a time parameter and the vector output as a position, we were able to associate a path with each such function. Calculus again allowed us to investigate both geometrical properties of the path's image curve as well as physical notions such as velocity and acceleration.

In the remainder of this book, we proceed to consider *multivariable functions*—that is, functions having more than one independent variable. In this chapter, we concentrate on *scalar-valued* multivariable functions, which are often called **scalar fields**. We discuss some techniques for visualizing and interpreting their graphical representations, and we examine the special examples of linear and quadratic multivariable functions in detail. Then, in the next two chapters, we show how derivatives and integrals can be extended as tools for analyzing multivariable functions.

13.1 MULTIVARIABLE FUNCTIONS—EXAMPLES AND TERMINOLOGY

In this section, we introduce some examples of multivariable functions and some of the language and notation we need to discuss them.

Functions of several variables arise naturally everywhere we look. Indeed, it is common for us to find that a quantity depends on the values of several other variable quantities rather than just one. The very first functions you ever studied were all two-input functions: addition, subtraction, multiplication, and division!

Many formulas from elementary geometry can be thought of as defining functions of several variables. For example, the Pythagorean theorem's statement that

$$z = \sqrt{x^2 + y^2}$$

expresses the length of the hypotenuse z in terms of x and y, the lengths of the legs. By varying x and y, we can think of z as a function of two variables. To emphasize this functional relationship, we write

$$z(x, y) = \sqrt{x^2 + y^2}$$

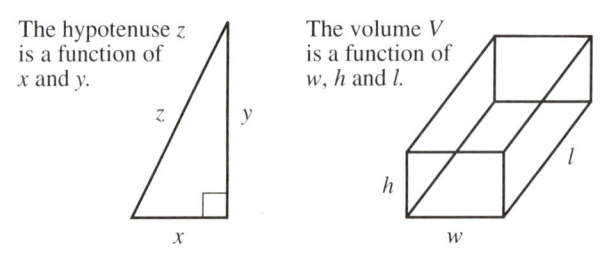

(see Figure 13.1).

Figure 13.1 Multivariable functions: $z = \sqrt{x^2 + y^2}$ and $V = w\ell h$.

The volume V of a box (rectangular parallelepiped) can be written

$$V(w, \ell, h) = w\ell h,$$

where w is the width, ℓ the length, and h the height of the box. Thus, the volume V is a function of three variables (see Figure 13.1). In each of these examples, the output is a single number that is uniquely determined by the values of more than one input.

The term **scalar field** is often used in the context of physics to refer to a scalar-valued multivariable function. If we think of three inputs x, y, and z as specifying the location (x, y, z) of an object in space, then there are a variety of scalar quantities that can be associated with this location, including temperature, humidity, barometric pressure, light intensity, and

13.1 MULTIVARIABLE FUNCTIONS—EXAMPLES AND TERMINOLOGY

others. The values of each of these quantities may well depend on the precise location of the object. Each quantity defines a *field* of scalar values. For example, we think of a temperature field as "tagging" a specific temperature

$$T(x, y, z)$$

to each position in space. In other words, the temperature at a point is a function of the three variables that specify its position.

Notation for multivariable functions

The inputs to a multivariable function can specify an ordered pair (x, y) (for 2 variables), an ordered triple (x, y, z) (for 3 variables), or, in general, an n-tuple $(x_1, x_2, \ldots, x_n)$ (for n variables), depending on how many real number inputs are required. As such, we think of the domain of a multivariable function as a subset of $\mathbb{R}^n$, where n denotes the number of independent variables. If the domain of a multivariable function f includes all real numbers for each of its n inputs, we can write

$$f : \mathbb{R}^n \longrightarrow \mathbb{R}.$$

If the domain D is some subset of $\mathbb{R}^n$, then we write

$$f : D \longrightarrow \mathbb{R}.$$

EXAMPLE 1 Describe suitable domains for the multivariable functions that express the hypotenuse of a right triangle as a function of the leg lengths and the volume of a rectangular box as a function of its three dimensions:

$$z = \sqrt{x^2 + y^2} \quad \text{and} \quad V = w\ell h.$$

Solution Because of the physical circumstances involved, only positive values make sense as inputs. In other words, to indicate their domains, we could write

$$z : \{(x, y) : x > 0, y > 0\} \longrightarrow \mathbb{R}$$

and

$$V : \{(w, \ell, h) : w > 0, \ell > 0, h > 0\} \longrightarrow \mathbb{R}$$

for these two functions. ■

Graph of $z = f(x, y)$—surfaces

Because we live in three spatial dimensions, we'll devote a good deal of our attention to scalar-valued functions of two variables (two inputs and one output), for they still provide a ready means of visualization. If we denote the two independent variables as x and y, and the dependent variable as z, then the graph of the function f is the set of points (x, y, z) in space satisfying the equation

$$z = f(x, y).$$

Generally, we can think of this graph as describing a *surface* in space (see Figure 13.2).

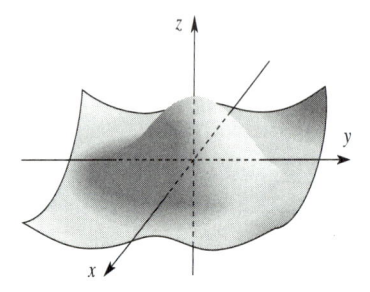

Figure 13.2 The graph of $z = f(x, y)$ is a surface.

For each ordered pair (x, y) in the domain of the function f, we plot the point

$$(x, y, z) = (x, y, f(x, y))$$

on the surface. In $\mathbb{R}^3$, we can think of locating the input pair (x, y) in the xy-plane (actually the point $(x, y, 0)$), and then move vertically up or down to the appropriate level for $z = f(x, y)$ to plot the point on the graph.

EXAMPLE 2 Consider the multivariable function $f : \mathbb{R}^2 \longrightarrow \mathbb{R}$, where

$$z = f(x, y) = \sqrt{x^2 + y^2}.$$

Describe its graph.

Solution The graph of $z = f(x, y)$ has the shape of a cone with its vertex at the origin and opening upwards. One point on the graph is $(3, 4, 5)$, since

$$f(3, 4) = \sqrt{3^2 + 4^2} = \sqrt{25} = 5.$$

Figure 13.3 illustrates how this point is located on the graph. ■

13.1 MULTIVARIABLE FUNCTIONS—EXAMPLES AND TERMINOLOGY

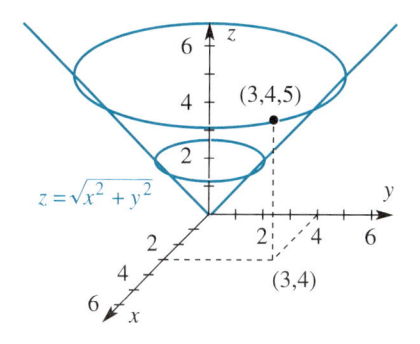

Figure 13.3 Locating a point on the graph of $f(x,y) = \sqrt{x^2+y^2}$.

▶ ▶ ▶ **The surface representing the graph of a function of two variables will satisfy the "vertical line test." That is, any vertical line (parallel to the z-axis) will pierce the surface in at most one point.**

Although we "run out" of dimensions for pictorially representing the graph of a function with more than two inputs, we can still use the idea of a surface to motivate our thinking. Some of the other visualization techniques we discuss in the next section extend well to functions of any number of inputs, while others work well only for two-input functions.

Examples of surfaces

Let's start out by taking a look at several examples of surfaces. By way of comparison, let's note that a *curve* in the Cartesian plane is often described as the solution set of an equation in two variables x and y. While this equation may represent a functional relationship between the two variables, such as

$$y = x^2,$$

an equation in two variables can also represent some non-functional relationship, such as

$$x^2 + y^2 = 9,$$

whose solution set is a circle of radius 3 centered at the origin.

Similarly, a surface in space can be described as the solution set to an equation in the three variables x, y, and z.

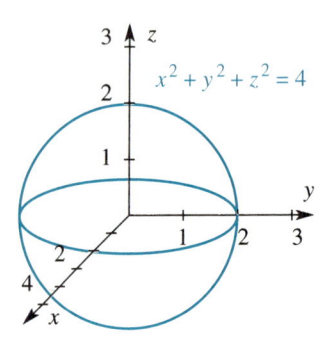

Figure 13.4 A sphere of radius 2 centered at the origin.

EXAMPLE 3 Describe the solution set of the equation

$$x^2 + y^2 + z^2 = 4.$$

Solution Since the distance between any point (x, y, z) in space and the origin is $\sqrt{x^2 + y^2 + z^2}$, the solution set of the equation

$$x^2 + y^2 + z^2 = 4$$

describes a *sphere* of radius 2, centered at the origin (see Figure 13.4). ∎

Note that this sphere is not the graph of a function of two variables x and y, since there can be two different values z corresponding to a single input pair (x, y). If we solve the equation for z, we have

$$z = \pm\sqrt{4 - x^2 - y^2}.$$

The graph of $z = \sqrt{4 - x^2 - y^2}$ is the upper hemisphere and the graph of $z = -\sqrt{4 - x^2 - y^2}$ is the lower hemisphere.

On the other hand, we could view the left-hand side of the original equation as a function of all three variables x, y, and z:

$$f(x, y, z) = x^2 + y^2 + z^2.$$

The sphere of radius 2 centered at the origin is now exactly the set of *all* inputs in $\mathbb{R}^3$ that produce the output

$$f(x, y, z) = 4.$$

EXAMPLE 4 Describe the solution set of the equation

$$x + 2y + 3z = 6.$$

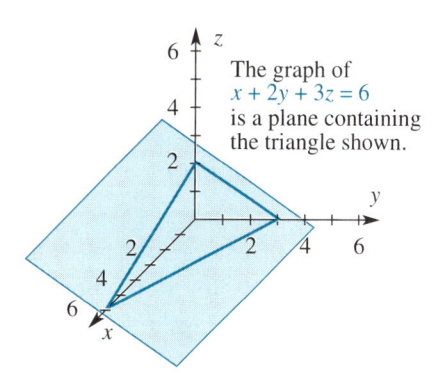

13.1 MULTIVARIABLE FUNCTIONS—EXAMPLES AND TERMINOLOGY

Figure 13.5 Three coordinate axis intercepts of the plane $x + 2y + 3z = 6$.

Solution The solution set of the linear equation

$$x + 2y + 3z = 6$$

describes a *plane*. Three points in this plane that we can identify readily are the *intercepts* with each of the three coordinate axes:

$$(6, 0, 0) \quad (0, 3, 0) \quad (0, 0, 2).$$

The triangle determined by these three points is shown in Figure 13.5.

This plane can be thought of as the graph of the function

$$z = f(x, y) = \frac{6 - x - 2y}{3}.$$

EXAMPLE 5 Describe the solution set of the equation $x^2 + y^2 + z^2 = 0$.

Solution The solution set of an equation in three variables could consist of a single point. The only point whose coordinates satisfy $x^2 + y^2 + z^2 = 0$ is the origin $(0, 0, 0)$.

A solution set could even be empty, as is the case with $x^2 + y^2 + z^2 = -1$, which has no solution in real numbers x, y, and z.

Cylindrical surfaces

A cylinder can be thought of as the surface generated by translating or sliding a circle along an axis. In general, a *cylindrical surface* is generated by translating a curve along some line. The solution set of an equation involving only 2 of the 3 variables x, y, and z will be a cylindrical surface in space. To visualize the surface, we can simply graph the curve in the coordinate plane corresponding to the two variables present in the equation,

EXAMPLE 6 Describe the solution set of $x^2 + y^2 = 4$.

Solution The solution set of $x^2 + y^2 = 4$ is an infinite cylinder of radius 2 with the z-axis running down its center. (Note that z is not involved in the equation.) Part of the surface is depicted in Figure 13.6. ∎

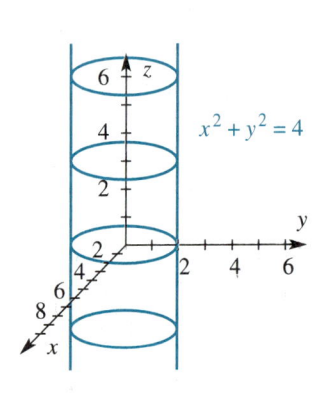

Figure 13.6 The infinite cylinder $x^2 + y^2 = 4$.

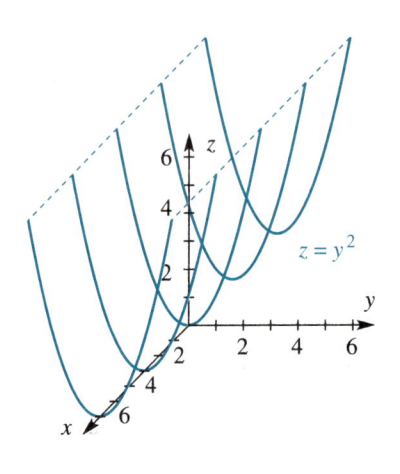

Figure 13.7 The infinite "trough" $z = y^2$.

EXAMPLE 7 Describe the solution set of $z = y^2$.

Solution The solution set of $z = y^2$ is the infinite "trough" generated by translating the appropriate parabola parallel to the x-axis. (Note that x is not involved in the equation.) Part of the surface is shown in Figure 13.7. ∎

Quadric surfaces

The sphere, the cylinder, and the parabolic "trough" all belong to a special class of surfaces known as the **quadric surfaces**. A quadric surface is the solution set of a quadratic equation in three variables—that is, an equation involving only terms of degree 2 and lower.

You are familiar with quadratic equations in two variables. Their graphs in the Cartesian plane are known as **conic sections**: *ellipses* (in-

13.1 MULTIVARIABLE FUNCTIONS—EXAMPLES AND TERMINOLOGY

cluding circles), *hyperbolas*, and *parabolas*. Each of the conic sections is a curve with very special geometric properties.

Quadric surfaces provide an important set of examples, because close inspection reveals that many surfaces behave very much like quadric surfaces. Below, we look at several quadric surfaces. Take special note of the *cross-sections*, the curves representing the intersection of the surface with a plane parallel to one of the coordinate planes. The cross-sections of quadric surfaces are conic sections.

As already mentioned, a **sphere** is a quadric surface. In general, a sphere of radius R centered at the origin has equation $x^2 + y^2 + z^2 = R^2$. All cross-sections of a sphere are circles (see Figure 13.8).

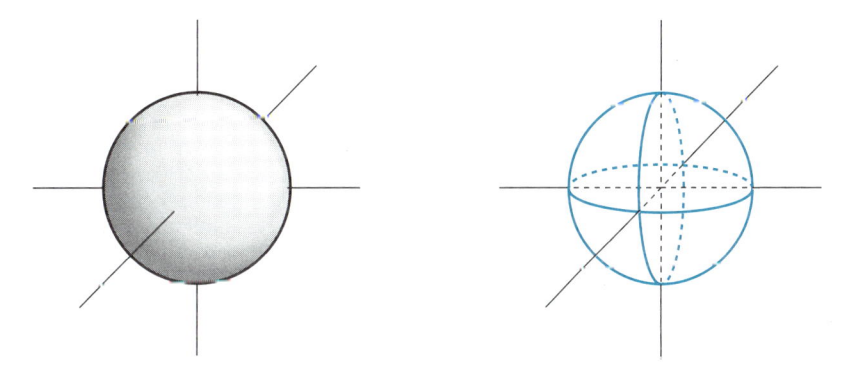

Figure 13.8 The sphere $x^2 + y^2 + z^2 = R^2$ and some cross-sections.

The sphere is a special case of an **ellipsoid**. An ellipsoid centered at the origin has an equation of the form $\frac{x^2}{a^2} + \frac{y^2}{b^2} + \frac{z^2}{c^2} = 1$, where a, b, and c are the distances from the origin to the intercepts with the x-axis, y-axis, and z-axis, respectively. As the name might suggest, the cross-sections of an ellipsoid are ellipses (see Figure 13.9).

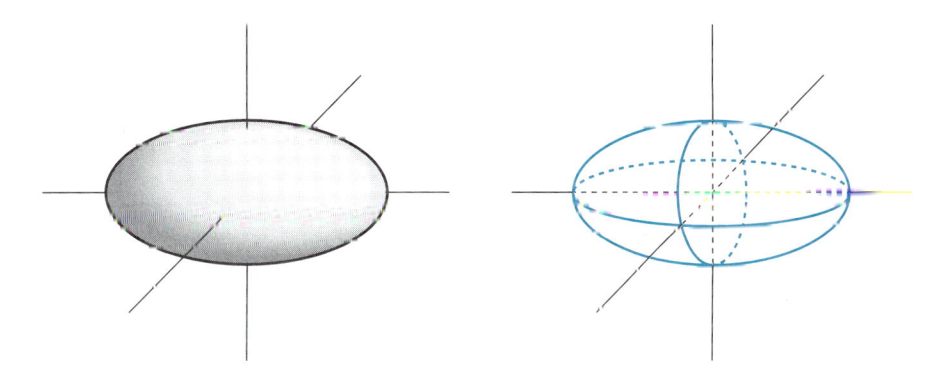

Figure 13.9 The ellipsoid $\frac{x^2}{a^2} + \frac{y^2}{b^2} + \frac{z^2}{c^2} = 1$ and some cross-sections.

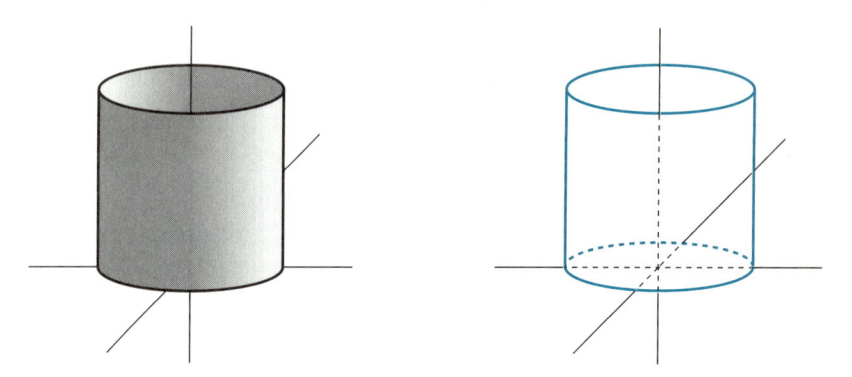

Figure 13.10 The elliptic cylinder $\frac{x^2}{a^2} + \frac{y^2}{b^2} = 1$ and some cross-sections.

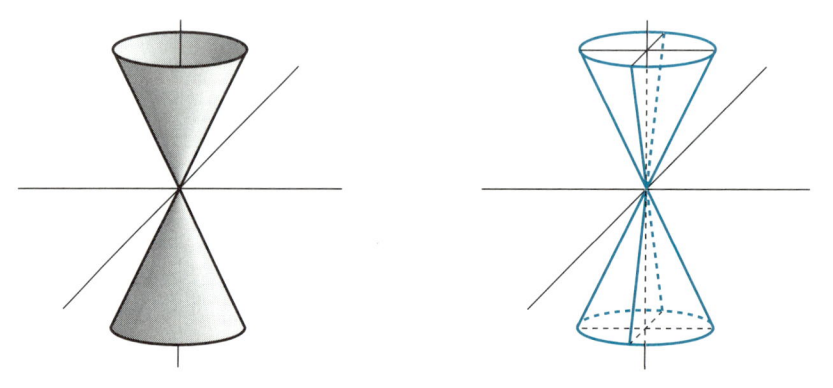

Figure 13.11 The elliptic cone $\frac{z^2}{c^2} = \frac{x^2}{a^2} + \frac{y^2}{b^2}$ and some cross-sections.

An **elliptic cylinder** having the z-axis running down its center has an equation of the form

$$\frac{x^2}{a^2} + \frac{y^2}{b^2} = 1.$$

Each horizontal cross-section is an ellipse (see Figure 13.10).

An **elliptic cone** having the z-axis running down its center has an equation of the form

$$\frac{z^2}{c^2} = \frac{x^2}{a^2} + \frac{y^2}{b^2}.$$

Again, note that each horizontal cross-section is an ellipse (see Figure 13.11). The vertical cross-sections are hyperbolas, with the exception of the xz-plane and the yz-plane. In these two cases, the vertical cross-sections are two intersecting lines.

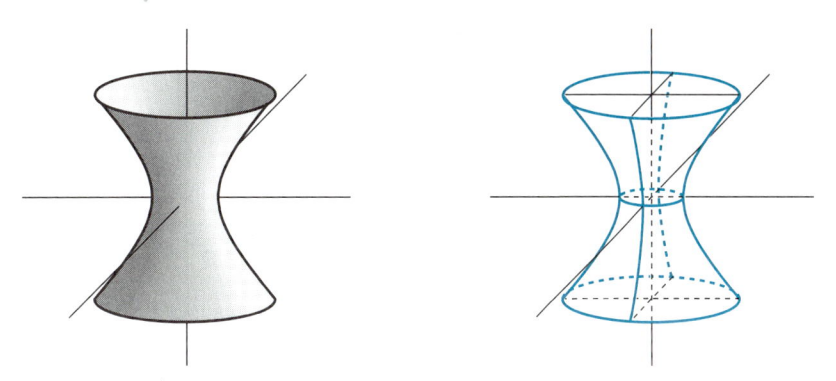

Figure 13.12 The hyperboloid of one sheet $\frac{z^2}{c^2} = \frac{x^2}{a^2} + \frac{y^2}{b^2} - 1$ and some cross-sections.

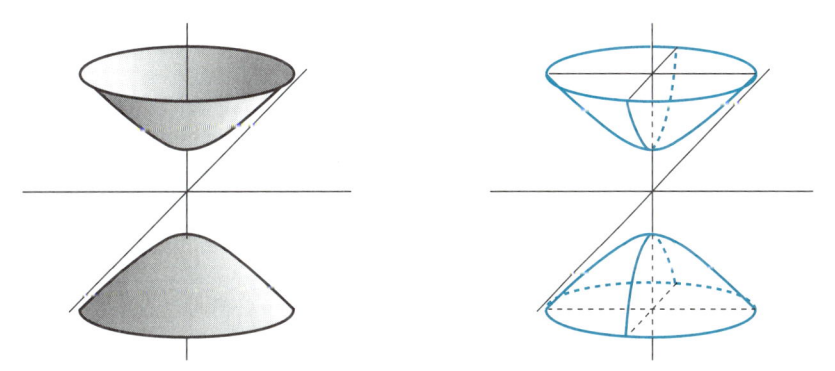

Figure 13.13 The hyperboloid of two sheets $\frac{z^2}{c^2} = \frac{x^2}{a^2} + \frac{y^2}{b^2} + 1$ and some cross-sections.

A **hyperboloid of one sheet** having the z-axis running down its center has an equation of the form

$$\frac{z^2}{c^2} = \frac{x^2}{a^2} + \frac{y^2}{b^2} - 1,$$

and a **hyperboloid of two sheets** has an equation of the form

$$\frac{z^2}{c^2} = \frac{x^2}{a^2} + \frac{y^2}{b^2} + 1.$$

The horizontal cross-sections of both hyperboloids are ellipses, while the vertical cross-sections are hyperbolas (see Figures 13.12 and 13.13).

An **elliptic paraboloid** having the z-axis running down its center has an equation of the form

$$cz = \frac{x^2}{a^2} + \frac{y^2}{b^2}.$$

The horizontal cross-sections are ellipses, while the vertical cross-sections are parabolas. The sign of c determines whether the surface is "bowl up" ($c > 0$) or "bowl down" ($c < 0$).

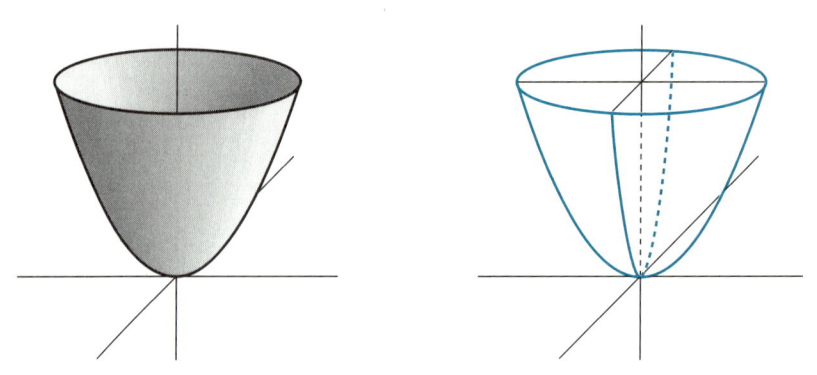

Figure 13.14 The elliptic paraboloid $cz = \dfrac{x^2}{a^2} + \dfrac{y^2}{b^2}$ ($c > 0$) and some cross-sections.

Figure 13.14 illustrates an elliptic paraboloid with $c > 0$. By changing a sign in the previous equation, we obtain the equation of a **hyperbolic paraboloid**:

$$cz = \frac{x^2}{a^2} - \frac{y^2}{b^2}.$$

Now, the horizontal cross-sections are hyperbolas, while the vertical cross-sections are parabolas. If $c < 0$, cross-sections parallel to the xz-plane open down while cross-sections parallel to the yz-plane open up (see Figure 13.15). If $c > 0$, the situation is reversed. In either case, the surface has the shape of a "saddle."

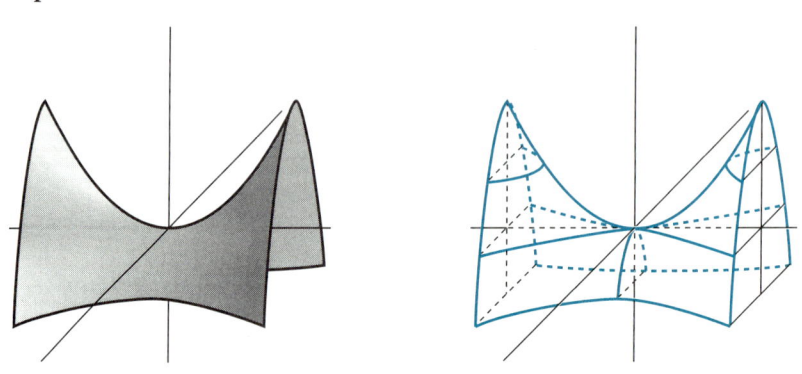

Figure 13.15 The hyperbolic paraboloid $cz = \dfrac{x^2}{a^2} - \dfrac{y^2}{b^2}$ ($c < 0$) and some cross-sections.

The origin $(0, 0, 0)$ is a key point for many of the quadric surfaces described above. The sphere and ellipsoid have the origin at their centers. The elliptic cone has the origin at its vertex, as does the elliptic paraboloid. The origin is a point of symmetry for both hyperboloids, and it is the "saddle point" of the hyperbolic paraboloid.

Given any point (x_0, y_0, z_0), we can translate any of these surfaces so that (x_0, y_0, z_0) acts as the new "origin" simply by replacing x by $x - x_0$, y by $y - y_0$, and z by $z - z_0$. The new "axes" all pass through this point but are still parallel with the original coordinate axes.

13.1 MULTIVARIABLE FUNCTIONS—EXAMPLES AND TERMINOLOGY

EXAMPLE 8 What is the equation of the ellipsoid whose center is $(1, -2, 3)$ and whose three "radii" are 4, 5, and 2 in the directions parallel to the x-axis, y-axis, and z-axis, respectively?

Solution We have $a = 4$, $b = 5$, $c = 2$, and $(x_0, y_0, z_0) = (1, -2, 3)$. The equation of the ellipsoid is

$$\frac{(x-1)^2}{16} + \frac{(y+2)^2}{25} + \frac{(z-3)^2}{4} = 1.$$

After expanding and collecting terms, we can arrive at an equivalent equation for this ellipsoid:

$$25x^2 + 16y^2 + 100z^2 - 50x + 64y - 600z + 589 = 0.$$

There are still no mixed terms (xy, xz, or yz) involved. Quadric surfaces whose equations include mixed terms can have axes with directions not parallel to any of the coordinate axes.

EXERCISES for Section 13.1

1. The graphs of the addition and subtraction functions

$$z = a(x, y) = x + y \quad \text{and} \quad z = s(x, y) = x - y$$

are two planes, each containing the origin $(0, 0, 0)$. Describe the line of intersection of these two planes.

2. The graph of the multiplication function

$$z = m(x, y) = xy$$

also contains the origin $(0, 0, 0)$. Is this graph also a plane?

3. What is the domain of the division function

$$z = d(x, y) = x/y,$$

that is, what is the set of ordered pairs (x, y) that are acceptable inputs for this function?

For exercises 4-9: Sketch each of the cylindrical surfaces whose equations are given.

4. $xy = 1$.

5. $xz = 1$.

6. $yz = 1$.

7. $x^2 + z^2 = 9$.

8. $\dfrac{y^2}{9} + \dfrac{z^2}{16} = 1$.

9. $|x| + |y| = 1$.

10. When is the graph of a function $z = f(x,y)$ a cylindrical surface? Give an example.

11. What is the equation of a sphere of radius 7, centered at the point $(-2, 4, -7)$?

12. What is the equation of a circular cylinder of radius 7, with a vertical axis containing the point $(-2, 4, -7)$?

13. For what values a, b, and c is the ellipsoid of Figure 13.9 the same as the sphere of Figure 13.8?

For exercises 14-20: For each of the quadric surfaces described, indicate whether or not it could be the graph of a function

$$z = f(x,y)$$

by using the vertical line test.

14. the ellipsoid shown in Figure 13.9

15. the elliptic cylinder shown in Figure 13.10

16. the elliptic cone shown in Figure 13.11

17. the hyperboloid of one sheet shown in Figure 13.12

18. the hyperboloid of two sheets shown in Figure 13.13

19. the elliptic paraboloid shown in Figure 13.14

20. the hyperbolic paraboloid shown in Figure 13.15

For exercises 21-26: Each of the quadric surfaces whose equation is given has the z-axis as its "central axis." In each exercise, make the necessary change to the equation so that the type of quadric surface is the same, but the central axis is now the x-axis. Then make the change so that the central axis is the y-axis.

21. the elliptic cylinder shown in Figure 13.10

22. the elliptic cone shown in Figure 13.11

23. the hyperboloid of one sheet shown in Figure 13.12

24. the hyperboloid of two sheets shown in Figure 13.13

25. the elliptic paraboloid shown in Figure 13.14

26. the hyperbolic paraboloid shown in Figure 13.15

13.2 VISUALIZING AND INTERPRETING MULTIVARIABLE FUNCTIONS

All of our visual communications media (written material, computer and calculator screens, and even the retinal surfaces of our eyes, for that matter) present at most two dimensions, so multivariable functions require more effort to be represented completely and unambiguously. In this section we explore a number of methods for graphically representing multivariable functions and for interpreting these graphs. Although no single technique can capture all of a multivariable function's behavior, together they provide an important aid to understanding.

Plotting individual points in a haphazard or random manner is usually not a good way to graph an equation in three variables, because it's so hard to figure out how to "connect the dots." It is much more sensible to graph these equations in some systematic manner. Let's turn now to some strategies for graphing multivariable functions of the form

$$z = f(x, y).$$

Slicing

Suppose we *freeze* one of the inputs of $z = f(x, y)$. The result can be thought of as a single-input function that can be graphed, visualized, and interpreted like any other function of one variable. The graph of this new function is a cross-sectional *slice* of the surface graph of the original function.

EXAMPLE 9 Find and graph the slices of the two-input function

$$f(x, y) = x^2 + y^2$$

obtained by freezing the second input at $y = -2, -1, 0, 1,$ and 2.

Solution   If we *freeze* the second input at $y = -2$, the result is a function of x alone:

$$f(x, -2) = x^2 + 4.$$

If, instead, we freeze the second input at $y = -1$, the result is a different but related function of x:

$$f(x, -1) = x^2 + 1.$$

Continuing in this manner, we also have

$$f(x, 0) = x^2, \quad f(x, 1) = x^2 + 1, \quad f(x, 2) = x^2 + 4.$$

The graphs of these "slice functions" are shown in Figure 13.16. ■

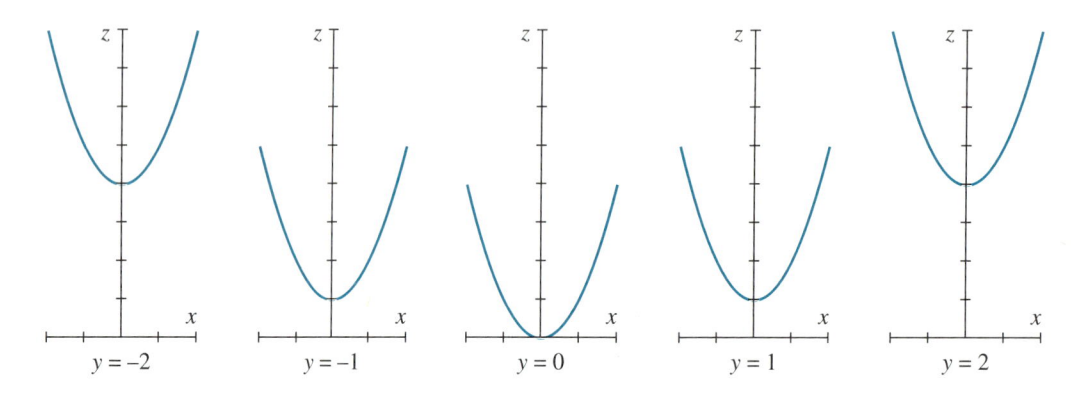

Figure 13.16 Graphs of slice functions.

The function $z = f(x, c)$, for each value $y = c$, corresponds to restricting the domain of f to the set of inputs

$$\{(x, y) : x \in \mathbb{R}, y = c\}.$$

The graphs of these slice functions can now be placed in the appropriate plane, corresponding to the value of the frozen input. The graph of $f(x, -2) = x^2 + 4$ is placed in the vertical plane $y = -2$, the graph of $f(x, -1) = x^2 + 1$ is placed in the vertical plane $y = -1$, and so on.

Figure 13.17 shows the relative positions of the slices in three dimensions. Each slice is a parabola, with the vertex height depending on the value y. You can think of the vertical planes $y = c$ as panes of glass (or transparency slides) with the graphs drawn on them. The graph of a slice gives us an image along a single direction of the shape of the graph of $z = f(x, y)$ in the same way that the edge of a slice of bread gives us an image (along one direction) of the shape of the loaf.

Of course, we can also slice the same surface with the first input x frozen. Now we plot the function of the form $z = f(c, y)$ for some fixed value $x = c$. This corresponds to restricting the domain of f to the set of inputs

$$\{(x, y) : y \in \mathbb{R}, x = c\}.$$

Each slice can then be placed in the vertical plane $x = c$.

EXAMPLE 10 Graph the slices of $f(x, y) = x^2 + y^2$ resulting from freezing $x = -2, -1, 0, 1,$ and 2.

Solution Figure 13.18 illustrates the slices of the two-input function

$$f(x, y) = x^2 + y^2$$

13.2 VISUALIZING AND INTERPRETING MULTIVARIABLE FUNCTIONS

obtained by freezing the first input at $x = -2, -1, 0, 1,$ and 2. In order from smallest value of x to largest, these graphs are of

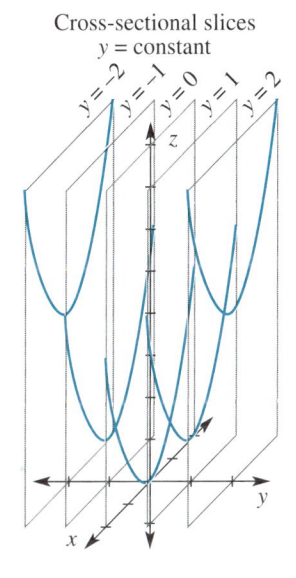

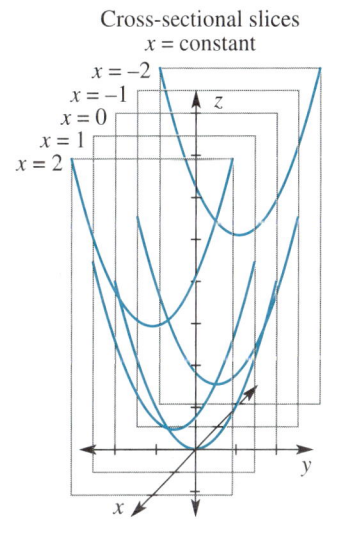

$$z = 4 + y^2, \quad z = 1 + y^2, \quad z = y^2, \quad z = 1 + y^2, \quad z = 4 + y^2.$$ ∎

Figure 13.17 Visualizing the slices (for y frozen) of $z = x^2 + y^2$ in $\mathbb{R}^3$.

Figure 13.18 Visualizing the slices (for x frozen) of $z = x^2 + y^2$ in $\mathbb{R}^3$.

If we are especially interested in the "shape" of a function graph near a particular point, it is often useful to graph each of the cross-sectional slices at that point. Doing this is analogous to slicing a loaf of bread first cross-wise and then length-wise to examine the shape of the loaf near the intersection. Figure 13.19 shows several slices of $z = x^2 + y^2$ for both x and y. The "fishnet" has a shape known as a *paraboloid of revolution*, since the graph can be generated by rotating a parabola about the z-axis.

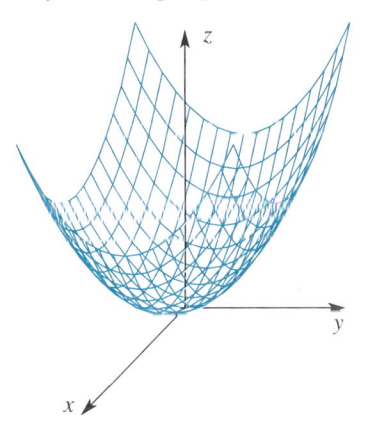

Figure 13.19 Several cross-sections of the paraboloid $z = x^2 + y^2$.

Repeating this process of setting one of the variables equal to a constant, graphing the result, and placing the result in the appropriate plane for several values of both inputs can give us a good visual image of the graph of the surface $z = f(x, y)$.

Slicing functions with more than two inputs

We cannot graph a function of three variables

$$w = f(x, y, z),$$

because we need four dimensions (three inputs, one output). However, slicing is one of the few methods that also works reasonably well for functions with three or more inputs.

We freeze all but one of the inputs and graph the resulting single-input function, repeating this process for various choices of the frozen inputs. It is helpful to arrange the resulting graphs in an orderly manner reflecting the choice of frozen inputs.

EXAMPLE 11   Graph the y-slices of the function

$$w = f(x, y, z) = 2.62 + 0.6x + 0.4y + z - x^2 - y^2 - z^2$$

over the region $[0, 0.6] \times [0, 0.6] \times [0, 0.6]$ obtained by freezing both x and z at all possible combinations of $x = 0, 0.2, 0.4, 0.6$ and $z = 0, 0.2, 0.4, 0.6$ and graphing the resulting single variable functions of y.

Solution   The region $[0, 0.6] \times [0, 0.6] \times [0, 0.6]$ represents all points (x, y, z) satisfying

$$0 \leq x, y, z \leq 0.6$$

(the cube with one corner at the origin $(0, 0, 0)$ and the opposite corner at $(0.6, 0.6, 0.6)$). Figure 13.20 shows the sixteen different slices arranged by the values of x and z. ■

Note the scaling on each window (shown on the lower left-hand window): along the horizontal, the y-range runs from 0 to 0.6; along the vertical, the w-range runs from 2.5 to 3.0. The dotted "crosshairs" provide a visual reference and do *not* represent the y- and w-axes.

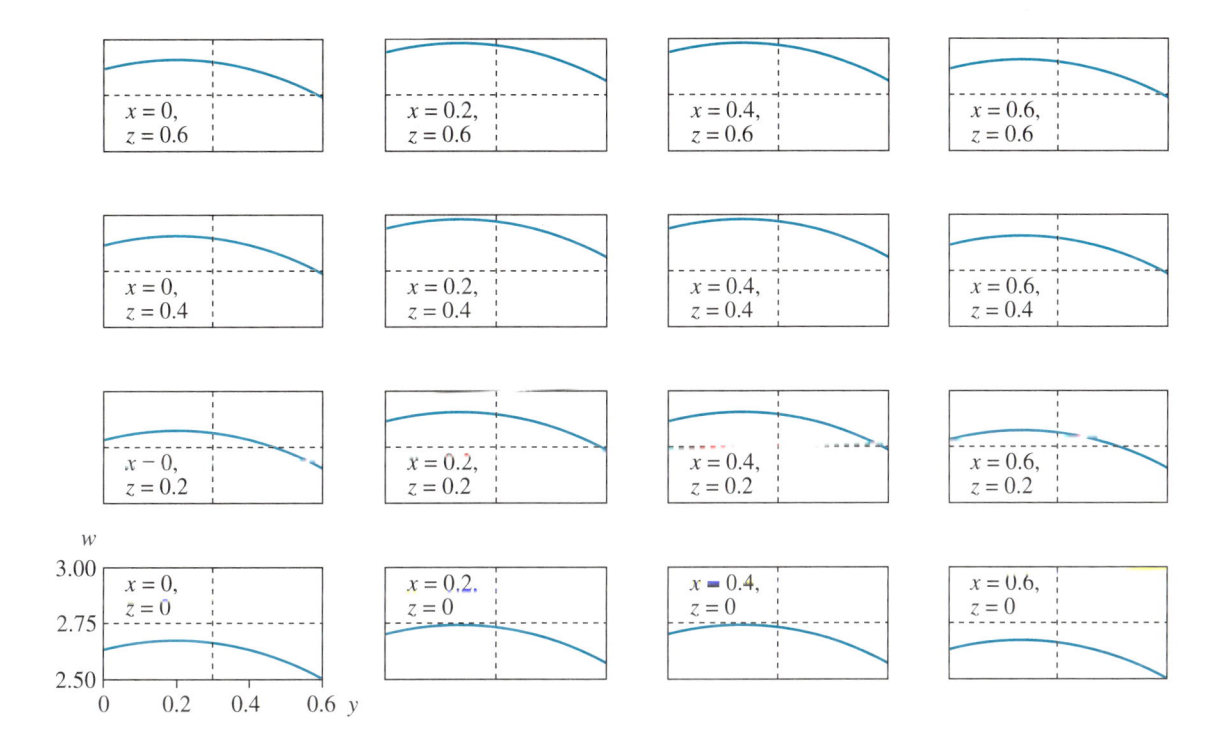

13.2 VISUALIZING AND INTERPRETING MULTIVARIABLE FUNCTIONS

Figure 13.20 Slices of a function graph $w = f(x, y, z)$ with respect to y.

To find the output value w corresponding to

$$(x, y, z) = (0.4, 0.6, 0.2)$$

we need to refer to the third picture of the third row from the top. At the far right-hand edge of the window, we can see that

$$f(0.4, 0.6, 0.2) \approx 2.75,$$

and you can check that the actual function value at this point is 2.74.

A scientist may use this method when studying a quantity that depends on several variables. By fixing the values of all the variables except one, she can isolate the effect of varying the lone remaining independent variable on the value of the dependent variable. This process of replacing all but one of the inputs to a function with fixed values to produce a single-input slice function is one important enough to be named. It is called *Currying*, in honor of the mathematician Haskel Curry.

EXAMPLE 12 Use the slices in Figure 13.20 to determine approximately where the relative maximum output of f occurs over the region $[0, 0.6] \times [0, 0.6] \times [0, 0.6]$.

Solution   In just the few y-slices taken, the very highest w values appear to be approximately 3.0 in four different windows, namely the second and third windows of the top two rows. The highest point in each of these four windows appears to occur near $y = 0.2$. In fact, the highest point in each of the sixteen windows appears to occur near $y = 0.2$. The top two rows correspond to $z = 0.6$ and $z = 0.4$, while the middle two columns correspond to $x = 0.2$ and $x = 0.4$. Averaging the z-values and x-values, we might make a reasonable guess that a maximum occurs near $(x = 0.3, y = 0.2, z = 0.5)$.

You can verify that

$$2.62 + 0.6x + 0.4y + z - x^2 - y^2 - z^2 = (1 - (x - .03)^2) + (1 - (y - .02)^2) + (1 - (z - .05)^2)$$

by expanding out the right-hand side and collecting like terms. Using this form of the function, we can see that the maximum value of $w = 3$ actually does occur exactly at $(x = 0.3, y = 0.2, z = 0.5)$. ∎

Contour plots

A variant of the slicing method is widely used in graphing elevation (that is, height above sea level) of regions of the earth. A **contour** is a "line of constant elevation." A contour may also be called a **level curve**. A contour map of a region shows and labels several level curves. Figure 13.21 shows a landscape with two tall hills and a dry lake bed nearby; Figure 13.22 is a contour map of the same region.

Near the hilltops there are small, roughly circular contours. Similarly, near the bottom of the lake bed are more roughly circular contours. Between two given contour lines, the land is steepest where the curves are closest together.

Figure 13.21 A landscape.

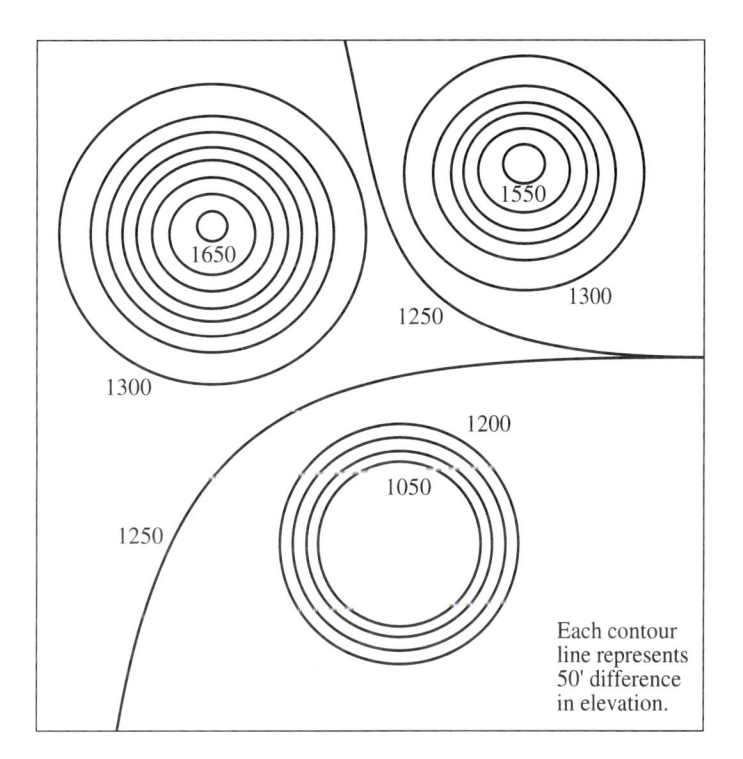

13.2 VISUALIZING AND INTERPRETING MULTIVARIABLE FUNCTIONS

Figure 13.22 Contour map: each line represents 50 feet of elevation change.

Instead of the cross-sectional slices made by vertical planes, contours represent the level curves of a surface sliced along horizontal planes. For a function

$$z = f(x, y),$$

a level curve is found by setting $z = c$ for a constant value c and graphing the result. This curve (or curves) represent the set of inputs (x, y) that produce the output

$$f(x, y) = c.$$

For three-dimensional graphing, the level curve can be placed in the horizontal plane $z = c$.

EXAMPLE 13 Graph the level curves of $z = f(x, y) = x^2 + y^2$ corresponding to $z = 0, 1, 2, 3,$ and 4. (The graph has no points corresponding to negative values of z.)

Solution The level curve corresponding to $z = 0$ is a single point $(0, 0)$, since $x = 0$ and $y = 0$ are the only pair of real values satisfying

$$0 = x^2 + y^2.$$

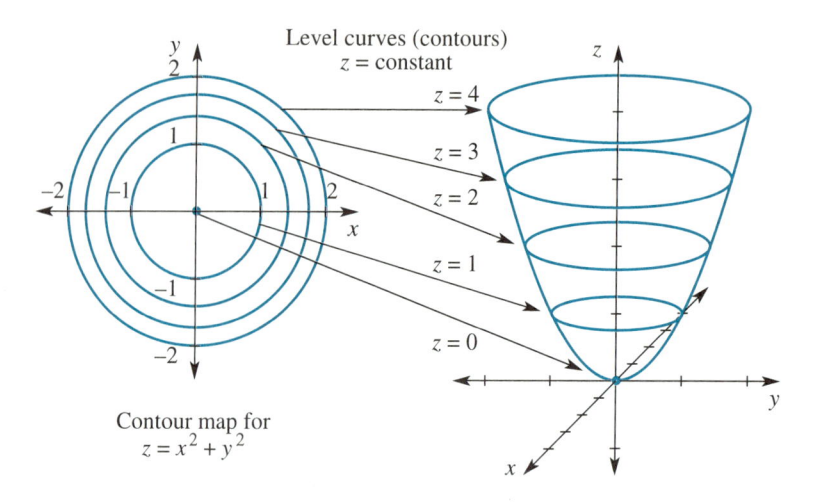

Figure 13.23 Graph of $z = x^2 + y^2$ showing level curves.

The other level curves are circles with centers at $(0,0)$. The contours are shown in Figure 13.23, along with their relationship to the graph of the surface. ∎

▶▶▶ **When viewed from above, contours of a function graph**

$$c = f(x, y)$$

representing different values $z = c$ cannot overlap, because only one output is associated with any given input pair (x, y).

The "mountaintops" and the "lake bottoms" of a surface represent its *local maxima* and *local minima*. For the paraboloid $z = x^2 + y^2$, there is a single local extremum (a minimum) at $(0,0)$. In fact, the absolute minimum output is 0, and there is no maximum output.

EXAMPLE 14 The contours of a function plotted over the domain $[0,3] \times [0,3]$ for values

$$c = \ldots, \ -1.0, \ -0.5, \ 0, \ 0.5, \ 1.0, \ 1.5, \ \ldots$$

are shown in Figure 13.24. Use them to determine the approximate locations of the extrema (maxima and minima) and the location of the steepest region of the graph over this region.

Solution From this picture we see that the extrema must be located near $(0.5, 1)$ and $(2, 2)$. The steep regions surround these extrema, and the flattest regions are off in the opposite corners. (To identify each extremum as either a relative maximum or minimum, we need to know the actual value c associated to each level curve shown.) ∎

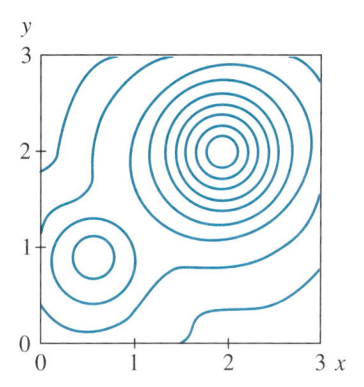

13.2 VISUALIZING AND INTERPRETING MULTIVARIABLE FUNCTIONS

Figure 13.24 Contour plot of the function f.

The cross-sections of a function graph $z = f(x,y)$ are usually easy to graph with most graphing software or graphing calculators. Just substitute the appropriate frozen value $x = c$ or $y = c$ and graph the one-input function (of the other variable) that results.

Level curves can be more difficult to plot, because the graph of

$$c = f(x,y)$$

is a more general equation in two variables. In the exercises, we discuss a way you can use slope fields to obtain contour plots.

Level surfaces

For a function of three variables $w = f(x,y,z)$, we have **level surfaces** determined by setting the output w equal to a constant. The level surface $c = f(x,y,z)$ gives us the set of inputs (x,y,z) in space that all produce the same output $w = c$.

EXAMPLE 15 A heat source is placed at the origin. After some time elapses, the temperature at point (x,y,z) is given by the function

$$T(x,y,z) = \frac{300}{1+x^2+y^2+z^2},$$

where the temperature is 300 degrees Kelvin at the heat source. Describe the level surfaces of the function T for temperatures $c \leq 300$.

Solution Setting $T(x,y,z) = c$, we have

$$\frac{300}{1+x^2+y^2+z^2} = c,$$

or equivalently,

$$\frac{300}{c} - 1 = x^2 + y^2 + z^2.$$

The level surfaces are concentric spheres centered at the origin. The set of points having temperature $T = c$ lie on the sphere of radius $\sqrt{\frac{300}{c} - 1}$ for $c < 300$ (and just the origin $(0,0,0)$ for $T = 300$). ■

Perspective

Imagine the drawing of a graph of a single-input function standing on edge, and observing it from various locations in the room. As you move away from it, the image you see gets smaller. As you move your head up, the image seems to move down; as you move left, the image moves right. These simple observations are the basis for graphing a two-input function in *perspective*. The technique of drawing in perspective was discovered and explored in the Renaissance, and it quickly became an important element of graphical artistry.

If we take the set of graphs produced by slicing the graph of a function of two variables $z = f(x,y)$ and shrink them in proportion to their distance from an imaginary "eye point," offsetting them left-to-right and up and down in the same manner, and then combine the resulting images, we will have produced a *perspective view* of the graph of the function.

For example, the right-hand side of Figure 13.25 illustrates a collection of slices: near to far, from bottom to top. On the left, these slices have first been stacked and offset vertically to show their order, and then shrunk in proportion to their distance from the eye point.

Figure 13.25 Putting slices in perspective.

13.2 VISUALIZING AND INTERPRETING MULTIVARIABLE FUNCTIONS

While this is a simple combination of the pictures produced by slicing, its three-dimensional effect can be of benefit in grasping the shape of a surface.

Wireframe plotting

A variant of the perspective graphing technique is quite simple but often adds a considerable amount of realism to the picture. First we make a rectangular grid in the domain of the function. Then, we plot the graph points corresponding to the grid intersections. Graph points plotted from adjacent grid points are connected in space with straight line segments.

Instead of plotting the smooth curves of the several slices corresponding to the grid lines, we are essentially plotting their *piecewise linear* approximations. The result resembles a piece of wire screening molded into the shape of the surface defined by the function's graph. For this reason, it is often called a *wireframe plot* of the function. If the grid is not too fine, a machine can generate a wireframe plot fairly quickly, since there are relatively few "true" graph points to be computed. However, even a coarse grid can often give us a reasonably good feel for the shape of the graph.

Figure 13.26 shows the grid used to plot part of the paraboloid

$$z = x^2 + y^2.$$

Less than 300 graph points were actually computed. (To compare, most graphing calculators plot well over 100 points to graph a function of one variable $y = f(x)$.) The straight line segments connecting these points are short enough that they give us a feeling for the shape of the paraboloid.

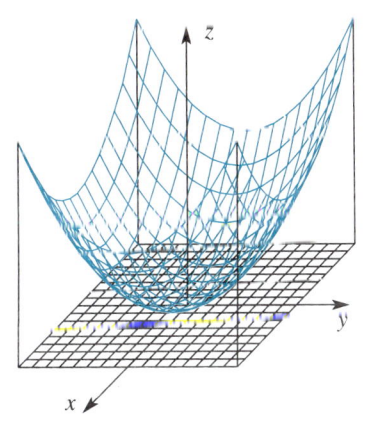

Figure 13.26 Points on a grid determine this wireframe plot of $z = x^2 + y^2$.

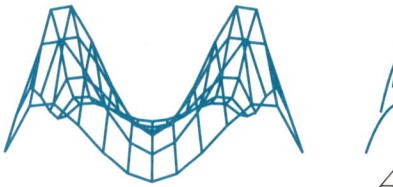

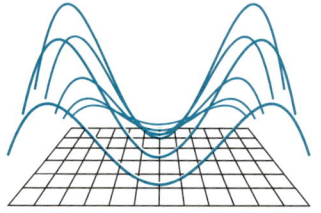

Figure 13.27 $f(x,y) = \dfrac{1 - x(x-1)(x-2)(x-3)}{(y-1)^2 + 1}$ plotted over $[0,3] \times [0,3]$.

Figure 13.27 shows a wireframe plot of the function

$$f(x,y) = \frac{1 - x(x-1)(x-2)(x-3)}{(y-1)^2 + 1}$$

as well as its slices plotted in perspective over the input region $[0,3] \times [0,3]$.

The graph of f has two peaks: one near $(0.5, 1)$ and one near $(2.5, 1)$. Between these two, at approximately $(1.5, 1)$, is an input at which the graph of the function resembles a "mountain pass" —locally flat, downhill in two directions, uphill in two other directions. Such a point is called a **saddle point**, so named because of the saddle-shape of the surface.

EXERCISES for Section 13.2

For exercises 1-5: Graphically determine the approximate locations of all extrema of the given function in the input region $[0,3] \times [0,3]$.

1. $f(x,y) = (x-1)^2 + (x-0.3)(y-0.7) + (y-1)^2$.

2. $f(x,y) = \dfrac{x^3 + x^2 y - y^2}{(x-1.5)^2 + (y-2.3)^2}$.

3. $f(x,y) = x^{(y+1)^2}$.

4. $f(x,y) = x + \ln(x^2 + y^2 + 1)$.

5. $f(x,y) = 3.5x - 7.2y$.

For exercises 6-13: Describe in your own words the shape of the graph of the given function near the input $(0,0)$. (Words you may consider using are: "bowl up," "bowl down," "saddle," "trough," or "flat," etc.)

6. $f(x,y) = x^2 + y^2$.

7. $f(x,y) = x^3 + y^3$.

8. $f(x,y) = x^3 - y^2$.

9. $f(x,y) = x^4 - y^4$.

10. $f(x,y) = xy$.

11. $f(x,y) = x^2 y$.

12. $f(x,y) = xy^3$.

13. $f(x,y) = x^2 y^2$.

13.2 VISUALIZING AND INTERPRETING MULTIVARIABLE FUNCTIONS

For exercises 14-21: An ant is standing on a hill whose height is given by

$$H(x,y) = x^4 + y^4 - 4x^2y^2.$$

Take "north" to be the direction of the positive y-axis and "east" to be the direction of the positive x-axis, and use graphical methods to answer these exercises.

14. If the ant is standing at the origin and begins walking north, will it soon be going uphill, downhill, or neither?

15. If the ant is standing at the origin and begins walking south, will it soon be going uphill, downhill, or neither?

16. If the ant is standing at the origin and begins walking east, will it soon be going uphill, downhill, or neither?

17. If the ant is standing at the origin and begins walking west, will it soon be going uphill, downhill, or neither?

18. If the ant is standing at the origin and begins walking northeast, will it soon be going uphill, downhill, or neither?

19. If the ant is standing at the origin and begins walking northwest, will it soon be going uphill, downhill, or neither?

20. If the ant is standing at the origin and begins walking southeast, will it soon be going uphill, downhill, or neither?

21. If the ant is standing at the origin and begins walking southwest, will it soon be going uphill, downhill, or neither?

For exercises 22-24: Suppose now that the height of the hill is given by

$$H(x,y) = x/y.$$

22. If the ant is standing at $x = 1$, $y = 1$, in which directions can it walk staying at the same height?

23. If the ant is standing at $x = 0$, $y = 1$, in which directions can it walk staying at the same height?

24. If the ant is standing at $x = -1$, $y = 1$, in which directions can it walk staying at the same height?

25. Graph the function $f(x,y) = \dfrac{x^2 \sin(xy)}{x^2 + y^2 + 1}$ on the region $[0,3] \times [0,3]$ by *freezing* the second input at $0, 0.3, 0.6, \ldots, 3.0$ and graphing the resulting slices. Then use these to approximate the inputs for f where $f(x,y)$ achieves its maximum and minimum values on this region.

26. Use the graphical slicing technique to determine whether the input $(1, -3, 2)$ to the function $f(x, y, z) = x^2yz^2 - 3xyz + y^2$ is a maximum, minimum, or neither.

There is a way to use slope fields to get an idea of the shape of the contours of a function of two variables. We illustrate the technique with the following example:

$$f(x, y) = x^2 + \sin(\pi y)x + y.$$

First, note that the level curves have an equation of the form

$$f(x, y) = c.$$

For our example, the level curves have equations of the form

$$x^2 + \sin(\pi y)x + y = c.$$

Recall that we can use implicit differentiation to find the slope of such a curve. For our example, we treat y implicitly as a function of x, and differentiate with respect to x:

$$2x + \pi y' \cos(\pi y)x + \sin(\pi y) + y' = 0.$$

Now we solve for y' in terms of both x and y:

$$y' = -\frac{2x + \sin(\pi y)}{1 + \pi \cos(\pi y)}.$$

We plot the slope field generated by this slope function of two variables in the input region $[0, 3] \times [0, 3]$ and obtain the window illustrated below:

Using this we can rough sketch the shape of some of the level curves:

Note that this method of plotting level curves gives us no information about what level $z = c$ is represented by each contour. For this, we need to sample values to identify the level curves; otherwise, there is no reason to believe that the curves we sketch correspond to *equally spaced* values of c, so be careful in using the closeness of the contours to judge how steep the surface is at any given input. Try the method in exercises 6-13 above.

13.3 LINEAR AND QUADRATIC MULTIVARIABLE FUNCTIONS

In this section, we take time to discuss two simple but very important types of multivariable functions in more detail: linear functions and quadratic functions.

Linear functions

A linear function of one variable has a straight line for a graph and a formula of the form

$$y = f(x) = mx + b,$$

where m is the slope of the line and b is the y-intercept.

A two-input function f is called a **linear function of two variables**, provided that its formula can be put in the form

$$f(x, y) = m_1 x + m_2 y + c.$$

If f has such a formula, then its graph is a *plane* with z-intercept $c = f(0,0)$ (see Figure 13.28).

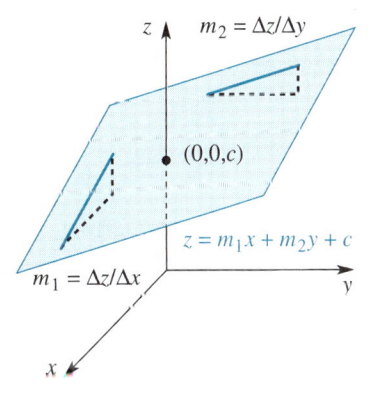

Figure 13.28 The graph of a linear two-input function is a plane.

Planes are important surfaces to study because, upon close inspection, many multivariable function graphs also look planar (just as close-up, many graphs of a single variable look straight). Indeed, for centuries the world was thought to be flat because people could visually grasp only relatively tiny regions of it.

We can think of m_1 and m_2 as slopes in much the same way we think of m representing the slope of the line $y = mx + b$. If we pick any two input pairs (x_1, y_0) and (x_2, y_0) having the same y-coordinate y_0 and calculate the corresponding outputs $z_1 = f(x_1, y_0)$, and $z_2 = f(x_2, y_0)$, then

$$m_1 = \frac{f(x_2, y_0) - f(x_1, y_0)}{x_2 - x_1} = \frac{z_2 - z_1}{x_2 - x_1}.$$

In other words, m_1 is the "rise over run" of the graph plane where we measure the rise vertically in the z-direction and the run in the x-direction. We write

$$m_1 = \frac{\Delta f}{\Delta x} = \frac{\Delta z}{\Delta x}$$

to denote that m_1 is the ratio of the change in function output $z = f(x, y)$ to the change in input x, and we call m_1 the **slope with respect to** x.

Similarly, if we pick any two input pairs (x_0, y_1) and (x_0, y_2) having the same x-coordinate x_0 and calculate the corresponding outputs $z_1 = f(x_0, y_1)$, and $z_2 = f(x_0, y_2)$, then we have

$$m_2 = \frac{f(x_0, y_2) - f(x_0, y_1)}{y_2 - y_1} = \frac{z_2 - z_1}{y_2 - y_1}.$$

In other words, m_2 is the "rise over run" of the graph plane where we measure the rise vertically in the z-direction and the run in the y-direction. We write

$$m_2 = \frac{\Delta f}{\Delta y} = \frac{\Delta z}{\Delta y}$$

to denote that m_2 is the ratio of the change in function output $z = f(x, y)$ to the change in input y, and we call m_2 the **slope with respect to** y.

The cross-sectional slices of the graph of a linear function will be straight lines. Figure 13.29 shows several slices of a linear function taken at equally spaced intervals along the y-direction, each representing a fixed value of y. Note that each slice is a straight line, all with the same slope with respect to x, and each is translated up or down by the same amount from the previous slice. Similarly, any slice for a fixed value of x will have the same slope with respect to y.

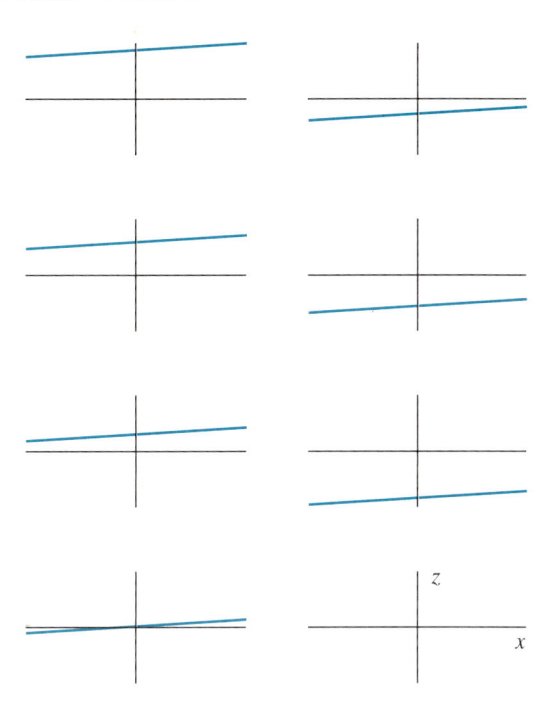

13.3 LINEAR AND QUADRATIC MULTIVARIABLE FUNCTIONS

Figure 13.29 Slices of a linear function $z = f(x, y)$ with y fixed.

EXAMPLE 16 Consider the graph of the linear function $f(x, y) = 3x + 7y + 1$. Find the slope with respect to x, the slope with respect to y, and the z-intercept.

Solution   We can see that the slope with respect to x is

$$m_1 = 3,$$

the slope with respect to y is $m_2 = 7,$

and the z-intercept is $c = 1$.

Let's verify these computations directly. First, we choose any two distinct ordered pairs with the same y-coordinate—say $(1, 2)$ and $(4, 2)$. We can use these to calculate the slope with respect to x:

$$\frac{\Delta f}{\Delta x} = \frac{f(4,2) - f(1,2)}{4 - 1} = \frac{(12 + 14 + 1) - (3 + 14 + 1)}{3} = \frac{9}{3} = 3$$

To calculate the slope with respect to y, we choose any two distinct ordered pairs with the same x-coordinate—say $(-2, 3)$ and $(-2, -5)$:

$$\frac{\Delta f}{\Delta y} = \frac{f(-2,-5) - f(-2,3)}{-5 - 3} = \frac{(-6 - 35 + 1) - (-6 + 21 + 1)}{-8} = \frac{-56}{-8} = 7.$$

Finally, the z-intercept is $f(0,0) = 3(0) + 7(0) + 1 = 1$. ■

EXAMPLE 17 Find the slope with respect to x, the slope with respect to y, and the z-intercept of the plane with equation

$$x + 2y + 3z = 6.$$

Solution The equation of this plane is not written in a form expressing z as a function of x and y. If we solve for z, we obtain

$$z = \frac{-x - 2y + 6}{3},$$

from which we can see that the slope with respect to x is $-1/3$, the slope with respect to y is $-2/3$, and the z-intercept is $6/3 = 2$. ∎

EXAMPLE 18 Find the slope with respect to x, the slope with respect to y, and the z-intercept of the plane with equation $z = c$, where c is a constant.

Solution This is a *horizontal* plane. The slopes with respect to x and with respect to y of a horizontal plane are both 0, and the z-intercept is c. ∎

In contrast, a *vertical* plane has an equation involving only x and y, and therefore cannot represent the graph of a function $z = f(x, y)$. Examples of equations of vertical planes include

$$x = 2, \quad y = -3, \quad \text{and} \quad 2x - 3y = 5.$$

The slopes with respect to x and y of a vertical plane are *undefined*. The notion of z-intercept does not make sense either, for a vertical plane will either contain the entire z-axis, or not intersect it at all.

Taylor form of a linear function

The point-slope form of the equation for a line having slope m and passing through the point (x_0, y_0) is

$$y - y_0 = m(x - x_0).$$

Writing in function notation (with $x_0 = a$, $y = f(x)$ and $y_0 = f(a)$) and rearranging the terms gives us the **Taylor form** for the linear function f about $x = a$:

$$f(x) = f(a) + m(x - a).$$

13.3 LINEAR AND QUADRATIC MULTIVARIABLE FUNCTIONS

Similarly, there is a Taylor form for a linear function in two variables about a point (a, b):

$$f(x, y) = f(a, b) + m_1(x - a) + m_2(y - b),$$

where (a, b) is a specific input pair, m_1 is the x-slope, and m_2 is the y-slope.

EXAMPLE 19 Find the Taylor form of the linear function $f(x, y) = 3x + 7y + 1$ about the point $(4, -2)$.

Solution Since $f(4, -2) = 3(4) + 7(-2) + 1 = -1$, we have

$$f(x, y) = 3(x - 4) + 7(y + 2) - 1$$

as the Taylor form of f. (You can expand and collect terms on the left-hand side of this equation to verify that it is equivalent to the original formula.) ∎

A linear function of three variables has the form

$$w = f(x, y, z) = m_1 x + m_2 y + m_3 z + c,$$

where

$$m_1 = \frac{\Delta w}{\Delta x}, \qquad m_2 = \frac{\Delta w}{\Delta y}, \qquad m_3 = \frac{\Delta w}{\Delta z}$$

are the three slopes with respect to x, y, and z, respectively. To compute the x-slope, choose two distinct ordered triples (x_1, y_0, z_0) and (x_2, y_0, z_0) with the same y and z coordinates and compute the difference quotient

$$m_1 = \frac{\Delta w}{\Delta x} = \frac{f(x_2, y_0, z_0) - f(x_1, y_0, z_0)}{x_2 - x_1}.$$

This value m_1 is the slope of every straight line slice obtained by freezing the values of y and z. The other two slopes are computed in a similar manner: m_2 using two points with the same x and z coordinates; m_3 using two points with the same x and y coordinates.

In general, a linear function of n variables has the form

$$f(x_1, x_2, \ldots, x_n) = m_1 x_1 + m_2 x_2 + \cdots + m_n x_n + c,$$

where each m_i ($1 \leq i \leq n$) is the slope with respect to x_i and represents the "rise/run" of the slice obtained by freezing the values of all the other variables and plotting the resulting line.

Quadratic functions

A **quadratic function** has a polynomial form involving only terms of degree 2 or less. (Linear functions involve only terms of degree 1 or less.) A quadratic function of two variables has the form

$$f(x,y) = c_{11}x^2 + c_{12}xy + c_{22}y^2 + c_1 x + c_2 y + c_0.$$

Note that we use the double-subscripted coefficients c_{11}, c_{12}, and c_{22} for the second-degree terms x^2, xy, and y^2, respectively, and the single subscripted coefficients c_1, c_2, and c_0 for the linear and constant terms.

The graph of a quadratic function will be a quadric surface, of which several examples were discussed in the first section of the chapter (see Figures 13.8 through 13.15). Here we want to take a close look at how the coefficients of the quadratic affect the shape of the graph of

$$z = f(x,y).$$

Let's look at some examples.

EXAMPLE 20 Examine the vertical cross-sections at the input $(0,0)$ for each of the following quadratic functions:

$$f(x,y) = 2x^2 + y^2,$$
$$g(x,y) = -x^2 - 2y^2,$$
$$h(x,y) = y^2 - 2x^2,$$
$$j(x,y) = 2xy - x^2 - y^2.$$

Determine which has a strict maximum at $(0,0)$, which has a strict minimum, and which must have neither. Describe the graph of each quadratic function.

Solution The slices determined by setting $y = 0$ and $x = 0$ have equations:

$$f(x,0) = 2x^2 \quad f(0,y) = y^2$$
$$g(x,0) = -x^2 \quad g(0,y) = -2y^2$$
$$h(x,0) = -2x^2 \quad h(0,y) = y^2$$
$$j(x,0) = -x^2 \quad j(0,y) = -y^2$$

13.3 LINEAR AND QUADRATIC MULTIVARIABLE FUNCTIONS

We can see in Figure 13.30 that f *could* have a minimum at $(0,0)$, since both $f(x,0)$ and $f(0,y)$ have a minimum there. Indeed, f *must* have a strict minimum at $(0,0)$, since $2x^2 + y^2 > 0$ for all other ordered pairs (x,y). The graph of $z = f(x,y)$ is a "bowl up" elliptic paraboloid.

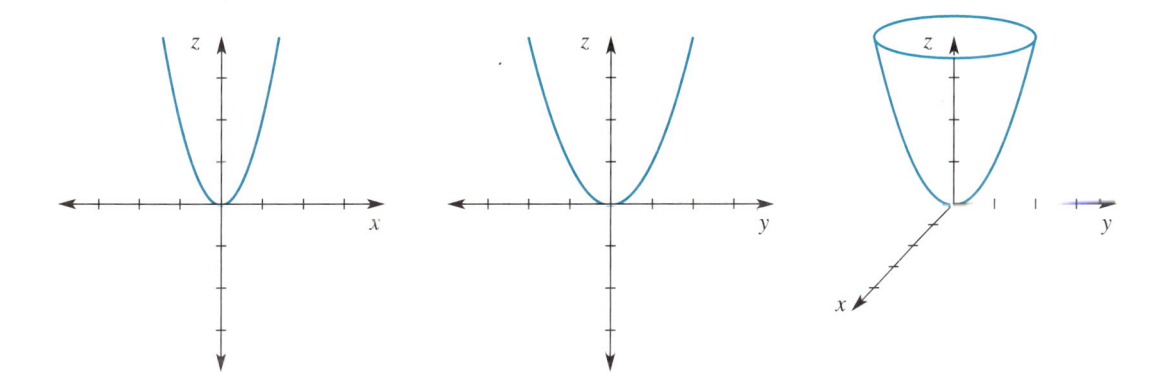

Figure 13.30 Vertical slices and graph of $z = f(x,y)$.

The slices of g suggest, however, that a maximum occurs at $(0,0)$ (see Figure 13.31). It is the case that g has a strict maximum at $(0,0)$, since $-x^2 - y^2 < 0$ except when $x = 0$ and $y = 0$. The graph of $z = g(x,y)$ is a "bowl down" elliptic paraboloid.

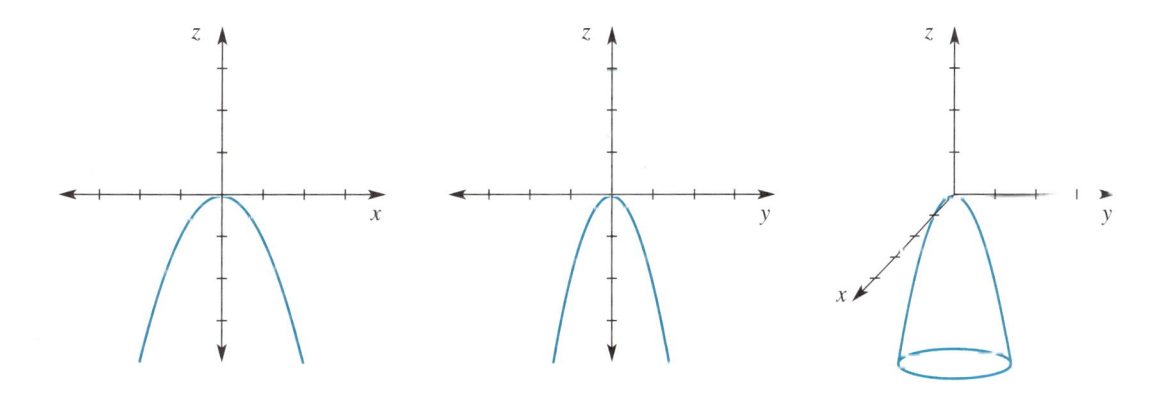

Figure 13.31 Vertical slices and graph of $z = g(x,y)$.

The function h must have neither a maximum nor a minimum at $(0,0)$, since it has a maximum along one slice but a minimum along the other! The graph of $z = h(x,y)$ is a "saddle," or hyperbolic, paraboloid, and the origin is called the saddle point (see Figure 13.32).

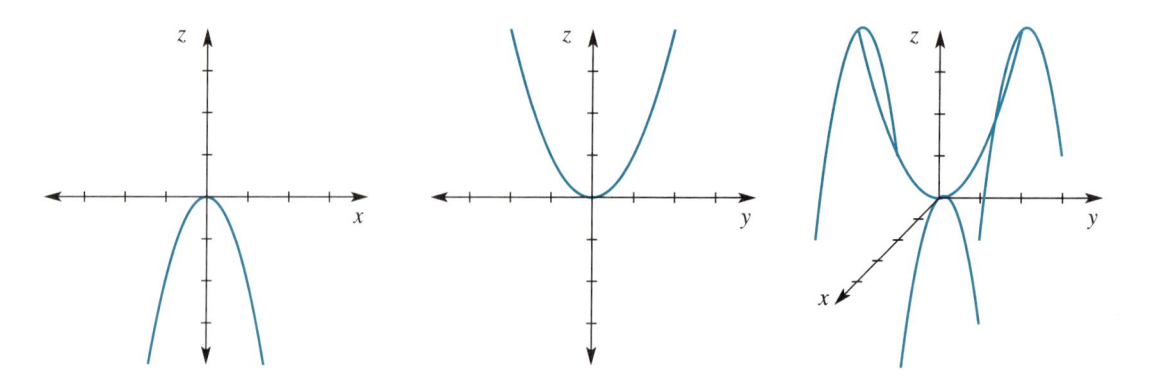

Figure 13.32 Vertical slices and graph of $z = h(x,y)$.

Finally, the slices of j also suggest that a maximum occurs at $(0,0)$, but this is not a strict maximum, since

$$j(x,y) = 2xy - x^2 - y^2 = -(x-y)^2,$$

which means that $j(x,y) = 0$ whenever $x = y$. The graph of j is an inverted "trough" whose crest runs along the line $x = y$ in the xy-plane (see Figure 13.33). ■

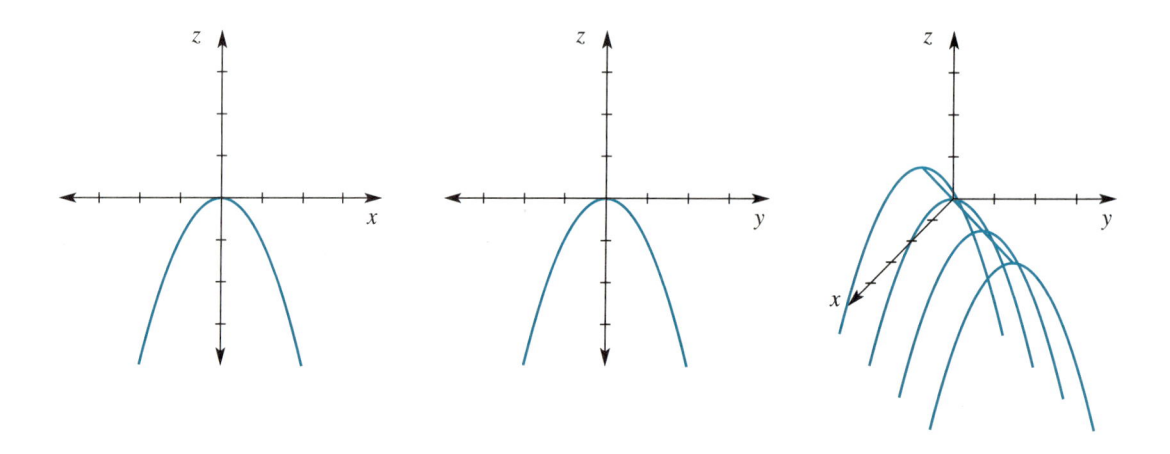

Figure 13.33 Vertical slices and graph of $z = j(x,y)$.

Slicing $z = Ax^2 + 2Bxy + Cy^2$ at the origin

The behavior we have just seen in these examples is typical of quadratic functions. To justify this statement, we'll find it convenient to write the quadratic coefficients as follows:

$$c_{11} = A, \qquad c_{12} = 2B, \qquad c_{22} = C,$$

13.3 LINEAR AND QUADRATIC MULTIVARIABLE FUNCTIONS

so that our quadratic has the form

$$z = Ax^2 + 2Bxy + Cy^2,$$

whose graph contains the origin $(0, 0, 0)$. Here's the main observation:

▶▶▶ **A vertical plane containing the origin will intersect the surface**

$$z = Ax^2 + 2Bxy + Cy^2$$

in either a parabola with its vertex at the origin or in a horizontal line lying in the xy-plane.

For instance, the intersection of the xz-plane (the plane $y = 0$) and the surface $z = Ax^2 + 2Bxy + Cy^2$ is

$$z = Ax^2.$$

If $A \neq 0$, this is a parabola with its vertex at the origin. If $A = 0$, the intersection is simply the x-axis.

However, the intersection of the yz-plane (the plane $x = 0$) and the surface $z = Ax^2 + 2Bxy + Cy^2$ is

$$z = Cy^2,$$

which is either a parabola with its vertex at the origin ($C \neq 0$) or simply the y-axis ($C = 0$). Any other vertical plane containing the origin has an equation of the form $y = mx$ ($m \neq 0$). Substituting mx for y, we see that the intersection of this plane with the surface has equation

$$z = Ax^2 + 2Bxy + Cy^2 = Ax^2 + 2Bx(mx) + C(mx)^2 = (A + 2Bm + Cm^2)x^2.$$

So, again, we can see that this is either a parabola with its vertex at the origin (if $A + 2Bm + Cm^2 \neq 0$) or the line $y = mx$ lying in the xy-plane (if $A + 2Bm + Cm^2 = 0$). With these observations in mind, we can list the possibilities:

1. If *all* the cross-sections at the origin are parabolas opening up, then the shape of the surface will be "bowl up" (an elliptic paraboloid with minimum at the origin).

2. If all the cross-sections at the origin are parabolas opening down, the shape of the surface will be "bowl down" (an elliptic paraboloid with maximum at the origin).

3. If there is a mix of parabolas, some opening up and some opening down, then the shape is a "saddle" (hyperbolic paraboloid with saddle point at the origin).

4. It is also possible that the graph could be a parabolic "trough" with its bottom (or crest if inverted) running along a line in the xy-plane.

5. The extreme case is the horizontal plane $z = 0$ (if $A = B = C = 0$).

The discriminant $D = AC - B^2$

Fortunately, there is often an easy way to detect the shape of the graph of $z = Ax^2 + 2Bxy + Cy^2$ by examining the coefficients. Using some algebra, we can rewrite the form of the quadratic function $z = Ax^2 + 2Bxy + Cy^2$ in a way that gives us useful information about the shape of the graph. Assuming $A \neq 0$, we can rewrite the equation in the form

$$z = \frac{1}{A}[(Ax + By)^2 + (AC - B^2)y^2].$$

The quantity
$$D = AC - B^2$$

is called the **discriminant** of this quadratic function, and substituting D into the equation gives us

$$z = \frac{1}{A}[(Ax + By)^2 + Dy^2].$$

The *sign* of the discriminant D tells us much about the shape of the graph.

First, note that if $D < 0$, then the quantity in the square brackets can be either negative (substitute $x = -B/A$ and $y = 1$) or positive (substitute $x = 1$ and $y = 0$). Hence, the vertical cross-sections of the graph at the origin will include a mix of open-up and open-down parabolas with the origin at the vertex. Therefore, the graph must be saddle-shaped at the origin.

Now, if $D > 0$, then the quantity in the square brackets is positive at every point (x, y) except the origin. Hence, the factor $1/A$ determines whether all the cross-sections are open-up ($A > 0$) or open-down ($A < 0$) parabolas with their vertex at the origin. In either case, the graph must be bowl-shaped, with a strict minimum or strict maximum at the origin.

If $D = 0$, then the quantity in the square brackets is simply $(Ax + By)^2$, which is 0 all along the line $Ax + By = 0$, and positive off this line. The graph is *trough-shaped* (opening up or down depending on the sign of A).

All this analysis was done assuming $A \neq 0$. However, if $A = 0$, then the original function simplifies to

$$z = 2Bxy + Cy^2$$

and the discriminant simplifies to $D = -B^2$. If $B \neq 0$, then $D < 0$ and the graph is again saddle-shaped. To see this, note that the cross-section with the plane $x = 0$ is $z = Cy^2$, while the cross-section with the plane

13.3 LINEAR AND QUADRATIC MULTIVARIABLE FUNCTIONS

$x = -Cy/B$ is $z = -Cy^2$. If $C \neq 0$, then these two slices are parabolas opening in opposite directions. If $C = 0$, then the graph of $z = 2Bxy$ is still saddle-shaped. Finally, if both $A = 0$ and $B = 0$, then $D = 0$ and the graph of $z = Cy^2$ is either trough-shaped ($C \neq 0$) or a horizontal plane ($C = 0$).

We summarize these observations in a theorem.

Theorem 13.1

Hypothesis: Suppose that $z = Ax^2 + 2Bxy + Cy^2$. Let $D = AC - B^2$.

Case 1: $D < 0$.
Conclusion: A saddle point occurs at $(0,0)$.

Case 2: $D > 0$ and $A > 0$.
Conclusion: A strict minimum occurs at $(0,0)$.

Case 3: $D > 0$ and $A < 0$.
Conclusion: A strict maximum occurs at $(0,0)$.

Case 4: $D = 0$.
Conclusion: No strict maximum, minimum, or saddle point occurs at $(0,0)$.

EXAMPLE 21 Use the theorem to analyze the five quadratic functions:

$$f(x,y) = 2x^2 + y^2,$$
$$g(x,y) = -x^2 - 2y^2,$$
$$h(x,y) = y^2 - 2x^2,$$
$$j(x,y) = 2xy - x^2 - y^2,$$
$$k(x,y) = 4xy.$$

Solution All of these functions satisfy the hypothesis of the theorem, so for each we evaluate the discriminant D and make the conclusion indicated:

- For $f(x,y) = 2x^2 + y^2$, we have $A = 2$, $B = 0$, and $C = 1$, so the discriminant is $AC - B^2 = 2 > 0$. Since $A > 0$, the function has a *strict minimum* at $(0,0)$.

- For $g(x,y) = -x^2 - 2y^2$, we have $A = -1$, $B = 0$, and $C = -2$, so the discriminant is $D = AC - B^2 = 2 > 0$. Since $A < 0$, the function has a *strict maximum* at $(0,0)$.

- For $h(x,y) = y^2 - 2x^2$, we have $A = -2$, $B = 0$, and $C = 1$, so the discriminant is $D = AC - B^2 = -2 < 0$. The function has a *saddle point* at $(0,0)$.

- For $j(x,y) = 2xy - x^2 - y^2$, we have $A = -1$, $B = 1$, and $C = -1$, so the discriminant is $D = AC - B^2 = 0$. Recall that this function has no strict minimum, maximum, or saddle point at $(0,0)$.

- For $k(x,y) = 4xy$, we have $A = 0$, $B = 2$, and $C = 0$, so the discriminant is $D = AC - B^2 = -4$. There is a *saddle point* at $(0,0)$. ∎

If we replace x by $x - x_0$, y by $y - y_0$, and z by $z - z_0$ in any equation of three variables, the geometric effect on the graph is the same as relabelling the origin (x_0, y_0, z_0). Hence, the discriminant D can also be used to judge the behavior of a quadric surface

$$z - z_0 = A(x - x_0)^2 + 2B(x - x_0)(y - y_0) + C(y - y_0)^2$$

at the point (x_0, y_0, z_0).

Taylor form for quadratic functions

If

$$f(x,y) = c_{11}x^2 + c_{12}xy + c_{22}y^2 + c_1 x + c_2 y + c_0,$$

then the Taylor form of f about the point (a,b) is

$$f(x,y) = f(a,b) + d_1(x-a) + d_2(y-b) + c_{11}(x-a)^2 + c_{12}(x-a)(y-b) + c_{22}(y-b)^2.$$

Note that the quadratic coefficients are the same, but the linear and constant coefficients may change. In particular, the constant term is $f(a,b)$.

EXAMPLE 22 Write the quadratic function

$$f(x,y) = x^2 - 2xy - y^2 + 2x + 4y - 5$$

in Taylor form about the point $(-1, 2)$.

Solution Since $f(-1,2) = 1 + 4 - 4 - 2 + 8 - 5 = 2$, we can write

$$f(x,y) = 2 + d_1(x+1) + d_2(y-2) + (x+1)^2 - 2(x+1)(y-2) - (y-2)^2,$$

and we need only determine d_1 and d_2. Perhaps the easiest way to accomplish this is to substitute values for (x,y) into both the original form and the Taylor form and set them equal to each other.

13.3 LINEAR AND QUADRATIC MULTIVARIABLE FUNCTIONS

For instance, if we substitute $(0, 2)$ into the original form, we have

$$f(0, 2) = 0 - 0 - 4 + 0 + 8 - 5 = -1.$$

If we substitute the same point into the Taylor form we have

$$f(0, 2) = 2 + d_1 + 0 + 1 - 0 - 0 = 3 + d_1.$$

Equating these, we conclude that $d_1 = -4$.

We chose the point $(0, 2)$ for no other reason than to simplify the computation. (Each term in the Taylor form having $(y - 2)$ as a factor has value 0 at this point.) Now, if we substitute $(-1, 0)$ into the original form, we have

$$f(-1, 0) = 1 - 0 - 0 - 2 + 0 - 5 = -6,$$

and in the Taylor form:

$$f(-1, 0) = 2 + 0 - 2d_2 + 0 + 0 - 4 = -2d_2 - 2.$$

Equating these, we conclude that $d_2 = 2$. Now we can write the complete Taylor form of f about the point $(-1, 2)$:

$$f(x, y) = 2 - 4(x + 1) + 2(y - 2) + (x + 1)^2 - 2(x + 1)(y - 2) - (y - 2)^2.$$

As one more check, we substitute $(0, 0)$ into the Taylor form and find that $f(0, 0) = 2 - 4 - 4 + 1 + 4 - 4 = -5$, which is the correct value of the function at this point. ■

The Taylor form is useful when we are interested in studying a function's behavior around a specific point (a, b). In particular, if the Taylor form of a quadratic function has no linear terms, we can use the discriminant to judge the shape of the graph at that point.

EXAMPLE 23 The Taylor form of a quadratic function f about the point $(3, -4)$ is found to have the form

$$z = f(x, y) = -7 + 5(x - 3)^2 - 2(x - 3)(y + 4) + 6(y + 4)^2.$$

Characterize the behavior of the graph at $(3, -4, -7)$.

Solution There are no linear terms. Note that the equation can be written

$$z + 7 = 5(x - 3)^2 - 2(x - 3)(y + 4) + 6(y + 4)^2,$$

so we can make use of the discriminant. In this form, we have $A = 5$, $B = -1$, and $C = 6$, so the discriminant is

$$D = AC - B^2 = 30 - (-1)^2 = 29 > 0.$$

We also have $A > 0$, so the function f must have a *strict minimum* at $(x, y) = (3, -4)$. The graph of $z = f(x, y)$ is a bowl-up elliptic paraboloid with a strict minimum at $(3, -4, -7)$. ∎

EXERCISES for Section 13.3

For exercises 1-5: Suppose that the graph of a two-input linear function $z = f(x, y)$ is a plane containing the three points $P = (1, 1, 1)$, $Q = (-1, 1, -3)$, and $R = (1, 3, 2)$.

1. Use the points P and Q to compute m_1, the x-slope of the plane.

2. Use the points P and R to compute m_2, the y-slope of the plane.

3. Express the equation of the plane in the form $z = m_1 x + m_2 y + c$.

4. Find the points at which the plane intersects each coordinate axis.

5. Find the Taylor form of f at $(-3, 2)$.

For exercises 6-10: For each of the planes whose equations are given, find the slope with respect to x, the slope with respect to y, and the z-intercept, if they exist. If the plane is the graph of a function, find its Taylor form at $(-1, 4)$.

6. $z = 1 - 5y + 2x$.

7. $z = y$.

8. $z = \sqrt{17}$.

9. $x = y$.

10. $2x - 5y + 8z - 9 = 0$.

For exercises 11-15: For each of the quadratic functions given, describe the behavior of the graph at the origin.

11. $f(x, y) = 3x^2 - y^2$.

12. $g(x, y) = -x^2 + 6xy - 2y^2$.

13. $h(x, y) = x^2 - 4xy + 4y^2$.

14. $j(x, y) = 3xy - 5x^2 - y^2$.

15. $k(x, y) = 4xy + (x - y)^2 - (x + y)^2$.

For exercises 16-20: For each of the quadratic functions given, find its Taylor form about the point $(-1, 3)$.

16. $z = x^2 + xy + y^2 + x + y + 1$.

13.3 LINEAR AND QUADRATIC MULTIVARIABLE FUNCTIONS

17. $z = x^2 - y^2 + 5$.

18. $z = xy$.

19. $z = x^2 + 2y^2 + 2x - 12y + 7$.

20. $z = -8$.

---

21. A quadratic function is found to have the Taylor form
$$z = 17 - (x+1)^2 + 8(x+1)(y-3) - 4(y-3)^2$$
about the point $(-1, 3)$. Characterize the behavior of the graph at the point $(-1, 3, 17)$.

22. What is the Taylor form of
$$z = 17 - (x+1)^2 + 8(x+1)(y-3) - 4(y-3)^2$$
at $(0,0)$?

23. We know that three noncollinear points determine a unique plane in space. What is the least number of points on the graph of the quadratic function
$$z = c_{11}x^2 + c_{12}xy + c_{22}y^2 + c_1 x + c_2 y + c_0$$
that uniquely determine its equation?

For exercises 24-30: Find an example of a function whose graph fits the description.

24. The graph is a bowl-up elliptic paraboloid with a strict minimum at the point $(1, -2, 3)$.

25. The graph is a bowl-down elliptic paraboloid with a strict maximum at the point $(0, 1, -5)$.

26. The graph is a hyperbolic paraboloid with a saddle point at $(-1, 3, -5)$.

27. The graph is an open-up parabolic trough whose bottom is the intersection of the vertical plane $2x - y + 3 = 0$ and the horizontal plane $z = 5$.

28. The graph is an inverted parabolic trough whose crest is the y-axis.

29. The graph is a plane that intersects the y-axis and the z-axis, but not the x-axis

30. Every vertical cross-section made parallel to the xz-plane or yz-plane is a straight line, but the graph is not a plane.

14

Differential Calculus of Multivariable Functions

We have seen how the derivative can be a powerful tool for analyzing functions of a single variable. The derivative

$$f'(x) = \frac{dy}{dx}$$

has both geometrical and physical interpretations: (1) geometrically, the derivative tells us the local slope of the graph of $y = f(x)$; and (2) physically, it tells us the instantaneous rate of change of output $f(x)$ per unit change in input x.

The derivative allows us to detect the relative extrema (maxima and minima) of a function and determines the *best linear approximation* (first-degree Taylor polynomial) at any input $x = a$. The graph of this best linear approximation is the tangent line to the graph. Higher order derivatives allow us to measure concavity and detect inflection points and to determine the best quadratic (second-degree Taylor polynomial) as well as higher degree approximations of the function.

In Chapter 12, we saw how to extend the idea of derivative to study *vector valued functions* $\mathbf{r}(t)$ of a single variable t. In this chapter, we extend the idea of derivative to scalar-valued functions of several variables (scalar fields).

A function of several variables has several derivatives, one for each input. These *partial derivatives* can be used to study the various vertical slices of a multivariable function graph in much the same way that the derivative of a single-variable function is used.

Taken together, the partial derivatives provide a *total derivative*, which describes the locally linear behavior of a multivariable function. One form of total derivative we'll find particularly useful is called the *gradient* vector, which can be used to measure rate of change, detect relative extrema, and form the best linear approximation of a scalar field.

Higher order partial derivatives give us additional information regarding the shape of the graph of a multivariable function and can be used to find the best quadratic approximation of a scalar field. The best quadratic approximation, in turn, affords a powerful test for classifying the critical points of a function of two variables.

14.1 PARTIAL DERIVATIVES

The derivative of a single-variable function $y = f(x)$ is defined as

$$f'(x) = \frac{dy}{dx} = \lim_{\Delta x \to 0} \frac{\Delta y}{\Delta x} = \lim_{\Delta x \to 0} \frac{f(x + \Delta x) - f(x)}{\Delta x}.$$

The difference quotient $\dfrac{\Delta y}{\Delta x}$ measures the *average rate of change*, while its limiting value $\dfrac{dy}{dx}$ tells us the *instantaneous rate of change* of the output y per unit change in input x.

For a multivariable function, we can consider the rate of change of output with respect to *each* input variable. For example, suppose that we have a function of two variables

$$z = f(x, y).$$

If we hold the input variable y fixed, we can study the rate of change of output z with respect to change in x alone. The difference quotient

$$\frac{\Delta z}{\Delta x} = \frac{f(x + \Delta x, y) - f(x, y)}{\Delta x}$$

gives us the ratio of change in output z to change in the input x (the two points $(x + \Delta x, y)$ and (x, y) have the same y-coordinate). We call $\dfrac{\Delta z}{\Delta x}$ the *average rate of change of z with respect to x*. Similarly,

$$\frac{\Delta z}{\Delta y} = \frac{f(x, y + \Delta y) - f(x, y)}{\Delta y}$$

gives us the ratio of change in output z to change in the input y (the two points $(x, y + \Delta y)$ and (x, y) have the same x-coordinate). We call $\dfrac{\Delta z}{\Delta y}$ the *average rate of change of z with respect to y*. The limiting values of these difference quotients are called **partial derivatives.**

14.1 PARTIAL DERIVATIVES

Definition 1

The **partial derivative of** $z = f(x,y)$ **with respect to** x is
$$\frac{\partial z}{\partial x} = \lim_{\Delta x \to 0} \frac{\Delta z}{\Delta x} = \lim_{\Delta x \to 0} \frac{f(x + \Delta x, y) - f(x,y)}{\Delta x},$$
and the **partial derivative of** $z = f(x,y)$ **with respect to** y is
$$\frac{\partial z}{\partial y} = \lim_{\Delta y \to 0} \frac{\Delta z}{\Delta y} = \lim_{\Delta y \to 0} \frac{f(x, y + \Delta y) - f(x,y)}{\Delta y},$$
provided that these limits exist.

Note that the partial derivative symbol ∂ is different than the Leibniz d used for single-variable functions. Other notations for $\frac{\partial z}{\partial x}$ include

$$\frac{\partial f}{\partial x} \qquad f_x \qquad z_x \qquad D_x f \qquad D_x z.$$

Similarly, other notations for $\frac{\partial z}{\partial y}$ include

$$\frac{\partial f}{\partial y} \qquad f_y \qquad z_y \qquad D_y f \qquad D_y z.$$

The partial derivatives $\frac{\partial f}{\partial x}$ and $\frac{\partial f}{\partial y}$ give us the *instantaneous rate of change* of $z = f(x,y)$ with respect to x and y, respectively.

For a function of three variables $w = f(x,y,z)$, we have three **partial derivatives**:

$$\frac{\partial f}{\partial x} = \lim_{\Delta x \to 0} \frac{f(x + \Delta x, y, z) - f(x,y,z)}{\Delta x}$$
$$\frac{\partial f}{\partial y} = \lim_{\Delta y \to 0} \frac{f(x, y + \Delta y, z) - f(x,y,z)}{\Delta y}$$
$$\frac{\partial f}{\partial z} = \lim_{\Delta z \to 0} \frac{f(x, y, z + \Delta z) - f(x,y,z)}{\Delta z}.$$

In general, a function of n variables has n partial derivatives. Each is computed by taking the limit of the appropriate difference quotient, with all variables but one fixed. If $f(x_1, x_2, ... x_i, ... x_n)$ is a function of n variables, then

$$\frac{\partial f}{\partial x_i} = \lim_{\Delta x_i \to 0} \frac{f(x_1, x_2, ..., x_i + \Delta x_i, ..., x_n) - f(x_1, x_2, ..., x_i, ..., x_n)}{\Delta x_i}.$$

Calculating partial derivatives

In computing a partial derivative, all but one of the variables are held fixed when taking the limit of difference quotients. For the purposes of partial differentiation, we can treat all but one of the variables as *constants*. Doing so makes it a fairly simple matter to compute partial derivatives of a multivariable function built up from the standard functions.

▶▶▶ **For partial differentiation, consider all the variables except the one of differentiation as constants and use the usual rules of differentiation for single-variable functions (product rule, quotient rule, chain rule, and so forth).**

EXAMPLE 1 Calculate $\dfrac{\partial f}{\partial x}$ and $\dfrac{\partial f}{\partial y}$ and evaluate them at the point $(-2,3)$ if

$$f(x,y) = x^2 + 5xy - y^3.$$

Solution Treating y as a constant and differentiating with respect to x, we have

$$\frac{\partial f}{\partial x} = 2x + 5y.$$

(Note that if y is considered constant, then y^3 is also a constant.) Treating x as a constant and differentiating with respect to y, we have

$$\frac{\partial f}{\partial y} = 5x - 3y^2.$$

The values of the partial derivatives at $(-2,3)$ are:

$$\left.\frac{\partial f}{\partial x}\right|_{(-2,3)} = 2(-2)+5(3) = 11 \quad \text{and} \quad \left.\frac{\partial f}{\partial y}\right|_{(-2,3)} = 5\cdot(-2)-3(3)^2 = -37. \;\blacksquare$$

EXAMPLE 2 Calculate $\dfrac{\partial s}{\partial u}$ and $\dfrac{\partial s}{\partial v}$ and evaluate them at the point $(0,0)$ if $s = \sin(2u - 3v)$.

Solution

$$\frac{\partial s}{\partial u} = 2\cos(2u - 3v) \quad \Longrightarrow \quad \left.\frac{\partial s}{\partial u}\right|_{(0,0)} = 2\cos 0 = 2.$$

$$\frac{\partial s}{\partial v} = -3\cos(2u - 3v) \quad \Longrightarrow \quad \left.\frac{\partial s}{\partial v}\right|_{(0,0)} = -3\cos 0 = -3. \;\blacksquare$$

EXAMPLE 3 Find f_x, f_y, and f_z for the function

$$f(x,y,z) = \frac{x^2 e^{yz}}{y}.$$

Solution $\quad f_x = \dfrac{2xe^{yz}}{y}, \quad f_y = \dfrac{zx^2 e^{yz} \cdot y - x^2 e^{yz} \cdot 1}{y^2}, \quad f_z = \dfrac{yx^2 e^{yz}}{y} = x^2 e^{yz}. \;\blacksquare$

14.1 PARTIAL DERIVATIVES

▶ ▶ ▶ **Note that each partial derivative is itself a function of several variables.**

Geometrical and physical interpretations

In the last chapter, we discussed how freezing the values of all but one variable of a multivariable function f creates a *slice function* that can be graphed like any function of one variable. The partial derivative with respect to this variable gives us the local slope of this slice function. Figures 14.1 and 14.2 illustrate the idea for a function of two variables.

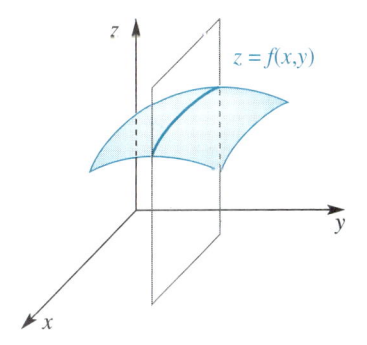

Figure 14.1 $\partial f / \partial x$ gives the local slope of a slice with y fixed.

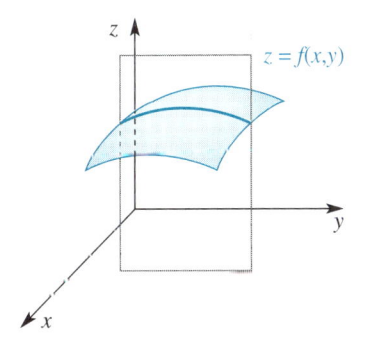

Figure 14.2 $\partial f / \partial y$ gives the local slope of a slice with x fixed.

Geometrically, you can think of a partial derivative as measuring the local slope of a slice of the surface as we move parallel to the coordinate axis represented by the variable of differentiation.

The y-slices of the function

$$w = f(x,y,z) = 2.62 + 0.6x + 0.4y + z - x^2 - y^2 - z^2$$

over the region $[0, 0.6] \times [0, 0.6] \times [0, 0.6]$ are shown in Figure 14.3 for all possible combinations of $x = 0, 0.2, 0.4, 0.6$ and $z = 0, 0.2, 0.4, 0.6$. The partial derivative

$$\frac{\partial w}{\partial y} = 0.4 - 2y$$

gives us the local slope at each point y on every one of the graphs shown.

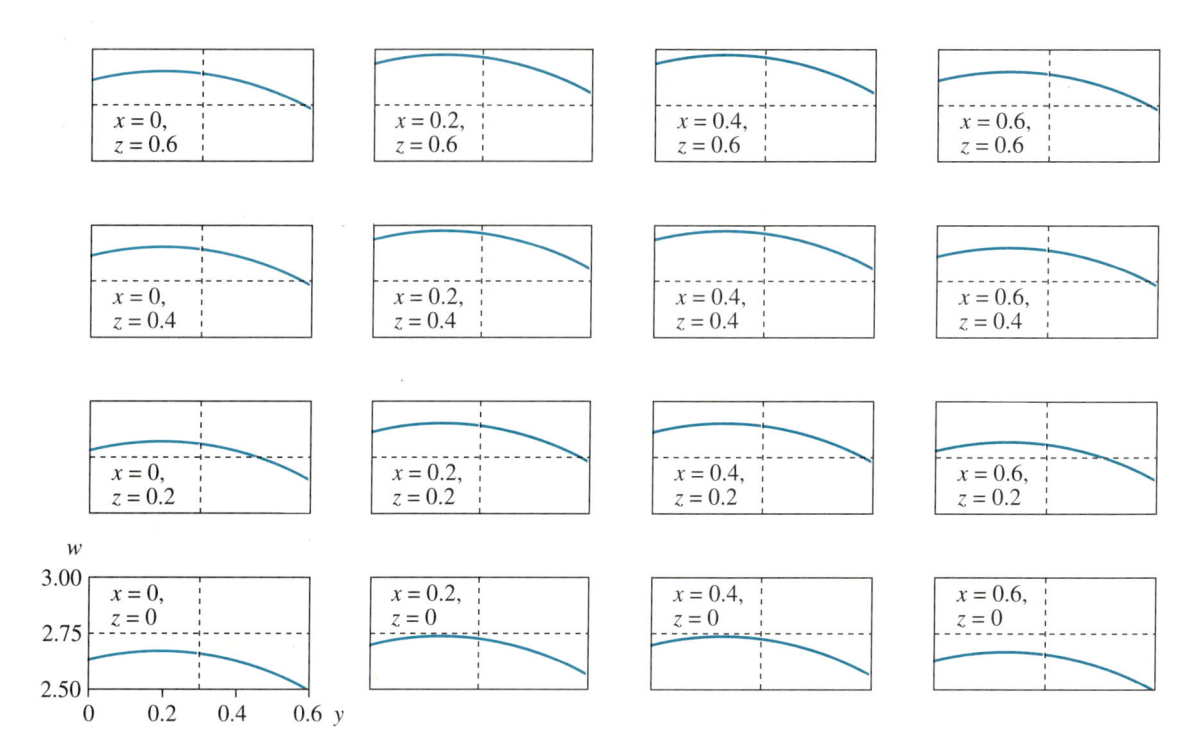

Figure 14.3 Slices of a function graph $w = f(x, y, z)$ with respect to y.

In this particular case, note that $\frac{\partial w}{\partial y}$ depends only on y, and not on the values of x and z. This means that all the slices of the graph should be "parallel," since they must all have the same local slope with respect to y. Note that the slices in Figure 14.3 illustrate this phenomenon.

Physically, a partial derivative measures the instantaneous rate of change of output per unit change when we move the input point in a direction parallel to one of the coordinate axes.

EXAMPLE 4 The temperature in °Kelvin at a point (x, y, z) is given by the function

$$T(x, y, z) = \frac{300}{1 + x^2 + y^2 + z^2}.$$

(A 300° heat source is located at the origin.) Find the instantaneous rate of temperature change per unit when we move vertically up from the point $(1, 2, 3)$.

14.1 PARTIAL DERIVATIVES

Solution A vertical move up is parallel to the z-axis, with z increasing. Therefore, we need the partial derivative with respect to z:

$$\frac{\partial T}{\partial z} = \frac{-600z}{(1+x^2+y^2+z^2)^2}.$$

Evaluating at the point $(1, 2, 3)$ we have

$$\left.\frac{\partial T}{\partial z}\right|_{(1,2,3)} = \frac{-1800}{(1+1+4+9)^2} = \frac{-1800}{225} = -8.$$

This means that we will experience a $-8°$ change in temperature per unit as we move *up* from the point $(1, 2, 3)$. If we move *down*, then we will experience a $+8°$ change per unit. This makes sense, as a move up from $(1, 2, 3)$ takes us a bit farther from the heat source, but a move down takes us a bit closer to it. ∎

EXERCISES for Section 14.1

For exercises 1-10: For each function, find

$$\frac{\partial f}{\partial x} \quad \text{and} \quad \frac{\partial f}{\partial y}$$

and evaluate each at the indicated point (a, b).

1. $f(x, y) = \dfrac{x+y}{x-y}$, $(a, b) = (1, 2)$.
2. $f(x, y) = xy + x/y$, $(a, b) = (-1, 3)$.
3. $f(x, y) = e^{x^2+y^2}$, $(a, b) = (1, -2)$.
4. $f(x, y) = 2x - 3y$, $(a, b) = (0, 0)$.
5. $f(x, y) = \sin(x^2 y)$, $(a, b) = (2, 0)$.
6. $f(x, y) = \arctan(y/x)$, $(a, b) = (-3, -3)$.
7. $f(x, y) = \tan(y/x) - \cot(y)$, $(a, b) = (2, \pi/2)$.
8. $f(x, y) = x + \ln(x^2 + y^2 + 1)$, $(a, b) = (-1, -2)$.
9. $f(x, y) = \sqrt{x^2 + y^2}$, $(a, b) = (-3, 4)$.
10. $f(x, y) = x^y$, $(a, b) = (e, 2)$.

For exercises 11-18: For each function, find $\dfrac{\partial f}{\partial x}$ and $\dfrac{\partial f}{\partial y}$ and evaluate each at the indicated point (a, b).

11. $f(x, y) = e^{x/y}$, $(a, b) = (1, 2)$
12. $f(x, y) = \cos(x + y^2)$, $(a, b) = (\frac{\pi}{2}, 0)$
13. $f(x, y) = \cos^2(xy)$, $(a, b) = (\frac{1}{2}, \frac{\pi}{2})$
14. $f(x, y) = e^{-5x} \sin y$, $(a, b) = (1, 0)$

15. $f(x,y) = \ln x + e^{-y}$, $\quad (a,b) = (1,1)$

16. $f(x,y) = y^2 e^{\cos x}$, $\quad (a,b) = 0, -2)$

17. $f(x,y) = \dfrac{x^2}{x+y}$, $\quad (a,b) = (-1,-1)$

18. $f(x,y) = \dfrac{2xy}{x-y}$, $\quad (a,b) = (1,-2)$

For exercises 19-31: For each function, find

$$f_x, \quad f_y, \quad f_z$$

and evaluate each at the indicated point (a,b,c).

19. $f(x,y,z) = \dfrac{z}{x+y}$, $\quad (a,b,c) = (1,2,3)$.

20. $f(x,y,z) = x^3 y^4 z^5$, $\quad (a,b,c) = (1,-2,-1)$.

21. $f(x,y,z) = 5x - 4y + 3z$, $\quad (a,b,c) = (-3,2,4)$.

22. $f(x,y,z) = \exp 1 - x^2 - y^2 - z^2$, $\quad (a,b,c) = (0,1,-1)$.

23. $f(x,y,z) = x \arcsin z - y \arctan z$, $\quad (a,b,c) = (.5,.5,.5)$.

24. $f(x,y,z) = \ln(1 + x^2 + y^2 - z^2)$, $\quad (a,b) = (1,1,1)$

25. $f(x,y,z) = \sin(xyz)$, $\quad (a,b,c) = (1,-\pi,0)$

26. $f(x,y,z) = e^{2x+y^2+z}$, $\quad (a,b,c) = (0,1,-1)$

27. $f(x,y,z) = \dfrac{xyz}{x^2+y^2-z^2}$, $\quad (a,b,c) = (1,1,-1)$

28. $f(x,y,z) = e^{xy} \cos(yz)$, $\quad (a,b,c) = (1,1,\tfrac{\pi}{4})$

29. $f(x,y,z) = \dfrac{x}{x+y^2+z^3}$, $\quad (a,b,c) = (3,2,1)$

30. $f(x,y,z) = \sin x \sin y \sin z$, $\quad (a,b,c) = (\tfrac{\pi}{4}, \tfrac{\pi}{2}, \pi)$

31. $f(x,y,z) = \dfrac{x+y}{x^3+y^2+z}$, $\quad (a,b,c) = (1,-1,1)$

For exercises 32-35: An ant is standing on a hill whose height is given by

$$H(x,y) = x^4 + y^4 - 4x^2 y^2.$$

Take "north" to be the direction of the positive y-axis and "east" to be the direction of the positive x-axis and use partial derivatives to answer these questions.

32. If the ant is standing at the point $(2,3)$ and begins walking north, is it going uphill, downhill, or neither?

14.1 PARTIAL DERIVATIVES

33. If the ant is standing at the point $(-1, 2)$ and begins walking south, is it going uphill, downhill, or neither?

34. If the ant is standing at the point $(3, -2)$ and begins walking east, is it going uphill, downhill, or neither?

35. If the ant is standing at the point $(-2, -1)$ and begins walking west, is it going uphill, downhill, or neither?

For exercises 36-39: Suppose now that the height of the hill is given by

$$H(x, y) = x/y.$$

36. If the ant is standing at the point $(2, 3)$ and begins walking north, is it going uphill, downhill, or neither?

37. If the ant is standing at the point $(-1, 2)$ and begins walking south, is it going uphill, downhill, or neither?

38. If the ant is standing at the point $(3, -2)$ and begins walking east, is it going uphill, downhill, or neither?

39. If the ant is standing at the point $(-2, -1)$ and begins walking west, is it going uphill, downhill, or neither?

For exercises 40-43: Two charged objects with charge 2 and -1 units, respectively, located in the plane at locations $(0.2, 0.01)$ and $(0.8, 0.01)$ produce a potential function on the plane given by

$$P(x, y) = \frac{2}{\sqrt{(x - 0.2)^2 + (y - 0.01)^2}} + \frac{-1}{\sqrt{(x - 0.8)^2 + (y - 0.01)^2}}.$$

40. If a test particle is located at the origin $(0, 0)$, how fast will the potential change if it starts moving along the positive x-axis?

41. If a test particle is located at the origin $(0, 0)$, how fast will the potential change if it starts moving along the positive y-axis?

42. If a test particle is located at $(0.5, 0.12)$, how fast will the potential change if it starts moving straight toward the y-axis along the line $y = -0.12$?

43. If a test particle is located at $(0.5, -0.12)$, how fast will the potential change if it starts moving straight toward the x-axis along the line $x = 0.5$?

For exercises 44-47: Suppose now that the field-producing objects had charge 2 and 1 instead of 2 and -1. The potential would then be given by

$$P(x,y) = \frac{2}{\sqrt{(x-0.2)^2 + (y-0.01)^2}} + \frac{1}{\sqrt{(x-0.8)^2 + (y-0.01)^2}}.$$

44. If a test particle is located at the origin $(0,0)$, how fast will the potential change if it starts moving along the positive x-axis?

45. If a test particle is located at the origin $(0,0)$, how fast will the potential change if it starts moving along the positive y-axis?

46. If a test particle is located at $(-0.35, -0.42)$, how fast will the potential change if it starts moving straight toward the y-axis along the line $y = -0.42$?

47. If a test particle is located at $(-0.35, -0.42)$, how fast will the potential change if it starts moving straight toward the x-axis along the line $x = -0.35$?

For exercises 48-51: The temperature in °Kelvin at a point (x, y, z) is given by the function

$$T(x,y,z) = \frac{300}{1 + x^2 + y^2 + z^2}.$$

(A 300° heat source is located at the origin.) Use this information to answer these questions.

48. Find the instantaneous rate of temperature change per unit when we move from the point $(1, 0, 0)$ straight toward the origin.

49. Find the instantaneous rate of temperature change per unit when we move from the point $(0, -3, 0)$ straight toward the origin.

50. Find the instantaneous rate of temperature change per unit when we move from the point $(1, 2, 3)$ straight toward the point $(1, 5, 3)$.

51. At which points in space will the temperature decrease regardless of whether we move straight up or straight down?

14.2 THE TOTAL DERIVATIVE

In the last section, we discussed *partial* derivatives of multivariable functions. In this section we talk about the **total derivative** of a multivariable function.

14.2 THE TOTAL DERIVATIVE

Limits and continuity of multivariable functions

Limits provide a convenient language for talking about continuity and differentiability of multivariable functions, just as they do for single-variable functions. A few comments are in order regarding the notion of limit as it applies to multivariable functions. Let's start by reviewing the basic ideas for single-variable functions.

For a function of a single variable $y = f(x)$, we write

$$\lim_{x \to a} f(x) = L.$$

This just means that the difference $|f(x) - L|$ can be made as small as we wish, provided that the input x is chosen sufficiently close (but not equal) to a. Recall that the actual function output $f(a)$ need not equal L (nor even be defined) for the limit to exist. However, if we *do* have

$$\lim_{x \to a} f(x) = f(a),$$

then we say that f is **continuous at** $x = a$.

We can make a similar definition for the limit of a multivariable function. If $z = f(x, y)$ is a function of two variables, we write

$$\lim_{(x,y) \to (a,b)} f(x, y) = L$$

to mean that the difference $|f(x, y) - L|$ can be made as small as we wish by choosing (x, y) sufficiently close (but not equal) to (a, b). A function of two variables is said to be **continuous at** (a, b) if

$$\lim_{(x,y) \to (a,b)} f(x, y) = f(a, b).$$

For a function of three variables $w = f(x, y, z)$ we write

$$\lim_{(x,y,z) \to (a,b,c)} f(x, y, z) = L$$

to indicate that $|f(x, y, z) - L|$ approaches 0 as (x, y, z) approaches (a, b, c). The function f is said to be **continuous at** (a, b, c) if

$$\lim_{(x,y,z) \to (a,b,c)} f(x, y, z) = f(a, b, c).$$

▶▶▶ **Virtually all of the standard functions (algebraic, trigonometric, inverse trigonometric, exponential, logarithmic) are continuous wherever they are defined. Multivariable functions built up from these through algebraic operations or composition will also be continuous wherever defined.**

For example, the function $z = f(x,y) = \dfrac{e^{xy}}{\sin(x^2+y^2)}$ is continuous at every point $(x,y) \neq (0,0)$. At $(0,0)$ the function is *undefined* since the denominator $\sin(0^2+0^2) = 0$. No limit exists at $(0,0)$, since the numerator approaches 1 and the denominator approaches 0 as (x,y) approaches $(0,0)$.

We have already noted that the partial derivatives of a multivariable function are themselves multivariable functions. When all these partial derivatives are continuous, the function deserves a special term.

Definition 2

> If all the partial derivatives of a multivariable function f are *continuous* at a point, we say that the function f is C^1-**differentiable** at that point.

For example, the function $z = f(x,y) = \dfrac{e^{xy}}{\sin(x^2+y^2)}$ is C^1-differentiable at every point except $(0,0)$, since its partial derivatives

$$\frac{\partial z}{\partial x} = \frac{ye^{xy}\sin(x^2+y^2) - 2xe^{xy}\cos(x^2+y^2)}{\sin^2(x^2+y^2)}$$

$$\frac{\partial z}{\partial y} = \frac{xe^{xy}\sin(x^2+y^2) - 2ye^{xy}\cos(x^2+y^2)}{\sin^2(x^2+y^2)}$$

are both continuous at every point $(x,y) \neq (0,0)$.

What is a total derivative?

If a function $y = f(x)$ of one variable is differentiable at a point $x = a$, then we can think of the function as being approximately *locally linear*. The derivative $f'(a)$ gives us a simple multiplication factor (namely, the local slope) for measuring change in output for small changes in input. In other words, if we make a small change in input Δx from $x = a$, then we just multiply by $f'(a)$ to find the approximate change in output:

$$f(a+\Delta x) - f(a) \approx f'(a)\Delta x.$$

We want to use a similar criterion for differentiability of a function of several variables. If a function of *several* variables is approximately *locally linear*, then for small changes in the inputs, the total change in output must be approximately a simple *linear combination* of the small changes in the various inputs. By this, we mean that we can multiply each input's change by its own multiplication factor and add up the results to approximate the change in output.

The partial derivatives of a multivariable function give us the local slope multiplication factors for each individual input, with all other inputs

14.2 THE TOTAL DERIVATIVE

fixed. If all of the partial derivatives are *continuous* (in other words, f is C^1-differentiable), this guarantees that the function behaves approximately locally linear. In this case, we can talk about the **total derivative** of f.

Definition 3

> The **total derivative** Df of a C^1-differentiable scalar-valued function f is defined by the single-row matrix of partial derivative values.

The total derivative of a function of two variables $z = f(x, y)$ is

$$Df = \begin{bmatrix} \dfrac{\partial f}{\partial x} & \dfrac{\partial f}{\partial y} \end{bmatrix}.$$

For a function of three variables $w = f(x, y, z)$:

$$Df = \begin{bmatrix} \dfrac{\partial f}{\partial x} & \dfrac{\partial f}{\partial y} & \dfrac{\partial f}{\partial z} \end{bmatrix},$$

and, in general, for a function f of n variables $x_1, x_2, \ldots x_n$,

$$Df = \begin{bmatrix} \dfrac{\partial f}{\partial x_1} & \dfrac{\partial f}{\partial x_2} & \cdots & \dfrac{\partial f}{\partial x_n} \end{bmatrix}.$$

Calculating total derivatives

If we calculate the total derivative of a function of two variables

$$z = f(x, y)$$

at a point (a, b), we obtain the 1×2 matrix

$$Df(a, b) = \begin{bmatrix} \dfrac{\partial f}{\partial x}\bigg|_{(a,b)} & \dfrac{\partial f}{\partial y}\bigg|_{(a,b)} \end{bmatrix}.$$

EXAMPLE 5 Find the total derivative of the function

$$z = f(x, y) = x^2 + 5xy - y^3$$

at the points $(-2, 3)$ and $(1.2, -2.1)$.

Solution Since $\dfrac{\partial f}{\partial x} = 2x + 5y$ and $\dfrac{\partial f}{\partial y} = 5x - 3y^2$, we have

$$Df(-2, 3) = \begin{bmatrix} 11 & -37 \end{bmatrix}$$

and

$$Df(1.2, -2.1) = \begin{bmatrix} -8.1 & -7.23 \end{bmatrix}.$$

Note that we can get a *different* matrix at each point (a, b).

EXAMPLE 6 Find the total derivative of $w = g(x,y,z) = \ln(x^2 + y^2 + z^2)$ at the point $(-2, 1, 5)$.

Solution $Dg = \begin{bmatrix} \dfrac{\partial g}{\partial x} & \dfrac{\partial g}{\partial y} & \dfrac{\partial g}{\partial z} \end{bmatrix} = \begin{bmatrix} \dfrac{2x}{x^2+y^2+z^2} & \dfrac{2y}{x^2+y^2+z^2} & \dfrac{2z}{x^2+y^2+z^2} \end{bmatrix}.$

Hence, $Dg(-2, 1, 5) = \begin{bmatrix} \dfrac{-2}{15} & \dfrac{1}{15} & \dfrac{1}{3} \end{bmatrix}.$ ∎

For functions of a single variable, we usually think of the derivative $f'(a)$ as a number (the local slope at $x = a$). Strictly speaking, this number represents the 1×1 matrix of the total derivative.

Interpreting the total derivative

What exactly does the total derivative tell us? In a very real sense, Df captures the "total rate of change" of a multivariable function f. Let's illustrate this statement for a function f of two variables. If we move away from the point (x, y) a small amount in *any* direction, we arrive at a new point

$$(x + \Delta x, y + \Delta y).$$

The total derivative Df at (x, y) can tell us approximately the change in output of our function f by applying the total derivative to the *vector of changes* $\langle \Delta x, \Delta y \rangle$. In other words,

$$\Delta z = f(x + \Delta x, y + \Delta y) - f(x, y) \approx Df(x,y) \begin{bmatrix} \Delta x \\ \Delta y \end{bmatrix},$$

or written in terms of the partial derivatives:

$$\Delta z \approx \frac{\partial f}{\partial x} \Delta x + \frac{\partial f}{\partial y} \Delta y.$$

This approximation becomes better and better as $\langle \Delta x, \Delta y \rangle$ becomes smaller and smaller.

EXAMPLE 7 Use the total derivative to approximate the change in output

$$\Delta z = f(-2.05, +3.02) - f(-2, 3)$$

for the function $z = f(x,y) = x^2 + 5xy - y^3$.

Solution We have already computed

$$Df(-2, 3) = \begin{bmatrix} 11 & -37 \end{bmatrix}.$$

14.2 THE TOTAL DERIVATIVE

The vector of changes is

$$\langle \Delta x, \Delta y \rangle = \langle -2.05, 3.02 \rangle - \langle -2, 3 \rangle = \langle -.05, .02 \rangle.$$

Therefore,

$$\Delta z \approx \begin{bmatrix} 11 & -37 \end{bmatrix} \begin{bmatrix} -.05 \\ .02 \end{bmatrix} = 11(-.05) - 37(.02) = -.55 - .74 = -1.29.$$

Comparing this to the actual change in the output

$$\Delta z = f(-2.05, 3.02) - f(-2, 3) = -54.296108 - (-53) = -1.296108,$$

we see that using the total derivative approximates the change in output accurate to within .01 of the actual change in output. ∎

Best linear approximations of multivariable functions

If a function $y = f(x)$ of one variable is differentiable at a point $x = a$, then it has a best linear approximation at that point. Geometrically, the graph of this best linear approximation is the *tangent line* to the graph of $y = f(x)$ at the point $(a, f(a))$. The derivative value $f'(a)$ is the slope of this tangent line. The equation of the tangent line is called the *first-degree Taylor polynomial approximation of f about $x = a$*:

$$f(x) \approx f(a) + f'(a)(x - a).$$

In the last chapter, we discussed how a *linear* function of two variables

$$z = m_1 x + m_2 y + c$$

has a *plane* for a graph. This plane has one slope in the x-direction, computed by the difference quotient $m_1 = \Delta z / \Delta x$ for any two distinct points in the plane with the same y-coordinates. This plane also has a slope in the y-direction, computed by the difference quotient $m_2 = \Delta z / \Delta y$ for any two distinct points in the plane with the same x-coordinates. The slope with respect to x measures "rise/run" as we move parallel to the x-axis, and the slope with respect to y measures "rise/run" as we move parallel to the y axis.

▶▶▶ A function f of two variables has a **total derivative at a point** (a, b) if both its partial derivatives are continuous at that point. Geometrically, the surface $z = f(x, y)$ has a *tangent plane* at the point $(a, b, f(a, b))$, **whose slopes with respect to x and y are the** *partial derivatives* **of f at (a, b). The equation of the tangent plane to the graph of $z = f(x, y)$ at the point $(a, b, f(a, b))$ is**

$$z = f(a,b) + f_x(a,b)(x-a) + f_y(a,b)(y-b).$$

Figure 14.4 illustrates a function graph and a tangent plane.

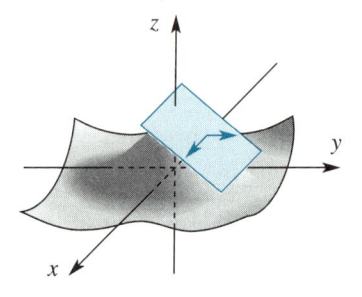

Figure 14.4 Tangent plane to the graph of $z = f(x, y)$.

Definition 4

For a C^1-differentiable function of two variables $f(x,y)$, the **first-degree Taylor polynomial approximation** about $(x, y) = (a, b)$ is

$$f(x,y) \approx f(a,b) + f_x(a,b)(x-a) + f_y(a,b)(y-b).$$

This is the best linear approximation of the function f at the point $(a, b, f(a, b))$. Written in terms of the total derivative, we have

$$f(x,y) \approx f(a,b) + Df(a,b) \begin{bmatrix} x-a \\ y-b \end{bmatrix}.$$

Note the similarity to the first-degree Taylor approximation for a function of one variable:

$$f(x) \approx f(a) + f'(a)(x-a).$$

EXAMPLE 8 Find the best linear approximation to the function

$$z = f(x,y) = x^2 + 5xy - y^3$$

at the point $(-2, 3)$.

14.2 THE TOTAL DERIVATIVE

Solution We have $f(-2, 3) = (-2)^2 + 5(-2)3 - 3^3 = 4 - 30 - 27 = -53$. We have already computed

$$Df(-2, 3) = \begin{bmatrix} 11 & -37 \end{bmatrix}.$$

Thus, the best linear approximation for the function is

$$f(x, y) \approx f(-2, 3) + \begin{bmatrix} 11 & -37 \end{bmatrix} \begin{bmatrix} x + 2 \\ y - 3 \end{bmatrix} = -53 + 11(x + 2) - 37(y - 3). \quad \blacksquare$$

Definition 5

For a C^1-differentiable function of three variables $f(x, y, z)$, the **first-degree Taylor polynomial approximation of f about** $(x, y, z) = (a, b, c)$ is

$$f(x, y, z) \approx f(a, b, c) + f_x(a, b, c)(x - a) + f_y(a, b, c)(y - b) + f_z(a, b, c)(z - c)$$

$$= f(a, b, c) + Df(a, b, c) \begin{bmatrix} x - a \\ y - b \\ z - c \end{bmatrix}.$$

EXAMPLE 9 Find the best linear approximation to the function $w = g(x, y, z) = \ln(x^2 + y^2 + z^2)$ at the point $(-2, 1, 5)$.

Solution We have $g(-2, 1, 5) = \ln((-2)^2 + 1^2 + 5^2) = \ln(30)$. We have already computed

$$Dg(-2, 1, 5) = \begin{bmatrix} \dfrac{-2}{15} & \dfrac{1}{15} & \dfrac{11}{3} \end{bmatrix}.$$

Thus, the best linear approximation for the function is

$$w = g(x, y, z) \approx \ln(30) + \begin{bmatrix} \dfrac{-2}{15} & \dfrac{1}{15} & \dfrac{1}{3} \end{bmatrix} \begin{bmatrix} x + 2 \\ y - 1 \\ z - 5 \end{bmatrix}$$

$$= \ln(30) - \dfrac{2}{15}(x + 2) + \dfrac{1}{15}(y - 1) + \dfrac{1}{3}(z - 5). \quad \blacksquare$$

EXERCISES for Section 14.2

For exercises 1-10. For each function, find the total derivative Df, and use it to find the equation of the best linear approximation to f about the indicated point (a, b).

1. $f(x, y) = \dfrac{x + y}{x - y}$, $(a, b) = (1, 2)$.
2. $f(x, y) = xy + x/y$, $(a, b) = (-1, 3)$.
3. $f(x, y) = e^{x^2 + y^2}$, $(a, b) = (1, -2)$.

4. $f(x,y) = 2x - 3y$, $\quad (a,b) = (0,0)$.

5. $f(x,y) = \sin(x^2 y)$, $\quad (a,b) = (2,0)$.

6. $f(x,y) = \arctan(y/x)$, $\quad (a,b) = (-3,-3)$.

7. $f(x,y) = \tan(y/x) - \cot(y)$, $\quad (a,b) = (2, \pi/2)$.

8. $f(x,y) = x + \ln(x^2 + y^2 + 1)$, $\quad (a,b) = (0,0)$.

9. $f(x,y) = \sqrt{x^2 + y^2}$, $\quad (a,b) = (-3, 4)$.

10. $f(x,y) = x^y$, $\quad (a,b) = (e, 2)$.

For exercises 11-15: For each function, find the total derivative Df and use it to find the equation of the best linear approximation to f at the indicated point (a, b, c).

11. $f(x,y,z) = \dfrac{z}{x+y}$, $\quad (a,b,c) = (1,2,3)$.

12. $f(x,y,z) = x^3 y^4 z^5$, $\quad (a,b,c) = (1,-2,-1)$.

13. $f(x,y,z) = 5x - 4y + 3z$, $\quad (a,b,c) = (-3, 2, 4)$.

14. $f(x,y,z) = \exp(1 - x^2 - y^2 - z^2)$, $\quad (a,b,c) = (0, 1, -1)$.

15. $f(x,y,z) = x \arcsin z - y \arctan z$, $\quad (a,b,c) = (.5, .5, .5)$.

For exercises 16-19: An ant is standing on a hill whose height is given by

$$H(x,y) = x^4 + y^4 - 4x^2 y^2.$$

Take "north" to be the direction of the positive y-axis and "east" to be the direction of the positive x-axis and use this information to answer these questions.

16. If the ant is standing at the point $(2, 3)$ and walks 0.1 units north, then 0.1 units east, what is its total change in elevation?

17. What is the total derivative of the height function at the point $(2, 3)$? Use it to approximate the change in elevation computed in the previous exercise, and compare your results.

18. If the ant is standing at the point $(-1, 2)$ and walks 0.2 units south, then 0.3 units west, what is the total change in elevation?

19. What is the total derivative of the height function at the point $(-1, 2)$? Use it to approximate the change in elevation computed in the previous exercise, and compare your results.

For exercises 20-23: Suppose now that the height of the hill is given by

$$H(x,y) = x/y.$$

14.3 THE GRADIENT

20. If the ant is standing at the point $(3, -2)$ and walks 0.2 units east, then 0.1 units south, what is its total change in elevation?

21. What is the total derivative of the height function at the point $(3, -2)$? Use it to approximate the change in elevation computed in the previous exercise, and compare your results.

22. If the ant is standing at the point $(-2, -1)$ and walks 0.1 units north, then 0.2 units west, what is its total change in elevation?

23. What is the total derivative of the height function at the point $(-2, -1)$? Use it to approximate the change in elevation computed in the previous exercise, and compare your results.

For exercises 24-25: The temperature in °Kelvin at a point (x, y, z) is given by the function

$$T(x, y, z) = \frac{300}{1 + x^2 + y^2 + z^2}.$$

(A $300°$ heat source is located at the origin.)

24. Find the total temperature change when we move from point $(1, 2, 3)$ to point $(1.1, 1.9, 3.2)$.

25. Find the total derivative of the temperature function. Use it to approximate the change in temperature computed in the previous exercise, and compare your results.

For exercises 26-30: You'll need 3D-graphing tools for these exercises, which are intended to investigate how well the tangent plane approximates the surface of a graph.

(a) Find the equation of the tangent plane to the graph of the indicated function at the point $(1, 2, f(1, 2))$.

(b) Plot the contours of the original function and the line contours of the tangent plane for the inputs (x, y) in $[0, 2] \times [1, 3]$. (Use level curves corresponding to values of z spaced 0.1 units apart.)

(c) Graph the wireframe plot of each over the same set of inputs.

26. $f(x, y) = \dfrac{x + y}{x - y}$.

27. $f(x, y) = xy + x/y$.

28. $f(x, y) = e^{x^2 + y^2}$.

29. $f(x, y) = \sqrt{x^2 + y^2}$.

30. $f(x, y) = x^y$.

14.3 THE GRADIENT

The total derivative Df is represented by a single-row matrix of partial derivatives. If we think of this matrix as a *vector*, we call it the **gradient** of the multivariable function f. The gradient is a versatile tool for analyzing the behavior of a multivariable function. In this section we'll see how the gradient can be used to calculate the rate of change of the outputs of a function f in any direction or along a path we may take through the domain of inputs.

Definition 6

> The **gradient** of a function f is the vector of partial derivatives of f. The gradient is denoted by ∇f and read as "grad f" or "del f."

Hence, the gradient of a function $f(x, y)$ of two variables is

$$\nabla f = \left\langle \frac{\partial f}{\partial x}, \frac{\partial f}{\partial y} \right\rangle;$$

the gradient of a function of three variables, $f(x, y, z)$ is

$$\nabla f = \left\langle \frac{\partial f}{\partial x}, \frac{\partial f}{\partial y}, \frac{\partial f}{\partial z} \right\rangle;$$

and, in general, the gradient of a function of n variables $f(x_1, x_2, \ldots, x_n)$ is

$$\nabla f = \left\langle \frac{\partial f}{\partial x_1}, \frac{\partial f}{\partial x_2}, \ldots, \frac{\partial f}{\partial x_n} \right\rangle.$$

EXAMPLE 10 Find the gradient of the function $z = f(x, y) = y \arctan x$ and evaluate it at $(x, y) = (1, 3)$.

Solution The gradient function is

$$\nabla f = \left\langle \frac{\partial f}{\partial x}, \frac{\partial f}{\partial y} \right\rangle = \left\langle \frac{y}{1 + x^2}, \arctan x \right\rangle.$$

Therefore,

$$\nabla f(1, 3) = \left\langle \frac{3}{1 + 1^2}, \arctan 1 \right\rangle = \left\langle \frac{3}{2}, \frac{\pi}{4} \right\rangle.$$ ∎

14.3 THE GRADIENT

EXAMPLE 11 Find $\nabla f(-3,-2,-1)$ if $f(x,y,z) = e^{xyz}$.

Solution The gradient function is

$$\nabla f = \left\langle \frac{\partial f}{\partial x}, \frac{\partial f}{\partial y}, \frac{\partial f}{\partial z} \right\rangle = \langle yze^{xyz}, xze^{xyz}, xye^{xyz} \rangle = e^{xyz}\langle yz, xz, xy \rangle.$$

Therefore, $\nabla f(-3,-2,-1) = e^{-6}\langle 2, 3, 6 \rangle$. ∎

Note that the gradient gives us a *vector value* at each point. For this reason, we can say that the gradient defines a *vector field*. In this section, we discuss some important interpretations and uses of the gradient vector.

We already know that the total derivative can be used to approximate the change in output of a C^1-differentiable function f. Written in terms of the gradient, this change is the *dot product of the gradient vector with the vector of changes in the inputs*. For example, if $z = f(x,y)$ is a C^1-differentiable function of two variables, then a change in input $\langle \Delta x, \Delta y \rangle$ results in a change in output

$$\Delta z \approx \frac{\partial f}{\partial x}\Delta x + \frac{\partial f}{\partial y}\Delta y = \nabla f \cdot \langle \Delta x, \Delta y \rangle.$$

For a C^1-differentiable function of three variables $w = f(x,y,z)$, a change in input of $\langle \Delta x, \Delta y, \Delta z \rangle$ results in the change in output

$$\Delta w \approx \frac{\partial f}{\partial x}\Delta x + \frac{\partial f}{\partial y}\Delta y + \frac{\partial f}{\partial z}\Delta z = \nabla f \cdot \langle \Delta x, \Delta y, \Delta z \rangle.$$

With this in mind, we will be able to use the gradient to measure the rate of change in *any* direction.

Directional derivatives

Let's consider the case of a function of two variables $z = f(x,y)$. Each partial derivative of f measures the instantaneous rate of change of output per unit change in a particular variable. Geometrically, we can think of $\frac{\partial f}{\partial x}$ as measuring the rate of output change of z per unit change in the direction parallel to the x-axis, while $\frac{\partial f}{\partial y}$ measures the rate of output change per unit change in the direction parallel to the y-axis.

These are not the only directions in which we can move in the xy-plane. Indeed, there are infinitely many directions one can move from any point. Suppose that we are at a given input (a,b) and we want to measure the instantaneous rate of change of $z = f(x,y)$ when the input is shifted in the direction

$$\mathbf{u} = \langle u_1, u_2 \rangle,$$

where **u** is a *unit vector* ($\|\mathbf{u}\| = \sqrt{u_1^2 + u_2^2} = 1$).

If we move Δt units in this direction, then the change in input is

$$\langle u_1 \Delta t, u_2 \Delta t \rangle,$$

and the corresponding change in output is

$$\Delta z = f(a + u_1 \Delta t, b + u_2 \Delta t) - f(a,b).$$

Thus, the *average rate of change* in the direction **u** is

$$\frac{\Delta z}{\Delta t} = \frac{f(a + u_1 \Delta t, b + u_2 \Delta t) - f(a,b)}{\Delta t}.$$

The limit of this difference quotient as Δt approaches 0 gives us the *instantaneous rate of change* per unit in the direction **u**, and this forms the basis of our definition of a **directional derivative**.

Definition 7

The **directional derivative** of f at (a,b) in the unit direction $\mathbf{u} = \langle u_1, u_2 \rangle$ is defined as

$$D_{\mathbf{u}} f(a,b) = \lim_{\Delta t \to 0} \frac{f(a + u_1 \Delta t, b + u_2 \Delta t) - f(a,b)}{\Delta t},$$

provided that this limit exists.

▶▶▶ **Don't forget that a directional derivative $D_{\mathbf{u}} f$ is always with respect to a *unit vector* u.**

Now, the gradient gives us a very handy way of computing this directional derivative. If f is a C^1-differentiable function, then the approximate change in output is simply

$$\Delta z \approx \nabla f(a,b) \cdot \langle u_1 \Delta t, u_2 \Delta t \rangle.$$

Dividing by Δt, we have

$$\frac{\Delta z}{\Delta t} \approx \frac{\nabla f(a,b) \cdot \langle u_1 \Delta t, u_2 \Delta t \rangle}{\Delta t} = \nabla f(a,b) \cdot \langle u_1, u_2 \rangle.$$

This approximation becomes better and better as Δt approaches 0. In fact,

$$D_{\mathbf{u}} f(a,b) = \nabla f(a,b) \cdot \mathbf{u}.$$

If $\mathbf{u} = \mathbf{i}$ or $\mathbf{u} = \mathbf{j}$, then the directional derivative is exactly the partial derivative with respect to x or y, respectively. That is,

14.3 THE GRADIENT

$$D_\mathbf{i} f(a,b) = \left.\frac{\partial f}{\partial x}\right|_{(a,b)} \quad \text{and} \quad D_\mathbf{j} f(a,b) = \left.\frac{\partial f}{\partial y}\right|_{(a,b)}.$$

EXAMPLE 12 Suppose that $z = f(x,y) = x^2 y + 2xy - y^3$. Find the instantaneous rate of change of $f(x,y)$ per unit if we move from the input point $(-2,3)$ straight toward the point $(1,7)$ in the xy-plane.

Solution First, we need to determine a unit vector describing the direction in which we're moving. The direction vector is given by

$$\langle 1, 7 \rangle - \langle -2, 3 \rangle = \langle 3, 4 \rangle.$$

A unit vector in this same direction is given by

$$\mathbf{u} = \frac{\langle 3, 4 \rangle}{\|\langle 3, 4 \rangle\|} = \frac{\langle 3, 4 \rangle}{\sqrt{3^2 + 4^2}} = \frac{\langle 3, 4 \rangle}{5} = \left\langle \frac{3}{5}, \frac{4}{5} \right\rangle.$$

The gradient of f is

$$\nabla f = \langle 2xy + 2y, x^2 + 2x - 3y^2 \rangle.$$

Hence,

$$D_\mathbf{u} f(-2,3) = \nabla f(-2,3) \cdot \langle 3/5, 4/5 \rangle = \langle -6, -27 \rangle \cdot \langle 3/5, 4/5 \rangle = -25.2.$$

This tells us that as we move from $(-2,3)$ towards $(1,7)$, the output $f(x,y)$ is *decreasing* at an instantaneous rate of 25.2 units per unit traveled. ∎

The definition of directional derivative is similar for functions of three or more variables.

Definition 8

The **directional derivative** of f at (a,b,c) with respect to the direction specified by unit vector $\mathbf{u} = \langle u_1, u_2, u_3 \rangle$ is

$$D_\mathbf{u} f(a,b,c) = \lim_{\Delta t \to 0} \frac{f(a + u_1 \Delta t, b + u_2 \Delta t, c + u_3 \Delta t) - f(a,b,c)}{\Delta t}.$$

For C^1-differentiable functions, the same observation regarding the gradient carries through. If f is a C^1-differentiable function of three variables and $\mathbf{u} = \langle u_1, u_2, u_3 \rangle$ is a unit vector, then

$$D_\mathbf{u} f(a,b,c) = \nabla f(a,b,c) \cdot \langle u_1, u_2, u_3 \rangle.$$

▶▶▶ **In general, for any C^1-differentiable function f of n variables, and any unit vector $\mathbf{u} \in \mathbb{R}^n$, we have**

$$D_\mathbf{u} f = \nabla f \cdot \mathbf{u}.$$

EXAMPLE 13 The temperature in °Kelvin at a point (x, y, z) is given by the function

$$T(x, y, z) = \frac{300}{1 + x^2 + y^2 + z^2}.$$

(A 300° heat source is located at the origin.) Find the instantaneous rate of temperature change per unit when we move from the point $(1, 2, 3)$ in the direction specified by the unit vector $\mathbf{u} = \langle -1/3, -2/3, 2/3 \rangle$.

Solution   First we calculate the gradient of the temperature function:

$$\nabla T = \left\langle \frac{-600x}{(1 + x^2 + y^2 + z^2)^2}, \frac{-600y}{(1 + x^2 + y^2 + z^2)^2}, \frac{-600z}{(1 + x^2 + y^2 + z^2)^2} \right\rangle.$$

At the point $(1, 2, 3)$, the gradient has the value

$$\nabla T(1, 2, 3) = \left\langle \frac{-600}{225}, \frac{-1200}{225}, \frac{-1800}{225} \right\rangle = \langle -8/3, -16/3, -8 \rangle.$$

Taking the dot product with the given direction, we have

$$D_\mathbf{u} T = \nabla T(1, 2, 3) \cdot \mathbf{u} = \left\langle \frac{-8}{3}, \frac{-16}{3}, -8 \right\rangle \cdot \left\langle \frac{-1}{3}, \frac{-2}{3}, \frac{2}{3} \right\rangle = \frac{8}{9} + \frac{32}{9} + \frac{-16}{3} = \frac{-8}{9}.$$

Therefore, we experience a $(-8/9)°$ change in temperature per unit of distance as we move from $(1, 2, 3)$ in the direction $\langle -1/3, -2/3, 2/3 \rangle$. ■

Direction of fastest change

In what direction should we move in the domain of a function to realize the *fastest* rate of increase in output of a function f? In terms of the directional derivative, we are asking: *What unit vector $\mathbf{u}$ maximizes $D_\mathbf{u} f$?*

For a C^1-differentiable function, we have just seen that

$$D_\mathbf{u} f = \nabla f \cdot \mathbf{u}.$$

If we interpret the dot product geometrically, we have

$$\nabla f \cdot \mathbf{u} = ||\nabla f|| \, ||\mathbf{u}|| \cos \theta = ||\nabla f|| \cos \theta,$$

where θ is the smallest angle between ∇f and $\mathbf{u}$ (provided $\nabla f \neq 0$). Note that we have used the fact that $||\mathbf{u}|| = 1$ for a unit vector.

14.3 THE GRADIENT

Since $-1 \leq \cos\theta \leq 1$ for all values θ, we can see that the maximum rate of change occurs when $\cos\theta = 1$. But this happens precisely when the two vectors ∇f and $\mathbf{u}$ have exactly the same direction. How remarkable! At every point, the gradient points exactly in the direction we should move to achieve the greatest rate of increase. Furthermore, this maximum rate of change is precisely $\|\nabla f\|$.

▶▶▶ **When the gradient $\nabla f \neq 0$, it points in the direction of fastest rate of increase of the function f. If $\mathbf{u}$ is a unit vector having this direction (namely $\mathbf{u} = \nabla f / \|\nabla f\|$), then**

$$D_{\mathbf{u}} f = \|\nabla f\|.$$

Note that this means that $-\nabla f$ must point in the direction of fastest rate of *decrease* of the function f.

EXAMPLE 14 The temperature in °Kelvin at a point (x, y, z) is given by the function

$$T(x, y, z) = \frac{300}{1 + x^2 + y^2 + z^2}.$$

(A 300° heat source is located at the origin.) Find a unit vector $\mathbf{u}$ that specifies the direction of the fastest increase in temperature change per unit when we move from the point $(1, 2, 3)$. What is the rate of temperature change when we move in this direction?

Solution First, we have already calculated the gradient of this temperature function:

$$\nabla T = \left\langle \frac{-600x}{(1+x^2+y^2+z^2)^2}, \frac{-600y}{(1+x^2+y^2+z^2)^2}, \frac{-600z}{(1+x^2+y^2+z^2)^2} \right\rangle.$$

At the point $(1, 2, 3)$, the gradient has the value

$$\nabla T(1, 2, 3) = \left\langle \frac{-8}{3}, \frac{-16}{3}, -8 \right\rangle.$$

A unit vector having the same direction is

$$\mathbf{u} = \frac{\nabla T(1, 2, 3)}{\|\nabla T(1, 2, 3)\|} = \frac{\langle -8/3, -16/3, -8 \rangle}{8\sqrt{14}/3} = \frac{1}{\sqrt{14}} \langle -1, -2, -3 \rangle.$$

Note that this vector points from $(1, 2, 3)$ directly back toward the origin, where the heat source is located.

The maximum instantaneous rate of temperature increase in this direction is

$$\|\nabla T(1, 2, 3)\| = \frac{8\sqrt{14}}{3} \approx 9.98.$$

In other words, if we head straight for the heat source from the point $(1, 2, 3)$, we'll experience an approximate $10°$ per unit instantaneous rate of temperature increase. ∎

Rate of change along a path—chain rule

So far, we've been using the gradient to measure rate of change in straight line directions. Now we want to investigate how the gradient can be used to measure rate of change along a curved path. For example, suppose that we have a temperature scalar field in space, and we have an object moving through space along some path. How can we measure the rate of change of temperature along this path? Clearly several things will affect this rate of change:

1. the distribution of temperatures in space represented by the scalar field,
2. the exact positions of the points on the path, and
3. how quickly we traverse the path.

Now, the temperature at each point in space (x, y, z) is given by some scalar field

$$f(x, y, z).$$

If the path is described by the vector position function

$$\mathbf{r}(t) = \langle x(t), y(t), z(t) \rangle,$$

then the temperature at each point along the path is given by composing f with $\mathbf{r}$:

$$f(\mathbf{r}(t)) = f(x(t), y(t), z(t)).$$

This is a scalar-valued function of a single variable t, and the rate of change with respect to t is simply

$$\frac{df(\mathbf{r}(t))}{dt} = \lim_{\Delta t \to 0} \frac{f(\mathbf{r}(t + \Delta t)) - f(\mathbf{r}(t))}{\Delta t}.$$

The gradient again gives us a handy way of computing this derivative. In the numerator of the difference quotient, we have the difference in outputs of f at the two points represented by $\mathbf{r}(t + \Delta t)$ and $\mathbf{r}(t)$. When Δt is small, the change in input is

$$\mathbf{r}(t + \Delta t) - \mathbf{r}(t) \approx \mathbf{r}'(t) \Delta t,$$

and the corresponding change in output is therefore

$$f(\mathbf{r}(t + \Delta t)) - f(\mathbf{r}(t)) \approx \nabla f(\mathbf{r}(t)) \cdot \mathbf{r}'(t) \Delta t.$$

14.3 THE GRADIENT

We can approximate the difference quotient by

$$\frac{f(\mathbf{r}(t+\Delta t)) - f(\mathbf{r}(t))}{\Delta t} \approx \nabla f(\mathbf{r}(t)) \cdot \mathbf{r}'(t),$$

and this approximation becomes better and better as Δt approaches 0.

This final result is known as the gradient form of the **chain rule**:

$$\frac{d}{dt}(f(\mathbf{r}(t))) = \nabla f(\mathbf{r}(t)) \cdot \mathbf{r}'(t).$$

Note that the value of the derivative is written in terms of:

1. the gradient of the scalar field f,
2. the position $\mathbf{r}(t)$, and
3. the velocity $\mathbf{r}'(t)$.

EXAMPLE 15 The temperature in °Kelvin at a point (x, y, z) is given by the function

$$T(x, y, z) = \frac{300}{1 + x^2 + y^2 + z^2}.$$

(A 300° heat source is located at the origin.) A particle is moving along the path

$$\mathbf{r}(t) = \langle t, t^2, t^3 \rangle.$$

Find the instantaneous rate of temperature change of the particle at time $t = 2$ where t is measured in seconds.

Solution We will compute this rate of change in two ways. First, we note that the temperature at any time t (in seconds) along the path is given by

$$T(\mathbf{r}(t)) = \frac{300}{1 + (x(t))^2 + (y(t))^2 + (z(t))^2} = \frac{300}{1 + t^2 + t^4 + t^6}.$$

The derivative of this function is

$$\frac{d}{dt}T(\mathbf{r}(t)) = \frac{-300(2t + 4t^3 + 6t^5)}{(1 + t^2 + t^4 + t^6)^2}.$$

At time $t = 2$, we have

$$\left.\frac{d}{dt}T(\mathbf{r})\right|_{t=2} = \frac{-300(4 + 32 + 192)}{(1 + 4 + 16 + 64)^2} = \frac{-68400}{7225} \approx -9.467.$$

Now, let's use the gradient form of the chain rule to obtain the same result. First, the gradient of the temperature field is given by

$$\nabla T = \left\langle \frac{-600x}{(1 + x^2 + y^2 + z^2)^2}, \frac{-600y}{(1 + x^2 + y^2 + z^2)^2}, \frac{-600z}{(1 + x^2 + y^2 + z^2)^2} \right\rangle.$$

The velocity vector of the path is given by

$$\mathbf{r}'(t) = \langle 1, 2t, 3t^2 \rangle.$$

At time $t = 2$, we are at the point $\mathbf{r}(2) = \langle 2, 4, 8 \rangle$, and the temperature gradient there is

$$\nabla T(\mathbf{r}(2)) = \frac{\langle -600(2), -600(4), -600(8) \rangle}{(1 + 2^2 + 4^2 + 8^2)^2} = \frac{\langle -1200, -2400, -4800 \rangle}{7225}.$$

The velocity is $\mathbf{r}'(2) = \langle 1, 4, 12 \rangle$. Hence, the desired rate of change is given by

$$\frac{d}{dt}T(\mathbf{r})\bigg|_{t=2} = \nabla T(\mathbf{r}(2)) \cdot \mathbf{r}'(2) = \left\langle \frac{-1200}{7225}, \frac{-2400}{7225}, \frac{-4800}{7225} \right\rangle \cdot \langle 1, 4, 12 \rangle$$

$$= \frac{-1200}{7225} + \frac{-9600}{7225} + \frac{-57600}{7225} = \frac{-68400}{7225} \approx -9.467.$$

We see again that at time $t = 2$, the temperature along the path is *decreasing* at an instantaneous rate of approximately $9.467°$ per second. ∎

Directional derivative of f along a path $\mathbf{r}$

Definition 9

> The **directional derivative** of f along the path $\mathbf{r}$ at any time t is obtained by computing the directional derivative of f with respect to the *unit tangent vector* $\mathbf{T}(t)$. We write
>
> $$D_\mathbf{r} f(t) = \frac{\nabla f(\mathbf{r}(t)) \cdot \mathbf{r}'(t)}{\|\mathbf{r}'(t)\|} = \nabla f(\mathbf{r}(t)) \cdot \mathbf{T}(t).$$

Note that the directional derivative of f along a path $\mathbf{r}$ is simply the rate of change along the path divided by the speed $\|\mathbf{r}'(t)\|$.

EXAMPLE 16 The temperature in °Kelvin at a point (x, y, z) is given by the function

$$T(x, y, z) = \frac{300}{1 + x^2 + y^2 + z^2}.$$

(A $300°$ heat source is located at the origin.) A particle is moving along the path

$$\mathbf{r}(t) = \langle t, t^2, t^3 \rangle.$$

Find the directional derivative of T along $\mathbf{r}$ at $t = 2$.

14.3 THE GRADIENT

Solution In the last example, we computed the rate of temperature change at $t = 2$ to be

$$\nabla T(\mathbf{r}(2)) \cdot \mathbf{r}'(2) \approx -9.467.$$

The speed at $t = 2$ is

$$\|\mathbf{r}'(2)\| = \sqrt{1^2 + 4^2 + 12^2} = \sqrt{161} \approx 12.689.$$

Hence, the directional derivative of T along $\mathbf{r}$ at $t = 2$ is

$$D_\mathbf{r} f \bigg|_{t=2} \approx \frac{-9.467}{12.689} \approx -0.746.$$

So, at time $t = 2$, the temperature along the path is decreasing at an instantaneous rate of approximately $-0.746°$ per unit of distance. ■

Note that the derivative of $f(\mathbf{r}(t))$ with respect to t is measured in output units per time unit (or whatever other unit t represents), while the directional derivative $D_\mathbf{r}(f)$ is measured in output units per unit of distance. This certainly makes sense in terms of dimensional analysis. In our examples using a temperature field T,

$$\frac{d}{dt} f(\mathbf{r}(t)) \text{ is measured in units of } \frac{degrees}{time}.$$

The speed $\|\mathbf{r}'(t)\|$ is measured in units of $distance/time$, so when we divide by the speed, we have

$$D_\mathbf{r} f \text{ is measured in units of } \frac{degrees}{distance}.$$

The relationship between the gradient and level curves and surfaces

In the last chapter we noted that given a contour map showing the level curves of a function of two variables, the direction of greatest increase or of greatest decrease at a point is naturally perpendicular to the level curve through that point. We have noted earlier in this section that the gradient gives the direction of greatest rate of increase. That fact suggests that the gradient vector at a point should be *orthogonal to the level curve through that same point.*

To see why this is reasonable, suppose that we have a level curve with equation

$$f(x, y) = c$$

for some constant c. Now, suppose that the differentiable path $\mathbf{r}(t)$ traces out the same curve. If we evaluate the function f along this path, we have

$$f(\mathbf{r}(t)) = c$$

for each t.

Taking the derivative with respect to t:

$$\nabla f(\mathbf{r}(t)) \cdot \mathbf{r}'(t) = 0.$$

The left-hand side is the dot product of the gradient ∇f at the point $\mathbf{r}(t)$ with the velocity vector $\mathbf{r}'(t)$. This tells us that $\nabla f(\mathbf{r}(t))$ is orthogonal to $\mathbf{r}'(t)$, which is *tangent* to the curve. Hence, $\nabla f(\mathbf{r}(t))$ is orthogonal to $\mathbf{r}'(t)$.

The same goes for functions of three variables. Now there could be several differentiable paths $\mathbf{r}(t)$ on a level surface

$$f(x, y, z) = c$$

that pass through a given point on the surface. For each one, $\nabla f(\mathbf{r}(t)) \cdot \mathbf{r}'(t) = 0$. If $\nabla f \neq \mathbf{0}$, then (since all the vectors $\mathbf{r}'(t)$ are tangent to the surface) we must have ∇f normal to the tangent plane of the level surface $f(x, y, z) = c$.

Definition 10

> A vector normal to the tangent plane of a surface at a point is said to be **normal** to the surface at that same point.

In other words, we have just established that a nonzero gradient ∇f is normal to a level surface $f(x, y, z) = c$.

Using ∇f to find the equation of a tangent plane.

If $f : \mathbb{R}^2 \to \mathbb{R}$, then the tangent plane to the graph of $z = f(x, y)$ at the point $(a, b, f(a, b))$ is given by the equation of the best linear approximation. We can write this in terms of the gradient as follows:

$$z = f(a, b) + \nabla f(a, b) \cdot \langle x - a, y - b \rangle.$$

EXAMPLE 17 Given the function
$$f(x, y) = y^3 e^x,$$
Find the equation of the tangent plane to the graph of $z = f(x, y)$ at $(0, 2)$.

Solution First, we compute the gradient:

$$\nabla f = \left\langle \frac{\partial f}{\partial x}, \frac{\partial f}{\partial y} \right\rangle = \langle y^3 e^x, 3y^2 e^x \rangle$$

14.3 THE GRADIENT

and evaluate it at the input point $(a, b) = (0, 2)$:

$$\nabla f(a, b) = \nabla f(0, 2) = (8, 12).$$

The equation of the tangent plane is

$$\begin{aligned} z &= f(a, b) + \nabla f(a, b) \cdot (x - a, y - b) \\ &= 8 + (8, 12) \cdot (x - 0, y - 2) \\ &= 8 + (8x) + (12y - 24) \\ &= 8x + 12y - 16. \end{aligned}$$

Recall that if $\langle a, b, c \rangle$ is a vector normal to a plane containing the point (x_0, y_0, z_0), then every point (x, y, z) in the plane satisfies the equation

$$a(x - x_0) + b(y - y_0) + c(z - z_0) = 0.$$

This follows from the observation that the vector

$$\langle x - x_0, y - y_0, z - z_0 \rangle$$

is parallel to the plane, so its dot product with the perpendicular normal vector must be 0:

$$\langle a, b, c \rangle \cdot \langle x - x_0, y - y_0, z - z_0 \rangle = 0.$$

Many surfaces have equations in three variables that can be written in the form

$$f(x, y, z) = 0.$$

This is simply a level surface for the function f. If (x_0, y_0, z_0) is a point on the surface, then the gradient $\nabla f(x_0, y_0, z_0)$ is normal to the level surface at that point (provided that the gradient is not the zero vector). We can use it to find the equation of the tangent plane to the level surface.

EXAMPLE 18 What is the equation of the tangent plane to $\sqrt{x^2 + y^2 + z^2} = 5$ at the point $(0, 3, 4)$?

Solution The surface described is a level surface for the function

$$f(x, y, z) = \sqrt{x^2 + y^2 + z^2} - 5.$$

The gradient of f is

$$\nabla f = \left\langle \frac{x}{\sqrt{x^2 + y^2 + z^2}}, \frac{y}{\sqrt{x^2 + y^2 + z^2}}, \frac{z}{\sqrt{x^2 + y^2 + z^2}} \right\rangle.$$

At the point $(x_0, y_0, z_0) = (0, 3, 4)$, we have a normal vector given by

$$\nabla f(0, 3, 4) = \left\langle 0, \frac{3}{5}, \frac{4}{5} \right\rangle.$$

Every point in the tangent plane must satisfy

$$\left\langle 0, \frac{3}{5}, \frac{4}{5} \right\rangle \cdot (x - 0, y - 3, z - 4) = \frac{3}{5}(y - 3) + \frac{4}{5}(z - 4) = 0.$$

Therefore, the equation of the tangent plane to the surface at the point $(0, 3, 4)$ is

$$\frac{3}{5}y + \frac{4}{5}z - 5 = 0. \qquad \blacksquare$$

EXERCISES for Section 14.3

For exercises 1-6: For each function, find $D_\mathbf{u} f$ at indicated point (a, b) and in the direction of the indicated unit vector **u**.

1. $f(x, y) = \dfrac{x+y}{x-y}$, $(a, b) = (1, 2)$, $\mathbf{u} = \mathbf{i}/\sqrt{2} + \mathbf{j}/\sqrt{2}$.
2. $f(x, y) = xy + x/y$, $(a, b) = (-1, 3)$, $\mathbf{u} = \langle 3/5, 4/5 \rangle$.
3. $f(x, y) = \sin(x^2 y)$, $(a, b) = (2, 0)$, $\mathbf{u} = \mathbf{j}$.
4. $f(x, y) = \arctan(y/x)$, $(a, b) = (-3, -3)$, $\mathbf{u} = \mathbf{i}$.
5. $f(x, y) = x + \ln(x^2 + y^2 + 1)$, $(a, b) = (0, 0)$, $\mathbf{u} = \langle -0.8, 0.6 \rangle$.
6. $f(x, y) = \sqrt{x^2 + y^2}$, $(a, b) = (-3, 4)$, $\mathbf{u} = \langle 1/\sqrt{2}, -1/\sqrt{2} \rangle$.

For exercises 7-11: For each function, find the unit vector **u** that gives the direction of the fastest rate of increase from the indicated point.

7. $f(x, y, z) = \dfrac{z}{x+y}$, $(a, b, c) = (1, 2, 3)$.
8. $f(x, y, z) = x^3 y^4 z^5$, $(a, b, c) = (1, -2, -1)$.
9. $f(x, y, z) = 5x - 4y + 3z$, $(a, b, c) = (-3, 2, 4)$.
10. $f(x, y, z) = \exp 1 - x^2 - y^2 - z^2$, $(a, b, c) = (0, 1, -1)$.
11. $f(x, y, z) = x \arcsin z - y \arctan z$, $(a, b, c) = (.5, .5, .5)$.

For exercises 12-17: The temperature at the point (x, y, z) is given by the scalar field

$$T(x, y, z) = 3x^2 + 2y^2 - 4z,$$

where the temperature is measured in °Celsius.

12. Find the gradient function ∇T.

14.3 THE GRADIENT

13. What is the rate of change of temperature at the point $(-1, -3, 2)$ if one moves directly toward the origin?

14. What is the rate of change of temperature at the point $(5, -3, -4)$ if one moves in the direction $2\mathbf{i} - 6\mathbf{j} + 3\mathbf{k}$?

15. What unit vector gives the direction of greatest rate of increase of temperature at the point $(4, -1, 0)$?

16. A particle is moving along the path described by
$$\mathbf{r}(t) = \langle \cos t, \sin t, t \rangle,$$
where t is measured in seconds. Find how fast the temperature is changing (in degrees per second) when $t = 0$.

17. Find the directional derivative of the temperature (in degrees per distance unit) along the same path at time $t = 4$.

For exercises 18-21: A scalar field $f : \mathbb{R}^3 \to \mathbb{R}$ is given by
$$f(x, y, z) = \arctan(x^2 + y^2 + z^2).$$

18. Find ∇f.

19. Verify that $\quad D_{\mathbf{i}} f = f_x \quad D_{\mathbf{j}} f = f_y \quad D_{\mathbf{k}} f = f_z.$

20. If $\mathbf{r}(t) = \langle \cos 8t, \sin 6t, e^{4t} \rangle$, find $D_{\mathbf{r}} f$ at $t = 0$.

21. Find the best linear approximation to f at the point $(1, 0, 0)$.

For exercises 22-25: An ant is standing on a hill whose height is given by
$$H(x, y) = x^4 + y^4 - 4x^2 y^2.$$

Take "north" to be the direction of the positive y-axis and "east" to be the direction of the positive x-axis.

22. If the ant is standing at the point $(2, 3)$ and walks northeast, how fast is the elevation changing? Describe the direction (by a unit vector) that the ant should move to go uphill fastest.

23. If the ant is standing at the point $(-1, 2)$ and walks southwest, how fast is the elevation changing? Describe the direction (by a unit vector) that the ant should move to go downhill fastest.

24. If the ant is standing at the point $(3, -2)$ and walks northwest, how fast is the elevation changing? Describe the direction (by a unit vector) that the ant should move to go uphill fastest.

25. If the ant is standing at the point $(-2, -1)$ and walks southeast, how fast is the elevation changing? Describe the direction (by a unit vector) that the ant should move to go downhill fastest.

For exercises 26-29: Suppose now that the height of the hill is given by

$$H(x,y) = x/y.$$

26. If the ant is standing at the point $(2, 3)$ and walks northwest, how fast is the elevation changing? Describe the direction (by a unit vector) that the ant should move to go downhill fastest.

27. If the ant is standing at the point $(-1, 2)$ and walks southeast, how fast is the elevation changing? Describe the direction (by a unit vector) that the ant should move to go uphill fastest.

28. If the ant is standing at the point $(3, -2)$ and walks northeast, how fast is the elevation changing? Describe the direction (by a unit vector) that the ant should move to go downhill fastest.

29. If the ant is standing at the point $(-2, -1)$ and walks southwest, how fast is the elevation changing? Describe the direction (by a unit vector) that the ant should move to go uphill fastest.

---

30. For each function f in exercises 1-6, find a unit normal vector and the equation of the tangent plane to the surface $z = f(x, y)$ at the indicated point (a, b).

14.4 HIGHER ORDER PARTIAL DERIVATIVES

In the calculus of single-variable functions, higher order derivatives give us additional details about the behavior of a function. For example, the second derivative tells us the concavity of a function's graph, which in turn tells us how fast the slope of the graph is changing. Higher order derivatives can be used to approximate the behavior of a function with a polynomial, namely the Taylor polynomial. We'll see in this section how the higher order partial derivatives of a multivariate function also give us more detailed information about the behavior of the function.

A partial derivative of a multivariate function f is also a function of several variables. Hence, each partial derivative itself has partial derivatives. We call these the **second-order partial derivatives** of f.

14.4 HIGHER ORDER PARTIAL DERIVATIVES

For example, if f is a function of two variables, we can consider

$$\frac{\partial}{\partial x}\left(\frac{\partial f}{\partial x}\right), \quad \frac{\partial}{\partial y}\left(\frac{\partial f}{\partial x}\right), \quad \frac{\partial}{\partial x}\left(\frac{\partial f}{\partial y}\right), \quad \frac{\partial}{\partial y}\left(\frac{\partial f}{\partial y}\right).$$

The shorthand notations for these are

$$\frac{\partial^2 f}{\partial x^2}, \quad \frac{\partial^2 f}{\partial y \partial x}, \quad \frac{\partial^2 f}{\partial x \partial y}, \quad \frac{\partial^2 f}{\partial y^2}.$$

EXAMPLE 19 Suppose $z = f(x, y) = x^2 y + 2xy - y^3$. Find the second-order partial derivatives of f.

Solution The first-order partial derivatives of f are:

$$\frac{\partial f}{\partial x} = 2xy + 2y \quad \text{and} \quad \frac{\partial f}{\partial y} = x^2 + 2x - 3y^2.$$

The second-order partial derivatives with respect to x are:

$$\frac{\partial^2 f}{\partial x^2} = 2y \quad \text{and} \quad \frac{\partial^2 f}{\partial x \partial y} = 2x + 2.$$

The second-order partial derivatives with respect to y are:

$$\frac{\partial^2 f}{\partial y \partial x} = 2x + 2 \quad \text{and} \quad \frac{\partial^2 f}{\partial y^2} = -6y.$$

Note that the "mixed" second-order partial derivatives in this example are equal. This need not be the case. However, we can state the following:

▶▶▶ **If either of the mixed derivatives is continuous, then so is the other, and they must match.**

An alternative notation for second-order partial derivatives is

$$f_{xx} = \frac{\partial^2 f}{\partial x^2}, \quad f_{yx} = \frac{\partial^2 f}{\partial x \partial y}, \quad f_{xy} = \frac{\partial^2 f}{\partial y \partial x}, \quad f_{yy} = \frac{\partial^2 f}{\partial y^2}.$$

Note carefully the notation for the mixed partial derivatives, for it is an easy matter to confuse the order. Perhaps the best way to remember the notation is to consider a second-order derivative as the derivative of a derivative. For example,

$$\frac{\partial^2 f}{\partial x \partial y} = \frac{\partial}{\partial x}\left(\frac{\partial f}{\partial y}\right) = (f_y)_x = f_{yx}.$$

Fortunately, the mixed partial derivatives match anyway when they are continuous.

For a function of three variables $f(x, y, z)$, there are a total of *nine* second-order partial derivatives:

$$\frac{\partial^2 f}{\partial x^2}, \quad \frac{\partial^2 f}{\partial x \partial y}, \quad \frac{\partial^2 f}{\partial x \partial z}$$

$$\frac{\partial^2 f}{\partial y \partial x}, \quad \frac{\partial^2 f}{\partial y^2}, \quad \frac{\partial^2 f}{\partial y \partial z}$$

$$\frac{\partial^2 f}{\partial z \partial x}, \quad \frac{\partial^2 f}{\partial z \partial y}, \quad \frac{\partial^2 f}{\partial z^2}.$$

EXAMPLE 20 Suppose that $f(x, y, z) = e^{xyz}$. Find all the second-order partial derivatives of f.

Solution The first-order partial derivatives of f are:

$$\frac{\partial f}{\partial x} = yze^{xyz}, \quad \frac{\partial f}{\partial y} = xze^{xyz}, \quad \frac{\partial f}{\partial z} = xye^{xyz}.$$

The second-order partial derivatives of f are:

$$\frac{\partial^2 f}{\partial x^2} = y^2 z^2 e^{xyz}, \quad \frac{\partial^2 f}{\partial x \partial y} = ze^{xyz} + xyz^2 e^{xyz}, \quad \frac{\partial^2 f}{\partial x \partial z} = ye^{xyz} + xy^2 z e^{xyz},$$

$$\frac{\partial^2 f}{\partial y \partial x} = ze^{xyz} + xyz^2 e^{xyz}, \quad \frac{\partial^2 f}{\partial y^2} = x^2 z^2 e^{xyz}, \quad \frac{\partial^2 f}{\partial y \partial z} = xe^{xyz} + x^2 yz e^{xyz},$$

$$\frac{\partial^2 f}{\partial z \partial x} = ye^{xyz} + xy^2 z e^{xyz}, \quad \frac{\partial^2 f}{\partial z \partial y} = xe^{xyz} + x^2 yz e^{xyz}, \quad \frac{\partial^2 f}{\partial z^2} = x^2 y^2 e^{xyz}.$$

Again, note that the mixed partial derivatives involving the same variables of differentiation match. ∎

For a function of n variables $f(x_1, x_2, \ldots, x_n)$, the second-order partial derivative

$$\frac{\partial^2 f}{\partial x_i \partial x_j} = \frac{\partial}{\partial x_i} \left(\frac{\partial f}{\partial x_j} \right)$$

denotes the partial derivative with respect to x_i of the partial derivative of f with respect to x_j.

If all its second-order partial derivatives are continuous, a function f is called C^2-**differentiable**. This is a nice property for a function to have, since it guarantees that the mixed partial derivatives match. For example, knowing that a function $f : \mathbb{R}^3 \to \mathbb{R}$ is C^2-differentiable means that only six of the nine second-order partial derivatives need to be calculated.

We can compute third-order and higher order partial derivatives. The number of partial derivatives increases exponentially with the order. There

14.4 HIGHER ORDER PARTIAL DERIVATIVES

are $8 = 2^3$ third-order partial derivatives of a function of two variables:

$$\frac{\partial^3 f}{\partial x^3}, \quad \frac{\partial^3 f}{\partial x^2 \partial y}, \quad \frac{\partial^3 f}{\partial x \partial y \partial x}, \quad \frac{\partial^3 f}{\partial x \partial y^2}, \quad \frac{\partial^3 f}{\partial y \partial x^2}, \quad \frac{\partial^3 f}{\partial y \partial x \partial y}, \quad \frac{\partial^3 f}{\partial y^2 \partial x}, \quad \frac{\partial^3 f}{\partial y^3},$$

and $27 = 3^3$ third-order partial derivatives of a function of three variables!

Functions built up out of the standard algebraic, trigonometric, inverse trigonometric, exponential, and logarithmic functions are continuously differentiable *infinitely many times* at most points at which they are defined. We could say that they are C^∞-differentiable. Most of the functions we'll discuss are certainly at least C^2-differentiable.

Geometrical and physical interpretations

The second derivative $\dfrac{d^2 y}{dx^2}$ gives us information about the concavity of the graph of a function of one variable $y = f(x)$. The second-order partial derivative of a function of two variables gives us information about the shape of the graph of the function $z = f(x, y)$.

If $\dfrac{\partial^2 f}{\partial x^2} > 0$, then the slice of $z = f(x, y)$ with y fixed is concave up.

If $\dfrac{\partial^2 f}{\partial x^2} < 0$, then the slice of $z = f(x, y)$ with y fixed is concave down.

Similarly, the sign of $\dfrac{\partial^2 f}{\partial y^2}$ tells us whether the slice with x fixed is concave up or down.

The mixed partials give us information about how slopes change from slice to slice. For example, if $\dfrac{\partial^2 f}{\partial x \partial y} > 0$, this means that the y-slopes are increasing as we move from slice to slice in the direction of increasing x.

Perhaps the most effective way to incorporate all the information that the second derivatives supply is to use them to find the **best quadratic approximation** to the function at that point. By definition, the best quadratic approximation to a C^2-differentiable function $f(x, y)$ at a point (a, b) is the quadratic function q satisfying

$$q(a,b) = f(a,b), \qquad q_x(a,b) = f_x(a,b), \qquad q_y(a,b) = f_y(a,b),$$

$$q_{xx}(a,b) = f_{xx}(a,b), \qquad q_{xy}(a,b) = f_{xy}(a,b), \qquad q_{yy}(a,b) = f_{yy}(a,b).$$

In other words, q and f and all their corresponding first- and second-order derivatives must match at the point (a, b). These six requirements completely determine the quadratic approximation q, as illustrated in the following example.

EXAMPLE 21 Find the best quadratic approximation to
$$f(x,y) = \sin(xy) + e^{2x} + e^y$$
at the point $(0,0)$.

Solution We need to find a quadratic function
$$q(x,y) = c_{11}x^2 + c_{12}xy + c_{22}y^2 + c_1 x + c_2 y + c_0$$
such that
$$q(0,0) = f(0,0), \qquad q_x(0,0) = f_x(0,0), \qquad q_y(0,0) = f_y(0,0),$$
$$q_{xx}(0,0) = f_{xx}(0,0), \qquad q_{xy}(0,0) = f_{xy}(0,0), \qquad q_{yy}(0,0) = f_{yy}(0,0).$$

First, we gather the needed information regarding the function f:

$$f(x,y) = \sin(xy) + e^{2x} - 2e^y \quad \Longrightarrow \quad f(0,0) = -1$$

$$f_x(x,y) = y\cos(xy) + 2e^{2x} \quad \Longrightarrow \quad f_x(0,0) = 2$$

$$f_y(x,y) = x\cos(xy) - 2e^y \quad \Longrightarrow \quad f_y(0,0) = -2$$

$$f_{xx}(x,y) = -y^2 \sin(xy) + 4e^{2x} \quad \Longrightarrow \quad f_{xx}(0,0) = 4$$

$$f_{xy}(x,y) = f_{yx}(x,y) = \cos(xy) - xy\sin(xy) \quad \Longrightarrow \quad f_{xy}(0,0) = 1$$

$$f_{yy}(x,y) = -x^2 \sin(xy) - 2e^y \quad \Longrightarrow \quad f_{yy}(0,0) = -2$$

Now, we do the same for the quadratic q and solve for the coefficients that guarantee matching values with f:

$$q(0,0) = c_0 = -1$$

$$q_x(x,y) = 2c_{11}x + c_{12}y + c_1 \quad \Longrightarrow \quad q_x(0,0) = c_1 = 2$$

$$q_y(x,y) = c_{12}x + 2c_{22}y + c_2 \quad \Longrightarrow \quad q_y(0,0) = c_2 = -2$$

$$q_{xx}(x,y) = 2c_{11} \quad \Longrightarrow \quad q_{xx}(0,0) = 2c_{11} = 4 \quad \Longrightarrow \quad c_{11} = 2$$

$$q_{xy}(x,y) = q_{yx}(x,y) = c_{12} \quad \Longrightarrow \quad c_{12} = 1$$

$$q_{yy}(x,y) = 2c_{22} = -2 \quad \Longrightarrow \quad c_{22} = -1$$

Hence the best quadratic approximation to f at $(0,0)$ is
$$2x^2 + xy - y^2 + 2x - 2y - 1.$$
∎

14.4 HIGHER ORDER PARTIAL DERIVATIVES

The technique used in this example can be applied to any C^2- differentiable function f. The coefficients of the best quadratic approximation to $f(x, y)$ at the point $(0, 0)$ are

$$c_0 = f(0, 0)$$

$$c_1 = f_x(0, 0) \qquad c_2 = f_y(0, 0)$$

$$c_{11} = \frac{1}{2} f_{xx}(0, 0) \qquad c_{12} = f_{xy}(0, 0) \qquad c_{22} = \frac{1}{2} f_{yy}(0, 0)$$

Definition 11

This best quadratic approximation is called the **second-degree Taylor polynomial approximation to f at** $(0, 0)$. For (x, y) sufficiently close to $(0, 0)$ we have

$$f(x, y) \approx f(0, 0) + f_x(0, 0)x + f_y(0, 0)y$$
$$+ \frac{1}{2} \left(f_{xx}(0, 0)x^2 + 2f_{xy}(0, 0)xy + f_{yy}(0, 0)y^2 \right).$$

We can find the best quadratic approximation of f at a point other than the origin.

Definition 12

The **second-degree Taylor polynomial approximation to f at** (a, b) is

$$f(x, y) \approx f(a, b) + f_x(a, b)(x - a) + f_y(a, b)(y - b)$$
$$+ \frac{1}{2} (f_{xx}(a, b)(x - a)^2 + 2f_{xy}(a, b)(x - a)(y - b) + f_{yy}(a, b)(y - b)^2).$$

EXAMPLE 22 Find the second-degree Taylor polynomial approximation to $f(x, y) = x^2 \ln y$ at $(-2, 1)$.

Solution First, we compute the first- and second-order derivatives:

$$f_x = 2x \ln y \qquad f_y = \frac{x^2}{y}$$

$$f_{xx} = 2 \ln y \qquad f_{xy} = \frac{2x}{y} \qquad f_{yy} = \frac{-x^2}{y^2}$$

Therefore, $f(x, y) \approx f(-2, 1) + f_x(-2, 1)(x + 2) + f_y(-2, 1)(y - 1)$

$$+ \frac{1}{2} \left(f_{xx}(-2, 1)(x + 2)^2 + 2f_{xy}(-2, 1)(x + 2)(y - 1) + f_{yy}(-2, 1)(y - 1)^2 \right)$$

$$= 0 + 0(x+2) + 4(y-1) + \frac{1}{2}\left(0(x+2)^2 + 2(-4)(x+2)(y-1) + (-4)(y-1)^2\right)$$

$$= 4(y-1) - 4(x+2)(y-1) - 2(y-1)^2. \qquad \blacksquare$$

The second-degree Taylor polynomial approximation is very useful in helping us analyze extrema of functions of two variables, as we'll see in the next section.

EXERCISES for Section 14.4

For exercises 1-8: First find the second-order partial derivatives, and then use these to find the best quadratic approximation to each function at $(0,0)$. Use the discriminant (see Chapter 13) to determine the shape ("bowl up," "bowl down," "saddle," "trough," or "flat,") of the quadratic approximation function near the input $(0,0)$.

1. $f(x,y) = x^2 + y^2$.
2. $f(x,y) = x^3 + y^3$.
3. $f(x,y) = x^3 - y^2$.
4. $f(x,y) = x^4 - y^4$.
5. $f(x,y) = xy$.
6. $f(x,y) = x^2 y$.
7. $f(x,y) = xy^3$.
8. $f(x,y) = x^2 y^2$.

9. At a point (a,b), a student finds the following for a particular function f:

$$f_x(a,b) > 0 \qquad f_y(a,b) < 0$$

$$f_{xx}(a,b) > 0 \qquad f_{xy}(a,b) < 0 \qquad f_{yy}(a,b) > 0.$$

Describe as completely as you can what this information means in terms of the slices of the graph of $z = f(x,y)$ at the point $(a, b, f(a,b))$.

10. For a function of two variables $z = f(x,y)$, how many fourth-degree partial derivatives are there? How about for a function of three variables?

For exercises 11-15: For each function, find all the second-order partial derivatives of the function f.

11. $f(x,y,z) = \dfrac{z}{x+y}$.
12. $f(x,y,z) = x^3 y^4 z^5$.
13. $f(x,y,z) = 5x - 4y + 3z$.
14. $f(x,y,z) = \exp 1 - x^2 - y^2 - z^2$.

14.5 FINDING EXTREMA OF MULTIVARIABLE FUNCTIONS—OPTIMIZATION

15. $f(x, y, z) = x \arcsin z - y \arctan z$.

For exercises 16-20: For each function, find all the second-order partial derivatives, then use them to find the best quadratic approximation to the function at the indicated point (a, b).

16. $f(x, y) = \dfrac{x+y}{x-y}$, $(a, b) = (1, 2)$.

17. $f(x, y) = xy + x/y$, $(a, b) = (-1, 3)$.

18. $f(x, y) = e^{x^2+y^2}$, $(a, b) = (1, -2)$.

19. $f(x, y) = \sin(x^2 y)$, $(a, b) = (2, 0)$.

20. $f(x, y) = \arctan(y/x)$, $(a, b) = (-3, -3)$.

For exercises 21-40: For each function, find all the second-order partial derivatives.

21. $f(x, y) = e^{x/y}$

22. $f(x, y) = \cos(x + y^2)$

23. $f(x, y) = \cos^2(xy)$

24. $f(x, y) = \dfrac{x^2}{x+y}$

25. $f(x, y) = \dfrac{2xy}{x-y}$

26. $f(x, y) = e^{x^2+y^2}$

27. $f(x, y) = (x^2 + y^2)^3$

28. $f(x, y) = \dfrac{x+y}{x-y}$

29. $f(x, y) = \ln(x^2 + y^2)$

30. $f(x, y) = x^2 y + \dfrac{x}{y^2}$

31. $f(x, y) = (x^2 + 1)\sin y$

32. $f(x, y) = \dfrac{x}{y^2} + \dfrac{y}{x^2}$

33. $f(x, y, z) = \ln(1 + x^2 + y^2 - z^2)$

34. $f(x, y, z) = \sin(xyz)$

35. $f(x, y, z) = e^{2x+y^2+z}$

36. $f(x, y, z) = \dfrac{x}{x+y^2+z^3}$

37. $f(x, y, z) = \sin x \sin y \sin z$

38. $f(x, y, z) = \sqrt{x^2 + y^2 + z^2}$

39. $f(x, y, z) = \dfrac{xyz}{x+y+z}$

40. $f(x, y, z) = (z^3 + y^2 + x)^3$

14.5 FINDING EXTREMA OF MULTIVARIABLE FUNCTIONS—OPTIMIZATION

In this section we show how the derivatives of multivariable functions can be used to detect the locations of and analyze its extrema (maxima and minima). First, let's review some of the basic terminology regarding extrema of single-variable functions.

An input x_0 is the location of a *relative* or *local maximum* for $y = f(x)$ if there is a neighborhood of x_0 (an open interval centered at x_0) such that $f(x_0) \geq f(x)$ for all x in the neighborhood. The input x_0 is the location of a *relative* or *local minimum* for $y = f(x)$ if $f(x_0) \leq f(x)$ for all x in some neighborhood.

We can make very similar definitions for continuous multivariable functions.

Definition 13

> The input (x_0, y_0) for a function f of two variables is the location of a **relative** or **local maximum** if there is some neighborhood (the *interior of a circle* centered at (x_0, y_0)) such that
>
> $$f(x_0, y_0) \geq f(x, y)$$
>
> for all (x, y) in the neighborhood. The input (x_0, y_0) is the location of a **relative** or **local minimum** if
>
> $$f(x_0, y_0) \leq f(x, y)$$
>
> for all (x, y) in some neighborhood of (x_0, y_0). For a function of three variables, a point (x_0, y_0, z_0) is the location of either a **relative maximum** or **relative minimum** if there is a neighborhood (the *interior of a sphere* centered at (x_0, y_0, z_0)) such that
>
> $$f(x_0, y_0, z_0) \geq f(x, y, z) \quad \text{or} \quad f(x_0, y_0, z_0) \leq f(x, y, z),$$
>
> respectively, for all points (x, y, z) in the neighborhood.

Critical points

The key observation for a continuous function f of a single variable is that its relative extrema can occur only at *critical points*. A critical point x_0 is an input in the domain of f such that

$$f'(x_0) = 0 \quad \text{or} \quad f'(x_0) \text{ is undefined.}$$

Now, for a continuous multivariable function, a relative extremum will also appear to be a relative extremum on every *slice* of the function at that point. Geometrically, all this means is that a high or low point on the surface graph must also be a high or low point on every vertical cross-section through that point. If we hold all the variables constant except one and graph the resulting single variable slice function, a relative extremum must be a *critical point* for this slice. The corresponding partial derivative must be zero or undefined at the extremum. This leads us to make the following definition.

14.5 FINDING EXTREMA OF MULTIVARIABLE FUNCTIONS—OPTIMIZATION

Definition 14

> If f is a continuous multivariable function, then an input point having a neighborhood in the domain of f is called a **critical point** if either
>
> $$\nabla f = 0 \quad \text{or} \quad \nabla f \text{ is undefined}$$
>
> at that point.

In other words, a critical point for f is some point where *all the partial derivatives are zero*, or where *at least one partial derivative is undefined*.

Definition 15

> If f is a C^1-differentiable function, then a critical point where $\nabla f = 0$ is called a **stationary point**.

This terminology is well-suited, since the instantaneous rate of change in *any* direction from a stationary point is 0 (because the gradient is 0 at a stationary point, its dot product with any unit vector is 0). For a C^1-differentiable function of two variables, a stationary point will be the location of a *horizontal tangent plane* $z = c$, where c is the function value at the stationary point.

EXAMPLE 23 Find the stationary critical points of the function $f(x,y) = x^2 + 5xy - y^3$. What is the equation of the tangent plane at each of these points?

Solution $\nabla f = \langle 2x + 5y, 5x - 3y^2 \rangle = \langle 0, 0 \rangle$ only when we simultaneously have

$$2x + 5y = 0$$
$$5x - 3y^2 = 0.$$

Solving the first equation for x gives us $x = -5y/2$. Substituting into the second equation gives us

$$5(-5y/2) - 3y^2 = (y/2)(-25 - 6y) = 0 \implies y = 0 \quad \text{or} \quad y = -25/6.$$

When $y = 0$, $x = -5(0)/2 = 0$. When $y = -25/6$, $x = 125/12$. The two stationary critical points are

$$(0,0) \quad \text{and} \quad (125/12, -25/6).$$

The horizontal tangent plane at $(0,0)$ is

$$z = f(0,0) = 0.$$

The horizontal tangent plane at $(125/12, -25/6)$ is

$$z = f(125/12, -25/6) = -15625/432 \approx -36.17.$$

EXAMPLE 24 Find the stationary critical points of the function $f(x,y,z) = e^{xyz}$.

Solution $\nabla f = \langle yze^{xyz}, xze^{xyz}, xye^{xyz}\rangle = \langle 0,0,0\rangle$ only when we simultaneously have

$$yze^{xyz} = 0$$
$$xze^{xyz} = 0$$
$$xye^{xyz} = 0.$$

The factor e^{xyz} is never 0. We can see that all three of the partial derivatives are 0 whenever at least two of x, y, and z are 0. The set of stationary critical points of f consists of every point on the three coordinate-axes. ∎

Second derivative test for extrema of functions of two variables

For a function of two variables, three possible shapes the graph of $z = f(x,y)$ could have at a stationary critical point are a local minimum, local maximum, or a saddle point. Figure 14.5 illustrates each of these shapes.

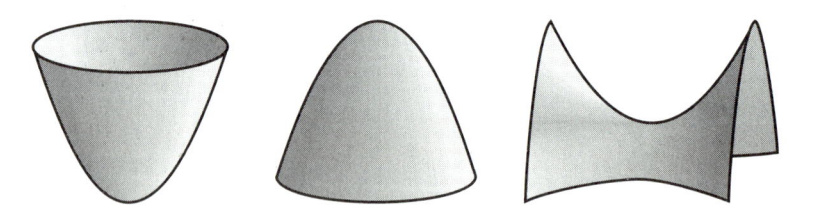

Figure 14.5 Stationary critical points: local minimum, local maximum, saddle point

At a stationary critical point, we know that both of the first-order partial derivatives are zero. Now, we'll show how the second-order partial derivatives can often be used to detect the shape of the graph.

If f is a C^2-differentiable function of two variables, then it has a second-degree Taylor approximation at any point (a,b) which can be written

$$f(x,y) \approx f(a,b) + \nabla f(a,b) \cdot \langle x-a, y-b\rangle$$
$$+ \frac{1}{2}(f_{xx}(a,b)(x-a)^2 + 2f_{xy}(a,b)(x-a)(y-b) + f_{yy}(a,b)(y-b)^2).$$

If (a,b) happens to be a stationary critical point for the function f, then

$$\nabla f(a,b) = \mathbf{0},$$

and the Taylor polynomial approximation simplifies to

14.5 FINDING EXTREMA OF MULTIVARIABLE FUNCTIONS—OPTIMIZATION

$$f(x,y) \approx f(a,b) + \frac{1}{2}(f_{xx}(a,b)(x-a)^2 + 2f_{xy}(a,b)(x-a)(y-b) + f_{yy}(a,b)(y-b)^2).$$

If we write

$$A = f_{xx}(a,b), \qquad B = f_{xy}(a,b), \qquad C = f_{yy}(a,b),$$

then we have

$$f(x,y) \approx f(a,b) + \frac{1}{2}(A(x-a)^2 + 2B(x-a)(y-b) + C(y-b)^2).$$

In the last chapter, we saw that the *discriminant* of a quadratic polynomial of the form

$$A(x-a)^2 + 2B(x-a)(y-b) + C(y-b)^2$$

is the quantity

$$D = AC - B^2.$$

The sign of this discriminant allows us to analyze the shape of the graph of the quadratic at (a,b).

The graph of our original function $z = f(x,y)$ has a shape very similar to this quadric surface near the point $(a, b, f(a,b))$. For this reason we say that the quantity

$$\Delta = AC - B^2,$$

where

$$A = \frac{\partial^2 f}{\partial x^2}, \qquad B = \frac{\partial^2 f}{\partial x \partial y}, \qquad C = \frac{\partial^2 f}{\partial y^2}$$

is the **discriminant** of the function $z = f(x,y)$. We can use this discriminant Δ for a C^2-differentiable function f in essentially the same way we use the discriminant D for quadratic functions, as summarized in the following theorem.

Theorem 14.1

Second derivative test for functions of two variables.

Suppose that $z = f(x,y)$ is a C^2-differentiable function, and let $\Delta = AC - B^2$, where

$$A = \frac{\partial^2 f}{\partial x^2}, \qquad B = \frac{\partial^2 f}{\partial x \partial y}, \qquad C = \frac{\partial^2 f}{\partial y^2}.$$

Hypothesis: The gradient $\nabla f(a,b) = \langle 0,0 \rangle$.
Case 1: $\Delta < 0$ at (a,b).
Conclusion: A saddle point occurs at (a,b).

Case 2: $\Delta > 0$ and $A > 0$ at (a,b).
Conclusion: A relative minimum occurs at (a,b).

Case 3: $\Delta > 0$ and $A < 0$ at (a,b).
Conclusion: A relative maximum occurs at (a,b).

Case 4: $\Delta = 0$ at (a,b).
Conclusion: None.

A brief comment is in order regarding Case 4 of the theorem. Here, "no conclusion" simply means that when $\Delta = 0$, we do not have enough information to identify the shape of the graph. It is still possible that the critical point is the location of a relative minimum, relative maximum, or a saddle point, but we will need to use some other method to make that identification (such as graphing a wireframe plot or examining a contour map of level curves).

If we form the 2×2 matrix of second-order partial derivatives

$$\begin{bmatrix} \frac{\partial^2 f}{\partial x^2} & \frac{\partial^2 f}{\partial x \partial y} \\ \frac{\partial^2 f}{\partial y \partial x} & \frac{\partial^2 f}{\partial y^2} \end{bmatrix},$$

then note that we can think of the discriminant Δ as the *determinant* of this matrix:

$$\Delta = \begin{vmatrix} \frac{\partial^2 f}{\partial x^2} & \frac{\partial^2 f}{\partial x \partial y} \\ \frac{\partial^2 f}{\partial y \partial x} & \frac{\partial^2 f}{\partial y^2} \end{vmatrix} = \begin{vmatrix} A & B \\ B & C \end{vmatrix} = AC - B^2.$$

Note that we are using the fact that $\frac{\partial^2 f}{\partial x \partial y} = \frac{\partial^2 f}{\partial y \partial x}$ for a C^2-differentiable function.

EXAMPLE 25 Use the second derivative test to classify each stationary critical point of the function $z = f(x,y) = x^2 + 5xy - y^3$ as a saddle point or relative extremum, if possible.

14.5 FINDING EXTREMA OF MULTIVARIABLE FUNCTIONS—OPTIMIZATION

Solution We found the stationary critical points to be $(0,0)$ and $(125/12, -25/6)$. The second-order partial derivatives are

$$\frac{\partial^2 f}{\partial x^2} = 2, \qquad \frac{\partial^2 f}{\partial x \partial y} = 5, \qquad \frac{\partial^2 f}{\partial y^2} = -6y.$$

At $(0,0)$, we have

$$A = 2, \qquad B = 5, \qquad C = 0,$$

and so

$$\Delta = AC - B^2 = 0 - 25 = -25 < 0.$$

We conclude that $(0,0)$ is the location of a saddle point.

At $(125/12, -25/6)$, we have

$$A = 2, \qquad B = 5, \qquad C = 25,$$

and so

$$\Delta = AC - B^2 = 50 - 25 = 25 > 0.$$

Since $A = 2 > 0$, we conclude that $(125/12, -25/6)$ is the location of a relative minimum. ∎

Absolute extrema over closed domains

An interval of real numbers $[a,b]$ is called *closed and bounded* since it has finite length and includes its endpoints. A region of the plane is called *closed and bounded* if it can be enclosed within a circle of finite area and the region includes all the points along its boundary. Similarly, a region of space is closed and bounded if it can be enclosed within a sphere of finite volume and the region includes all the points along its boundary.

Just as there is an extreme value theorem for single variable functions, we have an extreme value theorem for multivariable functions.

Theorem 14.2

Extreme value theorem for multivariable functions.
Hypothesis 1. The function $f : D \to \mathbb{R}$ is continuous over its domain D.
Hypothesis 2. D is closed and bounded.
Conclusion. The function f attains its extreme maximum and minimum values at some points within its domain D.

To find the **absolute extrema** of a function of two variables over some domain D in the xy-plane, we must check the output values at (1) the locations of all relative extrema that lie in D and (2) points along the boundary of D. Along the boundary of the domain, we may be able to write the function in terms of a single variable and use single-variable calculus to locate the relevant extrema along the boundary.

EXAMPLE 26 Find the absolute minimum and maximum of $z = f(x,y) = x^2 + 5xy - y^3$ over the domain $D = [0,2] \times [0,2]$.

Solution The domain in question is a square with corners at $(0,0)$, $(2,0)$, $(2,2)$, and $(0,2)$ (see Figure 14.6). In the previous example, we found that the only relative extremum occurs at $(125/12, -25/6)$, which lies outside the domain D. Hence, we need to examine the points along the boundary.

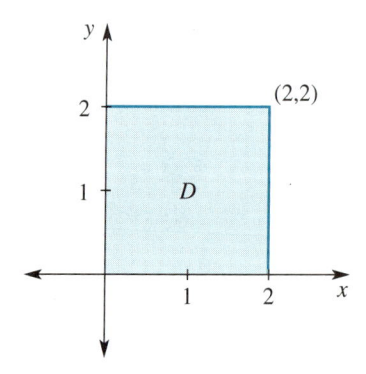

Figure 14.6 Domain D is a square.

First, points along the bottom edge of the square have coordinates

$$(x, 0) \quad \text{for } 0 \leq x \leq 2.$$

Substituting into the function, we have

$$z = f(x, 0) = x^2,$$

so the minimum value along the bottom edge is $f(0,0) = 0$ and the maximum value is $f(2,0) = 4$.

Along the left edge of the square, the points have coordinates

$$(0, y) \quad \text{for } 0 \leq y \leq 2.$$

Substituting into the function, we have

$$z = f(0, y) = -y^3,$$

so the minimum value along the left edge is $f(0,2) = -8$ and the maximum value is $f(0,0) = 0$.

Along the top edge of the square, the points have coordinates

$$(x, 2) \quad \text{for } 0 \leq x \leq 2.$$

14.5 FINDING EXTREMA OF MULTIVARIABLE FUNCTIONS—OPTIMIZATION

Substituting into the function, we have

$$z = f(x, 2) = x^2 + 10x - 8.$$

If we treat this as a function of a single variable x, we note that its derivative is $dz/dx = 2x + 10$. The critical point $x = -5$ lies outside the interval of x values, so the extrema must occur at the endpoints $x = 0$ and $x = 2$. The minimum value is $f(0, 2) = -8$ and the maximum value is $f(2, 2) = 16$.

Finally, along the right edge of the square, the points have coordinates

$$(2, y) \quad \text{for } 0 \leq y \leq 2.$$

Substituting into the function, we have

$$z = f(2, y) = 4 + 10y - y^3.$$

If we treat this as a function of a single variable y, we note that its derivative is $dz/dy = 10 - 3y^2$. Of the two critical points $y = \pm\sqrt{10/3}$, one *does* lie in the interval of y values, since

$$0 \leq \sqrt{10/3} \leq 2.$$

Hence, we check $y = 0$, $y = 2$, and $y = \sqrt{10/3}$. The corresponding output values are

$$f(2, 0) = 4, \quad f(2, \sqrt{10/3}) \approx 16.17, \quad f(2, 2) = 16.$$

Now we can see that the absolute minimum value $z = -8$ occurs at the corner $(0, 2)$, while the absolute maximum value $z \approx 16.17$ occurs at $(2, \sqrt{10/3})$ along the right edge of the square. ∎

EXAMPLE 27 Platypus Corporation must make a rectangular steel box whose three dimensions must have a sum of 120 cm. What is the maximum volume the box can have, and what are its dimensions?

Solution If we label the dimensions of the box x, y, and z, then we must maximize the volume

$$V = xyz,$$

given that

$$x + y + z = 120.$$

The constraint on the dimensions allows us to write the volume as a function of two variables. Since $z = 120 - x - y$, we have

$$V(x, y) = xy(120 - x - y) = 120xy - x^2 y - xy^2.$$

The critical points are where

$$\frac{\partial V}{\partial x} = 120y - 2xy - y^2 = y(120 - 2x - y) = 0,$$

and

$$\frac{\partial V}{\partial y} = 120x - x^2 - 2xy = x(120 - x - 2y) = 0.$$

We observe that both of these equations are satisfied if $(x, y) = (0, 0)$. (These are extremely small dimensions for a box, though, and certainly won't give us a maximum volume.)

Now, if $y \neq 0$, we can solve $\partial V/\partial x = 0$ for x,

$$x = \frac{y^2 - 120y}{-2y} = 60 - \frac{y}{2},$$

and substitute this result into the equation $\partial V/\partial y = 0$:

$$(60 - y/2)(120 - (60 - y/2) - 2y) = (60 - y/2)(60 - 3y/2) = 0.$$

This last equation is satisfied when $y = 120$ or when $y = 40$. If $y = 120$, then $x = z = 0$ and the box has zero volume. The value

$$y = 40 \text{ cm}$$

means that $\quad x = 60 - y/2 = 60 - 40/2 = 60 - 20 = 40$ cm.

If we check the second derivatives of V, we find

$$\frac{\partial^2 V}{\partial x^2} = -2y, \qquad \frac{\partial^2 V}{\partial x \partial y} = 120 - 2x - 2y, \qquad \frac{\partial^2 V}{\partial y^2} = -2x.$$

At $(x, y) = (40, 40)$, we have $A = -2(40) = -80$, and

$$\Delta = AC - B^2 = (-80)(-80) - (120 - 80 - 80)^2 = 6400 - 1600 = 4800 > 0.$$

The second derivative test tells us that a maximum occurs when $x = y = 40$ cm. The third dimension is

$$z = 120 - x - y = 120 - 40 - 40 = 40 \text{ cm},$$

and we conclude that the maximum volume of the box is

$$V = xyz = (40)(40)(40) = 64000 \text{ cm}^3,$$

obtained when the box is in the shape of a cube 40 cm on a side. ∎

EXERCISES for Section 14.5

For exercises 1-15: For each of the following functions of two variables, find all the stationary critical points (where $\nabla f = 0$), and then use the second derivative test to classify each point as a relative minimum, relative maximum, or saddle point. If no conclusion is possible, use graphical techniques to decide on the status of the critical point.

1. $f(x,y) = 3x^2 - 2y^2 + 3x - 2y + 5$.
2. $f(x,y) = 4xy - x^4 - 2y^2$.
3. $f(x,y) = 3xy - x^3 - y^3$.
4. $f(x,y) = 1 - x^2 + y^2$.
5. $f(x,y) = 9 + x^2 - y^2$.
6. $f(x,y) = 6x^2 - 2x^3 + 3y^2 + 6xy$.
7. $f(x,y) = x^4 + y^3 - 8x^2 - 3y + 5$.
8. $f(x,y) = x^2 - xy + y^2 - 2x + y$.
9. $f(x,y) = 3x^4 - 4x^2 y + y^2$.
10. $f(x,y) = \dfrac{x^2 + y^2}{e^y}$.
11. $f(x,y) = y^3 - 3x^2 y$.
12. $f(x,y) = 4xy^3 - 4x^3 y$.
13. $f(x,y) = \cos x \sinh y$.
14. $f(x,y) = x^4 - y^4$.
15. $f(x,y) = x^2 y^2$.

For exercises 16-20: For each of the functions of two variables, find the absolute maximum and minimum values of the function in the square region $D = [-1, 0] \times [-1, 0]$.

16. $f(x,y) = 3x^2 - 2y^2 + 3x - 2y + 5$.
17. $f(x,y) = 4xy - x^4 - 2y^2$.
18. $f(x,y) = 3xy - x^3 - y^3$.
19. $f(x,y) = 1 - x^2 + y^2$.
20. $f(x,y) = 9 + x^2 - y^2$.

For exercises 21-25: For each of the functions of two variables, find the absolute maximum and minimum values of the function in the triangular region bounded by the coordinate axes and the line $y = 1 - x$.

21. $f(x,y) = 6x^2 - 2x^3 + 3y^2 + 6xy$.
22. $f(x,y) = x^4 + y^3 - 8x^2 - 3y + 5$.
23. $f(x,y) = x^2 - xy + y^2 - 2x + y$.
24. $f(x,y) = 3x^4 - 4x^2 y + y^2$.
25. $f(x,y) = \dfrac{x^2 + y^2}{e^y}$.

14.6 THE METHOD OF LAGRANGE MULTIPLIERS—EXTREMA UNDER CONSTRAINTS

The second derivative test for functions of two variables is a handy tool. In this section we discuss a technique that can be used in some optimization problems involving functions of two, three, or more variables.

Sometimes we are faced with a problem of finding the maximum or minimum value of a multivariable function under some constraint on the possible inputs. That is, we're given some condition or conditions that the inputs must satisfy, and our task is to find an extremum of a function under those conditions. Here's an example of such a problem:

Problem: Find the maximum temperature that occurs at a point on a given surface in space.

The problem translates to finding the extreme value of a function $f(x,y,z)$ (in this case, temperature) under the constraint that $g(x,y,z) = c$ (the equation of the surface). Let's examine the geometric implications of the problem in terms of the gradients of both f and g.

First, let's assume that of the points on the surface described by $g(x,y,z) = c$, an extreme value of f is attained at (x_0, y_0, z_0), and that

$$\nabla g(x_0, y_0, z_0) \neq \mathbf{0}.$$

From properties of the gradient, we know that

▶▶▶ $\nabla g(x_0, y_0, z_0)$ **is normal to the level surface** $g(x,y,z) = c$.

Now, suppose that we have a differentiable path $\mathbf{r}(t)$ whose image curve lies completely on the surface $g(x,y,z) = c$, with

$$\mathbf{r}(t_0) = (x_0, y_0, z_0) \quad \text{and} \quad \mathbf{r}'(t_0) \neq \mathbf{0}.$$

In other words, you can imagine that we have drawn a smooth curve on the surface that passes through the point (x_0, y_0, z_0).

Since f attains an extreme value at (x_0, y_0, z_0), we must also have $f(\mathbf{r}(t))$ attaining an extreme value at $t = t_0$. This means that $t = t_0$ must be a *critical point* for $f(\mathbf{r}(t))$, so that

$$\frac{d}{dt} f(\mathbf{r}(t_0)) = 0.$$

The chain rule tells us that

$$\frac{d}{dt} f(\mathbf{r}(t)) = \nabla f(\mathbf{r}(t)) \cdot \mathbf{r}'(t),$$

so we must have $\quad \nabla f(\mathbf{r}(t_0)) \cdot \mathbf{r}'(t_0) = 0.$

Geometrically, this zero dot product tells us that $\nabla f(\mathbf{r}(t_0))$ is orthogonal to $\mathbf{r}'(t_0)$. But $\mathbf{r}'(t_0)$ is a vector that is tangent to the image curve of $\mathbf{r}$, and so it must also be tangent to the surface $g(x,y,z) = c$. Since $\nabla f(x_0, y_0, z_0)$ is perpendicular to every such image curve passing through (x_0, y_0, z_0), we conclude that

14.6 THE METHOD OF LAGRANGE MULTIPLIERS—EXTREMA UNDER CONSTRAINTS

▶▶▶ $\nabla f(x_0, y_0, z_0)$ **is normal to the level surface** $g(x, y, z) = c$.

The two gradients must point in the same or opposite directions! Put another way, there must be some nonzero scalar λ such that

$$\nabla f(x_0, y_0, z_0) = \lambda \nabla g(x_0, y_0, z_0).$$

Let's summarize the results of our reasoning in the statement of a theorem.

Theorem 14.3

Lagrange multiplier theorem (one constraint)
Hypothesis 1: The functions f and g are C^1-differentiable.
Hypothesis 2: The function f has an extremum under the constraint $g(x, y, z) = c$ at the point (x_0, y_0, z_0).
Hypothesis 3: The gradient $\nabla g(x_0, y_0, z_0) \neq 0$.
Conclusion: For some nonzero scalar λ, we must have

$$\nabla f(x_0, y_0, z_0) = \lambda \nabla g(x_0, y_0, z_0).$$

Note: The scalar λ is called a **Lagrange multiplier**, after the Italian-French mathematician Lagrange (1736-1813).

Note that the equation

$$\nabla f = \lambda \nabla g$$

actually represents three equations:

$$\frac{\partial f}{\partial x} = \lambda \frac{\partial g}{\partial x}, \qquad \frac{\partial f}{\partial y} = \lambda \frac{\partial g}{\partial y}, \qquad \frac{\partial f}{\partial z} = \lambda \frac{\partial g}{\partial z}.$$

Along with the original constraint $g(x, y, z) = c$, this gives us a total of four equations in four unknowns (x, y, z, and λ). The particular value of λ may not be of much interest, but its introduction as an *auxiliary variable* allows us to solve for $x = x_0$, $y = y_0$, and $z = z_0$, the coordinates of the potential location of an extremum. This strategy, a direct application of the theorem, is known as the **method of Lagrange multipliers**. We outline the method below:

Problem: Find the extreme values of $f(x, y, z)$ subject to the constraint $g(x, y, z) = c$, using the **method of Lagrange multipliers**:

Step 1. Compute the gradient ∇f.

Step 2. Compute the gradient ∇g.

Step 3. Solve the following system of equations for x, y, z, and λ:

$$\frac{\partial f}{\partial x} = \lambda \frac{\partial g}{\partial x}$$

$$\frac{\partial f}{\partial y} = \lambda \frac{\partial g}{\partial y}$$

$$\frac{\partial f}{\partial z} = \lambda \frac{\partial g}{\partial z}$$

$$g(x, y, z) = c.$$

Step 4. Check the values x_0, y_0, z_0 back in the original problem to determine whether $f(x_0, y_0, z_0)$ is an extreme value subject to the constraint.

EXAMPLE 28 Find the point on the plane $x - y + 2z = 5$ that is closest to the origin.

Solution The distance from any point (x, y, z) to the origin is

$$d = \sqrt{x^2 + y^2 + z^2}.$$

If we can minimize the function

$$f(x, y, z) = x^2 + y^2 + z^2$$

subject to the constraint $g(x, y, z) = x - y + 2z = 5$, we will have our solution.

The gradient of f is

$$\nabla f = \langle 2x, 2y, 2z \rangle,$$

and the gradient of our constraint function g is

$$\nabla g = \langle 1, -1, 2 \rangle.$$

The method of Lagrange multipliers leads us to solve

$$\nabla f = \lambda \nabla g$$

along with the original constraint. That is, we must solve

$$2x = \lambda, \quad 2y = -\lambda, \quad 2z = 2\lambda,$$

and

$$x - y + 2z = 5.$$

The three equations involving λ lead us to

$$x = \lambda/2, \quad y = -\lambda/2, \quad z = \lambda.$$

Substituting into the original constraint equation, we have

$$(\lambda/2) - (-\lambda/2) + 2\lambda = 5,$$

14.6 THE METHOD OF LAGRANGE MULTIPLIERS—EXTREMA UNDER CONSTRAINTS

which yields $3\lambda = 5$ or $\lambda = 5/3$. The coordinates of the point we seek are

$$x = 5/6, \qquad y = -5/6, \qquad z = 5/3.$$

In other words, $(5/6, -5/6, 5/3)$ is the point on the plane $x - y + 2z = 5$ closest to the origin. You can check that this point indeed satisfies the equation of the plane. (Also note that a normal vector to the plane is $\langle 1, -1, 2 \rangle$, while the vector from the origin to the point $(5/6, -5/6, 5/3)$ is

$$\langle 5/6, -5/6, 5/3 \rangle = (5/6)\langle 1, -1, 2 \rangle.$$

In other words, the point we found is on the perpendicular between the origin and the plane, and is therefore the closest point to the origin.) ■

EXAMPLE 29 Platypus Corporation must make a rectangular steel box whose three dimensions have a sum of 120 cm. What is the maximum volume the box can have, and what are its dimensions?

Solution We have solved this problem before, but this time we wish to illustrate the method of Lagrange multipliers. Our function to maximize is

$$V(x, y, z) = xyz$$

and our constraint is

$$g(x, y, z) = x + y + z = 120.$$

The gradient equation is

$$\nabla V = \langle yz, xz, xy \rangle = \lambda \langle 1, 1, 1 \rangle = \nabla g.$$

From this, we can see that x, y, and z must all be nonzero, and

$$yz = \lambda \quad \Longrightarrow \quad y = \frac{\lambda}{z},$$

$$xz = \lambda \quad \Longrightarrow \quad x = \frac{\lambda}{z},$$

$$xy = \lambda \quad \Longrightarrow \quad \frac{\lambda^2}{z^2} = \lambda \quad \Longrightarrow \quad \lambda = z^2.$$

If we substitute the third result into the first two, we see that

$$x = y = z.$$

The original constraint equation

$$x + y + z = 120$$

tells us that we must have $x = y = z = 40$ cm and that the maximum volume of the box is $V = xyz = 64000$ cm^3. ∎

▶▶▶ **The Lagrange multiplier theorem also holds for functions of two variables. That is, if f and g are C^1-differentiable functions of two variables and f has an extremum at (x_0, y_0) under the constraint $g(x, y) = c$, then**

$$\nabla f(x_0, y_0) = \lambda \nabla g(x_0, y_0)$$

for some Lagrange multiplier λ (provided that $\nabla g(x_0, y_0) \neq 0$).

EXERCISES for Section 14.6

For exercises 1-5: For each of these surfaces, use the technique of Lagrange multipliers to find the point on the surface closest to the origin.

1. $3x + 2y + z - 6 = 0$.
2. $(x-1)^2 + (y+2)^2 - (z-3)^2 = 4$.
3. $xyz = 8$.
4. $z = x^2 + y^2 - 4$. (Hint: rewrite the constraint in the form $g(x, y, z) = 0$ first.)
5. $xy + xz + yz = 12$.

6. Platypus Corporation is assigned to make a rectangular steel box with a total surface area of 1200 cm^2. What is the maximum volume of the box, and what are its dimensions?

7. Platypus Corporation is assigned to make a rectangular steel box with a fixed volume of 6000 cm^3. What is the minimum surface area of the box, and what are its dimensions?

8. Platypus Corporation is assigned to make a rectangular steel box *with an open top* and a fixed volume of 6000 cm^3. What is the minimum surface area of the box, and what are its dimensions?

9. Platypus Corporation must make a rectangular steel box with an open top of fixed volume 6000 cm^3 and with a reinforced bottom that costs twice as much per square centimeter as the four sides. What dimensions of the box will cost the least?

10. Platypus Corporation must make a closed rectangular steel box with a fixed volume of 6000 cm^3 out of material costing 20 cents per square centimeter for the four sides, 30 cents per square centimeter for the top, and 40 cents per square centimeter for the bottom. What dimensions of the box will cost the least?

14.6 THE METHOD OF LAGRANGE MULTIPLIERS—EXTREMA UNDER CONSTRAINTS

11. It is possible to have more than one constraint and still use the method of Lagrange multipliers. For example, we may be seeking the maximum temperature along a curve in space that is described as the intersection of two surfaces. In this situation, the gradient of the scalar field may be a *linear combination* of the constraint gradients at an extremum.

If we seek an extremum of the function $f(x, y, z)$ under the constraints

$$g(x, y, z) = c \quad \text{and} \quad h(x, y, z) = d,$$

then we look for points (x_0, y_0, z_0) satisfying the constraints as well as the gradient equation

$$\nabla f(x_0, y_0, z_0) = \lambda \nabla g(x_0, y_0, z_0) + \mu \nabla h(x_0, y_0, z_0)$$

for the Lagrange multipliers λ and μ. (However, the method may fail if one of the constraint gradients is a scalar multiple of the other.)

Use the technique of Lagrange multipliers to solve the following problem: Find the point that lies on both the paraboloid $z = x^2 + y^2$ and the plane $x + y + z = 6$ that is closest to the origin. Also find the point satisfying these constraints that lies farthest from the origin.

15

Integral Calculus of Multivariable Functions

The last chapter discussed the differential calculus of multivariable scalar-valued functions. To understand the principles of integration of functions of several variables, it will be wise for us to review the idea of definite integral for functions of a single variable.

If f is a real-valued function defined over the closed interval $[a,b]$, then we form a *Riemann sum* using the following steps:

Step 1. Form a regular partition of the interval $[a,b]$ by subdividing it into n equal-size subintervals of length Δx.

Step 2. Choose a single input x_i from each subinterval for $1 \leq i \leq n$.

Step 3. For each x_i, evaluate $f(x_i)$ and multiply by the length of the subinterval Δx.

Step 4. Sum up the results for all the subintervals:

$$\sum_{i=1}^{n} f(x_i)\,\Delta x.$$

Now, if a limiting value exists for these Riemann sums as we take finer and finer partitions (in other words, as n grows larger without bound and Δx approaches 0), then this limiting value is called the value of the definite integral

$$\int_a^b f(x)\,dx.$$

If $f(x) \geq 0$ for all $x \in [a,b]$, then this definite integral has a geometrical interpretation as the *area* under graph of $y = f(x)$ over this interval (see Figure 15.1).

A definite integral $\int_a^b f(x)\,dx$ can have a variety of *physical* meanings, depending on the interpretation of the function f. For example, we know that if $f(x)$ represents *force* at distance x, then $\int_a^b f(x)\,dx$ represents *work* performed over the interval $[a,b]$. In this chapter, we see how the notion of

903

definite integral can be extended to functions of several variables. These **multiple integrals** also have both geometric and physical interpretations.

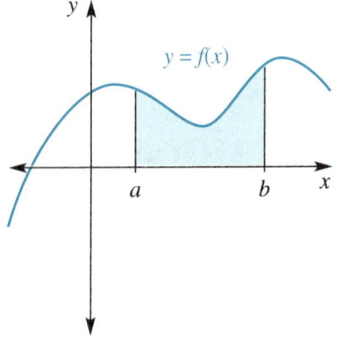

Figure 15.1 Geometric interpretation: $\int_a^b f(x)\,dx$ = area of shaded region.

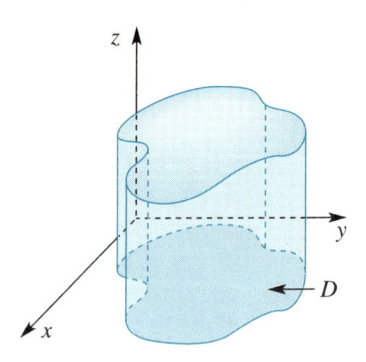

Figure 15.2 Geometric interpretation of $\iint_D f(x,y)\,dA$.

15.1 MULTIPLE INTEGRALS

Double integrals—notation and interpretation

Just as a definite integral of a function of one variable has a geometrical interpretation related to *area*, a **double integral** of a function of two variables has a geometrical interpretation related to *volume*. In short, if D is some region in the xy-plane, and if

$$f(x,y) \geq 0$$

for all $(x,y) \in D$, then

$$\iint_D f(x,y)\,dA$$

represents the *volume* between the graph of $z = f(x,y)$ and the xy-plane over the region D (see Figure 15.2).

Some comments regarding the notation need to be made here. The use of two integral signs $\iint$ reminds us that we're integrating over a 2-dimensional region D. The dA refers to an "area element." (We'll see that the precise form of dA will depend on how we choose to compute the value of the integral.)

15.1 MULTIPLE INTEGRALS

Calculating double integrals over rectangular regions

Let's start out by describing the method for the simplest type of region D, that is, a *rectangular region* whose sides lie parallel to the coordinate axes (see Figure 15.3).

In this case, we can represent D as a Cartesian product of closed intervals

$$D = [a, b] \times [c, d]$$
$$= \{(x, y) : a \leq x \leq b \text{ and } c \leq y \leq d\}.$$

Suppose that we form a regular partition of each of the intervals $[a, b]$ and $[c, d]$. That is, we subdivide $[a, b]$ into m equal-sized subintervals of length Δx and $[c, d]$ into n equal-sized subintervals of length Δy. (Note that it is *not* necessary to have either $m = n$ or $\Delta x = \Delta y$.) These two partitions create a grid of small rectangles, each of dimension Δx by Δy and having area

$$\Delta A = \Delta x \Delta y,$$

as shown in Figure 15.4.

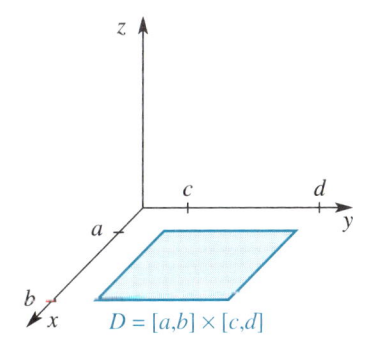

Figure 15.3 Rectangular region $D = [a, b] \times [c, d]$.

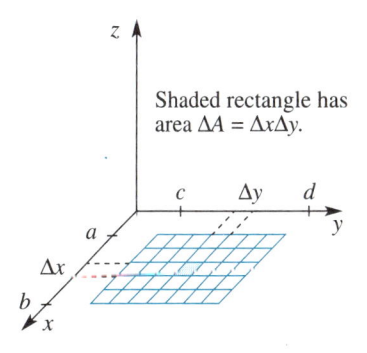

Figure 15.4 Grid partition of $D = [a, b] \times [c, d]$.

We have a total of $m \times n$ rectangles, which we could index with the letters i and j (i indexes a subinterval in the x-direction, so $1 \leq i \leq m$; j represents a subinterval in the y-direction, so $1 \leq j \leq n$). If we choose one point (x_i, y_j) from each rectangle, we can form the *double* Riemann sum

$$\sum_{j=1}^{n} \sum_{i=1}^{m} f(x_i, y_j) \Delta x \Delta y,$$

where the double summation signs simply indicate that we sum over all of the rectangles.

Geometrically, if (x, y) is a point in a small rectangle of dimensions Δx by Δy, then $f(x, y)\Delta x \Delta y$ is the volume of a box with the height $f(x, y)$ and

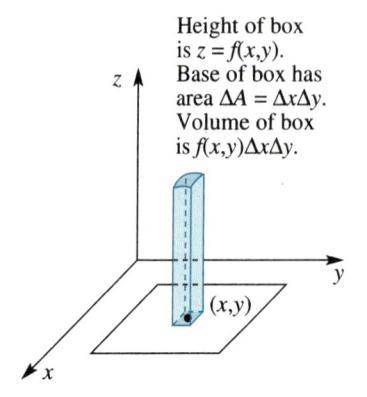

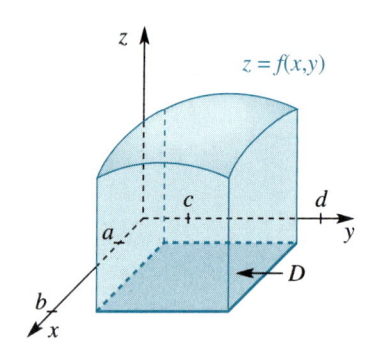

Figure 15.5 $f(x_i, y_j)\Delta x \Delta y$ approximates the volume over a small rectangle.

Figure 15.6 Interpreting a double integral as a volume.

base $\Delta A = \Delta x \Delta y$ (see Figure 15.5). This approximates the volume under the graph of $z = f(x, y)$ over this small rectangle.

Hence, the double Riemann sum gives us an approximation to the *total volume* under the graph of $z = f(x, y)$ over the region D. If the values of these double Riemann sum approximations approach a single limit as the grid partition becomes finer and finer, we call this limit the value of the double integral

$$\iint_D f(x, y)\, dA.$$

For a rectangular grid over the region $D = [a, b] \times [c, d]$, we write

$$\int_c^d \int_a^b f(x, y)\, dx\, dy,$$

where the area element

$$dA = dx\, dy$$

indicates that the Riemann sums have been formed with rectangular grid sections.

Computing double integrals iteratively

If f is a continuous function of two variables over the region D, then it's possible to reduce the calculation of the double integral

$$\iint_D f(x, y)\, dA$$

to two single integrals computed *iteratively*.

15.1 MULTIPLE INTEGRALS

The idea behind the method is perhaps best explained geometrically. For the time being, let's assume that $f(x,y) \geq 0$ over the entire region $D = [a,b] \times [c,d]$. The graph of $z = f(x,y)$ forms a "cap" over a solid with a rectangular base and vertical sides (see Figure 15.6).

Now, imagine that we select a particular $x_0 \in [a,b]$ and slice the solid with the vertical plane $x = x_0$ (see Figure 15.7). The continuity of f allows us to compute the area of this slice with the definite integral

$$\int_c^d f(x_0, y)\, dy.$$

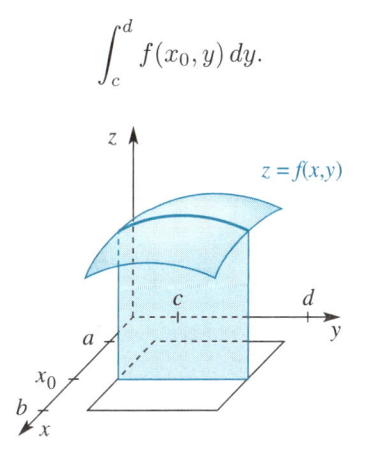

Figure 15.7 $\int_c^d f(x_0, y)\, dy$ is the area sliced by $x = x_0$.

EXAMPLE 1 Suppose that $f(x,y) = x^2 y^3$, and consider the solid bounded by the graph $z = x^2 y^3$ over the rectangular region $D = [-1, 3] \times [1, 4]$. Find the area of the slice of the solid cut out by the vertical plane $x = 2$.

Solution When $x = 2$, we have $f(2, y) = 4y^3$. The area of the slice is

$$\int_1^4 4y^3\, dy = y^4 \Big]_{y=1}^{y=4} = 256 - 1 = 255. \qquad \blacksquare$$

We can perform a similar computation to find the area of the slice for any value $x \in [a,b]$.

EXAMPLE 2 Using the same function as in the previous example, find the area of the slice for *any* value $x \in [-1, 3]$. Use your result to find the area of the slices cut by the end planes $x = -1$ and $x = 3$.

Solution We simply treat x as if it were a particular constant value and find

$$\int_1^4 x^2 y^3\, dy = \frac{x^2 y^4}{4}\Big]_{y=1}^{y=4} = 64x^2 - \frac{x^2}{4} = \frac{255 x^2}{4}.$$

For $x = 2$, note that we obtain the cross-sectional area 255, matching the result found in the previous example. At the two edges of the volume we have

$$\text{area of slice} = \frac{255}{4} \text{ at } x = -1$$

$$\text{area of slice} = \frac{2295}{4} \text{ at } x = 3.$$ ∎

The area $\int_c^d f(x,y)\,dy$ naturally depends on the value of x. Indeed, we can think of it as describing a *cross-sectional area function* that outputs the area of the slice corresponding to a value x. As such, if we integrate *this* function over the entire x-interval $[a,b]$, we obtain the *volume*

$$\int_a^b \left(\int_c^d f(x,y)\,dy \right) dx.$$

EXAMPLE 3 Find the volume of the solid bounded by the graph of $z = x^2 y^3$, the xy-plane, and the vertical planes $x = -1$, $x = 3$, $y = 1$, and $y = 4$.

Solution This is precisely the volume under the graph of $z = x^2 y^3$ over the region

$$D = [-1, 3] \times [1, 4]$$

(since $x^2 y^3 \geq 0$ for all $(x, y) \in D$). The volume is given by

$$\int_{-1}^{3} \left(\int_{1}^{4} x^2 y^3 \, dy \right) dx = \int_{-1}^{3} \frac{255 x^2}{4} dx = \frac{255 x^3}{12} \Big]_{x=-1}^{x=3} = \frac{2295}{4} + \frac{85}{4} = 595.$$ ∎

Changing the order of integration

Of course, we could also take vertical slices parallel to the x-axis (one for each possible value of $y = y_0$) as shown in Figure 15.8.

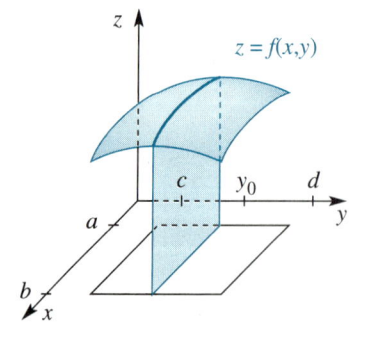

Figure 15.8 $\int_a^b f(x, y_0)\,dx$ is the area sliced by $y = y_0$.

15.1 MULTIPLE INTEGRALS

The volume in this case is given by

$$\int_c^d \left(\int_a^b f(x,y)\, dx \right) dy$$

EXAMPLE 4 Find the volume of the solid described in the previous example using the opposite order of integration.

Solution

$$\int_1^4 \left(\int_{-1}^3 x^2 y^3\, dx \right) dy = \int_1^4 \left(\frac{x^3 y^3}{3} \right)\bigg|_{x=-1}^{x=3} dy$$

$$= \int_1^4 \left(9y^3 + \frac{y^3}{3} \right) dy$$

$$= \int_1^4 \frac{28 y^3}{3}\, dy$$

$$= \frac{7 y^4}{3}\bigg|_{y=1}^{y=4}$$

$$= \frac{1792}{3} - \frac{7}{3} = \frac{1785}{3} = 595.$$

▶▶▶ As you can see, the order of integration is a matter of choice. This is not always the case, but for *continuous* functions f, we have

$$\int_c^d \int_a^b f(x,y)\, dx\, dy = \int_a^b \int_c^d f(x,y)\, dy\, dx.$$

Interpreting the double integral as signed volume

If $f(x,y) < 0$ for $(x,y) \in D$, then the double integral

$$\iint_D f(x,y)\, dA$$

represents the *negative* of the volume between the graph of $z = f(x,y)$ and the xy-plane under the region D. If $f(x,y)$ changes sign over D, then the integral $\iint_D f(x,y)\, dA$ represents the *net volume* above the xy-plane.

EXAMPLE 5 Find $\displaystyle\int_{-2}^1 \int_{-1}^3 xy\, dx\, dy$.

Solution   The function $f(x,y) = xy$ changes sign over the region

$$D = [-1, 3] \times [-2, 1],$$

and the sign of xy depends on which quadrant (x,y) lies in. Figure 15.9 shows D, the region of integration lying in the xy-plane.

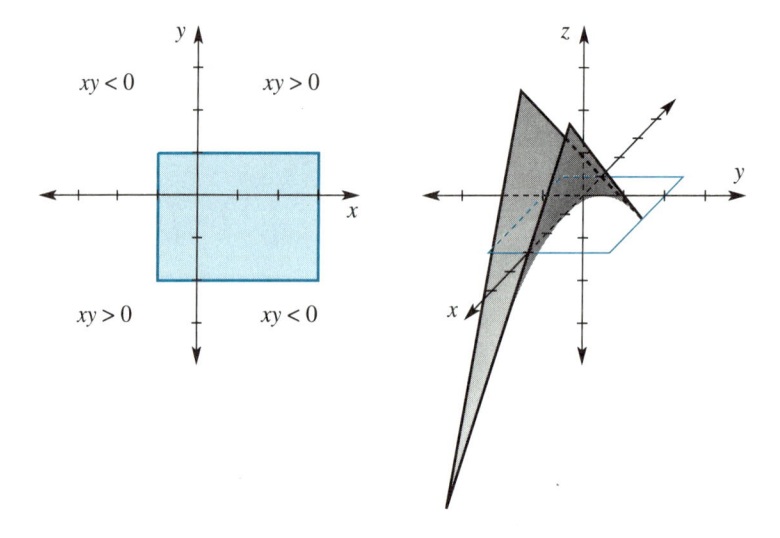

Figure 15.9 The sign of xy depends on the quadrant.

The graph of $z = xy$ lies below the xy-plane in the second and fourth quadrants and above the xy-plane in the first and third. The integral $\int_{-2}^{1} \int_{-1}^{3} xy \, dx \, dy$ represents the *net* volume between the xy-plane and the graph. We have

$$\int_{-2}^{1} \int_{-1}^{3} xy \, dx \, dy = \int_{-2}^{1} \frac{x^2 y}{2} \Big]_{x=-1}^{x=3} dy$$

$$= \int_{-2}^{1} \left(\frac{9y}{2} - \frac{y}{2} \right) dy$$

$$= \int_{-2}^{1} 4y \, dy$$

$$= 2y^2 \Big]_{y=-2}^{y=1}$$

$$= 2 - 8 = -6.$$

The negative value indicates that between the xy plane and the graph of $z = xy$, there are 6 more units of volume *under* the xy-plane than above. ∎

15.1 MULTIPLE INTEGRALS

You can check that

$$\int_{-2}^{0}\int_{-1}^{0} xy\, dx\, dy = 1, \qquad \int_{0}^{1}\int_{-1}^{0} xy\, dx\, dy = -\frac{1}{4},$$

$$\int_{-2}^{0}\int_{0}^{3} xy\, dx\, dy = -9, \qquad \int_{0}^{1}\int_{0}^{3} xy\, dx\, dy = \frac{9}{4},$$

and that these sum to the final value of the double integral.

EXAMPLE 6 Find $\int_{0}^{\pi/2}\int_{-1}^{2} y \cos x\, dy\, dx$.

Solution We have

$$\int_{0}^{\pi/2}\int_{-1}^{2} y \cos x\, dy\, dx = \int_{0}^{\pi/2} (y^2/2) \cos x \Big]_{-1}^{2} dx$$

$$= \int_{0}^{\pi/2} (2 \cos x - (1/2) \cos x)\, dx$$

$$= \int_{0}^{\pi/2} (3/2) \cos x\, dx$$

$$= (3/2) \sin x \Big]_{0}^{\pi/2} = 3/2 - 0 = 3/2. \qquad \blacksquare$$

Triple integrals

A triple integral is defined for a function of three variables over a region R in $\mathbb{R}^3$. The notation is

$$\iiint_{R} f(x, y, z)\, dV,$$

where dV is the notation for a "volume element," whose form will depend on our method of computation.

The simplest type of region R is a rectangular parallelepiped (box) with sides parallel with the coordinate planes. Such a region can be denoted as a Cartesian product of three closed intervals (see Figure 15.10):

$$R = [a, b] \times [c, d] \times [p, q]$$
$$= \{(x, y, z) : a \leq x \leq b \text{ and } c \leq y \leq d \text{ and } p \leq z \leq q\}.$$

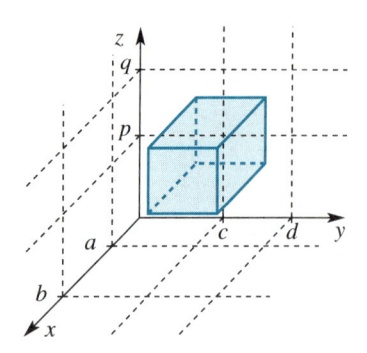

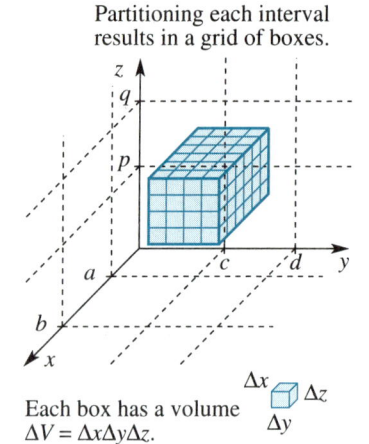

Figure 15.10 The box $[a,b] \times [c,d] \times [p,q]$.

Figure 15.11 Partitioning the box $[a,b] \times [c,d] \times [p,q]$.

Each box has a volume $\Delta V = \Delta x \Delta y \Delta z$.

Now we form a regular partition of each interval:

$[a,b]$ is subdivided into ℓ equal-sized subintervals of length Δx.

$[c,d]$ is subdivided into m equal-sized subintervals of length Δy.

$[p,q]$ is subdivided into n equal-sized subintervals of length Δz.

Figure 15.11 shows that these partitions create a 3-dimensional grid of rectangular boxes, each of dimension $\Delta x \times \Delta y \times \Delta z$ and having volume:

$$\Delta V = \Delta x \Delta y \Delta z.$$

Now, we choose one point (x_i, y_j, z_k) from each of the $\ell \times m \times n$ boxes ($1 \leq i \leq \ell, 1 \leq j \leq m, 1 \leq k \leq n$), evaluate the function f at each chosen point, and finally form the *triple* Reimann sum

$$\sum_{k=1}^{n} \sum_{j=1}^{m} \sum_{i=1}^{\ell} f(x_i, y_j, z_k) \Delta x \Delta y \Delta z.$$

If, as the partitions are made finer and finer, these sums approach a single limiting value, then that value is denoted by

$$\iiint_R f(x,y,z)\, dV$$

or

$$\int_p^q \int_c^d \int_a^b f(x,y,z)\, dx\, dy\, dz,$$

where $dV = dx\, dy\, dz$ represents the volume element corresponding to a rectangular grid.

Physical interpretation of the triple integral

While we don't have a purely geometric interpretation for this integral, we can provide a *physical* interpretation in terms of density and mass. If an object is made of a substance of *uniform* density, then one simply multiplies its volume to find its mass. For example, if an object of uniform density $\rho = 12$ g/cm^3 has volume $V = 350$ cm^3, then its

$$\text{total mass} = \rho V = (12 \text{ g/cm}^3)(350 \text{ cm}^3) = 4200 \text{ g}.$$

Now, suppose that we have a box R made of material with *variable* density. That is, if we sampled a cubic centimeter of the substance from different parts of the box, we would find that mass depended on the part of the box we took the sample from. We imagine the density

$$\rho = f(x, y, z)$$

to be a function of the location (x, y, z) of a point in the box.

If f is a continuous function, then dividing the box into an extremely fine grid will determine a collection of tiny boxes, each of approximately constant (uniform) density. Different tiny boxes will have different densities, depending on their location. If each tiny box has dimensions $\Delta x \times \Delta y \times \Delta z$, then its mass is approximately

$$f(x_i, y_j, z_k) \Delta x \Delta y \Delta z,$$

where (x_i, y_j, z_k) is a representative point of the box. Summing the masses (the triple Riemann sum) gives an approximation of the total mass of the box, and this approximation becomes better and better for finer and finer grids. Indeed, we have

$$\text{total mass} = \iiint_R \rho \, dV = \iiint_D f(x, y, z) \, dV.$$

Calculating triple integrals

Triple integrals can be computed iteratively in a manner entirely analogous to that used for double integrals. We'll illustrate with some examples.

EXAMPLE 7 Find $\int_{-1}^{0} \int_{1}^{2} \int_{0}^{1} (z + 2x + 3y) \, dy \, dx \, dz$.

Solution

$$\int_{-1}^{0}\int_{1}^{2}\int_{0}^{1}(z+2x+3y)\,dy\,dx\,dz = \int_{-1}^{0}\int_{1}^{2}\left(zy+2xy+\frac{3y^2}{2}\right)\bigg]_{y=0}^{y=1} dx\,dz$$

$$= \int_{-1}^{0}\int_{1}^{2}\left(z+2x+\frac{3}{2}\right)dx\,dz$$

$$= \int_{-1}^{0} zx+x^2+\frac{3x}{2}\bigg]_{x=1}^{x=2} dz$$

$$= \int_{-1}^{0}(2z+4+3)-(z+1+3/2)\,dz$$

$$= \int_{-1}^{0}(z+9/2)\,dz$$

$$= \frac{z^2+9z}{2}\bigg]_{z=-1}^{z=0}$$

$$= 0 - \frac{1-9}{2} = 4.$$ ∎

EXAMPLE 8 Find the mass of the unit cube $[0,1]\times[0,1]\times[0,1]$ if the density ρ at (x,y,z) is ye^{-xy} kilograms per cubic unit.

Solution The solution to the problem is given by the triple integral

$$\int_0^1\int_0^1\int_0^1 ye^{-xy}\,dx\,dz\,dy = \int_0^1\int_0^1 -e^{-xy}\bigg]_0^1 dz\,dy$$

$$= \int_0^1\int_0^1(-e^{-y}+1)\,dz\,dy = \int_0^1 z(-e^{-y}+1)\bigg]_0^1 dy$$

$$= \int_0^1(-e^{-y}+1)\,dy = e^{-y}+y\bigg]_0^1$$

$$= (e^{-1}+1)-(1+0) = e^{-1}.$$

The mass of the cube is $1/e \approx 0.368$ kilograms. ∎

EXERCISES for Section 15.1

For exercises 1-10: Calculate each of the double integrals.

1. $\int_1^2\int_1^4(x+y)\,dx\,dy$

2. $\int_0^1\int_0^1(1-x-y)\,dx\,dy$

3. $\int_0^1\int_0^1(x^2+y^2)\,dx\,dy$

4. $\int_1^e\int_1^2(x/y)\,dx\,dy$

5. $\int_0^1\int_1^3(x-3x^2y+\sqrt{y})\,dx\,dy$

6. $\int_0^{\pi/2}\int_0^{\pi/2}\sin(x+y)\,dx\,dy$

15.2 DOUBLE INTEGRALS OVER MORE GENERAL REGIONS

7. $\displaystyle\int_{-1}^{1}\int_{0}^{\pi/2}(x\sin(y)-ye^x)\,dy\,dx$

8. $\displaystyle\int_{1}^{4}\int_{0}^{1}y^2/\sqrt{x}\,dy\,dx$

9. $\displaystyle\int_{0}^{\pi/4}\int_{0}^{\pi/2}\frac{\cos(x)}{\cos^2(y)}\,dx\,dy$

10. $\displaystyle\int_{-2}^{-1}\int_{0}^{1}\frac{y}{1+x^2}\,dx\,dy$

11. Suppose that $f(x,y) = 4x+2y+1$ over the domain $[0,7]\times[-1,6]$. Suppose that the domain is partitioned into a 7 by 7 rectangular grid determined by vertical and horizontal lines with integer intercepts. If the center of each square is chosen as the point (x_i, y_j), calculate the corresponding Riemann sum.

12. Evaluate $\displaystyle\int_{1}^{4}\int_{-1}^{2}(2x+6x^2y)\,dy\,dx$.

13. Evaluate $\displaystyle\int_{-1}^{2}\int_{1}^{4}(2x+6x^2y)\,dx\,dy$.

14. Evaluate $\displaystyle\int_{0}^{2}\int_{-2}^{0}(x^3+4y)\,dy\,dx$.

15. Evaluate $\displaystyle\int_{1}^{3}\int_{\pi/6}^{2\pi}2y\cos x\,dx\,dy$.

16. Verify that $\int_{1}^{2}\int_{-1}^{2}(4xy^3+y)\,dy\,dx = \int_{-1}^{2}\int_{1}^{2}(4xy^3+y)\,dx\,dy$.

17. Find $\displaystyle\int_{0}^{3}\int_{1}^{2}\int_{1}^{4}x+y+z\,dx\,dy\,dz$.

18. Find $\displaystyle\int_{1}^{2}\int_{-1}^{2}\int_{0}^{2}xy^2-y/z\,dx\,dy\,dz$.

19. Suppose that the cube $[-1,1]\times[-1,1]\times[-1,1]$ is made up of a substance whose density ρ is given by $f(x,y,z) = \dfrac{y^2e^z}{1+x^2}$ grams per cubic unit at the point (x,y,z). Find the total mass of the cube.

20. Show that $\displaystyle\int_{c}^{d}\int_{a}^{b}1\,dx\,dy$ gives the area of the rectangle $[a,b]\times[c,d]$.

21. Show that $\displaystyle\int_{p}^{q}\int_{c}^{d}\int_{a}^{b}1\,dx\,dy\,dz$ gives the volume of the box $[a,b]\times[c,d]\times[p,q]$.

22. Evaluate $\displaystyle\iiint_{R}3xy^3z^2\,dV$, where $R = [-1,3]\times[1,4]\times[0,2]$.

23. Evaluate
$$\iiint_{R}(x+2y+4z)\,dV,$$
where $R = [1,2]\times[-1,0]\times[0,3]$ in six different ways.

15.2 DOUBLE INTEGRALS OVER MORE GENERAL REGIONS

In this section we'll discuss how to calculate double integrals when the region D is not a "nice" rectangle (that is, a Cartesian product of closed intervals). First, let's examine some important properties of multiple integrals.

Properties of multiple integrals

Many of the properties of multiple definite integrals are analogous to properties of single definite integrals.

> If c is a constant, then
> $$\iint_D cf(x,y)\,dA = c \iint_D f(x,y)\,dA$$

whenever the integrals on both sides of the equation exist. In other words, we can factor a *constant* through the double integral sign.

Similarly, the additive property extends to multiple integrals:

> $$\iint_D (f(x,y) + g(x,y))\,dA = \iint_D f(x,y)\,dA + \iint_D g(x,y)\,dA.$$

In other words, the multiple integral of a sum of two functions is the sum of their integrals.

> If the region of integration D is split into two non-overlapping regions D_1 and D_2, then
> $$\iint_D f(x,y)\,dA = \iint_{D_1} f(x,y)\,dA + \iint_{D_2} f(x,y)\,dA.$$

> If $f(x,y) \leq g(x,y)$ for every $(x,y) \in D$, then
> $$\iint_D f(x,y)\,dA \leq \iint_D g(x,y)\,dA.$$

All of these properties also hold for triple integrals.

15.2 DOUBLE INTEGRALS OVER MORE GENERAL REGIONS

Calculating double integrals by the slicing method

In the last section, we saw how we could think of the double integral

$$\int_a^b \int_c^d f(x,y)\,dy\,dx$$

in terms of *slices*. That is, for each $x \in [a,b]$ we can think of the "inner" integral

$$\int_c^d f(x,y)\,dy$$

as providing the signed area of the vertical cross-sectional slice of the graph of $z = f(x,y)$ over the y-interval $[c,d]$. We obtain such a value for each $x \in [a,b]$. When we integrate these values over the x-interval $[a,b]$, we obtain the signed volume under the graph of $z = f(x,y)$:

$$\int_a^b \left(\int_c^d f(x,y)\,dy \right) dx.$$

If f is continuous, then we can switch the order of integration to

$$\int_c^d \left(\int_a^b f(x,y)\,dx \right) dy.$$

Now, the inner integral represents a slice of the graph of $z = f(x,y)$ over the x-interval $[a,b]$ for each $y \in [c,d]$. We can integrate these values over the y-interval $[c,d]$ to obtain the same final double integral value.

Suppose that our region of integration D is *not* a rectangle. One possibility is shown in Figure 15.12.

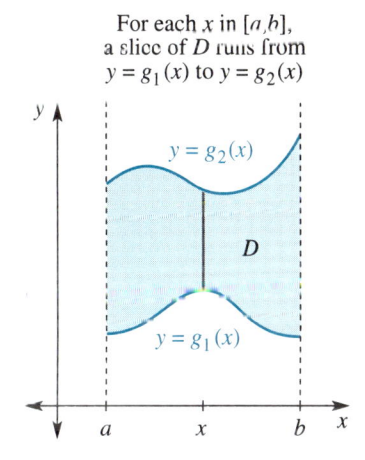

Figure 15.12 Region D is bounded by $y = g_1(x)$ and $y = g_2(x)$.

In this picture, our region D lies entirely between the graphs of $y = g_1(x)$ and $y = g_2(x)$ over the x-interval $[a, b]$. Such a region is sometimes described as being *vertically simple*.

To integrate $f(x, y)$ over this region, we can still use the method of slicing. For each value $x \in [a, b]$, a slice runs from $y = g_1(x)$ to $y = g_2(x)$. Hence, the inner integral will have the form

$$\int_{g_1(x)}^{g_2(x)} f(x, y) \, dy.$$

Integrating these cross-sections over the interval $[a, b]$ gives us

$$\int_a^b \int_{g_1(x)}^{g_2(x)} f(x, y) \, dy \, dx.$$

Another possibility is shown in Figure 15.13.

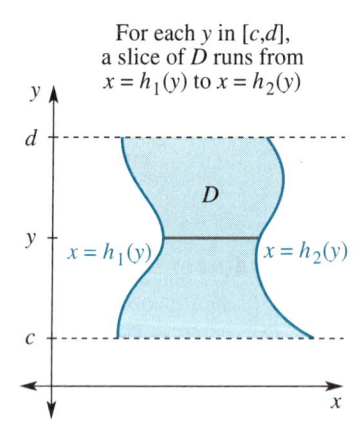

Figure 15.13 Region D is bounded by $x = h_1(y)$ and $x = h_2(y)$.

In this picture, our region D lies entirely between the graphs of $x = h_1(y)$ and $x = h_2(y)$ over the y-interval $[c, d]$. For each $y \in [c, d]$, our slice runs from $x = h_1(y)$ to $x = h_2(y)$. Such a region is sometimes described as being *horizontally simple*. Hence, the inner integral will have the form

$$\int_{h_1(y)}^{h_2(y)} f(x, y) \, dx.$$

Integrating these over the y-interval $[c, d]$ gives us the final result:

$$\int_c^d \int_{h_1(y)}^{h_2(y)} f(x, y) \, dx \, dy.$$

15.2 DOUBLE INTEGRALS OVER MORE GENERAL REGIONS

EXAMPLE 9 Find $\iint_D (x^2 + xy - y^2)\, dA$ where D is the region bounded by the graph of $y = x^2$ and the lines $y = 0$ and $x = 1$. Use both orders of integration.

Solution The region D is described in two ways in Figure 15.14. This particular region is both vertically and horizontally simple.

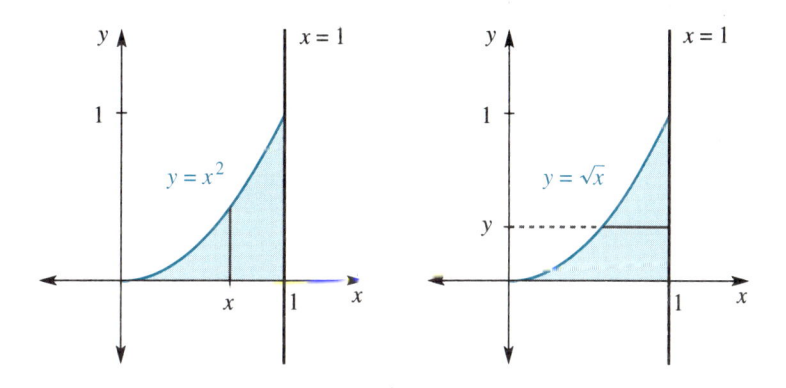

Figure 15.14 The region D bounded by $y = x^2$, $x = 1$, and $y = 0$.

In the picture on the left, the region is considered to lie between $y = x^2$ and $y = 0$ over the x-interval $[0, 1]$. Hence, for each slice corresponding to a value x between 0 and 1, the y values run from 0 to x^2.

Therefore, we have

$$\int_0^1 \int_0^{x^2} (x^2 + xy - y^2)\, dy\, dx = \int_0^1 \left[x^2 y + \frac{xy^2}{2} - \frac{y^3}{3} \right]_{y=0}^{y=x^2} dx$$

$$= \int_0^1 \left(x^4 + \frac{x^5}{2} - \frac{x^6}{3} \right) dx$$

$$= \left[\frac{x^5}{5} + \frac{x^6}{12} - \frac{x^7}{21} \right]_{x=0}^{x=1}$$

$$= \frac{1}{5} + \frac{1}{12} - \frac{1}{21} = \frac{33}{140}.$$

In the picture on the right, the region is considered to lie between $x = \sqrt{y}$ and $x = 1$ over the y interval $[0, 1]$. Hence, for each slice corresponding to a value y between 0 and 1, the x values run from $\sqrt{y}$ to 1.

Therefore, we have

$$\int_0^1 \int_{\sqrt{y}}^1 (x^2 + xy - y^2)\, dx\, dy = \int_0^1 \frac{x^3}{3} + \frac{x^2 y}{2} - xy^2 \Big]_{x=\sqrt{y}}^{x=1} dy$$

$$= \int_0^1 (1/3 + y/2 - y^2) - \left(\frac{y^{3/2}}{3} + \frac{y^2}{2} - y^{5/2}\right) dy$$

$$= (y/3 + y^2/4 - y^3/3) - \left(\frac{2y^{5/2}}{15} + \frac{y^3}{6} - \frac{2y^{7/2}}{7}\right)\Big]_{y=0}^{y=1}$$

$$= \left(\frac{1}{3} + \frac{1}{4} - \frac{1}{3}\right) - \left(\frac{2}{15} + \frac{1}{6} - \frac{2}{7}\right) = \frac{33}{140}.$$

With either order of integration, we obtain the same result. ■

EXAMPLE 10 Find $\iint_D (x^2 + y^2)\, dA$ where D is the region bounded by the coordinate axes and the line $y = (10 - 3x)/4$.

Solution The region D is the triangle pictured in Figure 15.15.

If we slice the region at each value $x \in [0, 10/3]$, we see that the corresponding y-values run from 0 to $(10 - 3x)/4$. Hence, we can write

$$\iint_D (x^2 + y^2)\, dA = \int_0^{10/3} \int_0^{(10-3x)/4} (x^2 + y^2)\, dy\, dx$$

$$= \int_0^{10/3} x^2 y + \frac{y^3}{3}\Big]_{y=0}^{y=(10-3x)/4} dx$$

$$= \int_0^{10/3} \left[\frac{x^2(10 - 3x)}{4} + \frac{((10 - 3x)/4)^3}{3}\right] dx.$$

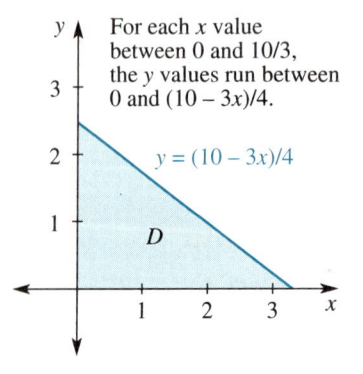

Figure 15.15 The triangular region D.

15.2 DOUBLE INTEGRALS OVER MORE GENERAL REGIONS

Now, to calculate the outer integral, we first multiply out the expression in square braces and collect terms:

$$\int_0^{10/3} \left(-\frac{57x^3}{64} + \frac{125x^2}{32} - \frac{-75x}{16} + \frac{125}{24} \right) dx$$

$$= \left(-\frac{57x^4}{256} + \frac{125x^3}{96} - \frac{-75x^2}{32} + \frac{125x}{24} \right) \Bigg]_0^{10/3}$$

$$= \left(-\frac{57(10/3)^4}{256} + \frac{125(10/3)^3}{96} - \frac{-75(10/3)^2}{32} + \frac{125(10/3)}{24} \right)$$

$$\approx 12.0563.$$

EXAMPLE 11   Find $\displaystyle\int_{-1}^0 \int_0^{2\sqrt{1-x^2}} x \, dy \, dx$.

Solution   We have

$$\int_{-1}^0 \int_0^{2\sqrt{1-x^2}} x \, dy \, dx = \int_{-1}^0 xy \Bigg]_{y=0}^{y=2\sqrt{1-x^2}} dx = \int_{-1}^0 2x\sqrt{1-x^2} \, dx.$$

To complete the computation, we make the following substitution: Let $u = 1 - x^2$. Differentiating, we have $du = -2x\,dx$. Substituting for the limits of integration, we have $u = 1$ when $x = 0$, and we have $u = 0$ when $x = -1$. Thus, our integral is transformed to

$$-\int_0^1 \sqrt{u}\,du = -\frac{2u^{3/2}}{3}\Bigg]_0^1 = -\frac{2}{3} - 0 = -\frac{2}{3}.$$

Subdividing the region of integration

The next two examples illustrate how it may be desirable to subdivide the region of integration.

EXAMPLE 12   Find $\displaystyle\int_{-2}^1 \int_{-2|x|}^{|x|} e^{x+y} \, dy \, dx$.

Solution   Note that we can graph the region of integration by examining the limits and order of integration carefully. The region must be bounded between the graphs of $y = |x|$ and $y = -2|x|$ between $x = -2$ and $x = 1$.

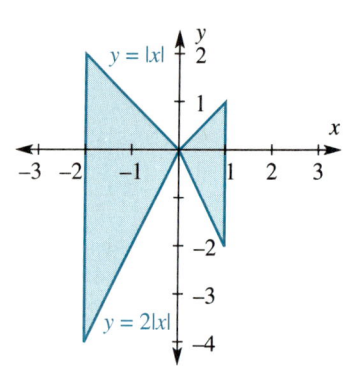

Figure 15.16 The region between $y = |x|$ and $y = -2|x|$ over $[-2, 1]$.

Figure 15.16 illustrates this region. If we use x-slices, it is convenient to break the region at $x = 0$ (since $|x|$ depends on the *sign* of x):

$$\int_{-2}^{1} \int_{-2|x|}^{|x|} e^{x+y} \, dy \, dx = \int_{-2}^{0} \int_{2x}^{-x} e^{x+y} \, dy \, dx + \int_{0}^{1} \int_{-2x}^{x} e^{x+y} \, dy \, dx$$

$$= \int_{-2}^{0} e^{x+y} \Big]_{y=2x}^{y=-x} dx + \int_{0}^{1} e^{x+y} \Big]_{y=-2x}^{y=x} dx$$

$$= \int_{-2}^{0} (1 - e^{3x}) \, dx + \int_{0}^{1} (e^{2x} - e^{-x}) \, dx$$

$$= (x - e^{3x}/3) \Big]_{x=-2}^{x=0} + (e^{2x}/2 + e^{-x}) \Big]_{x=0}^{x=1}$$

$$= \left[-\frac{1}{3} - \left(-2 - \frac{e^{-6}}{3} \right) \right] + \left[\frac{e^2}{2} + e^{-1} - \left(\frac{1}{2} + 1 \right) \right]$$

$$\approx 4.23.$$

EXAMPLE 13 Find $\iint_D e^{x-y} \, dA$ where D is the triangular region with vertices $(0, 0)$, $(1, 3)$, and $(2, 2)$.

Solution The region D is shown in Figure 15.17, where we have subdivided it into two subregions D_1 and D_2.

Each side of the triangle is part of a straight line. By finding the equation of each of these lines, we can note the following:

In region D_1, x runs from 0 to 1, while y runs from x to $3x$.

In region D_2, x runs from 1 to 2, while y runs from x to $-x + 4$.

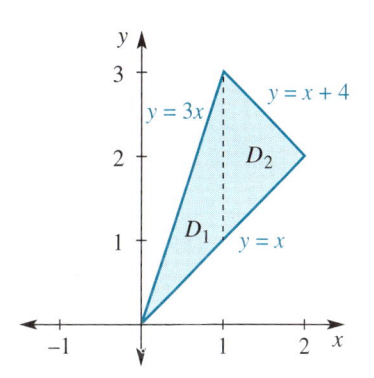

15.2 DOUBLE INTEGRALS OVER MORE GENERAL REGIONS

Figure 15.17 Subdividing the region of integration.

Thus, we can write

$$\iint_D e^{x-y}\,dy\,dx = \iint_{D_1} e^{x-y}\,dA + \iint_{D_2} e^{x-y}\,dA$$

$$= \int_0^1 \int_x^{3x} e^{x-y}\,dy\,dx + \int_1^2 \int_x^{-x+4} e^{x-y}\,dy\,dx$$

$$= \int_0^1 \left. -e^{x-y} \right]_{y=x}^{y=3x} dx + \int_1^2 \left. -e^{x-y} \right]_{y=x}^{y=-x+4} dx$$

$$= \int_0^1 (-e^{-2x} + e^0)\,dx + \int_1^2 (-e^{2x-4} + e^0)\,dx$$

$$= \left(\frac{e^{-2x}}{2} + x \right) \Big]_{x=0}^{x=1} + \left(\frac{-e^{2x-4}}{2} + x \right) \Big]_{x=1}^{x=2}$$

$$= \left[\left(\frac{e^{-2}}{2} + 1 \right) - \left(\frac{1}{2} + 0 \right) \right] + \left[\left(\frac{-1}{2} + 2 \right) - \left(\frac{-e^{-2}}{2} + 1 \right) \right]$$

$$= e^{-2} + 1 \approx 1.135.$$

Changing the order of integration

The choice of order of integration can be important. There are cases when one order of integration may simplify our computations over another or prevent us from having to subdivide a region.

EXAMPLE 14 Find $\int_{-1}^{1} \int_{|x|}^{1} (x+y)^2 \,dy\,dx$.

Solution First, let's look at the region under consideration (see Figure 15.18). The given order of integration indicates that the values x run over the interval $[-1, 1]$, and for each x-value in this interval, y runs from $|x|$ to 1.

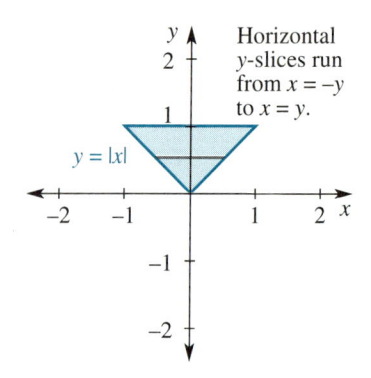

Figure 15.18 Region of integration for $\int_{-1}^{1} \int_{|x|}^{1} (x+y)^2 \, dy \, dx$.

If we integrate "$dx \, dy$" instead of "$dy \, dx$," we have the values y running over the interval $[0,1]$, and for each y-value, x runs from $-y$ to y. This gives us the equivalent integral

$$\int_0^1 \int_{-y}^{y} (x+y)^2 \, dx \, dy.$$

We compute this integral as follows:

$$\int_0^1 \int_{-y}^{y} (x+y)^2 \, dx \, dy = \int_0^1 \int_{-y}^{y} (x^2 + 2xy + y^2) \, dx \, dy$$

$$= \int_0^1 \left(\frac{x^3}{3} + x^2 y + xy^2 \right) \Bigg]_{x=-y}^{x=y} dy$$

$$= \int_0^1 \left[\left(y^3 + y^3 + \frac{y^3}{3}\right) - \left(-y^3 + y^3 - \frac{y^3}{3}\right) \right] dy$$

$$= \int_0^1 \frac{8}{3} y^3 \, dy = \frac{2}{3} y^4 \Bigg]_{y=0}^{y=1} = \frac{2}{3}. \quad \blacksquare$$

Following is another example of a double integral that is somewhat easier to compute if we change the order of integration.

EXAMPLE 15 Compute $\int_{-3}^{3} \int_{-\sqrt{9-y^2}}^{\sqrt{9-y^2}} x^2 \, dx \, dy$.

Solution Let's first look at the region of integration. Note that the values y run over the interval $[-3, +3]$ and that the region is bounded between the two curves $x = -\sqrt{9-y^2}$ and $x = \sqrt{9-y^2}$. Figure 15.19 illustrates that these two curves form the circle $x^2 + y^2 = 9$.

Of course, we could look at this same region as being bounded by the two curves $y = -\sqrt{9-x^2}$ and $y = \sqrt{9-x^2}$. Using this, we could rewrite the

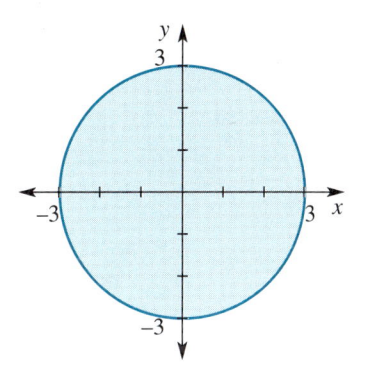

15.2 DOUBLE INTEGRALS OVER MORE GENERAL REGIONS

Figure 15.19 A circular region of integration.

integral:

$$\int_{-3}^{3}\int_{-\sqrt{9-y^2}}^{\sqrt{9-y^2}} x^2\,dx\,dy = \int_{-3}^{3}\int_{-\sqrt{9-x^2}}^{\sqrt{9-x^2}} x^2\,dy\,dx$$

$$= \int_{-3}^{3}\left(x^2 y\right)\Big]_{y=-\sqrt{9-x^2}}^{y=\sqrt{9-x^2}} dx$$

$$= \int_{-3}^{3}\left[x^2\sqrt{9-x^2} - x^2(-\sqrt{9-x^2})\right] dx$$

$$= \int_{-3}^{3} 2x^2\sqrt{9-x^2}\,dx.$$

Using an integral table or a numerical integrator, we find that this final single integral has the value $81\pi/4 \approx 63.617$. ∎

EXERCISES for Section 15.2

For exercises 1-8: Find the value of each integral.

1. $\displaystyle\int_{-3}^{2}\int_{0}^{y^2}(x^2+y)\,dx\,dy$

2. $\displaystyle\int_{0}^{1}\int_{y^2}^{y}(x^2+y^3)\,dx\,dy$

3. $\displaystyle\int_{0}^{\frac{\pi}{2}}\int_{\cos x}^{0} y\sin x\,dy\,dx$

4. $\displaystyle\int_{-1}^{1}\int_{|y|}^{1}(x+y)^2\,dx\,dy$

5. $\displaystyle\int_{1}^{2}\int_{-y}^{y}(x+y)\,dx\,dy$

6. $\displaystyle\int_{1}^{1}\int_{0}^{3}(y-x^2)\,dx\,dy$

7. $\displaystyle\int_{0}^{1}\int_{-\sqrt{y}}^{\sqrt{y}} yx^2\,dx\,dy$

8. $\displaystyle\int_{0}^{1}\int_{0}^{2} e^{x-y}\,dy\,dx$

9. Find $\displaystyle\iint_D f(x,y)\,dA$ where $f(x,y) = y^2\sqrt{x}$ and $D = \{(x,y) : x > 0,\ y > x^2,\ y < 10 - x^2\}$.

10. Sketch the region bounded by $y = \ln x$, the x-axis, and the line $x = e$ and integrate the function $f(x,y) = xe^y$ over this region.

11. Set up (but do not integrate) the previous exercise, reversing the order of integration.

12. Find $\iint_D xy\, dA$, where D is the region bounded by $y = x$, $y = 3x$ and $x = 1$.

13. Find $\iint_D (x+y)\, dA$, where D is the region bounded by $y = x$ and $y = x^3$. (Be careful!)

14. Find $\iint_D 2xy\, dx\, dy$, where D is the region in the xy-plane pictured below:

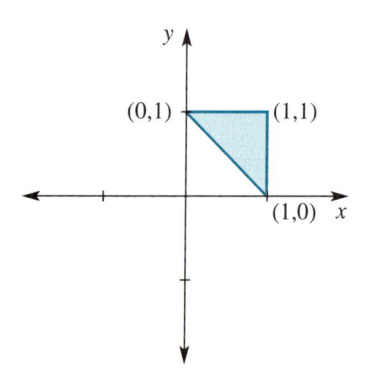

For Exercise 14

For exercises 15-20: Evaluate the following integrals.

15. $\displaystyle\int_1^2 \int_{1-x}^{\sqrt{x}} x^2 y\, dy\, dx$

16. $\displaystyle\int_{-1}^{1} \int_{x^3}^{x+1} f(x,y)\, dy\, dx$

17. $\displaystyle\int_1^2 \int_{x^3}^{x} e^{y/x}\, dy\, dx$

18. $\displaystyle\int_0^{\pi/6} \int_0^{\pi/2} (x\cos y - \cos x)\, dy\, dx$

19. $\displaystyle\int_{\pi/6}^{\pi/4} \int_{\tan x}^{\sec x} (y + \sin x)\, dy\, dx$

20. $\displaystyle\int_0^1 \int_y^1 \frac{1}{1+y^2}\, dx\, dy$

For exercises 21-24: Sketch the region D bounded by the graphs of the equations and express $\iint_D f(x,y)\, dA$ as an iterated integral.

21. $y = \sqrt{x}$, $x = 0$, $y = 2$
22. $y = x^3$, $x = 2$, $y = 0$
23. $y = \sqrt{x}$, $y = x^3$
24. $y = \sqrt{1-x^2}$, $y = 0$

For exercises 25-30: Sketch the region bounded by the graphs of the given equations and find its area by means of double integrals.

25. $y = \sqrt{x}$, $y = -x$, $x = 1$, $x = 4$

26. $y^2 = -x$, $x - y = 4$, $y = -1$, $y = 2$

27. $y = x$, $y = 3x$, $x + y = 4$

28. $y = e^x$, $y = \sin x$, $x = -\pi$, $x = \pi$

29. $y = \ln |x|$, $y = 0$, $y = 1$

30. $y = x^2$, $y = 1/(1 + x^2)$

---

31. Find the volume of the solid which lies under the graph of $z = 4x^2 + y^2$ and over the rectangular region R in the xy-plane having vertices $(0, 0, 0)$, $(0, 1, 0)$, $(2, 0, 0)$, and $(2, 1, 0)$.

32. Sketch the solid in the first octant that is bounded by the graphs of $x^2 + z^2 = 9$, $y = 2x$, $y = 0$, $z = 0$, and find its volume.

33. Find the volume of the solid bounded by the graphs of $z = x^2 + 4$, $y = 4 - x^2$, $x + y = 2$, $z = 0$.

34. The iterated integral $\int_0^1 \int_{3-x}^{3-x^2} \sqrt{25 - x^2 - y^2}\, dy\, dx$ represents the volume of a solid. Describe it.

15.3 DOUBLE INTEGRALS IN POLAR COORDINATES

▶▶▶ **If you wish to review polar coordinates, there is a section in the appendices that you will find useful.**

The region of integration D may be more simply described using polar rather than rectangular coordinates. For example, if the boundary of the region is made up of circular arcs, polar coordinates may be a convenient choice.

If we make a polar grid of a region (see Figure 15.20), one thing that we notice is that the grid sections are not the same size. The further away from the origin, the larger the section. Indeed, the area ΔA of a grid section is

$$\Delta A = r \Delta r \Delta \theta,$$

where r is the distance from the origin to the center of the section. This suggests a procedure for expressing a double integral in terms of polar coordinates.

$$\iint_D f(x, y)\, dA$$

Area of grid section is $\Delta A = r \Delta r \Delta \theta$, where r is the distance to the center of the section.

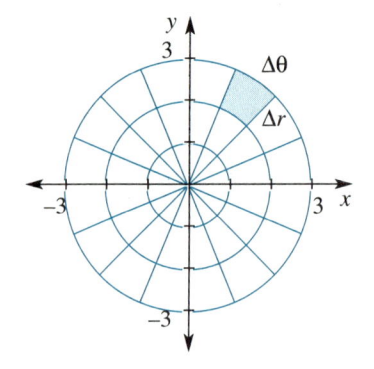

Figure 15.20 Polar grid of the plane.

Step 1. Describe the region D in terms of θ and r.

Step 2. Make the change of variables:

$$x = r\cos\theta \qquad y = r\sin\theta$$

Step 3. Use $dA = r\,dr\,d\theta$ as the area element in the integral.

EXAMPLE 16 Find $\iint_D (1 + x^2 + y^2)^{3/2} \, dA$, where D is the interior of the unit circle.

Solution We'll use a polar coordinate transformation to evaluate the integral. In terms of polar coordinates, D is the region of points (r, θ) satisfying

$$0 \leq r \leq 1 \quad \text{and} \quad 0 \leq \theta \leq 2\pi.$$

Substituting $x = r\cos\theta$, $y = r\sin\theta$, and $dA = r\,dr\,d\theta$, we can write

$$\iint_D (1 + x^2 + y^2)^{3/2} \, dA = \int_0^{2\pi} \int_0^1 (1 + r^2)^{3/2} r \, dr \, d\theta.$$

Let's make the substitution $u = 1 + r^2$; $du = 2r\,dr$. Our new limits of integration are $u = 1$ (when $r = 0$) and $u = 2$ (when $r = 1$). We have

$$\int_0^{2\pi} \frac{1}{2} \int_1^2 u^{3/2} \, du \, d\theta = \int_0^{2\pi} \frac{1}{5} (u)^{5/2} \Big]_1^2 \, d\theta$$

$$= \int_0^{2\pi} \left(\frac{2^{5/2}}{5} - \frac{1^{5/2}}{5} \right) d\theta = \left(\frac{4\sqrt{2}}{5} - \frac{1}{5} \right) \theta \Big]_0^{2\pi}$$

$$= \frac{8\sqrt{2}\pi}{5} - \frac{2\pi}{5}. \qquad \blacksquare$$

15.3 DOUBLE INTEGRALS IN POLAR COORDINATES

EXAMPLE 17 Find $\iint_D x^2 \, dx \, dy$, where D is the set of points (x,y) satisfying $0 \le x \le y$ and $x^2 + y^2 \le 1$.

Solution A picture of the region D is shown in Figure 15.21.

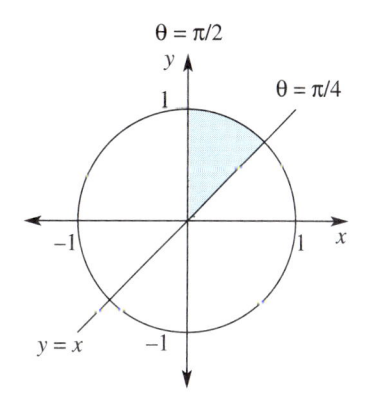

Figure 15.21 Region $D = \{(x,y) : 0 \le x \le y \text{ and } x^2 + y^2 \le 1\}$.

Using polar coordinates:

$$\int_{\pi/4}^{\frac{\pi}{2}} \int_0^1 r^2 \cos^2\theta \, r \, dr \, d\theta = \int_{\frac{\pi}{4}}^{\frac{\pi}{2}} \int_0^1 r^3 \cos^2\theta \, dr \, d\theta = \int_{\frac{\pi}{4}}^{\frac{\pi}{2}} \frac{r^4}{4} \cos^2\theta \bigg]_0^1 d\theta$$

$$= \frac{1}{4} \int_{\frac{\pi}{4}}^{\frac{\pi}{2}} \cos^2\theta \, d\theta = \frac{1}{4} \int_{\frac{\pi}{4}}^{\frac{\pi}{2}} \frac{1 + \cos 2\theta}{2} \, d\theta$$

$$= \frac{1}{8} \int_{\frac{\pi}{4}}^{\frac{\pi}{2}} d\theta + \frac{1}{8} \int_{\frac{\pi}{4}}^{\frac{\pi}{2}} \cos 2\theta \, d\theta$$

$$= \frac{1}{8} \theta \bigg]_{\frac{\pi}{4}}^{\frac{\pi}{2}} + \frac{1}{16} \sin 2\theta \bigg]_{\frac{\pi}{4}}^{\frac{\pi}{2}}$$

$$= \left(\frac{\pi}{16} - \frac{\pi}{32} \right) + \left(0 - \frac{1}{16} \right)$$

$$= \frac{\pi}{32} - \frac{1}{16}. \quad \blacksquare$$

▶▶▶ **The area of a region D can be found by simply integrating the constant function $f(x,y) = 1$ over the region:**

$$\text{area of } D = \iint_D 1 \, dA.$$

EXAMPLE 18 Find the area of the region bounded by $r = 1$ and $r = 1 + \sin\theta$ for $\theta \in [0, \pi]$.

Solution We have

$$\text{area of } D = \iint_D 1\, dA = \int_0^\pi \int_1^{1+\sin\theta} r\, dr\, d\theta$$

$$= \int_0^\pi r^2/2 \Big]_{r=1}^{r=1+\sin\theta} d\theta$$

$$= \int_0^\pi (1/2)((1+\sin\theta)^2 - 1)\, d\theta$$

$$= \int_0^\pi \left(\frac{\sin^2\theta}{2} + \sin\theta\right) d\theta$$

$$= \frac{1}{2}\int_0^\pi \sin^2\theta\, d\theta - \left(\cos\theta \Big]_0^\pi\right)$$

$$= \frac{1}{2}\int_0^\pi \sin^2\theta\, d\theta + 2.$$

Using an integral table, we find that

$$\frac{1}{2}\int_0^\pi \sin^2\theta\, d\theta = \pi/4 \approx 0.785,$$

so the total area of the region is approximately 2.785. ∎

EXERCISES for Section 15.3

1. Find $\iint_D (x^2 + y^2)\, dx\, dy$, where D is the region in the xy-plane pictured below:

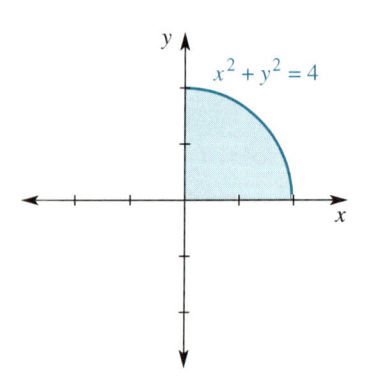

For Exercise 1

15.3 DOUBLE INTEGRALS IN POLAR COORDINATES

2. For the integral $\int_{-2}^{2} \int_{0}^{\sqrt{4-y^2}} xy\, dx\, dy$, sketch the region of integration, make a change of variables to polar coordinates, then perform the integration.

3. For the integral $\int_{-\sqrt{2}/2}^{\sqrt{2}/2} \int_{-\sqrt{1-x^2}}^{|x|} x^2 y^3\, dy\, dx$, sketch the region of integration, make a change of variables to polar coordinates, then perform the integration.

4. Compute $\iint_D (x^2 + y^2)\, dx\, dy$, where the region D is the unit circle.

For exercises 5-10: Find the area of the region D described by integrating
$$\iint_D 1\, dA.$$

5. D is the region enclosed by the polar graph $r = 1 + \sin(2\theta)$.

6. D is the region enclosed by the polar graph $r = 1 - \cos(2\theta)$.

7. D is the region enclosed by the inner loop of the polar graph $r = 1 + 2\sin(\theta)$.

8. D is the region enclosed by the inner loop of the polar graph $r = 1 - 2\sin(\theta)$.

9. D is the region enclosed by one "leaf" of the polar graph $r = \sin(3\theta)$.

10. D is the region enclosed by one "leaf" of the polar graph $r = \cos(2\theta)$.

---

11. Use a double integral to find the area of one loop of $r^2 = 8 \sin 2\theta$.

12. Use a double integral to find the area inside $r = 2 - 2\cos\theta$ and outside $r = 3$.

13. Use a double integral to find the area bounded by $r = 3 + 2\sin\theta$.

14. Find the volume of the solid that lies inside the sphere $x^2 + y^2 + z^2 = 25$ and outside the cylinder $x^2 + y^2 = 9$.

For exercises 15-18: Evaluate the following double integrals by changing to polar coordinates.

15. $\int_{-5}^{5} \int_{0}^{\sqrt{25-x^2}} e^{-(x^2+y^2)}\, dy\, dx$

16. $\int_{0}^{1} \int_{0}^{\sqrt{1-x^2}} e^{\sqrt{x^2+y^2}}\, dy\, dx$

17. $\int_{0}^{3} \int_{0}^{\sqrt{9-x^2}} (x^2 + y^2)^{3/2}\, dy\, dx$

18. $\int_{1}^{2} \int_{0}^{x} \frac{1}{\sqrt{x^2 + y^2}}\, dy\, dx$

15.4 TRIPLE INTEGRALS OVER MORE GENERAL REGIONS

Triple integrals over regions other than nice "boxes" are calculated in a way similar to double integrals. For example, suppose that we can describe the region of integration R as lying between two surfaces, $z = g_1(x, y)$ and $z = g_2(x, y)$, over some (2-dimensional) domain D in the xy-plane (see Figure 15.22).

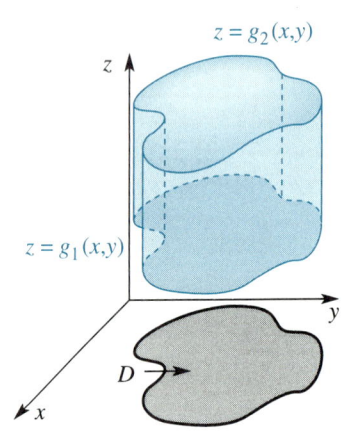

Figure 15.22 Region R is bounded by $z = g_1(x, y)$ and $z = g_2(x, y)$ over D.

The triple integral of a function $f(x, y, z)$ over R can be written in the form

$$\iiint_R f(x, y, z)\, dV = \iint_D \left(\int_{g_1(x,y)}^{g_2(x,y)} f(x, y, z)\, dz \right) dA.$$

Note that the value of the inner integral will be an expression in terms of x and y alone, so we have essentially reduced the problem to that of finding a double integral over the 2-dimensional domain D. This problem can then be treated like those of the previous section. (In this case, we can say that the region R is *simple* over the xy-plane.)

Similarly, if we can express the region of integration R as lying between two surfaces of the form $y = h_1(x, z)$ and $y = h_2(x, z)$ over a domain D in the xz-plane (see Figure 15.23), then we can write the triple integral in the form

$$\iiint_R f(x, y, z)\, dV = \iint_D \left(\int_{h_1(x,z)}^{h_2(x,z)} f(x, y, z)\, dy \right) dA,$$

where the area element dA refers to the xz-plane. (In this case, we can say that the region R is *simple* over the xz-plane.)

15.4 TRIPLE INTEGRALS OVER MORE GENERAL REGIONS

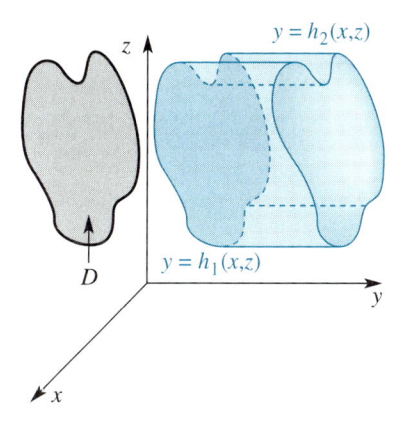

Figure 15.23 Region R is bounded by $y = h_1(x, z)$ and $y = h_2(x, z)$ over D.

A third possibility is that the region R is bounded by the two surfaces $x = k_1(y, z)$ and $x = k_2(y, z)$ over some domain D in the yz-plane. In this case, we could write the triple integral in the form

$$\iiint_R f(x, y, z)\, dV = \iint_D \left(\int_{k_1(y,z)}^{k_2(y,z)} f(x, y, z)\, dx \right) dA,$$

where the area element dA refers to the yz-plane. (In this case, we can say that the region R is *simple* over the yz-plane.)

EXAMPLE 19 Find $\int_0^1 \int_0^z \int_0^y xy^2 z^3 \, dx\, dy\, dz$.

Solution We have

$$\int_0^1 \int_0^z \int_0^y xy^2 z^3 \, dx\, dy\, dz = \int_0^1 \int_0^z \left[\frac{x^2 y^2 z^3}{2} \right]_0^y dy\, dz$$

$$= \int_0^1 \int_0^z \frac{y^4 z^3}{2} dy\, dz$$

$$= \int_0^1 \left[\frac{y^5 z^3}{10} \right]_0^z dz = \int_0^1 \frac{z^8}{10} dz = \left. \frac{z^9}{90} \right|_0^1 = \frac{1}{90}. \quad \blacksquare$$

EXAMPLE 20 Find $\iiint_R 2z \, dV$, where R is the region in space lying between the surfaces $z = 0$ and $z = \sqrt{xy}$ over the domain in the xy-plane bounded by $y = x^2 + 1$, $y = 1 - x^2$, and $x = 1$.

Solution The integral can be expressed as

$$\iiint_R 2z\,dV = \int_0^1 \int_{-x^2+1}^{x^2+1} \int_0^{\sqrt{xy}} 2z\,dz\,dy\,dx$$

$$= \int_0^1 \int_{-x^2+1}^{x^2+1} z^2 \Big]_{z=0}^{z=\sqrt{xy}} dy\,dx$$

$$= \int_0^1 \int_{-x^2+1}^{x^2+1} xy\,dy\,dx$$

$$= \int_0^1 (xy^2/2) \Big]_{y=-x^2+1}^{y=x^2+1} dx$$

$$= \int_0^1 \left(\frac{x(x^2+1)^2}{2} - \frac{x(-x^2+1)^2}{2} \right) dx$$

$$= \int_0^1 4x^3\,dx = x^4 \Big]_0^1 = 1. \quad \blacksquare$$

▶▶▶ The volume of a region R can be found by simply integrating the constant function $f(x,y,z) = 1$ over the region:

$$\text{volume of } R = \iiint_R 1\,dV.$$

EXAMPLE 21 Find the volume of the region R bounded by $z = x^2 + y^2$ and $z = -2xy$ over the domain $D = \{(x,y) : 0 \le x \le y^2,\ 0 \le y \le 1\}$.

Solution We have

$$\iiint_R 1\,dV = \int_0^1 \int_0^{y^2} \int_{-2xy}^{x^2+y^2} 1\,dz\,dx\,dy$$

$$= \int_0^1 \int_0^{y^2} z \Big]_{z=-2xy}^{z=x^2+y^2} dx\,dy$$

$$= \int_0^1 \int_0^{y^2} (x^2 + 2xy + y^2)\,dx\,dy$$

$$= \int_0^1 \left(\frac{x^3}{3} + x^2 y + xy^2 \right) \Big]_{x=0}^{x=y^2} dy$$

$$= \int_0^1 \left(\frac{y^6}{3} + y^5 + y^4 \right) dy$$

$$= \frac{y^7}{21} + \frac{y^6}{6} + \frac{y^5}{5} \Big]_0^1$$

$$= \frac{1}{21} + \frac{1}{6} + \frac{1}{5} = \frac{29}{70}. \quad \blacksquare$$

15.4 TRIPLE INTEGRALS OVER MORE GENERAL REGIONS

EXAMPLE 22 Find the volume of the region R bounded by the surface $x = z/y$ and the five planes: $x = 0$, $y = 1$, $y = e$, $z = 2$, and $z = 5$.

Solution The volume is given by

$$\iiint_R 1\, dV = \int_2^5 \int_1^e \int_0^{z/y} 1\, dx\, dy\, dz$$

$$= \int_2^5 \int_1^e x \Big]_{x=0}^{x=z/y} dy\, dz$$

$$= \int_2^5 \int_1^e (z/y)\, dy\, dz$$

$$= \int_2^5 z \ln y \Big]_{y=1}^{y=e} dz$$

$$= \int_2^5 z\, dz$$

$$= z^2/2 \Big]_2^5 = 25/2 - 4/2 = 21/2.$$

■

EXERCISES for Section 15.4

1. Find $\displaystyle\int_0^1 \int_{x^2}^x \int_0^{xy} (x + 2y + 3z)\, dz\, dy\, dx$.

2. Find $\displaystyle\int_{-3}^2 \int_{-y}^y \int_{-x}^x (xy^2 z^3)\, dz\, dx\, dy$.

3. A tetrahedron is bounded by the three coordinate planes and the plane $x + y + z = 1$. Find its volume.

4. Find the volume between the plane $z = 1$ and the cone $z = \sqrt{x^2 + y^2}$.

5. Find the volume between the plane $z = 4$ and the paraboloid $z = x^2 + y^2$.

For exercises 6-10: Consider the region

$$R = \{(x, y, z) : x^2 + y^2 + z^2 \leq 1,\ x \geq 0,\ y \geq 0,\ z \geq 0\}.$$

(The upper hemisphere of the unit sphere.)

6. Find the volume of the region R using geometry.

7. Calculate the volume of the region R as a triple integral.

8. The density of the region at any point (x, y, z) is given by $f(x, y, z) = xyz$ kilograms per cubic unit. Find the mass of the region R.

9. If $f(x,y,z)$ is the density at a point in a region R and M is the total mass of the region, then the **center of mass** of the region is the point

$$(\bar{x}, \bar{y}, \bar{z}),$$

whose coordinates are given by the formulas

$$\bar{x} = \frac{1}{M} \iiint_R x f(x,y,z)\, dV, \qquad \bar{y} = \frac{1}{M} \iiint_R y f(x,y,z)\, dV,$$

$$\text{and} \qquad \bar{z} = \frac{1}{M} \iiint_R z f(x,y,z)\, dV.$$

Find the center of mass for the region R if its density at any point is given by $f(x,y,z) = xyz$.

10. The **moment of inertia** of an object having point density $f(x,y,z)$ about the xy-plane is given by the formula

$$I_{xy} = \iiint_R z^2 f(x,y,z)\, dV.$$

Similarly, the moments of inertia of the object about the xz-plane and yz-plane are

$$I_{xz} = \iiint_R y^2 f(x,y,z)\, dV \quad \text{and} \quad I_{yz} = \iiint_R x^2 f(x,y,z)\, dV,$$

respectively. Find the moment of inertia of the region R about each of the coordinate planes if its density at any point is $f(x,y,z) = xyz$.

11. If the tetrahedron of exercise 3 has point density $x^2 + y^2 + z^2$, find its mass.

12. If the tetrahedron of exercise 3 has point density $x^2 + y^2 + z^2$, find its center of mass.

13. If the tetrahedron of exercise 3 has point density $x^2 + y^2 + z^2$, find its moment of inertia about each of the coordinate planes.

14. Find the volume of the region R bounded by the graphs of $z = 3x^2$, $z = 4 - x^2$, $y = 0$, and $z + y = 6$.

For exercises 15-18: Evaluate the following triple integrals.

15. $\displaystyle \int_0^1 \int_{1+x}^{2x} \int_z^{x+z} x\, dy\, dz\, dx$

16. $\displaystyle \int_1^2 \int_0^{z^2} \int_{x+z}^{x-z} 4\, dy\, dx\, dz$

17. $\displaystyle \int_{-1}^2 \int_1^{x^2} \int_0^{x+y} 2x^2 y\, dz\, dy\, dx$

18. $\displaystyle \int_{-2}^3 \int_0^{3y} \int_1^{yz} (2x + y + z)\, dx\, dz\, dy$

19. Sketch the region R bounded by the graphs of $z = 9 - 4x^2 - y^2$ and $z = 0$. Express its volume as a triple integral in six different ways.

15.5 CYLINDRICAL AND SPHERICAL COORDINATES

20. Sketch the region bounded by the graphs of $y = 2 - z^2$, $y = z^2$, $x + z = 4$, and $x = 0$. Find its volume using a triple integral.

21. The triple integral $\displaystyle\int_1^4 \int_{-z}^{z} \int_{-\sqrt{z^2-y^2}}^{\sqrt{z^2-y^2}} dx\, dy\, dz$ represents the volume of a solid region. Describe the region.

15.5 CYLINDRICAL AND SPHERICAL COORDINATES

Polar coordinates are a useful alternative to rectangular coordinates for describing certain curves in the plane, especially those involving circular arcs. There are also alternative coordinate systems for 3-dimensional space, and they are the subject of this section.

Cylindrical coordinates

Given a point expressed in rectangular coordinates (x, y, z), we obtain its representation in **cylindrical coordinates** (r, θ, z) by simply converting x and y to polar form:

$$x = r \cos \theta \qquad y = r \sin \theta.$$

In cylindrical coordinates,

- r is the horizontal distance from the z-axis;
- θ is the angle measured counterclockwise from the positive x-axis side of the xz-plane;
- z is still measured vertically from the xy-plane.

Given a point expressed in cylindrical coordinates (r, θ, z), we can imagine locating it on a vertical cylinder of radius r with the z-axis) running down its center (see Figure 15.24).

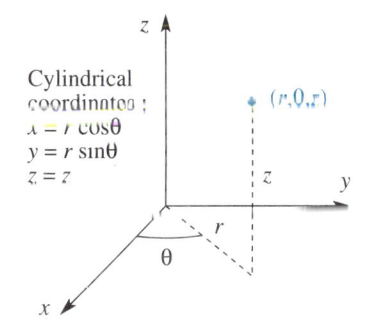

Figure 15.24 Cylindrical coordinates.

While we sometimes have occasion to consider $r < 0$ and arbitrary angles θ when using polar coordinates, we make the stipulation

$$r \geq 0 \quad \text{and} \quad 0 \leq \theta < 2\pi$$

for cylindrical coordinates. This gives each point in space (except those on the z-axis) exactly one representation in cylindrical coordinates.

EXAMPLE 23 A point in space has cylindrical coordinates $(r, \theta, z) = (4, \frac{\pi}{3}, -3)$. Find its rectangular coordinates.

Solution We have

$$x = 4\cos\frac{\pi}{3} = 4 \cdot \frac{1}{2} = 2$$

$$y = 4\sin\frac{\pi}{3} = 4 \cdot \frac{\sqrt{3}}{2} = 2\sqrt{3}$$

$$z = -3.$$

Therefore, $(x, y, z) = (2, 2\sqrt{3}, -3)$. ∎

To convert from rectangular coordinates to cylindrical coordinates, we make use of the formula

$$r = \sqrt{x^2 + y^2}$$

to determine r. The value θ is determined by

$$\theta = \begin{cases} \arctan(y/x) & \text{if } x > 0 \text{ and } y \geq 0 \\ \pi + \arctan(y/x) & \text{if } x < 0 \\ 2\pi + \arctan(y/x) & \text{if } x > 0 \text{ and } y < 0. \end{cases}$$

If $x = 0$, we can take $\theta = \pi/2$ for $y > 0$ or $\theta = 3\pi/2$ for $y < 0$.

EXAMPLE 24 A point in space has rectangular coordinates $(x, y, z) = (3, 4, -2)$. Find its cylindrical coordinates.

Solution We have $r = \sqrt{x^2 + y^2} = \sqrt{3^2 + 4^2} = \sqrt{25} = 5,$

and $\theta = \arctan(4/3).$

Therefore, $(r, \theta, z) = (5, \arctan(4/3), -2)$. ∎

15.5 CYLINDRICAL AND SPHERICAL COORDINATES

Some surfaces have particularly simple equations in terms of cylindrical coordinates. A cylinder of radius c and z-axis running down its center has equation

$$r = c.$$

A vertical "half-plane" with its edge along the z-axis has equation

$$\theta = c,$$

and a horizontal plane still has equation

$$z = c.$$

Spherical coordinates

If we imagine locating points on a sphere centered at the origin, we have the basic idea of **spherical coordinates**. For a point expressed in spherical coordinates (ρ, θ, φ):

- ρ is the distance from the origin;
- θ is the angle measured counterclockwise from the positive x-axis side of the xz-plane;
- φ is the angle measured down from the positive z-axis.

Figure 15.25 illustrates the location of a point by spherical coordinates.

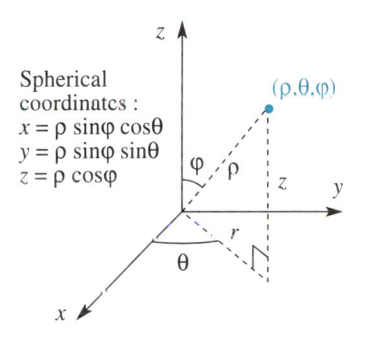

Spherical coordinates:
$x = \rho \sin\varphi \cos\theta$
$y = \rho \sin\varphi \sin\theta$
$z = \rho \cos\varphi$

Figure 15.25 Spherical coordinates.

To give each point in space except those on the z-axis a unique representation in spherical coordinates, we make the stipulations

$$\rho \geq 0, \quad 0 \leq \theta < 2\pi, \quad 0 \leq \varphi \leq \pi.$$

To convert from rectangular coordinates (x, y, z) to spherical coordinates, we make use of the formula

$$\rho = \sqrt{x^2 + y^2 + z^2}$$

to determine the spherical radius ρ. The coordinate θ is determined in the same way as for cylindrical coordinates. Using $r = \sqrt{x^2 + y^2}$, we can write

$$\varphi = \begin{cases} \arctan(r/z) = \arctan(\sqrt{x^2+y^2}/z) & \text{for } z > 0 \\ \pi + \arctan(r/z) = \pi + \arctan(\sqrt{x^2+y^2}/z) & \text{for } z < 0 \\ \pi/2 & \text{for } z = 0 \end{cases}$$

for points not on the z-axis. We have $\varphi = 0$ for points on the positive z-axis and $\varphi = \pi$ for points on the negative z-axis.

EXAMPLE 25 A point in space has rectangular coordinates $(x, y, z) = (3, 4, -2)$. Find its spherical coordinates (ρ, θ, φ).

Solution We have
$$\rho = \sqrt{x^2 + y^2 + z^2} = \sqrt{3^2 + 4^2 + (-2)^2} = \sqrt{29}.$$

The value θ is calculated as it was in the previous example to be

$$\theta = \arctan(4/3).$$

Finally, since $r = \sqrt{3^2 + 4^2} = \sqrt{25} = 5$, we have

$$\varphi = \pi + \arctan(5/(-2)).$$

Therefore, the spherical coordinates are

$$(\rho, \theta, \varphi) = (\sqrt{29}, \arctan(4/3), \pi + \arctan(-5/2)).\quad\blacksquare$$

We can convert directly from cylindrical coordinates (r, θ, z) to spherical coordinates by noting that

$$\rho = \sqrt{r^2 + z^2}.$$

The coordinate θ is the same, and φ can be calculated as discussed above using r and z.

EXAMPLE 26 A point in space has cylindrical coordinates $(r, \theta, z) = (4, \frac{\pi}{3}, -3)$. Find its spherical coordinates.

Solution We have $\rho = \sqrt{r^2 + z^2} = \sqrt{4^2 + (-3)^2} = \sqrt{25} = 5.$

Now, $\theta = \frac{\pi}{3}$ and $\varphi = \pi + \arctan(r/z) = \pi + \arctan(-\frac{4}{3}).$

Thus, the point has spherical coordinates $(\rho, \theta, \varphi) = (5, \frac{\pi}{3}, \pi + \arctan(-\frac{4}{3})).\quad\blacksquare$

15.5 CYLINDRICAL AND SPHERICAL COORDINATES

To change from spherical coordinates (ρ, θ, φ) to cylindrical coordinates (r, θ, z), we note that

$$r = \rho \sin \varphi,$$

$$\theta = \theta,$$

$$z = \rho \cos \varphi.$$

To change from spherical coordinates to rectangular, first convert to cylindrical coordinates, then convert these to rectangular coordinates:

$$x = \rho \sin \varphi \cos \theta,$$

$$y = \rho \sin \varphi \sin \theta,$$

$$z = \rho \cos \varphi.$$

EXAMPLE 27 A point has spherical coordinates $(\rho, \theta, \varphi) = (4, \pi/4, 2\pi/3)$. Express the point in both cylindrical and rectangular coordinates.

Solution For cylindrical coordinates, we have

$$r = \rho \sin \varphi = 4 \sin(2\pi/3) = 4(\sqrt{3}/2) = 2\sqrt{3},$$

$$\theta = \pi/4,$$

$$z = \rho \cos \varphi = 4 \cos(2\pi/3) = 4(-1/2) = -2.$$

Now, converting to rectangular coordinates, we have

$$x = \rho \sin \varphi \cos \theta = r \cos \theta = (2\sqrt{3}) \cos(\pi/4) = (2\sqrt{3})(\sqrt{2}/2) = \sqrt{6},$$

$$y = \rho \sin \varphi \sin \theta = r \sin \theta = (2\sqrt{3}) \cos(\pi/4) = (2\sqrt{3})(\sqrt{2}/2) = \sqrt{6},$$

$$z = \rho \cos \varphi = -2.$$

Hence, the cylindrical coordinates of the point are $(r, \theta, z) = (2\sqrt{3}, \pi/4, -2)$ and the rectangular coordinates are $(x, y, z) = (\sqrt{6}, \sqrt{6}, -2)$. ∎

Some surfaces have particularly simple equations in terms of spherical coordinates. A sphere of radius c has equation

$$\rho = c.$$

Again, a vertical "half-plane" with its edge along the z-axis has equation

$$\theta = c,$$

and a circular cone (actually the top or bottom half of a quadric surface) with its vertex at the origin has equation

$$\varphi = c.$$

Triple integrals in cylindrical coordinates

Some regions of integration for triple integrals may lend themselves to easier description by the use of *cylindrical* coordinates.

To form a cylindrical partition of a region of space, we first form a polar coordinate grid horizontally (parallel with the xy-plane) of grid dimension Δr by $\Delta \theta$, and then we partition vertically along the z-direction by Δz. The resulting small regions of space carved out by this method look like fragments of a cylindrical shell, each with volume

$$\Delta V \approx r \, \Delta r \, \Delta \theta \, \Delta z.$$

Figure 15.26 illustrates a typical cylindrical grid section. If the partition is fine enough, the sections approximate small rectangular boxes of dimensions Δr by $r\Delta\theta$ by Δz.

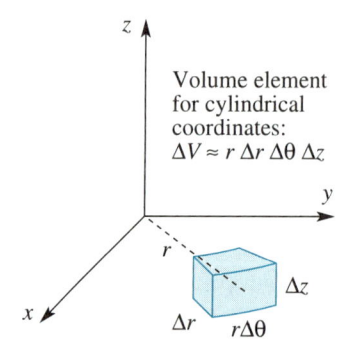

Figure 15.26 Forming a cylindrical partition of space.

To express a triple integral

$$\iiint_R f(x, y, z) \, dV$$

in terms of cylindrical coordinates:

15.5 CYLINDRICAL AND SPHERICAL COORDINATES

Step 1. Describe the region R in terms of θ, r, and z.

Step 2. Make the change of variables

$$x = r\cos\theta \qquad y = r\sin\theta \qquad z = z.$$

Step 3. Use $dV = r\,dr\,d\theta\,dz$ as the volume element in the integral.

EXAMPLE 28 A material has density given by $\rho = (x^2 z + y^2 z)$ g/cm³ at any point on or above the xy-plane. (Units of the three coordinate axes are in centimeters.) Find the mass of the cylindrical shell R with inner surface $x^2 + y^2 = 1$, outer surface $x^2 + y^2 = 4$, and lying between the horizontal planes $z = 0$ and $z = 2$.

Solution The shell is illustrated in Figure 15.27.

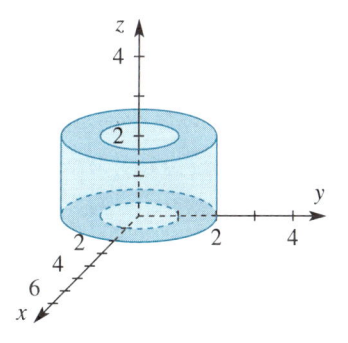

Figure 15.27 Finding the mass of a cylindrical shell.

In terms of cylindrical coordinates, the inner and outer surfaces of the cylindrical shell have the equations $r = 1$ and $r = 2$, respectively. The region R consists of the points

$$\{(r, \theta, z) : 1 \leq r \leq 2,\ 0 \leq \theta \leq 2\pi,\ 0 \leq z \leq 2\}.$$

The density function we are integrating can be written

$$x^2 z + y^2 z = z(x^2 + y^2) = zr^2$$

in cylindrical coordinates.

Hence, we can express the mass integral in cylindrical coordinates as follows:

$$\iiint_R (x^2 z + y^2 z) \, dV = \int_0^2 \int_0^{2\pi} \int_1^2 (zr^2) r \, dr \, d\theta \, dz$$

$$= \int_0^2 \int_0^{2\pi} \int_1^2 zr^3 \, dr \, d\theta \, dz.$$

Carrying out the integral computation gives us

$$\int_0^2 \int_0^{2\pi} zr^4/4 \Big]_{r=1}^{r=2} d\theta \, dz = \int_0^2 \int_0^{2\pi} (4z - z/4) \, d\theta \, dz$$

$$= \int_0^2 (15z/4)\theta \Big]_{\theta=0}^{\theta=2\pi} dz$$

$$= \int_0^2 (15\pi z/2) \, dz$$

$$= (15\pi z^2/4) \Big]_{z=0}^{z=2} = 15\pi.$$

The shell has a mass of $15\pi \approx 47.124$ grams. ∎

Triple integrals in spherical coordinates

Similarly, some regions of integration for triple integrals may lend themselves to easier descriptions by the use of *spherical* coordinates (ρ, θ, φ).

If we partition the distance from the origin by $\Delta \rho$, we obtain a set of concentric spheres. These spheres are then partitioned by $\Delta \theta$ and $\Delta \varphi$ like longitude and latitude lines on the face of the earth. The resulting small regions of space carved out by this method look like fragments of a spherical shell, each with volume

$$\Delta V \approx \rho^2 \sin \varphi \, \Delta \rho \, \Delta \theta \, \Delta \phi.$$

Figure 15.28 illustrates a typical spherical grid section. If the partition is fine enough, the sections approximate small rectangular boxes of dimensions $\Delta \rho$ by $\rho \sin \varphi \Delta \theta$ by $\rho \Delta \varphi$.

To express a triple integral $\iiint_R f(x, y, z) \, dV$ in terms of spherical coordinates:

15.5 CYLINDRICAL AND SPHERICAL COORDINATES

Step 1. Describe the region R in terms of ρ, θ, and φ.

Step 2. Make the change of variables

$$x = \rho \sin\varphi \cos\theta \qquad y = \rho \sin\varphi \sin\theta \qquad z = \rho \cos\varphi.$$

Step 3. Use $dV = \rho^2 \sin\varphi \, d\rho \, d\theta \, d\varphi$ as the volume element in the integral.

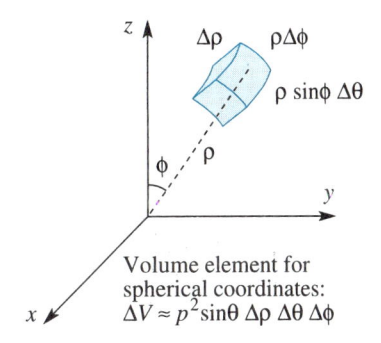

Figure 15.28 Forming a spherical partition of space.

EXAMPLE 29 Find $\iiint_R (x^2 + y^2 + z^2)^{3/2} \, dV$, where R is the region within the unit sphere having all three coordinates $x \geq 0$, $y \geq 0$, $z \geq 0$.

Solution Figure 15.29 illustrates the region in question.

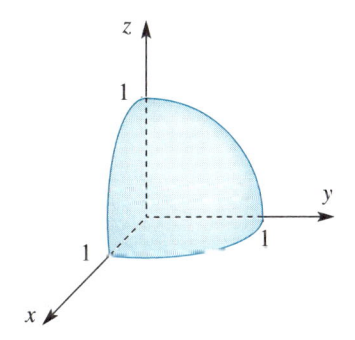

Figure 15.29 Region R is the interior of the unit sphere within the first octant.

In terms of spherical coordinates, R is the set of points satisfying

$$0 \leq \rho \leq 1, \qquad 0 \leq \theta \leq \pi/2, \qquad 0 \leq \varphi \leq \pi/2.$$

Since $x^2 + y^2 + z^2 = \rho^2$ and $dV = \rho^2 \sin\varphi \, d\rho \, d\theta \, d\varphi$, we can express the integral as

$$\iiint_R (x^2 + y^2 + z^2)^{3/2} \, dV = \int_0^{\pi/2} \int_0^{\pi/2} \int_0^1 (\rho^3)\rho^2 \sin\varphi \, d\rho \, d\theta \, d\varphi$$

$$= \int_0^{\pi/2} \int_0^{\pi/2} \int_0^1 \rho^5 \sin\varphi \, d\rho \, d\theta \, d\varphi$$

$$= \int_0^{\pi/2} \int_0^{\pi/2} (\rho^6/6) \sin\varphi \Big]_{\rho=0}^{\rho=1} d\theta \, d\varphi$$

$$= \int_0^{\pi/2} \int_0^{\pi/2} (1/6) \sin\varphi \, d\theta \, d\varphi$$

$$= \int_0^{\pi/2} (\theta/6) \sin\varphi \Big]_{\theta=0}^{\theta=\pi/2} d\varphi$$

$$= \int_0^{\pi/2} (\pi/12) \sin\varphi \, d\varphi$$

$$= (-\pi/12) \cos\varphi \Big]_{\varphi=0}^{\varphi=\pi/2} = 0 - (-\pi/12) = \pi/12. \blacksquare$$

EXERCISES for Section 15.5

For exercises 1-12: Fill in the table.

	Rectangular	Cylindrical	Spherical
1.	$(1, 2, -2)$		
2.		$(4, 5\pi/4, -3)$	
3.			$(5, 2\pi/3, 2\pi/3)$
4.	$(-3, -3, -3)$		
5.		$(7, 3\pi/4, 6)$	
6.			$(5, 11\pi/6, \pi/2)$
7.	$(0, 0, -5)$		
8.		$(2, 0, 0)$	
9.			$(4, \pi/2, \pi/2)$
10.	$(0, 6, 0)$		
11.		$(2, 330°, -1)$	
12.			$(6, 256°, 127°)$

13. What points in space have exactly the same coordinates in both the rectangular and cylindrical systems (angles measured in radians)?

14. What points in space have exactly the same coordinates in both the rectangular and spherical systems (angles measured in radians)?

15. What points in space have exactly the same coordinates in both the cylindrical and spherical systems (angles measured in radians)?

15.5 CYLINDRICAL AND SPHERICAL COORDINATES

16. What points in space have exactly the same coordinates in all three systems (angles measured in radians)?

17. What points in space have more than one representation in the cylindrical and spherical coordinate systems? Why?

For exercises 18-30: An equation in terms of the coordinates of one system can be written in terms of another by substitution, using the conversion formulas. For example, the equation of the plane

$$x + 2y - 3z = 6$$

in rectangular coordinates has the equation

$$r\cos\theta + 2r\sin\theta - 3z = 6$$

in cylindrical coordinates, and the equation

$$\rho\cos\theta\sin\varphi + 2\rho\sin\theta\sin\varphi - 3\rho\cos\varphi = 6$$

in spherical coordinates. Make the conversion of the given equation into cylindrical and spherical coordinates.

18. the elliptic paraboloid $z = x^2 + y^2$

19. the hyperbolic paraboloid $z = x^2 - y^2$

20. the sphere $x^2 + y^2 + z^2 = 16$

21. the plane $z = 5$

22. the plane $y = x$

23. the cylinder $x^2 + y^2 = 9$

24. the cone $z^2 = x^2 + y^2$

25. the parabolic trough $z = y^2$

26. the hyperbolic paraboloid $z = 2xy$

27. the surface $z = x/y$

28. the hyperboloid of one sheet $z^2 = x^2 + y^2 - 1$

29. the hyperboloid of two sheets $z^2 = x^2 + y^2 + 1$

30. the surface $z = \arctan(y/x)$

---

31. Find the volume of the upper hemisphere of the unit sphere using a triple integral in cylindrical coordinates.

32. Find the volume of the upper hemisphere of the unit sphere using a triple integral in spherical coordinates.

33. Evaluate $\displaystyle\int_0^{\pi/2}\int_0^{\pi/4}\int_0^{3\cos\theta} \rho^2 \sin\phi\, d\rho\, d\phi\, d\theta$.

34. Find the volume of the ellipsoid
$$\frac{x^2}{4} + \frac{y^2}{9} + \frac{z^2}{16} = 1.$$

15.6 NUMERICAL TECHNIQUES FOR MULTIPLE INTEGRALS

In this section, discuss some techniques for numerically approximating the value of a multiple integral. Recall that a single definite integral
$$\int_a^b f(x)\,dx$$
can be approximated by simply calculating a Reimann sum
$$\sum_{i=1}^n f(x_i)\Delta x$$
for a suitably fine partition of $[a, b]$ and some particular choice of points x_i $(1 \le i \le n)$ from the resulting subintervals. The left rectangle, right rectangle, and midpoint rules are all examples of this strategy, corresponding to choosing each point x_i as the left endpoint, right endpoint, or midpoint, respectively, of each subinterval.

A similar strategy can be used to approximate the value of a double integral
$$\iint_D f(x, y)\,dA.$$

If D is a rectangular region, then we can follow these steps:

Step 1. We subdivide the region D using a fine rectangular grid, where each rectangle has dimensions $\Delta x \times \Delta y$.

Step 2. For each rectangle in our grid, we choose a point (x_i, y_j).

Step 3. We calculate $f(x_i, y_j)\Delta x \Delta y$ for each of the rectangles.

Step 4. We sum them up to find
$$\iint_D f(x, y)\,dA \approx \sum_{j=1}^m \sum_{i=1}^n f(x_i, y_j)\,\Delta x \Delta y.$$

We'll illustrate the method by using it to approximate the volume between the paraboloid $z = x^2 + y^2$ and the xy-plane over the unit square $[0, 1] \times [0, 1]$.

15.6 NUMERICAL TECHNIQUES FOR MULTIPLE INTEGRALS

First, we note that in this case, we can calculate the volume exactly:

$$\int_0^1 \int_0^1 (x^2 + y^2)\, dx\, dy = \int_0^1 \left(\frac{x^3}{3} + xy^2\right)\Bigg]_{x=0}^{x=1} dy$$

$$= \int_0^1 \left(\frac{1}{3} + y^2\right) dy$$

$$= \frac{y + y^3}{3}\Bigg]_{y=0}^{y=1}$$

$$= 2/3.$$

We have chosen a simple integral for which we know the exact value for purposes of comparing our approximations. The importance of numerical techniques is more evident when we need to approximate the value of a definite integral that is difficult or impossible to calculate exactly.

Suppose that we partition $[0, 1] \times [0, 1]$ into 25 small squares by partitioning $[0, 1]$ into 5 equal subintervals along both the x and y axes. In other words, we have

$$\Delta x = 0.2 = \Delta y,$$

so that the area of each small square is

$$\Delta A = 0.04 = \Delta x \Delta y.$$

To use this grid to approximate the value of the integral

$$\int_0^1 \int_0^1 (x^2 + y^2)\, dx\, dy$$

we need to:

1. pick one point (x, y) from each small square,
2. evaluate $x^2 + y^2$ at each chosen point, and
3. add all these values and multiply by 0.04.

The choice of points to make from each rectangle in our grid is purely arbitrary. However, if we wanted to automate the procedure, we might make some systematic choices, as we do for the left rectangle or right rectangle numerical methods of integration. Figure 15.30 shows three possible systematic choices of points we could make for a rectangular grid.

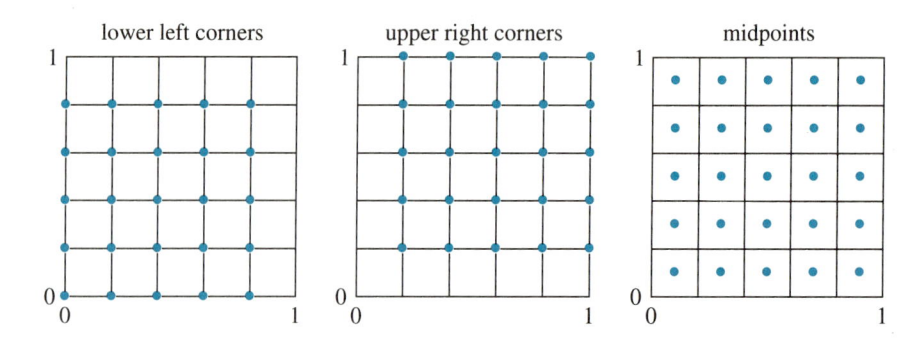

Figure 15.30 Three possible choices of points for a rectangular grid.

EXAMPLE 30 Approximate $\int_0^1 \int_0^1 x^2+y^2 \, dx \, dy$ using each of the choices of points indicated in Figure 15.30.

Solution Below, we have assembled the data we need to calculate each approximation, depending on the choice of points from the grid sections.

lower left corners		upper right corners		midpoints	
(x,y)	x^2+y^2	(x,y)	x^2+y^2	(x,y)	x^2+y^2
$(0.0, 0.0)$	0.00	$(0.2, 0.2)$	0.08	$(0.1, 0.1)$	0.02
$(0.2, 0.0)$	0.04	$(0.4, 0.2)$	0.20	$(0.1, 0.3)$	0.10
$(0.4, 0.0)$	0.16	$(0.6, 0.2)$	0.40	$(0.1, 0.5)$	0.26
$(0.6, 0.0)$	0.36	$(0.8, 0.2)$	0.68	$(0.1, 0.7)$	0.50
$(0.8, 0.0)$	0.64	$(1.0, 0.2)$	1.04	$(0.1, 0.9)$	0.82
$(0.0, 0.2)$	0.04	$(0.2, 0.4)$	0.20	$(0.3, 0.1)$	0.10
$(0.2, 0.2)$	0.08	$(0.4, 0.4)$	0.32	$(0.3, 0.3)$	0.18
$(0.4, 0.2)$	0.20	$(0.6, 0.4)$	0.52	$(0.3, 0.5)$	0.34
$(0.6, 0.2)$	0.40	$(0.8, 0.4)$	0.80	$(0.3, 0.7)$	0.58
$(0.8, 0.2)$	0.68	$(1.0, 0.4)$	1.16	$(0.3, 0.9)$	0.90
$(0.0, 0.4)$	0.16	$(0.2, 0.6)$	0.40	$(0.5, 0.1)$	0.26
$(0.2, 0.4)$	0.20	$(0.4, 0.6)$	0.52	$(0.5, 0.3)$	0.34
$(0.4, 0.4)$	0.32	$(0.6, 0.6)$	0.72	$(0.5, 0.5)$	0.50
$(0.6, 0.4)$	0.52	$(0.8, 0.6)$	1.00	$(0.5, 0.7)$	0.74
$(0.8, 0.4)$	0.80	$(1.0, 0.6)$	1.36	$(0.5, 0.9)$	1.06
$(0.0, 0.6)$	0.36	$(0.2, 0.8)$	0.68	$(0.7, 0.1)$	0.50
$(0.2, 0.6)$	0.40	$(0.4, 0.8)$	0.80	$(0.7, 0.3)$	0.58
$(0.4, 0.6)$	0.52	$(0.6, 0.8)$	1.00	$(0.7, 0.5)$	0.74
$(0.6, 0.6)$	0.72	$(0.8, 0.8)$	1.28	$(0.7, 0.7)$	0.98
$(0.8, 0.6)$	1.00	$(1.0, 0.8)$	1.64	$(0.7, 0.9)$	1.30
$(0.0, 0.8)$	0.64	$(0.2, 1.0)$	1.04	$(0.9, 0.1)$	0.82
$(0.2, 0.8)$	0.68	$(0.4, 1.0)$	1.16	$(0.9, 0.3)$	0.90
$(0.4, 0.8)$	0.80	$(0.6, 1.0)$	1.36	$(0.9, 0.5)$	1.06
$(0.6, 0.8)$	1.00	$(0.8, 1.0)$	1.64	$(0.9, 0.7)$	1.30
$(0.8, 0.8)$	1.28	$(1.0, 1.0)$	2.00	$(0.9, 0.9)$	1.62

If we total the values $x^2 + y^2$ for each choice and multiply by $\Delta x \Delta y = 0.04$, we obtain the corresponding approximations:

Using lower left corners: $\int_0^1 \int_0^1 (x^2 + y^2)\, dx\, dy \approx (12)(0.04) = 0.48.$

Using upper right corners: $\int_0^1 \int_0^1 (x^2 + y^2)\, dx\, dy \approx (22)(0.04) = 0.88.$

Using midpoints: $\int_0^1 \int_0^1 (x^2 + y^2)\, dx\, dy \approx (16.5)(0.04) = 0.66.$ ∎

Numerical techniques for general regions of integration

For a more general region D, we have an additional complication, since not all of the rectangles will fit entirely within the region. What do we do with those that overlap the boundary of the region? One idea might be to count only those within the region. Another would be to count all those rectangles that include any part of the region. Figure 15.31 shows which rectangles from our grid partition of $[0,1] \times [0,1]$ we would include if our region of integration had been the quarter of the unit circle in this square.

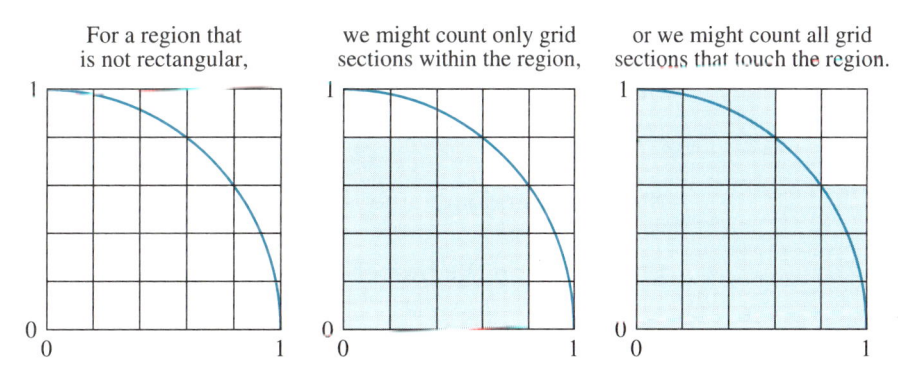

Figure 15.31 Choosing grid sections for a non-rectangular region.

If we choose only those rectangles completely within the region, we stand a good chance of underestimating the value of the integral. On the other hand, if we count *all* the rectangles that include some part of the region, then we stand a good chance of overestimating the area of the region. For a very irregularly shaped region, the ratio of the number of rectangles completely inside the region to the number overlapping any part of the region may be quite small, but for finer and finer grid partitions, we expect this ratio to be nearly 1. (There are, however, some pathologically bizarre regions for which this is not true.)

Another strategy for approximating a double integral

$$\iint_D f(x,y)\, dy\, dx$$

is outlined below:

Step 1. Partition the region D in strips of width Δx in the x-direction.

Step 2. Choose a slice of the solid from each strip.

Step 3. Approximate the area of each slice like any single integral, perhaps using the trapezoidal rule or Simpson's rule.

Step 4. Multiply the area of each slice by Δx and add the results.

(If the order of integration is $dx\,dy$ instead, we can partition D into strips of width Δy in the y-direction in Step 1, and then multiply the area of the chosen slices by Δy in Step 4.)

A technique similar to this one is used by many machine integrators to numerically approximate double integrals.

Monte Carlo method of approximating integrals

A **Monte Carlo** method is one making use of randomly generated numbers. Knowledge of the laws of probability can then be used to our advantage. Here's how random numbers could be used to estimate the area of a region:

Step 1. Completely enclose the region D with a rectangle R (or some other region of known area).

Step 2. Randomly choose several points within R.

Step 3. Find the proportion of points falling within region D.

Step 4. Multiply this proportion by the area of R.

Figure 15.32 illustrates this Monte Carlo method as applied to approximate the area of one quarter of the unit circle.

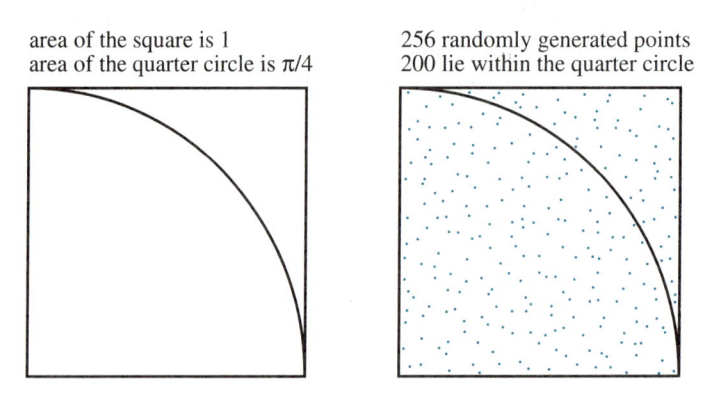

Figure 15.32 Using randomly selected points to approximate area.

15.6 NUMERICAL TECHNIQUES FOR MULTIPLE INTEGRALS

Of the 256 randomly generated points in the unit square R (of area 1), 200 of the points fell within the quarter circle D. This leads to the estimate:

$$\text{area of quarter circle} \approx (200/256) \cdot 1 = 0.78125.$$

In comparison, the exact area of the quarter circle is

$$\pi/4 \approx 0.7854.$$

A similar Monte Carlo method can be used to approximate the value of a double integral over an irregularly shaped region. To approximate

$$\iint_D f(x,y)\, dA,$$

we add two more steps:

Step 5. Evaluate $f(x,y)$ at each point that falls in the region D, and find the average of these function values.

Step 6. Multiply this average function value by the estimated area found in Step 4.

This is the approximate value of the integral.

EXAMPLE 31 The average value of $x^2 + y^2$ for the 200 points within the quarter circle D of Figure 15.32 was found to be 0.506. Use this to approximate

$$\iint_D (x^2 + y^2)\, dA,$$

and compare it with the actual integral value.

Solution The estimated area of the quarter circle was 0.78125. Therefore, the estimated value of the integral is

$$\iint_D (x^2 + y^2)\, dA \approx (0.78125)(0.506) = 0.3953125.$$

In comparison, we can use polar coordinates to find the actual integral value:

$$\iint_D (x^2 + y^2)\, dA = \int_0^{\pi/2} \int_0^1 (r^2) r\, dr\, d\theta$$

$$= \int_0^{\pi/2} \int_0^1 r^3\, dr\, d\theta$$

$$= \int_0^{\pi/2} r^4/4 \Big]_{r=0}^{r=1} d\theta$$

$$= \int_0^{\pi/2} 1/4\, d\theta$$

$$= \theta/4 \Big]_0^{\pi/2} = \pi/8 \approx 0.3927.$$

We can see that the Monte Carlo estimate compares very favorably with the actual value. ∎

With suitable adjustments, these Riemann sum, slicing, and Monte Carlo methods can also be used to approximate triple integrals.

EXERCISES for Section 15.6

1. Using the same grid as in Example 31, approximate the value of the integral

$$\int_0^1 \int_0^1 e^{x^2+y^2}\, dx\, dy$$

using the lower left corners of each grid section.

2. Using the same grid as in Example 31, approximate the value of the integral

$$\int_0^1 \int_0^1 e^{x^2+y^2}\, dx\, dy$$

using the upper right corners of each grid section.

3. Using the same grid as in Example 31, approximate the value of the integral

$$\int_0^1 \int_0^1 e^{x^2+y^2}\, dx\, dy$$

using the midpoints of each grid section.

15.6 NUMERICAL TECHNIQUES FOR MULTIPLE INTEGRALS

4. Approximate $\iint_D (x^2+y^2)\,dA$, where D is the interior of the quarter circle of Figure 15.31 by using only those grid sections completely inside D and choosing the lower left corner of each section.

5. Approximate $\iint_D (x^2+y^2)\,dA$, where D is the interior of the quarter circle of Figure 15.31 by using all those grid sections having any overlap with D and choosing the upper right corner of each section.

6. Average the results of the previous two exercises and compare with the actual value.

---

7. Approximate $\int_{0.2}^{0.6} \int_{0.4}^{1.0} (x^2 + y^2)\,dA$ using midpoints of appropriate grid sections.

8. Find the exact value of
$$\iint_D e^{x^2+y^2}\,dA,$$
where D is the same quarter circle. (Hint: change to polar coordinates!)

9. Generate 100 ordered pairs of numbers (x,y) such that $0 \leq x \leq 1$ and $0 \leq y \leq 1$. Keep only those ordered pairs satisfying $x^2 + y^2 \leq 1$. For each of these ordered pairs, evaluate $e^{x^2+y^2}$ and average the results. Finally, multiply this average by the percentage of the points that fell within 1 unit of the origin, and compare with the answer from the previous exercise.

---

10. A lake is roped off within a 100-meter by 200-meter rectangle. Fifty points are chosen at random within the rectangle, and 36 of these points fall within the lake. A depth reading is taken at each of the 36 points, and the average depth reading is 2.6 meters. What is the estimated volume of water in the lake in kiloliters?

16

Vector Analysis

In this chapter, we examine how differential and integral calculus can be used to analyze vector fields. We examine the different types of derivatives that can be formed on vector fields, then turn to some generalizations of integral calculus to curves and surfaces in space. In the next chapter, we'll see that each of these generalizations in turn gives rise to what could be called the *fundamental theorems of vector calculus*.

What is a vector field?

A vector-valued multivariable function is called a **vector field**. We can think of a vector-valued function as a vector of scalar-valued functions. Most often, we'll think of the inputs to a vector field as the coordinates specifying the location of an object, either in a plane (2 inputs) or in space (3 inputs).

If $\mathbf{F} : \mathbb{R}^2 \longrightarrow \mathbb{R}^2$ is a vector field having points in the plane as inputs and 2-dimensional vectors as outputs, we can write

$$\mathbf{F}(x,y) = \langle P(x,y),\ Q(x,y) \rangle$$
$$= P(x,y)\mathbf{i} + Q(x,y)\mathbf{j},$$

where the components P and Q are scalar-valued functions of two variables.

If $\mathbf{F} : \mathbb{R}^3 \longrightarrow \mathbb{R}^3$ is a vector field having points in space as inputs and 3-dimensional vectors as outputs, we can write

$$\mathbf{F}(x,y,z) = \langle P(x,y,z),\ Q(x,y,z),\ R(x,y,z) \rangle$$
$$= P(x,y,z)\mathbf{i} + Q(x,y,z)\mathbf{j} + R(x,y,z)\mathbf{k},$$

where the components P, Q, and R are scalar-valued functions of three variables.

Some authors restrict their attention to vector fields in space ($\mathbb{R}^3 \longrightarrow \mathbb{R}^3$). There really is no loss, for a vector field

$$\mathbf{F}(x,y) = \langle P(x,y),\ Q(x,y) \rangle$$

can also be thought of as representing

$$\mathbf{F}(x,y,z) = \langle P(x,y),\ Q(x,y),\ 0 \rangle,$$

where our attention is restricted to the xy-plane.

A vector field $\mathbf{F}$ is described as being *continuous, differentiable, C^1-differentiable,* or *C^2-differentiable,* provided that all of its scalar-valued components have that property.

Force fields and velocity fields

Much of vector calculus was invented as a tool for working with physics, and it is here that we see the most natural examples of vector fields. For motivation and interpretations, we'll appeal to two kinds of vector fields that arise often in physical situations: *force fields* and *velocity fields.*

Many forces may act on an object (gravitational, electrostatic, magnetic, and others). The direction and magnitude of these force vectors may well depend on the precise location of the object relative to another body or an electric charge; hence, each force can be thought of as a vector-valued function of the three variables that specify its position. If an object is immersed in a fluid that is flowing, then the *velocity* vector of the fluid at each location gives us the instantaneous direction and speed of the fluid flow at that point.

Consider a satellite in orbit around the earth. The magnitude of the gravitational force exerted on the satellite by the earth is proportional to the product of the masses of the earth and the satellite, and is inversely proportional to the square of the distance between the satellite and the center of the earth. Precisely, the magnitude of the gravitational force $\mathbf{F}$ is

$$||\mathbf{F}|| = \frac{GMm}{d^2},$$

where M is the (constant) mass of the earth, m is the (constant) mass of the satellite, d is the distance from the satellite to the center of the earth, and G is a gravitational constant.

If we designate the center of the earth as the origin of a rectangular coordinate system, then the position of the satellite can be indicated by some ordered triple of coordinates (x, y, z). The distance is given by

$$d = \sqrt{x^2 + y^2 + z^2}.$$

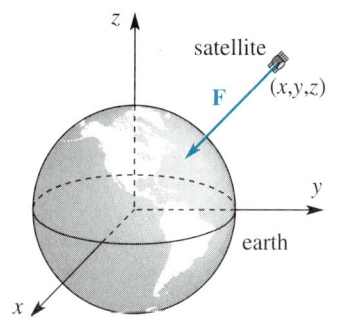

Figure 16.1 Gravitational forces form a vector field.

The direction of the gravitational force **F** is toward the center of the earth (see Figure 16.1).

Since the position vector of the satellite is $\langle x, y, z \rangle$, a *unit vector* pointing back toward the center of the earth is

$$\frac{-\langle x, y, z \rangle}{\sqrt{x^2 + y^2 + z^2}}.$$

Now we can write

$$\mathbf{F}(x, y, z) = \frac{-GMm}{(x^2 + y^2 + z^2)^{3/2}} \langle x, y, z \rangle,$$

so **F** is a vector-valued function of the satellite coordinates x, y, and z.

Visualizing vector fields

We can obtain a visual picture of a vector field by selecting several input points and drawing the output vector of appropriate length and direction emanating from each point. In the plane, this gives us a picture that is similar to a *slope* or *direction field*, except now our line segments have arrowheads and may have different lengths.

Suppose that a thin sheet of liquid swirling about a central point has a velocity field given by

$$\mathbf{F}(x, y) = \langle -y, x \rangle,$$

where we consider the central point as the origin $(0, 0)$. Here's a table of vectors at a selection of points:

point (x,y)	vector $\mathbf{F}(x,y)$	point (x,y)	vector $\mathbf{F}(x,y)$
$(0,0)$	$\langle 0,0 \rangle$	$(3,0)$	$\langle 0,3 \rangle$
$(1,0)$	$\langle 0,1 \rangle$	$(0,3)$	$\langle -3,0 \rangle$
$(0,1)$	$\langle -1,0 \rangle$	$(-3,0)$	$\langle 0,-3 \rangle$
$(-1,0)$	$\langle 0,-1 \rangle$	$(0,-3)$	$\langle 3,0 \rangle$
$(0,-1)$	$\langle 1,0 \rangle$	$(\sqrt{2},\sqrt{2})$	$\langle -\sqrt{2},\sqrt{2} \rangle$
$(2,0)$	$\langle 0,2 \rangle$	$(-\sqrt{2},\sqrt{2})$	$\langle -\sqrt{2},-\sqrt{2} \rangle$
$(0,2)$	$\langle -2,0 \rangle$	$(-\sqrt{2},-\sqrt{2})$	$\langle \sqrt{2},-\sqrt{2} \rangle$
$(-2,0)$	$\langle 0,-2 \rangle$	$(\sqrt{2},-\sqrt{2})$	$\langle \sqrt{2},\sqrt{2} \rangle$
$(0,-2)$	$\langle 2,0 \rangle$		

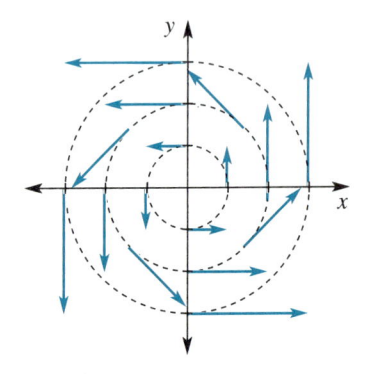

Figure 16.2 Illustration of the vector field $\mathbf{F}(x,y) = \langle -y, x \rangle$.

Figure 16.2 illustrates these velocity vectors along with some *flow lines* (shown in dashes). We can see that the flow lines are circles and that the direction of the flow is counterclockwise. Also, the further out the point from the center, the greater the velocity vector's magnitude.

16.1 DERIVATIVES OF VECTOR FIELDS—DIVERGENCE AND CURL

We have seen that a *scalar field* f has a *vector* derivative, as given by its gradient:

$$\mathbf{grad}\, f = \nabla f,$$

obtained by forming the vector of partial derivatives of f. For example, if we have a scalar function of three variables,

$$f(x, y, z),$$

then $\mathbf{grad}\, f$ is given by

$$\nabla f = \left\langle \frac{\partial f}{\partial x}, \frac{\partial f}{\partial y}, \frac{\partial f}{\partial z} \right\rangle.$$

16.1 DERIVATIVES OF VECTOR FIELDS—DIVERGENCE AND CURL

In this section we'll examine derivatives of vector fields. Because of their many physical interpretations, we'll find it useful to form different "types" of derivatives for different purposes.

Divergence of a vector field

One type of derivative of a vector field is known as the *divergence*, and it can be computed whenever all the components of the vector field are differentiable.

Definition 1

> The **divergence** of a vector field $\mathbf{F} : \mathbb{R}^2 \longrightarrow \mathbb{R}^2$, where
> $$\mathbf{F}(x,y) = \langle P(x,y), Q(x,y) \rangle,$$
> is defined as
> $$\operatorname{div} \mathbf{F} = \frac{\partial P}{\partial x} + \frac{\partial Q}{\partial y}.$$
> If $\mathbf{F} : \mathbb{R}^3 \longrightarrow \mathbb{R}^3$, where
> $$\mathbf{F}(x,y,z) = P(x,y,z)\mathbf{i} + Q(x,y,z)\mathbf{j} + R(x,y,z)\mathbf{k}$$
> then
> $$\operatorname{div} \mathbf{F} = \frac{\partial P}{\partial x} + \frac{\partial Q}{\partial y} + \frac{\partial R}{\partial z}.$$
> Note that the divergence of a vector field is always a *scalar field*.

In general, the divergence can be defined whenever $\mathbf{F} : \mathbb{R}^n \longrightarrow \mathbb{R}^n$, where

$$\mathbf{F} = \langle f_1, f_2, f_3, \ldots, f_n \rangle,$$

and each of the component functions f_i is a differentiable scalar field of the n variables $x_1, x_2, x_3, \ldots, x_n$. In this case, we have

$$\operatorname{div} \mathbf{F} = \frac{\partial f_1}{\partial x_1} + \frac{\partial f_2}{\partial x_2} + \frac{\partial f_3}{\partial x_3} + \cdots + \frac{\partial f_n}{\partial x_n}.$$

EXAMPLE 1 Find the divergence of each of the following vector fields:

$$\mathbf{F}(x,y,z) = \langle x, y, z \rangle$$
$$\mathbf{G}(x,y,z) = \langle y^2 + z^2, x^2 + z^2, x^2 + y^2 \rangle$$
$$\mathbf{H}(x,y,z) = \langle 3x^2, 4y^2, 5z^2 \rangle.$$

Solution
$$\text{div } \mathbf{F} = 1 + 1 + 1 = 3$$
$$\text{div } \mathbf{G} = 0 + 0 + 0 = 0$$
$$\text{div } \mathbf{H} = 6x + 8y + 10z.$$
∎

EXAMPLE 2   A thin sheet of fluid has a velocity field given by

$$\mathbf{F}(x, y) = \left\langle \frac{y}{x^2 + y^2}, \frac{-x}{x^2 + y^2} \right\rangle.$$

Find div **F**.

Solution
$$\text{div } \mathbf{F} = \frac{0(x^2 + y^2) - 2xy}{(x^2 + y^2)^2} + \frac{0(x^2 + y^2) - 2y(-x)}{(x^2 + y^2)^2} = \frac{-2xy + 2xy}{(x^2 + y^2)^2} = 0.$$
∎

Physical interpretation of divergence

What exactly does div **F** tell us about the vector field **F**? Let's think in terms of **F** representing the velocity field of a gas (a gas is a fluid that can compress or expand). At any point, imagine a small cubical volume V of the gas centered at the point. As time passes, the different particles of gas in this cube will move different distances, and the volume may change in both shape and size.

The instantaneous rate of change of the volume V with respect to time is dV/dt. A negative rate indicates compression, and a positive rate indicates expansion of the gas. If we divide this rate by the original volume V of the gas, then

$$\frac{dV/dt}{V}$$

gives us the *instantaneous rate of change of volume per unit of volume*.

As the cube of gas V centered at the point is taken smaller and smaller, the value of this quantity approaches the divergence of **F** at that point:

$$\text{div } \mathbf{F} = \lim_{V \to 0} \frac{dV/dt}{V}.$$

In other words, div **F** gives us a measure of the instantaneous rate of volume change *per unit of volume*. Roughly speaking, if the arrows of a vector field "converge," we'd expect the divergence of **F** to be negative; if the arrows "diverge," we'd expect the divergence to be positive (see Figure 16.3). However, even if the arrows neither converge nor diverge, a vector field can have nonzero divergence, as shown in the following example.

16.1 DERIVATIVES OF VECTOR FIELDS—DIVERGENCE AND CURL

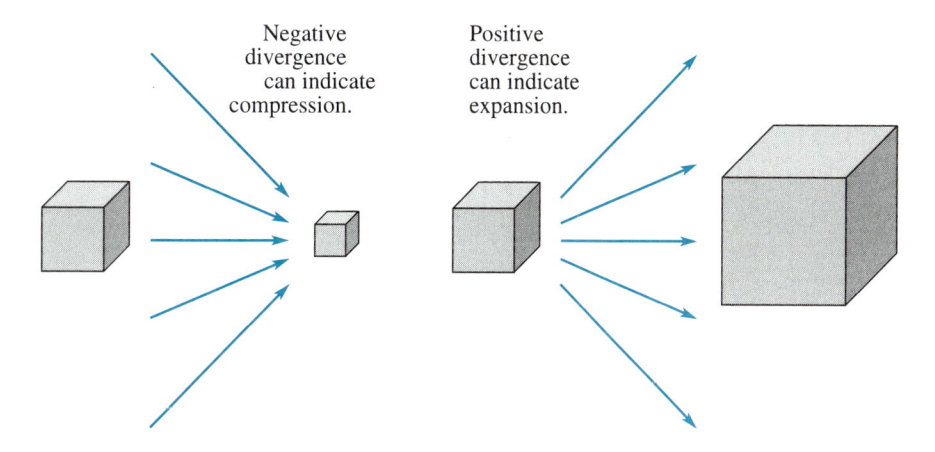

Figure 16.3 Negative and positive divergence.

EXAMPLE 3 Suppose that $\mathbf{F}(x,y) = \left(\dfrac{1}{1+y}\right)\mathbf{j}$ gives us the velocity field of a gas for all points (x, y, z) in space (except those in the plane $y = -1$). Illustrate this vector field and calculate its divergence.

Solution Since $\mathbf{F}$ does not depend on z, we can picture the flow of gas by looking at the vectors in the xy-plane. Figure 16.4 illustrates some of the vectors in this vector field.

Looking at this picture, we would not describe the arrows as converging or diverging. If, however, we calculate the divergence, we have

$$\text{div } \mathbf{F}(x, y, z) = 0 + \frac{-1}{(1+y)^2} + 0 = \frac{-1}{(1+y)^2},$$

which is negative at any point where $y \neq -1$. This indicates that a small cube of gas would compress in this field. (Can you see why from Figure 16.4?)

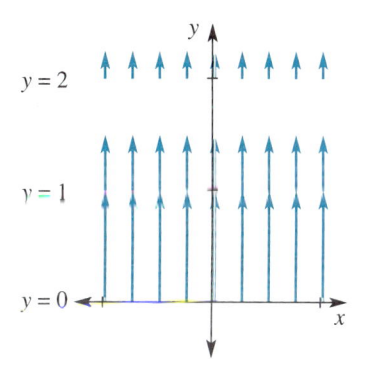

Figure 16.4 The vector field $\mathbf{F}(x,y) = \left(\dfrac{1}{1+y}\right)\mathbf{j}$.

Definition 2

> A vector field **F** is called **incompressible at a point** (x, y, z) if
> $$\text{div } \mathbf{F}(x, y, z) = 0.$$
> A vector field **F** is called **incompressible** if it is incompressible at every point.

Physically, if the vector field is incompressible, this simply means that a fluid moving in this vector field would neither compress nor expand.

EXAMPLE 4 The velocity field of a draining bathtub is approximated by
$$\mathbf{F}(x, y, z) = \frac{y}{x^2 + y^2}\mathbf{i} - \frac{x}{x^2 + y^2}\mathbf{j} - \mathbf{k}.$$
Is **F** an incompressible vector field?

Solution We have
$$\text{div } \mathbf{F} = \frac{\partial}{\partial x}\left[\frac{y}{x^2 + y^2}\right] + \frac{\partial}{\partial y}\left[\frac{-x}{x^2 + y^2}\right] + \frac{\partial}{\partial z}(-1)$$
$$= \frac{0 \cdot (x^2 + y^2) - 2xy}{(x^2 + y^2)^2} + \frac{0(x^2 + y^2) - 2y(-x)}{(x^2 + y^2)^2} + 0$$
$$= \frac{-2xy + 2xy}{(x^2 + y^2)^2} = 0.$$

Therefore, **F** is incompressible. (Fluid water, for all intents and purposes, does not compress or expand, so this makes sense.) ∎

Curl of a vector field

The divergence of a vector field of any dimension produces a new *scalar field*. In contrast, the *curl* of a vector field is defined only for $\mathbb{R}^3$, and it produces a new *vector field*, also of dimension three.

Definition 3

> Suppose that $\mathbf{F}: \mathbb{R}^3 \longrightarrow \mathbb{R}^3$ is a vector field, with
> $$\mathbf{F}(x, y, z) = P(x, y, z)\mathbf{i} + Q(x, y, z)\mathbf{j} + R(x, y, z)\mathbf{k}$$
> for some scalar components P, Q, and R. If **F** is differentiable, then we define the **curl** of **F** as
> $$\text{curl } \mathbf{F} = \left(\frac{\partial R}{\partial y} - \frac{\partial Q}{\partial z}\right)\mathbf{i} + \left(\frac{\partial P}{\partial z} - \frac{\partial R}{\partial x}\right)\mathbf{j} + \left(\frac{\partial Q}{\partial x} - \frac{\partial P}{\partial y}\right)\mathbf{k}.$$
> Note that curl **F** is also a vector field $\mathbb{R}^3 \longrightarrow \mathbb{R}^3$.

16.1 DERIVATIVES OF VECTOR FIELDS—DIVERGENCE AND CURL

The formula looks complicated, but fortunately there is a simple way to remember it. If we form a 3×3 "matrix" with the standard basis vectors **i**, **j**, and **k** in the first row, the partial derivative operators $\frac{\partial}{\partial x}$, $\frac{\partial}{\partial y}$, and $\frac{\partial}{\partial z}$ in the second row, and the three components of **F** (P, Q, and R) in the third row, then curl **F** is simply the "determinant":

$$\text{curl } \mathbf{F} = \begin{vmatrix} \mathbf{i} & \mathbf{j} & \mathbf{k} \\ \partial/\partial x & \partial/\partial y & \partial/\partial z \\ P & Q & R \end{vmatrix}.$$

Recall that we used a similar device for remembering the *cross product* of two 3-dimensional vectors.

EXAMPLE 5 Find curl $\mathbf{F}(0, 1, 0)$ if $\mathbf{F}(x, y, z) = \langle x, 3xy, z \rangle$.

Solution

$$\text{curl } \mathbf{F} = \begin{vmatrix} \mathbf{i} & \mathbf{j} & \mathbf{k} \\ \partial/\partial x & \partial/\partial y & \partial/\partial z \\ x & 3xy & z \end{vmatrix}$$

$$= 0\mathbf{i} - 0\mathbf{j} + (3y - 0)\mathbf{k} = \langle 0, 0, 3y \rangle.$$

Therefore, curl $\mathbf{F}(0, 1, 0) = \langle 0, 0, 3 \rangle$. ∎

EXAMPLE 6 Find curl $\mathbf{F}(1, 1, 1)$ if $\mathbf{F}(x, y, z) = \langle y, z, x \rangle$.

Solution We have

$$\text{curl } \mathbf{F} = \begin{vmatrix} \mathbf{i} & \mathbf{j} & \mathbf{k} \\ \partial/\partial x & \partial/\partial y & \partial/\partial z \\ y & z & x \end{vmatrix}$$

$$= -1\mathbf{i} - 1\mathbf{j} - 1\mathbf{k} = \langle -1, -1, -1 \rangle.$$

Therefore, curl $\mathbf{F}(1, 1, 1) = \langle -1, -1, -1 \rangle$. ∎

EXAMPLE 7 Find curl $\mathbf{F}(2, 0, 1)$ if

$$\mathbf{F}(x, y, z) = (x + y)^3 \mathbf{i} + (\sin xy)\mathbf{j} + (\cos xyz)\mathbf{k}.$$

Solution   We have

$$\text{curl } \mathbf{F} = \begin{vmatrix} \mathbf{i} & \mathbf{j} & \mathbf{k} \\ \partial/\partial x & \partial/\partial y & \partial/\partial z \\ (x+y)^3 & \sin xy & \cos xyz \end{vmatrix}$$

$$= (-xz \sin xyz)\mathbf{i} - (-yz \sin xyz)\mathbf{j} + (y \cos xy - 3(x+y)^2)\mathbf{k}.$$

Therefore, curl $\mathbf{F}(2,0,1) = \langle 0, 0, -12 \rangle$. ∎

Physical interpretation of curl

What exactly does curl $\mathbf{F}$ tell us about $\mathbf{F}$? Let's imagine that $\mathbf{F}$ describes the velocity field of a fluid. Now suppose that we drop a small paddle wheel like the one shown in Figure 16.5 into the fluid.

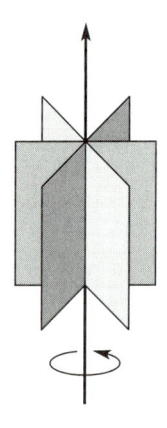

Figure 16.5 A paddle wheel.

Three questions we could ask in this situation are:

1. In what direction should the axis of the paddle wheel point if we are to achieve the fastest rate of rotation of the paddle wheel?

2. If the axis of the paddle wheel is aligned for maximum rotational velocity, just how fast does the paddle wheel rotate?

3. In what direction does the paddle wheel rotate?

The vector curl $\mathbf{F}$ provides us with answers to all of these questions!

1. At any point in the fluid, the paddle wheel will rotate fastest if its axis is aligned with the direction of curl $\mathbf{F}$.

2. If the axis of the paddle wheel is aligned in this way, then the magnitude of curl $\mathbf{F}$ is precisely *twice* the maximum angular velocity of the paddle wheel. The Greek letter omega (ω) is often used to indicate angular velocity. We can write

$$\|\text{curl } \mathbf{F}\| = 2\omega.$$

16.1 DERIVATIVES OF VECTOR FIELDS—DIVERGENCE AND CURL

3. If the axis of the paddle wheel is imagined to have *right-hand screw threads*, then the paddle wheel rotates in the direction that results in the "screw" moving in the direction of curl **F**. Figure 16.5 illustrates that the direction of rotation is counterclockwise (looking down on the paddle wheel) if the curl vector points straight up.

Definition 4

> A vector field **F** is called **irrotational** at a point (x, y, z) if
> $$\text{curl } \mathbf{F}(x, y, z) = \mathbf{0} = \langle 0, 0, 0 \rangle.$$
> We call the vector field **F** **irrotational** if it is irrotational at every point.

As the term suggests, a paddlewheel would not rotate in a fluid moving in an irrotational vector field.

Del Notation

We've already used the "del" notation for the gradient of a scalar field f:

$$\text{grad } f = \nabla f.$$

The ∇ notation is also convenient for expressing the divergence and curl of a vector field **F**. If we think of ∇ as the vector of partial derivative operators, then for functions of the three rectangular coordinate variables x, y, and z, we have

$$\nabla = \left\langle \frac{\partial}{\partial x}, \frac{\partial}{\partial y}, \frac{\partial}{\partial z} \right\rangle.$$

Given a vector field

$$\mathbf{F}(x, y, z) = P(x, y, z)\mathbf{i} + Q(x, y, z)\mathbf{j} + R(x, y, z)\mathbf{k},$$

we can write its divergence as the *dot product* of ∇ and **F**:

$$\text{div } \mathbf{F} = \nabla \cdot \mathbf{F} = \frac{\partial P}{\partial x} + \frac{\partial Q}{\partial y} + \frac{\partial R}{\partial z}.$$

Similarly, we can write the curl **F** as the *cross product* of ∇ and **F**:

$$\text{curl } \mathbf{F} = \nabla \times \mathbf{F} = \begin{vmatrix} \mathbf{i} & \mathbf{j} & \mathbf{k} \\ \partial/\partial x & \partial/\partial y & \partial/\partial z \\ P & Q & R \end{vmatrix}$$

$$= \left(\frac{\partial R}{\partial y} - \frac{\partial Q}{\partial z} \right)\mathbf{i} + \left(\frac{\partial P}{\partial z} - \frac{\partial R}{\partial x} \right)\mathbf{j} + \left(\frac{\partial Q}{\partial x} - \frac{\partial P}{\partial y} \right)\mathbf{k}.$$

EXAMPLE 8 Find the divergence and curl of each of the following vector fields:

$$\mathbf{F}(x, y, z) = \langle 2x, 3y, 4z \rangle$$

$$\mathbf{G}(x, y, z) = \langle x^2, y^2, z^2 \rangle$$

$$\mathbf{H}(x, y, z) = \langle x + y, y + z, z + x \rangle.$$

Identify the points at which each of the fields are irrotational or incompressible.

Solution $\nabla \cdot \mathbf{F} = 2 + 3 + 4 = 9$ at every point, so $\mathbf{F}$ is incompressible nowhere.

$$\nabla \times \mathbf{F} = \begin{vmatrix} \mathbf{i} & \mathbf{j} & \mathbf{k} \\ \partial/\partial x & \partial/\partial y & \partial/\partial z \\ 2x & 3y & 4z \end{vmatrix} = 0\mathbf{i} - 0\mathbf{j} + 0\mathbf{k} = \langle 0, 0, 0 \rangle,$$

so $\mathbf{F}$ is irrotational everywhere.

$\nabla \cdot \mathbf{G} = 2x + 2y + 2z$, so $\mathbf{G}$ is incompressible at every point in the plane $x + y + z = 0$.

$$\nabla \times \mathbf{G} = \begin{vmatrix} \mathbf{i} & \mathbf{j} & \mathbf{k} \\ \partial/\partial x & \partial/\partial y & \partial/\partial z \\ x^2 & y^2 & z^2 \end{vmatrix} = 0\mathbf{i} - 0\mathbf{j} + 0\mathbf{k} = \langle 0, 0, 0 \rangle,$$

so $\mathbf{G}$ is irrotational everywhere.

$\nabla \cdot \mathbf{H} = 1 + 1 + 1 = 3$ at every point, so $\mathbf{H}$ is incompressible nowhere.

$$\nabla \times \mathbf{H} = \begin{vmatrix} \mathbf{i} & \mathbf{j} & \mathbf{k} \\ \partial/\partial x & \partial/\partial y & \partial/\partial z \\ x+y & y+z & z+x \end{vmatrix} = (0-1)\mathbf{i} - (1-0)\mathbf{j} + (0-1)\mathbf{k} = \langle -1, -1, -1 \rangle,$$

so $\mathbf{H}$ is irrotational nowhere. ∎

The Laplacian

The del operator can be used as a building block to define other operators. For example, starting with a scalar field f, we can find its *gradient* ∇f. The *divergence* of this new vector field can be written

$$\text{div}(\text{grad } f) = \nabla \cdot (\nabla f) = \frac{\partial^2 f}{\partial x^2} + \frac{\partial^2 f}{\partial y^2} + \frac{\partial^2 f}{\partial z^2}.$$

This is called the **Laplacian.** A shorthand notation for the Laplacian of f is

$$\nabla^2 f.$$

16.1 DERIVATIVES OF VECTOR FIELDS—DIVERGENCE AND CURL

EXAMPLE 9 Compute the Laplacian of the scalar field $f(x, y, z) = x^2 + y^2 + z^2$.

Solution The gradient of f is $\text{grad } f = \nabla f = \langle 2x, 2y, 2z \rangle$. The Laplacian is

$$\text{div } (\text{grad } f) = \nabla \cdot (\nabla f) = \nabla^2 f = 2 + 2 + 2 = 6. \quad \blacksquare$$

The Laplacian $\nabla^2 f(x, y, z)$ can be thought of roughly as providing a measure of the difference between the *average value* of f at points near (x, y, z) and the *actual value* $f(x, y, z)$. Hence, if f is a temperature field and $\nabla^2 f(x, y, z)$ is negative, then the temperature at (x, y, z) is higher than the average temperature of the immediate surroundings.

The operator ∇^2 applied to a *vector field* $\mathbf{F}$ "distributes" over the components. For example, if

$$\mathbf{F}(x, y) = \langle P(x, y), \ Q(x, y) \rangle,$$

then

$$\nabla^2 \mathbf{F} = \left(\frac{\partial^2 P}{\partial x^2} + \frac{\partial^2 P}{\partial y^2} \right) \mathbf{i} \ + \ \left(\frac{\partial^2 Q}{\partial x^2} + \frac{\partial^2 Q}{\partial y^2} \right) \mathbf{j}.$$

Many properties relating div, grad, curl, and the Laplacian can be written down very conveniently using the del notation.

EXAMPLE 10 Suppose that $\mathbf{F}(x, y, z) = \langle P(x, y, z), Q(x, y, z), R(x, y, z) \rangle$ and that P, Q, and R are all C^2-differentiable. (In other words, the second-order partial derivatives are all continuous.) Show that

$$\text{div } (\text{curl } \mathbf{F}) = \nabla \cdot (\nabla \times \mathbf{F}) = 0.$$

Solution First, we calculate curl $\mathbf{F}$:

$$\nabla \times \mathbf{F} = \left\langle \frac{\partial R}{\partial y} - \frac{\partial Q}{\partial z}, \ \frac{\partial P}{\partial z} - \frac{\partial R}{\partial x}, \ \frac{\partial Q}{\partial x} - \frac{\partial P}{\partial y} \right\rangle.$$

Now,

$$\text{div } (\text{curl } \mathbf{F}) = \frac{\partial}{\partial x}\left(\frac{\partial R}{\partial y} - \frac{\partial Q}{\partial z} \right) + \frac{\partial}{\partial y}\left(\frac{\partial P}{\partial z} - \frac{\partial R}{\partial x} \right) + \frac{\partial}{\partial z}\left(\frac{\partial Q}{\partial x} - \frac{\partial P}{\partial y} \right)$$

$$= \left(\frac{\partial^2 R}{\partial x \partial y} - \frac{\partial^2 Q}{\partial x \partial z} \right) + \left(\frac{\partial^2 P}{\partial y \partial z} - \frac{\partial^2 R}{\partial y \partial x} \right) + \left(\frac{\partial^2 Q}{\partial z \partial x} - \frac{\partial^2 P}{\partial z \partial y} \right).$$

Since P, Q, and R are C^2-differentiable, the mixed partial derivatives of each must match. If we write div (curl $\mathbf{F}$), rearranging the terms, we have

$$\text{div } (\text{curl } \mathbf{F}) = \left(\frac{\partial^2 P}{\partial y \partial z} - \frac{\partial^2 P}{\partial z \partial y} \right) + \left(\frac{\partial^2 Q}{\partial z \partial x} - \frac{\partial^2 Q}{\partial x \partial z} \right) + \left(\frac{\partial^2 R}{\partial x \partial y} - \frac{\partial^2 R}{\partial y \partial x} \right) = 0. \quad \blacksquare$$

EXAMPLE 11 Suppose that f is a C^2-differentiable scalar field. Show that $\nabla \times (\nabla f) = 0$.

Solution

$$\nabla \times (\nabla f) = \begin{vmatrix} \mathbf{i} & \mathbf{j} & \mathbf{k} \\ \partial/\partial x & \partial/\partial y & \partial/\partial z \\ \frac{\partial f}{\partial x} & \frac{\partial f}{\partial y} & \frac{\partial f}{\partial z} \end{vmatrix}$$

$$= \left(\frac{\partial^2 f}{\partial y \partial z} - \frac{\partial^2 f}{\partial z \partial y}\right)\mathbf{i} - \left(\frac{\partial^2 f}{\partial x \partial z} - \frac{\partial^2 f}{\partial z \partial x}\right)\mathbf{j} + \left(\frac{\partial^2 f}{\partial x \partial y} - \frac{\partial^2 f}{\partial y \partial x}\right)\mathbf{k}$$

$$= \langle 0, 0, 0 \rangle,$$

since the mixed partial derivatives of a C^2-differentiable function match. Hence, the curl of a gradient vector field f is always the zero vector, provided that f is C^2-differentiable. ∎

EXERCISES for Section 16.1

For exercises 1-10: Sketch several vectors for each of the following vector fields.

1. $F(x,y) = 2\mathbf{i} + \mathbf{j}$
2. $F(x,y) = 2\mathbf{i} - \mathbf{j}$
3. $F(x,y) = -2\mathbf{i} - \mathbf{j}$
4. $F(x,y) = 2\mathbf{i} + x\mathbf{j}$
5. $F(x,y) = 2\mathbf{i} + y\mathbf{j}$
6. $F(x,y) = \dfrac{-x\mathbf{i}}{\sqrt{x^2+y^2}} - \dfrac{y\mathbf{j}}{\sqrt{x^2+y^2}}$
7. $F(x,y) = x\mathbf{i} + (x-y)\mathbf{j}$
8. $F(x,y) = \mathbf{i} + (x+y)\mathbf{j}$
9. $F(x,y) = \dfrac{1}{x}\mathbf{i} + \dfrac{1}{y}\mathbf{j}$
10. $F(x,y) = \dfrac{1}{y}\mathbf{i} + \dfrac{1}{x}\mathbf{j}$

For exercises 11-30: Refer to the following scalar and vector fields:

$f(x,y,z) = x^2 y^3 z^4$ $\mathbf{F}(x,y,z) = \langle y^2 + z^2, x^2 + z^2, x^2 + y^2 \rangle$

$g(x,y,z) = \ln(xyz)$ $\mathbf{G}(x,y,z) = \langle xyz, xy^2 z^3, x^3 y^2 z \rangle$

$h(x,y,z) = e^{xyz}$ $\mathbf{H}(x,y,z) = \langle xyz, x^2 y^2 z^2, x^2 + y^2 + z^2 \rangle$

$p(x,y,z) = x + y - z$ $\mathbf{P}(x,y,z) = \langle x + y + z, x + y^2 + z^3, x^3 - y^2 - z \rangle$

16.1 DERIVATIVES OF VECTOR FIELDS—DIVERGENCE AND CURL

Compute the indicated scalar or vector field.

11. grad $f = \nabla f$
12. grad $g = \nabla g$
13. grad $h = \nabla h$
14. grad $p = \nabla p$
15. div $\mathbf{F} = \nabla \cdot \mathbf{F}$
16. div $\mathbf{G} = \nabla \cdot \mathbf{G}$
17. div $\mathbf{H} = \nabla \cdot \mathbf{H}$
18. div $\mathbf{P} = \nabla \cdot \mathbf{P}$
19. curl $\mathbf{F} = \nabla \times \mathbf{F}$
20. curl $\mathbf{G} = \nabla \times \mathbf{G}$
21. curl $\mathbf{H} = \nabla \times \mathbf{H}$
22. curl $\mathbf{P} = \nabla \times \mathbf{P}$
23. $\nabla^2 f$
24. $\nabla^2 g$
25. $\nabla^2 h$
26. $\nabla^2 p$
27. $\nabla^2 \mathbf{F}$
28. $\nabla^2 \mathbf{G}$
29. $\nabla^2 \mathbf{H}$
30. $\nabla^2 \mathbf{P}$

For exercises 31-35: Suppose that $f(x, y, z)$ and $g(x, y, z)$ are any two C^2-differentiable scalar fields and that $\mathbf{F}(x, y, z)$ and $\mathbf{G}(x, y, z)$ are any two C^2-differentiable vector fields. Verify each of the statements.

31. $\nabla \cdot (\nabla f \times \nabla g) = 0$.
32. $\nabla \cdot (\mathbf{F} \times \mathbf{G}) = \mathbf{G} \cdot (\nabla \times \mathbf{F}) - \mathbf{F} \cdot (\nabla \times \mathbf{G})$.
33. curl $(f\mathbf{F}) = f$ curl $\mathbf{F} + \nabla f \times \mathbf{F}$.
34. $\nabla \times (\nabla \times \mathbf{G}) = \nabla(\nabla \cdot \mathbf{G}) - (\nabla^2 \mathbf{G})$.
35. div $(f\mathbf{F}) = f$ div $\mathbf{F} + \mathbf{F} \cdot \nabla f$.

36. A scalar field f that satisfies *Laplace's equation*,
$$\nabla^2 f = 0,$$
is called **harmonic**. Show that
$$f(x, y, z) = \sin x \sinh y + \cos x \cosh z$$
is harmonic.

37. Which of the scalar fields f, g, h, and p of exercises 11-30 are harmonic?

38. Which of the vector fields $\mathbf{F}$, $\mathbf{G}$, $\mathbf{H}$, and $\mathbf{P}$ of exercises 11-30 are incompressible everywhere?

39. Which of the vector fields $\mathbf{F}$, $\mathbf{G}$, $\mathbf{H}$, and $\mathbf{P}$ of exercises 11-30 are irrotational everywhere?

40. A student says that the curl of any vector field is incompressible and that the gradient of any scalar field is irrotational. Are these statements accurate?

41. Some people describe the curl as measuring the "swirl" of a vector field. However, that's a bit misleading, as shown in this exercise. Suppose that a fluid has a velocity field given by
$$\mathbf{F}(x, y, z) = \frac{y}{4}\mathbf{i}.$$

Sketch some of the vectors in the xy-plane and note that they do not "swirl." Then find curl **F** and note that it is not zero. Does this make sense in terms of the paddle wheel interpretation? Explain why.

16.2 TOTAL DERIVATIVE OF A VECTOR FIELD—JACOBIANS

So far, we've mentioned the divergence and curl as special types of derivatives for vector fields. In this section, we'll turn to the notion of a *total derivative* of a vector field.

The Jacobian

The **total derivative** of a C^1-differentiable vector field **F** is represented by the matrix of partial derivatives of its components. This matrix is known as the **Jacobian matrix**.

Definition 5

> The **Jacobian matrix** of a vector field $\mathbf{F} : \mathbb{R}^2 \longrightarrow \mathbb{R}^2$, where
> $$\mathbf{F}(x,y) = \langle P(x,y),\ Q(x,y) \rangle,$$
> is defined as
> $$D\mathbf{F} = \begin{bmatrix} \partial P/\partial x & \partial P/\partial y \\ \partial Q/\partial x & \partial Q/\partial y \end{bmatrix}.$$
> If $\mathbf{F} : \mathbb{R}^3 \longrightarrow \mathbb{R}^3$, where
> $$\mathbf{F}(x,y,z) = P(x,y,z)\mathbf{i} + Q(x,y,z)\mathbf{j} + R(x,y,z)\mathbf{k},$$
> then
> $$D\mathbf{F} = \begin{bmatrix} \partial P/\partial x & \partial P/\partial y & \partial P/\partial z \\ \partial Q/\partial x & \partial Q/\partial y & \partial Q/\partial z \\ \partial R/\partial x & \partial R/\partial y & \partial R/\partial z \end{bmatrix}.$$

Let's make some observations regarding the Jacobian matrix.

Observation 1. Each row of the Jacobian matrix $D\mathbf{F}$ can be viewed as the *gradient* of a scalar component of **F**. For example, if
$$\mathbf{F}(x,y,z) = \langle P(x,y,z), Q(x,y,z), R(x,y,z) \rangle,$$
then
$$D\mathbf{F} = \begin{bmatrix} \nabla P \\ \nabla Q \\ \nabla R \end{bmatrix} = \begin{bmatrix} \partial P/\partial x & \partial P/\partial y & \partial P/\partial z \\ \partial Q/\partial x & \partial Q/\partial y & \partial Q/\partial z \\ \partial R/\partial x & \partial R/\partial y & \partial R/\partial z \end{bmatrix}.$$

16.2 TOTAL DERIVATIVE OF A VECTOR FIELD—JACOBIANS

Observation 2. The sum of the diagonal entries of DF (called the **trace** of a matrix) is just the divergence of $\mathbf{F}$:

$$\text{div } \mathbf{F} = \frac{\partial P}{\partial x} + \frac{\partial Q}{\partial y} + \frac{\partial R}{\partial z}.$$

Observation 3. If $\mathbf{F} : \mathbb{R}^3 \longrightarrow \mathbb{R}^3$, the determinant of the Jacobian matrix DF is the triple scalar product of the gradients of its components:

$$|D\mathbf{F}| = \begin{vmatrix} \partial P/\partial x & \partial P/\partial y & \partial P/\partial z \\ \partial Q/\partial x & \partial Q/\partial y & \partial Q/\partial z \\ \partial R/\partial x & \partial R/\partial y & \partial R/\partial z \end{vmatrix} = \nabla P \cdot (\nabla Q \times \nabla R).$$

The Jacobian matrix gives us a way of measuring the instantaneous rate of change in the vector output of a vector field with respect to change in the inputs. For example, suppose that

$$\mathbf{F}(x, y, z) = \langle P(x, y, z), Q(x, y, z), R(x, y, z) \rangle,$$

and we want to measure the change in vector output when we move from the input point (x_0, y_0, z_0) to the point $(x_0+\Delta x, y_0+\Delta y, z_0+\Delta z)$. The resulting changes in each component of $\mathbf{F}$ are

$$\Delta P = P(x_0 + \Delta x, y_0 + \Delta y, z_0 + \Delta z) - P(x_0, y_0, z_0)$$
$$\approx \nabla P(x_0, y_0, z_0) \cdot \langle \Delta x, \Delta y, \Delta z \rangle,$$

$$\Delta Q = Q(x_0 + \Delta x, y_0 + \Delta y, z_0 + \Delta z) - Q(x_0, y_0, z_0)$$
$$\approx \nabla Q(x_0, y_0, z_0) \cdot \langle \Delta x, \Delta y, \Delta z \rangle,$$

$$\Delta R = R(x_0 + \Delta x, y_0 + \Delta y, z_0 + \Delta z) - R(x_0, y_0, z_0)$$
$$\approx \nabla R(x_0, y_0, z_0) \cdot \langle \Delta x, \Delta y, \Delta z \rangle.$$

These three component changes can be conveniently expressed in terms of the Jacobian matrix. We can write

$$\Delta \mathbf{F} = \mathbf{F}(x_0 + \Delta x, y_0 + \Delta y, z_0 + \Delta z) - \mathbf{F}(x_0, y_0, z_0)$$

$$\approx \begin{bmatrix} \nabla P(x_0, y_0, z_0) \cdot \langle \Delta x, \Delta y, \Delta z \rangle \\ \nabla Q(x_0, y_0, z_0) \cdot \langle \Delta x, \Delta y, \Delta z \rangle \\ \nabla R(x_0, y_0, z_0) \cdot \langle \Delta x, \Delta y, \Delta z \rangle \end{bmatrix}$$

$$= D\mathbf{F}(x_0, y_0, z_0) \begin{bmatrix} \Delta x \\ \Delta y \\ \Delta z \end{bmatrix}.$$

Similarly, if

$$\mathbf{F}(x, y) = \langle P(x, y), Q(x, y) \rangle,$$

then
$$\Delta \mathbf{F} = \mathbf{F}(x_0 + \Delta x, y_0 + \Delta y) - \mathbf{F}(x_0, y_0)$$
$$\approx \begin{bmatrix} \nabla P(x_0, y_0) \cdot \langle \Delta x, \Delta y \rangle \\ \nabla Q(x_0, y_0) \cdot \langle \Delta x, \Delta y \rangle \end{bmatrix}$$
$$= D\mathbf{F}(x_0, y_0) \begin{bmatrix} \Delta x \\ \Delta y \end{bmatrix}.$$

Jacobian determinant as scale factor

There is a very nice geometric interpretation of the determinant of the Jacobian matrix. Let's consider the 2-dimensional case first. Think of the points (x, y) and $(x + \Delta x, y + \Delta y)$ as the opposite corners of a rectangle of dimensions Δx by Δy. If $\mathbf{F} : \mathbb{R}^2 \to \mathbb{R}^2$ is a C^1-differentiable vector field, then for small Δx and Δy, the image of this rectangle under $\mathbf{F}$ will be approximately a parallelogram (see Figure 16.6).

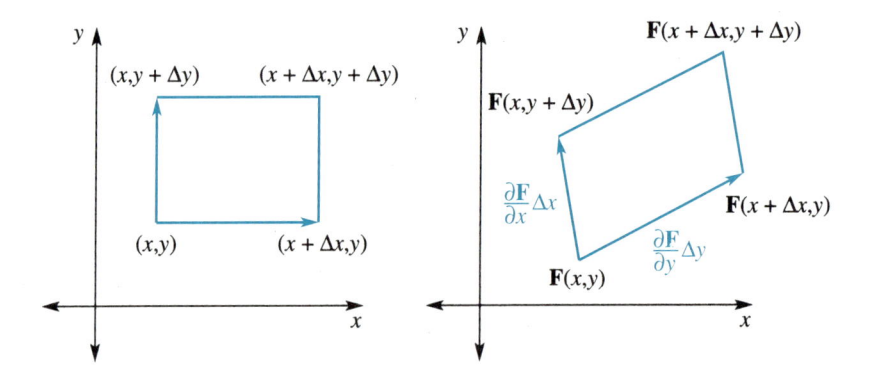

Figure 16.6 A rectangle and its image under $\mathbf{F}$.

The sides of this parallelogram are the vectors
$$\frac{\partial \mathbf{F}}{\partial x} \Delta x = \left\langle \frac{\partial P}{\partial x} \Delta x, \frac{\partial Q}{\partial x} \Delta x \right\rangle \quad \text{and} \quad \frac{\partial \mathbf{F}}{\partial y} \Delta y = \left\langle \frac{\partial P}{\partial y} \Delta y, \frac{\partial Q}{\partial y} \Delta y \right\rangle,$$

and the area of the parallelogram is given by
$$\left| \frac{\partial P}{\partial x} \Delta x \frac{\partial Q}{\partial y} \Delta y - \frac{\partial P}{\partial y} \Delta y \frac{\partial Q}{\partial x} \Delta x \right| = \left| \frac{\partial P}{\partial x} \frac{\partial Q}{\partial y} - \frac{\partial P}{\partial y} \frac{\partial Q}{\partial x} \right| \Delta x \Delta y.$$

The quantity in the absolute value signs on the right-hand side of this equation is the determinant of the Jacobian matrix:
$$\begin{vmatrix} \partial P / \partial x & \partial P / \partial y \\ \partial Q / \partial x & \partial Q / \partial y \end{vmatrix}.$$

If we denote this determinant by $J(x, y)$, we have the
$$\text{area of the parallelogram} = |J(x, y)| \Delta x \Delta y.$$

16.2 TOTAL DERIVATIVE OF A VECTOR FIELD—JACOBIANS

In other words, the absolute value of the determinant of the Jacobian matrix is the scale factor relating the area of the original rectangle to its image parallelogram under **F**.

In three dimensions, if we think of (x, y, z) and $(x + \Delta x, y + \Delta y, z + \Delta z)$ as the opposite corners of a small rectangular box, then Δx, Δy, and Δz are the dimensions, and the

$$\text{volume of the box } = \Delta x \Delta y \Delta z.$$

For very small dimensions, the image of this box under a differentiable vector field

$$\mathbf{F}(x, y, z) = \langle P(x, y, z), Q(x, y, z), R(x, y, z) \rangle,$$

is approximately a parallelepiped, and the

$$\text{volume of the parallelepiped } = |J(x, y, z)| \Delta x \Delta y \Delta z,$$

where

$$J(x, y, z) = \begin{vmatrix} \partial P/\partial x & \partial P/\partial y & \partial P/\partial z \\ \partial Q/\partial x & \partial Q/\partial y & \partial Q/\partial z \\ \partial R/\partial x & \partial R/\partial y & \partial R/\partial z \end{vmatrix}$$

is again the determinant of the Jacobian matrix. In this case, its absolute value provides the scale factor relating the volume of the original box to its image under **F**.

Application to change of variables in double integration

Given a double integral

$$\iint_D f(x, y) \, dx \, dy$$

over a domain D in the xy-plane, we can rewrite the integral in the form

$$\iint_D f(r \cos \theta, r \sin \theta) r \, dr \, d\theta,$$

where the region D is now described in polar coordinates, and the factor r arises in expressing the area element dA in polar coordinates.

This change of variables from (x, y) to (r, θ) can actually be thought of as describing a function from $\mathbb{R}^2$ to $\mathbb{R}^2$, where

$$x = P(r, \theta) = r \cos \theta \quad \text{and} \quad y = Q(r, \theta) = r \sin \theta.$$

In fact, if we label the coordinate axes with r and θ, we can think of the region D as the image of some region D^* in the $r\theta$-plane.

For example, the region

$$D = \{(x,y) : 0 \le x,\ 0 \le y,\ x^2 + y^2 \le 4\}$$

can be described as the image of a rectangle

$$D^* = \{(r,\theta) : 0 \le r \le 2,\ 0 \le \theta \le \pi/2\}$$

under the mapping that takes (r,θ) to (x,y):

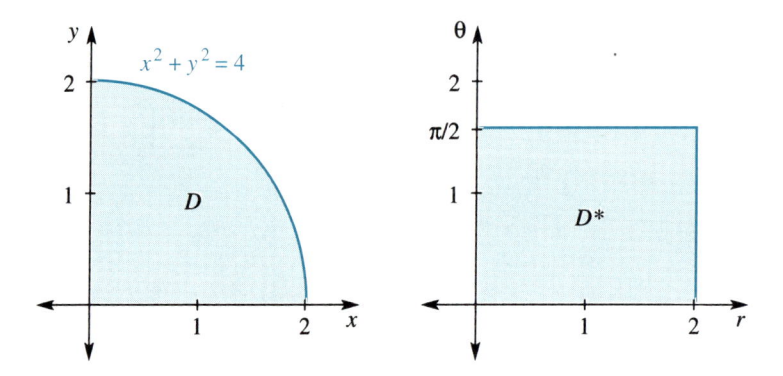

$$x = r\cos\theta \quad \text{and} \quad y = r\sin\theta).$$

(See Figure 16.7.)

Figure 16.7 The region D is the image of D^* when $x = r\cos\theta$ and $y = r\sin\theta$.

What is the Jacobian scale factor of this particular mapping? If we compute the determinant of the Jacobian matrix, we find

$$J(r,\theta) = \begin{vmatrix} \partial x/\partial r & \partial x/\partial \theta \\ \partial y/\partial r & \partial y/\partial \theta \end{vmatrix} = \begin{vmatrix} \cos\theta & -r\sin\theta \\ \sin\theta & r\cos\theta \end{vmatrix}$$

$$= r\cos^2\theta + r\sin^2\theta = r(\cos^2\theta + \sin^2\theta) = r.$$

Since $r \ge 0$, we have $|J(r,\theta)| = r$, which is precisely the scale factor we should expect in changing to polar coordinates in double integration.

A Jacobian scale factor arises when we make a change of variables in integration. In general, if we wish to compute

$$\iint_D f(x,y)\,dx\,dy$$

by changing variables from (x,y) to (u,v) via the mapping

$$x = P(u,v) \quad \text{and} \quad y = Q(u,v),$$

we must take the following steps:

16.2 TOTAL DERIVATIVE OF A VECTOR FIELD—JACOBIANS

1. Compute the Jacobian determinant

$$J(u,v) = \begin{vmatrix} \partial P/\partial u & \partial P/\partial v \\ \partial Q/\partial u & \partial Q/\partial v \end{vmatrix} = \frac{\partial P}{\partial u}\frac{\partial Q}{\partial v} - \frac{\partial P}{\partial v}\frac{\partial Q}{\partial u}.$$

2. Find the region D^* in the uv-plane whose image is D under the mapping. You may find it useful to solve

$$x = P(u,v) \quad \text{and} \quad y = Q(u,v)$$

for u and v in terms of x and y. Rewriting the description of the original region D in terms of u and v provides a description of D^* in the uv-plane.

3. Calculate

$$\iint_{D^*} f(P(u,v), Q(u,v)) |J(u,v)|\, du\, dv.$$

We'll illustrate the method with an example.

EXAMPLE 12 Calculate

$$\int_0^2 \int_0^x (x+y)\, dy\, dx$$

both directly and by making the change of variables:

$$x = u + v \quad \text{and} \quad y = u - v.$$

Solution We first calculate the double integral in the original variables:

$$\int_0^2 \int_0^x (x+y)\, dy\, dx = \int_0^2 \left(xy + \frac{y^2}{2}\right)\Big]_0^x dx$$

$$= \int_0^2 \left(x^2 + \frac{x^2}{2}\right) dx$$

$$= \int_0^2 \frac{3x^2}{2}\, dx$$

$$= \frac{x^3}{2}\Big]_0^2 = \frac{8}{2} - 0 = 4.$$

Now, we'll calculate the integral again, using the change of variables indicated.

1. The Jacobian determinant is

$$J(u,v) = \begin{vmatrix} \dfrac{\partial x}{\partial u} & \dfrac{\partial x}{\partial v} \\ \dfrac{\partial y}{\partial u} & \dfrac{\partial y}{\partial v} \end{vmatrix} = \begin{vmatrix} 1 & 1 \\ 1 & -1 \end{vmatrix} = -1 - 1 = -2.$$

Our conversion scale factor will be $|J(u,v)| = 2$.

2. The original region of integration D is defined by $0 \leq y \leq x$ and $0 \leq x \leq 2$. If we solve $x = u + v$, $y = u - v$ for u and v, we find:

$$x + y = 2u \qquad x - y = 2v,$$

so

$$u = \frac{x+y}{2} \qquad v = \frac{x-y}{2}.$$

The borderlines of the original region D lie on the lines $y = 0$, $x = 2$, and $y = x$. Therefore, the borderlines of the region D^* lie on the lines $u - v = 0$, $u + v = 2$, and $v = 0$. Figure 16.8 illustrates both the original region D and the new region of integration D^*.

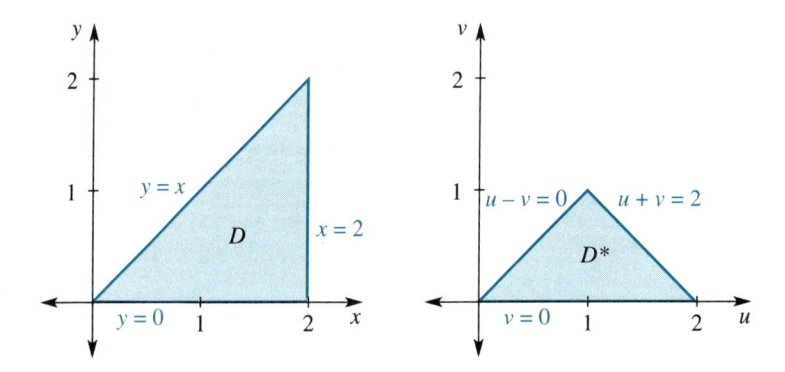

Figure 16.8 The region D is the image of D^* when $x = u + v$ and $y = u - v$.

3. We now calculate the double integral in terms of the new variables u and v:

16.2 TOTAL DERIVATIVE OF A VECTOR FIELD—JACOBIANS

$$\iint_D (x+y)\,dy\,dx = \iint_{D^*} ((u+v)+(u-v))2\,du\,dv$$

$$= \iint_{D^*} 4u\,du\,dv$$

$$= 4\int_0^1 \int_v^{2-v} u\,du\,dv$$

$$= 4\int_0^1 \frac{u^2}{2}\Big]_v^{2-v} dv$$

$$= 4\int_0^1 \left(\frac{(2-v)^2}{2} - \frac{v^2}{2}\right) dv$$

$$= 4\int_0^1 \left(\frac{4}{2} - \frac{4v}{2} + \frac{v^2}{2} - \frac{v^2}{2}\right) dv$$

$$= 4\int_0^1 (2-2v)\,dv$$

$$= 4(2v - v^2)\Big]_0^1$$

$$= (8v - 4v^2)\Big]_0^1 = (8-4) - (0-0) = 4.$$

We can see that the two calculations match exactly. ■

The chain rule for total derivatives

We have talked about vector fields having the same number of components as inputs, but we can also talk about functions

$$\mathbf{f}: \mathbb{R}^m \to \mathbb{R}^n,$$

where $m \neq n$, and

$$\mathbf{f}(x_1, x_2, \ldots, x_m) = \langle f_1(x_1, x_2, \ldots, x_m), \ldots, f_n(x_1, x_2, \ldots, x_m)\rangle,$$

and each component f_i ($1 \leq i \leq n$) is a scalar-valued function $\mathbb{R}^m \to \mathbb{R}$.

Provided that all the functions f_i are C^1-differentiable, the total derivative of **f** is the $n \times m$ matrix of partial derivatives:

$$D\mathbf{f} = \begin{bmatrix} \frac{\partial f_1}{\partial x_1} & \frac{\partial f_1}{\partial x_2} & \frac{\partial f_1}{\partial x_3} & \cdots & \frac{\partial f_1}{\partial x_m} \\ \frac{\partial f_2}{\partial x_1} & \frac{\partial f_2}{\partial x_2} & \frac{\partial f_1}{\partial x_3} & \cdots & \frac{\partial f_2}{\partial x_m} \\ \vdots & \vdots & \vdots & \ddots & \vdots \\ \frac{\partial f_n}{\partial x_1} & \frac{\partial f_n}{\partial x_2} & \frac{\partial f_n}{\partial x_3} & \cdots & \frac{\partial f_n}{\partial x_m} \end{bmatrix}$$

In other words, the ith row of $D\mathbf{f}$ consists of the components of the gradient of the ith component function f_i.

EXAMPLE 13 Find the total derivative of $\mathbf{h}: \mathbb{R}^3 \to \mathbb{R}^2$, where

$$\mathbf{h}(x, y, z) = \langle xyz, \sqrt{x^2 + y^2 + z^2} \rangle.$$

Evaluate $D\mathbf{h}(0, 3, 4)$.

Solution $D\mathbf{h}$ is represented by a 2×3 matrix. If

$$h_1(x, y, z) = xyz \quad \text{and} \quad h_2(x, y, z) = \sqrt{x^2 + y^2 + z^2},$$

then

$$D\mathbf{h} = \begin{bmatrix} \frac{\partial h_1}{\partial x} & \frac{\partial h_1}{\partial y} & \frac{\partial h_1}{\partial z} \\ \frac{\partial h_2}{\partial x} & \frac{\partial h_2}{\partial y} & \frac{\partial h_2}{\partial z} \end{bmatrix} = \begin{bmatrix} yz & xz & xy \\ \frac{x}{\sqrt{x^2+y^2+z^2}} & \frac{y}{\sqrt{x^2+y^2+z^2}} & \frac{z}{\sqrt{x^2+y^2+z^2}} \end{bmatrix}.$$

So, we have

$$D\mathbf{h}(0, 3, 4) = \begin{bmatrix} 12 & 0 & 0 \\ 0 & \frac{3}{5} & \frac{4}{5} \end{bmatrix}. \blacksquare$$

Properties of the total derivative

1. If $\mathbf{f}: \mathbb{R}^m \to \mathbb{R}^n$, then for any scalar c

$$D(c\mathbf{f}) = cD\mathbf{f}.$$

2. If **f** and **g** are functions $\mathbb{R}^m \to \mathbb{R}^n$, then

$$D(\mathbf{f} + \mathbf{g}) = D\mathbf{f} + D\mathbf{g}.$$

3. In the special case that $f: \mathbb{R}^n \to \mathbb{R}$, then Df is a $1 \times n$ matrix

$$Df = [\,\frac{\partial f}{\partial x_1} \quad \frac{\partial f}{\partial x_2} \quad \frac{\partial f}{\partial x_3} \quad \cdots \quad \frac{\partial f}{\partial x_n}\,],$$

and we can think of Df as the *gradient*:

$$\nabla f = \left\langle \frac{\partial f}{\partial x_1}, \frac{\partial f}{\partial x_2}, \cdots, \frac{\partial f}{\partial x_n} \right\rangle.$$

16.2 TOTAL DERIVATIVE OF A VECTOR FIELD—JACOBIANS

If f and g are functions $\mathbb{R}^n \to \mathbb{R}$, then

$$D(fg) = gDf + fDg = g\nabla f + f\nabla g,$$

and

$$D\left(\frac{f}{g}\right) = \frac{gDf - fDg}{g^2} = \frac{g\nabla f + f\nabla g}{g^2}$$

whenever $g \neq 0$.

Note: This product rule and quotient rule only hold for the special case of functions $\mathbb{R}^n \to \mathbb{R}$.

4. If $\mathbf{f} : \mathbb{R}^m \to \mathbb{R}^n$ and $\mathbf{g} : \mathbb{R}^n \to \mathbb{R}^p$, then $\mathbf{g} \circ \mathbf{f} : \mathbb{R}^m \to \mathbb{R}^p$, and at any point $\mathbf{x} = (x_1, x_2, \ldots, x_n)$ in $\mathbb{R}^n$, we have

$$D(\mathbf{g} \circ \mathbf{f})(\mathbf{x}) = D\mathbf{g}(\mathbf{f}(\mathbf{x})) D\mathbf{f}(\mathbf{x}).$$

This is the **chain rule**. Note how it appears virtually identical to the chain rule for single-variable calculus, except that the product indicated on the right-hand side of the equation is a *matrix product*.

EXAMPLE 14 Suppose that

$$f(u, v) = \frac{u^2 + v^2}{u^2 - v^2}$$

and $\qquad u(x, y) = e^{-x-y}, \qquad v(x, y) = e^{xy}.$

Find $D(f \circ \mathbf{g})(x, y)$, where $\mathbf{g}(x, y) = \langle u(x, y), v(x, y)\rangle$.

Solution We'll compute $D(f \circ \mathbf{g})$ using the chain rule:

$$Df(u, v) = \left[\begin{array}{cc} \dfrac{2u(u^2 - v^2) - 2u(u^2 + v^2)}{(u^2 - v^2)^2} & \dfrac{2v(u^2 - v^2) - (-2v(u^2 + v^2))}{(u^2 - v^2)^2} \end{array}\right]$$

$$= \left[\begin{array}{cc} \dfrac{-4uv^2}{(u^2 - v^2)^2} & \dfrac{4u^2 v}{(u^2 - v^2)^2} \end{array}\right]$$

and

$$D\mathbf{g}(x, y) = \left[\begin{array}{cc} -e^{-x-y} & -e^{-x-y} \\ ye^{xy} & xe^{xy} \end{array}\right].$$

The chain rule says:

$$D(f \circ \mathbf{g})(x, y) = Df(\mathbf{g}(x, y)) D\mathbf{g}(x, y)$$

$$= \left[\begin{array}{cc} \dfrac{-4e^{-x-y}e^{2xy}}{(e^{2(-x-y)} - e^{2xy})^2} & \dfrac{4e^{2(-x-y)}e^{xy}}{(e^{2(-x-y)} - e^{2xy})^2} \end{array}\right] \left[\begin{array}{cc} -e^{-x-y} & -e^{-x-y} \\ ye^{xy} & xe^{xy} \end{array}\right]$$

$$= \left[\begin{array}{cc} \dfrac{4e^{2(-x-y)}e^{2xy} + 4ye^{2(-x-y)}e^{2xy}}{(e^{2(-x-y)} - e^{2xy})^2} & \dfrac{4e^{2(-x-y)}e^{2xy} + 4xe^{2(-x-y)}e^{2xy}}{(e^{2(-x-y)} - e^{2xy})^2} \end{array}\right]$$

Now, if we substitute for u and v in terms of x and y in the composition $f \circ g$, we have a function of x and y:

$$h(x,y) = f \circ g(x,y) = \frac{e^{2(-x-y)} + e^{2xy}}{e^{2(-x-y)} - e^{2xy}}.$$

You can check directly that the matrix $Dh = [\partial h/\partial x \quad \partial h/\partial y]$ is precisely the one we found above. ∎

EXAMPLE 15 Suppose that $\mathbf{f}(x,y) = \langle e^x, x+y \rangle$ and $\mathbf{g}(u,v) = \langle u, \cos v, v+u \rangle$. Find $D(\mathbf{g} \circ \mathbf{f})(0,0)$ in two ways: by direct substitution and differentiation and by the chain rule.

Solution By direct substitution, we have

$$\mathbf{g} \circ \mathbf{f}(x,y) = \mathbf{g}(\mathbf{f}(x,y)) = \mathbf{g}(e^x, x+y) = \langle e^x, \cos(x+y), x+y+e^x \rangle.$$

So,

$$D(\mathbf{g} \circ \mathbf{f}(x,y)) = \begin{bmatrix} e^x & 0 \\ -\sin(x+y) & -\sin(x+y) \\ 1+e^x & 1 \end{bmatrix}.$$

At $(x,y) = (0,0)$ we have

$$D(\mathbf{g} \circ \mathbf{f})(0,0) = \begin{bmatrix} 1 & 0 \\ 0 & 0 \\ 2 & 1 \end{bmatrix}.$$

Now, by the chain rule, we have

$$D(\mathbf{g} \circ \mathbf{f})(0,0) = D\mathbf{g}(\mathbf{f}(0,0))D\mathbf{f}(0,0).$$

We compute

$$D\mathbf{g}(u,v) = \begin{bmatrix} 1 & 0 \\ 0 & -\sin v \\ 1 & 1 \end{bmatrix} \quad \text{and} \quad D\mathbf{f}(x,y) = \begin{bmatrix} e^x & 0 \\ 1 & 1 \end{bmatrix}.$$

16.2 TOTAL DERIVATIVE OF A VECTOR FIELD—JACOBIANS

Since $\mathbf{f}(0,0) = \langle 1,0 \rangle$, we have

$$D(\mathbf{g} \circ \mathbf{f})(0,0) = D\mathbf{g}(1,0)D\mathbf{f}(0,0)$$

$$= \begin{bmatrix} 1 & 0 \\ 0 & 0 \\ 1 & 1 \end{bmatrix} \begin{bmatrix} 1 & 0 \\ 1 & 1 \end{bmatrix}$$

$$= \begin{bmatrix} 1 & 0 \\ 0 & 0 \\ 2 & 1 \end{bmatrix},$$

which matches exactly the first computation. ∎

Using the same functions **f** and **g** in the previous example, note that the composition $\mathbf{f} \circ \mathbf{g}$ does not make sense, since the output of **g** has *three* components, but **f** requires *two* inputs.

EXERCISES for Section 16.2

For exercises 1-16: Refer to the following functions:

$$\mathbf{F}(u,v) = \langle 2u - 3v, 3u + v \rangle \qquad \mathbf{G}(u,v) = \langle uv, u/v, v/u \rangle$$

$$\mathbf{H}(x,y,z) = \langle xyz, xy^2z^3 \rangle \qquad \mathbf{P}(x,y,z) = \langle x+y+z, x+y^2+z^3, x^3-y^2-z \rangle.$$

Compute the indicated total derivative. If an indicated composition makes sense, then verify the chain rule by direct substitution and differentiation. If an indicated composition does not make sense, explain why.

1. $D\mathbf{F}(u,v)$
2. $D\mathbf{G}(u,v)$
3. $D\mathbf{H}(x,y,z)$
4. $D\mathbf{P}(x,y,z)$
5. $D(\mathbf{F} \circ \mathbf{G})(u,v)$
6. $D(\mathbf{G} \circ \mathbf{F})(u,v)$
7. $D(\mathbf{H} \circ \mathbf{P})(x,y,z)$
8. $D(\mathbf{P} \circ \mathbf{H})(x,y,z)$
9. $D(\mathbf{F} \circ \mathbf{P})(x,y,z)$
10. $D(\mathbf{F} \circ \mathbf{H})(x,y,z)$
11. $D(\mathbf{H} \circ \mathbf{G})(u,v)$
12. $D(\mathbf{H} \circ \mathbf{F})(u,v)$
13. $D(\mathbf{G} \circ \mathbf{P})(x,y,z)$
14. $D(\mathbf{G} \circ \mathbf{H})(x,y,z)$
15. $D(\mathbf{P} \circ \mathbf{G})(u,v)$
16. $D(\mathbf{P} \circ \mathbf{F})(u,v)$

17. Substitute $u = xy$, $v = y/x$ to find the area of the first quadrant region bounded by the lines $y = x$, $y = 2x$ and the hyperbolas $xy = 1$, $xy = 2$ (see picture below).

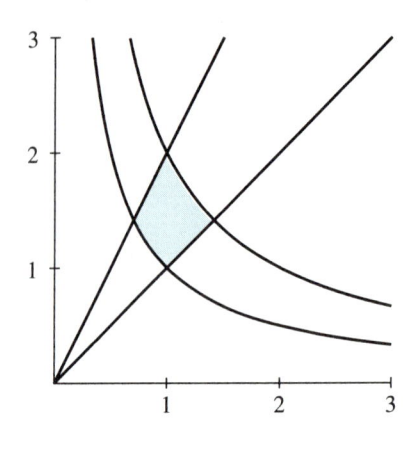

For Exercise 17

18. Elliptical coordinates are given by $x = ar\cos\theta$, $y = br\sin\theta$, where a and b are constants. What is the Jacobian scale factor for this change of variables?

A Jacobian scale factor also arises in triple integration. To change variables in the integral

$$\iiint_R f(x, y, z)\, dx\, dy\, dz$$

to u, v, and w via

$$x = P(u, v, w), \qquad y = Q(u, v, w), \qquad z = R(u, v, w),$$

one must:

1. calculate the Jacobian determinant

$$J(u, v, w) = \begin{vmatrix} \dfrac{\partial x}{\partial u} & \dfrac{\partial x}{\partial v} & \dfrac{\partial x}{\partial w} \\ \dfrac{\partial y}{\partial u} & \dfrac{\partial y}{\partial v} & \dfrac{\partial y}{\partial w} \\ \dfrac{\partial z}{\partial u} & \dfrac{\partial z}{\partial v} & \dfrac{\partial z}{\partial w} \end{vmatrix};$$

2. find the region R^* in uvw-space whose image is R under the given transformation; and

3. compute

$$\iiint_{R^*} f(P(u,v,w), Q(u,v,w), R(u,v,w)) |J(u,v,w)|\, du\, dv\, dw.$$

19. The transformation from Cartesian to cylindrical coordinates is via

$$x = r\cos\theta, \qquad y = r\sin\theta, \qquad z = z.$$

Verify that the Jacobian determinant $J(r, \theta, z) = r$.

20. The transformation from Cartesian to spherical coordinates is via

$$x = \rho\cos\theta\sin\varphi, \qquad y = \rho\sin\theta\sin\varphi, \qquad z = \rho\cos\varphi.$$

Verify that the Jacobian determinant $J(\rho, \theta, \varphi) = -\rho^2 \sin\varphi$.

16.3 LINE INTEGRALS

Now we turn our attention to integration of both scalar and vector fields. Let's recall the situation for single-variable calculus again. A definite integral

$$\int_a^b f(x)\,dx$$

measures the net effect of the real-valued function f over the interval $[a, b]$. We can interpret this value in a variety of ways, depending on the physical meaning of the function f. For example, if $f(x)$ represents the *force* acting on an object at distance x, then $\int_a^b f(x)\,dx$ represents the *work* performed as the object moves over the distance interval $[a, b]$.

In this example, we can consider the interval as a simple straight line path of the object. Our interest in this section turns to integrals over much more general paths. The traditional name for such an integral is *line integral*, though perhaps a better name (used occasionally) would be a *path integral*. You can think of a line integral as measuring the net effect of a scalar field or vector field on an object as it moves along some path. In this section we define and illustrate how line integrals are calculated and discuss some of their physical properties and interpretations.

Smooth and piece-wise smooth curves—arc length

A vector-valued function $\mathbf{r}(t)$ of one real variable t describes a *path*. The image of a continuous path is a *curve* in the plane ($\mathbb{R}^2$) if

$$\mathbf{r}(t) = \langle x(t), y(t) \rangle$$

or in space ($\mathbb{R}^3$) if

$$\mathbf{r}(t) = \langle x(t), y(t), z(t) \rangle.$$

In either case, we say **r** parametrizes the curve. A single curve can have infinitely many different parametrizations.

A continuous path **r** is **smooth** over the interval $a \leq t \leq b$ if $\mathbf{r}'(t)$ is *continuous* and *nonzero* for $a < t < b$. If we can break the interval $[a, b]$ into a finite number of subintervals, and if **r** is smooth over each subinterval then we say that **r** is **piece-wise smooth** over $[a, b]$. Figure 16.9 illustrates the image curves of a smooth and a piece-wise smooth path.

smooth path   piece-wise smooth path

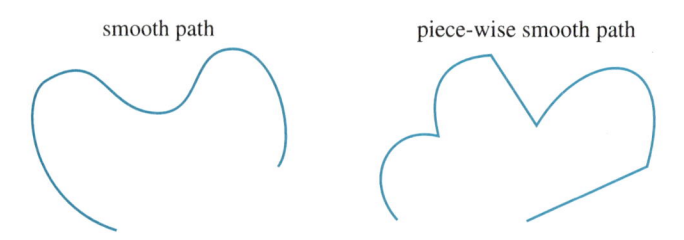

Figure 16.9 A smooth and piece-wise smooth path.

Recall (Section 12.4) that we can find the arc length of a smooth curve by integrating the speed $\|\mathbf{r}'(t)\|$ (magnitude of the velocity) over the interval $[a, b]$:

$$\text{arc length} = \int_a^b \|\mathbf{r}'(t)\|\, dt.$$

The arc length of a piece-wise smooth curve can be found by finding the lengths of each of its smooth pieces and adding the results.

We can use this result to form the arc length function for the path **r**:

$$s(t) = \int_a^t \|\mathbf{r}'(u)\|\, du.$$

The function s simply gives us the length of the path **r** over the interval $[a, t]$, and we have

$$\frac{ds}{dt} = \|\mathbf{r}'(t)\|.$$

In other words, the derivative of the arc length function with respect to t is just the *speed*. We call

$$ds = \|\mathbf{r}'(t)\|\, dt$$

the arc length element, so that

$$\int_a^b ds$$

gives the arc length of a curve with parametrization **r** over the interval $a \leq t \leq b$.

16.3 LINE INTEGRALS

A **closed** curve C is one whose initial and terminal points are the same. If $\mathbf{r}$ is a parametrization for a closed curve C defined on $[a, b]$, then

$$\mathbf{r}(a) = \mathbf{r}(b).$$

A **simple** curve has a parametrization $\mathbf{r}$ such that $\mathbf{r}(t_1) \neq \mathbf{r}(t_2)$ whenever $t_1 \neq t_2$, with the only possible exception allowed being the endpoints (see Figure 16.10).

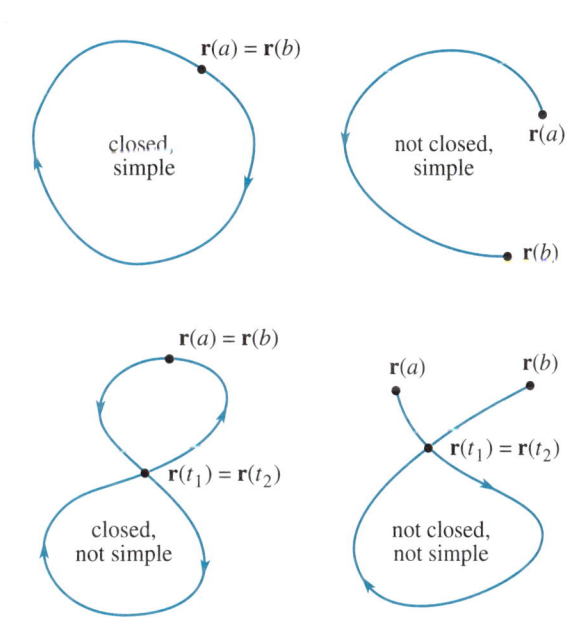

Figure 16.10 Types of curves.

A parametrized curve comes with an *orientation* (direction) determined by the position function $\mathbf{r}$. In Figure 16.10, the orientation of each curve is indicated by small arrowheads showing the direction in which the velocity vector $\mathbf{r}'$ points. Note that for an intersection point of a non-simple curve (where $\mathbf{r}(t_1) = \mathbf{r}(t_2)$, but $t_1 \neq t_2$), the idea of a tangent vector becomes ambiguous, for the direction of the velocity vector depends on the time t we cross that point.

If $\mathbf{r}(t)$ is a parametrization defined over $[a, b]$, we can easily obtain another parametrization that reverses the orientation of the curve:

$$\mathbf{q}(t) = \mathbf{r}(b + a - t), \qquad \text{for } a \leq t \leq b.$$

Note that $\mathbf{q}(a) = \mathbf{r}(b)$, $\mathbf{q}(b) = \mathbf{r}(a)$, and

$$\mathbf{q}'(t) = -\mathbf{r}'(b + a - t).$$

Line integral of a scalar field f with respect to arc length

If f is a scalar field defined in a region of the plane $\mathbb{R}^2$ or space $\mathbb{R}^3$, and $\mathbf{r}$ is a path whose image curve C lies in the same region, we can define a line integral of f with respect to arc length.

Definition 6

> The **line integral** of f with respect to arc length along C is
> $$\oint_C f\, ds = \int_a^b f(\mathbf{r}(t))\left(\frac{ds}{dt}\right) dt = \int_a^b f(\mathbf{r}(t)) \|\mathbf{r}'(t)\| \, dt.$$

▶▶▶ **The small circle in the symbol $\oint$ indicates that we are calculating a line integral, but not necessarily over a *closed* curve.**

If we think of C as a *wire* in space and $f(x, y, z)$ measures the mass per unit length (density) at each point (x, y, z), then we could approximate the entire mass of the wire with a Riemann sum. That is, we could partition the curve C into n tiny pieces of length Δs such that the density of each piece is relatively uniform. The total mass of the wire is given by

$$\sum_{i=1}^n f(x_i, y_i, z_i) \Delta s,$$

where (x_i, y_i, z_i) is a point chosen from the ith piece of wire. The limit of these Riemann sums as the partition is made finer and finer is

$$\oint_C f\, ds \quad = \quad \text{mass of the wire}.$$

We note that the line integral

$$\oint_C 1\, ds \quad = \quad \text{length of the wire}.$$

EXAMPLE 16 A straight wire has endpoints $(1, 2, 3)$ and $(-2, 3, -1)$ (units measured in cm). Its density at any point is given by $f(x, y, z) = y^2$ grams per cm. Find the mass of the wire.

Solution We can parametrize the line segment represented by the wire as

$$\mathbf{r}(t) = \langle 1, 2, 3 \rangle + t(\langle -2, 3, -1 \rangle - \langle 1, 2, 3 \rangle)$$
$$= \langle 1, 2, 3 \rangle + t\langle -3, 1, -4 \rangle$$
$$= \langle 1 - 3t,\ 2 + t,\ 3 - 4t \rangle.$$

16.3 LINE INTEGRALS

Note that $\mathbf{r}(0) = \langle 1, 2, 3 \rangle$ and $\mathbf{r}(1) = \langle -2, 3, -1 \rangle$. For $0 \leq t \leq 1$, $\mathbf{r}(t)$ traces out each point along the length of the wire.

The derivative is simply the constant vector $\mathbf{r}'(t) = \langle -3, 1, -4 \rangle$, so

$$\|\mathbf{r}'(t)\| = \sqrt{(-3)^2 + 1^2 + (-4)^2} = \sqrt{26},$$

and the mass is given by

$$\oint_C y^2 \, ds = \int_0^1 \sqrt{26}(2+t)^2 \, dt = \frac{\sqrt{26}}{3}(2+t)^3 \Big]_0^1 = 9\sqrt{26} - \frac{8}{3}\sqrt{26} = \frac{19}{3}\sqrt{26} \approx 32.29.$$

The wire has a mass of approximately 32.29 grams. ∎

EXAMPLE 17 Find $\oint_C f \, ds$, where $f(x, y, z) = x + y + yz$ and C is the helix parametrized by

$$\mathbf{r}(t) = \langle \sin t, \cos t, t \rangle \quad \text{for } 0 \leq t \leq 2\pi.$$

Solution The velocity and speed are

$$\mathbf{r}'(t) = \langle \cos t, -\sin t, 1 \rangle \quad \text{and} \quad \|\mathbf{r}'(t)\| = \sqrt{\cos^2 t + \sin^2 t + 1} = \sqrt{2}.$$

Since $x = \sin t$, $y = \cos t$, and $z = t$, we have

$$\oint_C f \, ds = \int_0^{2\pi} f(\sin t, \cos t, t) \sqrt{2} \, dt$$

$$= \int_0^{2\pi} (\sin t + \cos t + t \cos t) \sqrt{2} \, dt$$

$$= \sqrt{2} [-\cos t + \sin t + (t \sin t - \cos t)] \Big]_0^{2\pi}$$

$$= \sqrt{2}(-1 + 0 + (0 - 1)) - \sqrt{2}(-1 + 0 + (0 - 1))$$

$$= -2\sqrt{2} + 2\sqrt{2} = 0.$$
∎

EXAMPLE 18 Find $\oint_C f \, ds$, where $f(x, y, z) = x + \cos^2 z$ over the same curve as the previous example.

Solution The velocity and speed are the same as in the previous example, so

$$\oint_C f\,ds = \int_0^{2\pi}(\sin t + \cos^2 t)\sqrt{2}\,dt$$

$$= \sqrt{2}\left(-\cos t + \frac{t}{2} + \frac{\sin 2t}{4}\right)\bigg]_0^{2\pi}$$

$$= \sqrt{2}(-1 + \pi + 0) - \sqrt{2}(-1 + 0 + 0)$$

$$= \sqrt{2}\pi \approx 4.44.$$

■

EXAMPLE 19 Find $\oint_C f\,ds$, where $f(x, y, z) = x + y + z$ and C is parametrized by

$$\mathbf{r}(t) = \langle t, t^2, \tfrac{2}{3}t^3\rangle \qquad \text{for } 0 \leq t \leq 1.$$

Solution The velocity and speed are

$$\mathbf{r}'(t) = \langle 1, 2t, 2t^2\rangle \qquad \text{and} \qquad \|\mathbf{r}'(t)\| = \sqrt{1 + 4t^2 + 4t^4} = \sqrt{(1 + 2t^2)^2} = 1 + 2t^2.$$

Therefore,

$$\oint_C f\,ds = \int_0^1 (t + t^2 + \tfrac{2}{3}t^3)(1 + 2t^2)\,dt$$

$$= \int_0^1 (t + t^2 + \tfrac{2}{3}t^3 + 2t^3 + 2t^4 + \tfrac{4}{3}t^5)\,dt$$

$$= \left(\frac{t^2}{2} + \frac{t^3}{3} + \frac{1}{6}t^4 + \frac{1}{2}t^4 + \frac{2}{5}t^5 + \frac{2}{9}t^6\right)\bigg]_0^1$$

$$= \left(\frac{1}{2} + \frac{1}{3} + \frac{1}{6} + \frac{1}{2} + \frac{2}{5} + \frac{2}{9}\right) - (0)$$

$$= \frac{191}{90}.$$

■

Line integral of a vector field F

If $\mathbf{F}$ is a vector field defined in some region of the plane $\mathbb{R}^2$ or space $\mathbb{R}^3$, we can define a line integral of $\mathbf{F}$ as follows.

16.3 LINE INTEGRALS

Definition 7
> If C is a curve parametrized by a smooth path $\mathbf{r}$ defined over the interval $[a,b]$, and if $\mathbf{F}(\mathbf{r}(t))$ is defined for $a \leq t \leq b$, then
> $$\oint_C \mathbf{F} \cdot d\mathbf{s} = \int_a^b \mathbf{F}(\mathbf{r}(t)) \cdot \mathbf{r}'(t) \, dt$$
> is the **line integral** of $\mathbf{F}$ along $\mathbf{r}$. If $\mathbf{r}$ is piece-wise smooth, then the line integral of $\mathbf{F}$ along C is obtained by adding the line integrals over the smooth pieces of $\mathbf{r}$.

The notation $\oint_C$ reminds us again that this is a line integral over a curve C. The notation $\mathbf{F} \cdot d\mathbf{s}$ is meant to remind us that a dot product $\mathbf{F}(\mathbf{r}(t)) \cdot \mathbf{r}'(t)$ is involved in the calculation of the line integral.

If $\mathbf{F}$ represents a *vector force field* and $\mathbf{r}$ is the path of an object moving through the force field, then

$$\oint_C \mathbf{F} \cdot d\mathbf{s}$$

measures the work performed on the object over the path. Let's see why this makes sense. At any time t, the vector $\mathbf{F}(\mathbf{r}(t))$ gives us the force vector at a point on our path $\mathbf{r}$. The velocity vector $\mathbf{r}'(t)$ points in the tangent direction to the curve. Therefore, the dot product

$$\mathbf{F}(\mathbf{r}(t)) \cdot \mathbf{T}(t)$$

gives us exactly the component of the force that points in the same direction as the instantaneous motion of the object at time t (see Figure 16.11). Integrating this force component over the interval $[a,b]$ measures the work.

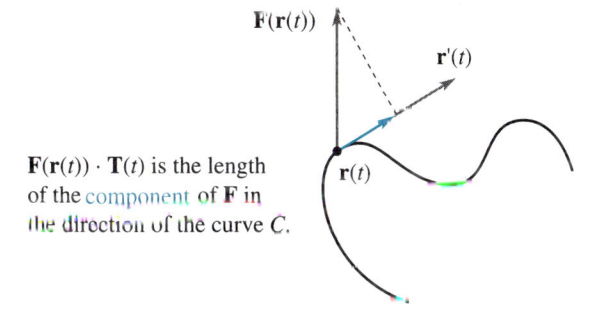

Figure 16.11 The component of a force field along a curve.

If **F** is the *velocity vector field* for a fluid and **r** describes a curve C through this fluid, then at any point $\mathbf{r}(t)$ on the curve we can take the dot product of **F** with the unit tangent vector **T**, and this gives us the tangential component of the fluid flow along the curve. In other words, at any time t,

$$\mathbf{F} \cdot \mathbf{T} = \mathbf{F}(\mathbf{r}(t)) \cdot \frac{\mathbf{r}'(t)}{\|\mathbf{r}'(t)\|}$$

measures the flow in the same direction as the curve at that point.

Note that $\mathbf{F} \cdot \mathbf{T}$ is a scalar (it's a dot product). If we take the line integral of this scalar over the curve C with respect to arc length, we have a measure of the total fluid flow or *circulation* along the curve:

$$\oint_C \mathbf{F} \cdot \mathbf{T} \, ds = \int_a^b \mathbf{F}(\mathbf{r}(t)) \cdot \frac{\mathbf{r}'(t)}{\|\mathbf{r}'(t)\|} \|\mathbf{r}'(t)\| \, dt.$$

Note that the speed factor $\|\mathbf{r}'(t)\|$ cancels out, so that

$$\oint_C \mathbf{F} \cdot \mathbf{T} \, ds = \int_a^b \mathbf{F}(\mathbf{r}(t)) \cdot \mathbf{r}'(t) \, dt = \oint_C \mathbf{F} \cdot d\mathbf{s}.$$

This is simply our definition of the line integral of the original vector field over the curve.

EXAMPLE 20 Find $\oint_C \mathbf{F} \cdot \mathbf{T} \, ds$ where $\mathbf{F}(x, y, z) = \langle y, 2x, y \rangle$ and the curve C is parametrized by

$$\mathbf{r}(t) = \langle t, t^2, t^3 \rangle \qquad \text{for } 0 \le t \le 1.$$

Solution We compute $\mathbf{r}'(t) = \langle 1, 2t, 3t^2 \rangle$, so

$$\oint_C \mathbf{F} \cdot \mathbf{T} \, ds = \int_0^1 \mathbf{F}(\mathbf{r}(t)) \cdot \mathbf{r}'(t) \, dt$$

$$= \int_0^1 \langle t^2, 2t, t^2 \rangle \cdot \langle 1, 2t, 3t^2 \rangle \, dt = \int_0^1 (t^2 + 4t^2 + 3t^4) \, dt$$

$$= \frac{t^3}{3} + \frac{4t^3}{3} + \frac{3t^5}{5} \bigg]_0^1 = \frac{1}{3} + \frac{4}{3} + \frac{3}{5} = \frac{34}{15}.$$ ∎

Other notations for line integrals

If **r** is a path $[a, b] \longrightarrow \mathbb{R}^2$, then we have

$$\mathbf{r}(t) = \langle x(t), y(t) \rangle$$

and

$$\mathbf{r}'(t) = \left\langle \frac{dx}{dt}, \frac{dy}{dt} \right\rangle.$$

16.3 LINE INTEGRALS

Now, if our vector field **F** has the form

$$\mathbf{F}(x,y) = \langle P(x,y), Q(x,y) \rangle,$$

then the line integral of **F** along the image curve C can be written

$$\oint_C \mathbf{F} \cdot d\mathbf{s} = \int_a^b \langle P(x,y), Q(x,y) \rangle \cdot \left\langle \frac{dx}{dt}, \frac{dy}{dt} \right\rangle dt$$

$$= \int_a^b \left[P(x,y) \frac{dx}{dt} + Q(x,y) \frac{dy}{dt} \right] dt.$$

A shorthand notation for this form is simply

$$\oint_C P\,dx + Q\,dy.$$

Similarly, if $\mathbf{r}(t) = \langle x(t), y(t), z(t) \rangle$, and

$$\mathbf{F}(x,y,z) = \langle P(x,y,z),\ Q(x,y,z),\ R(x,y,z) \rangle,$$

then

$$\oint_C \mathbf{F} \cdot d\mathbf{s} = \oint_C P\,dx + Q\,dy + R\,dz.$$

EXAMPLE 21 Find $\oint_C x\,dy - y\,dx$, where C is parametrized by

$$\mathbf{r}(t) = \langle \cos t, \sin t \rangle \qquad \text{for } 0 \leq t \leq 2\pi.$$

Solution We have $x = \cos t$, $\dfrac{dx}{dt} = -\sin t$, $y = \sin t$, $\dfrac{dy}{dt} = \cos t$.

Therefore, $\oint_C x\,dy - y\,dx = \int_0^{2\pi} [\cos^2 t + \sin^2 t]\,dt = \int_0^{2\pi} 1\,dt = t\Big|_0^{2\pi} = 2\pi.$ ∎

EXAMPLE 22 Find $\oint_C yz\,dx + xz\,dy + xy\,dz$, where C consists of straight line segments joining $(1,0,0)$ to $(0,1,0)$ to $(0,0,1)$.

Solution We will consider C in two parts. We can parametrize the line segment C_1 by

$$\mathbf{r}_1 : [0,1] \longrightarrow \mathbb{R}^3 \qquad \text{where } \mathbf{r}_1(t) = \langle 1-t, t, 0 \rangle.$$

Note that $\mathbf{r}_1(0) = \langle 1, 0, 0 \rangle$, $\mathbf{r}_1(1) = \langle 0, 1, 0 \rangle$, and $\mathbf{r}_1'(t) = \langle -1, 1, 0 \rangle$.

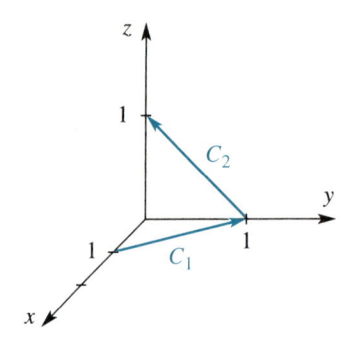

Figure 16.12 The smooth pieces of C are C_1 and C_2.

We can parametrize the line segment C_2 by

$$\mathbf{r}_2 : [0,1] \longrightarrow \mathbb{R}^3 \quad \text{where } \mathbf{r}_2(t) = \langle 0, 1-t, t\rangle.$$

Note that $\mathbf{r}_2(0) = \langle 0,1,0\rangle$, $\mathbf{r}_2(1) = \langle 0,0,1\rangle$, and $\mathbf{r}_2'(t) = \langle 0,-1,1\rangle$. (See Figure 16.12.) Thus,

$$\oint_C yz\,dx + xz\,dy + xy\,dz = \oint_{C_1} yz\,dx + xz\,dy + xy\,dz + \oint_{C_2} yz\,dx + xz\,dy + xy\,dz$$

$$= \int_0^1 [t(0)(-1) + (1-t)(0)(1) + (1-t)(t)(0)]\,dt$$

$$+ \int_0^1 [(1-t)(t)(0) + (0)(t)(-1) + (0)(1-t)(1)]\,dt$$

$$= \int_0^1 0\,dt + \int_0^1 0\,dt = 0 + 0 = 0. \qquad \blacksquare$$

Effects of orientation on line integrals

A single curve C can have infinitely many different parametrizations. What effect does the choice of parametrization have on the value of a line integral over a curve?

Suppose that a curve C has a smooth parametrization $\mathbf{r}$ defined over the interval $[a,b]$. If $u(t)$ is a function such that $u(c) = a$, $u(d) = b$ and $u'(t) > 0$, then the new parametrization

$$\mathbf{p}(t) = \mathbf{r}(u(t)), \quad \text{for } c \leq t \leq d$$

will trace out the curve C in the same direction as $\mathbf{r}$. We say $\mathbf{p}$ **preserves** the original orientation of C.

If $u(t)$ is a function such that $u(c) = b$, $u(d) = a$ and $u'(t) < 0$, then the new parametrization

$$\mathbf{q}(t) = \mathbf{r}(u(t)), \quad \text{for } c \leq t \leq d$$

16.3 LINE INTEGRALS

will trace out the curve in the *opposite* direction of r. We say that q **reverses** the original orientation of C. The effect of such a change in parametrization is different for line integrals of *scalar* and *vector* fields, as documented in the following theorem.

Theorem 16.1

Hypothesis 1: Curve C has smooth parametrization r defined on $[a, b]$.

Hypothesis 2: f is a differentiable scalar field defined at each point of C, and **F** is a differentiable vector field defined at each point of C.

Hypothesis 3: p and q are smooth parametrizations of C defined on $[c, d]$.

Conclusion 1: If p preserves the orientation of C given by r, then

$$\int_a^b f(\mathbf{r}(t))\|\mathbf{r}'(t)\|\,dt = \int_c^d f(\mathbf{p}(t))\|\mathbf{p}'(t)\|\,dt$$

and

$$\int_a^b \mathbf{F}(\mathbf{r}(t)) \cdot \mathbf{r}'(t)\,dt = \int_c^d \mathbf{F}(\mathbf{p}(t)) \cdot \mathbf{p}'(t)\,dt.$$

Conclusion 2: If q reverses the orientation of C given by r, then

$$\int_a^b f(\mathbf{r}(t))\|\mathbf{r}'(t)\|\,dt = \int_c^d f(\mathbf{q}(t))\|\mathbf{q}'(t)\|\,dt$$

and

$$\int_a^b \mathbf{F}(\mathbf{r}(t)) \cdot \mathbf{r}'(t)\,dt = -\int_c^d \mathbf{F}(\mathbf{q}(t)) \cdot \mathbf{q}'(t)\,dt.$$

In other words, $\oint_C f\,ds$ does not depend on the the parametrization of C, regardless of the orientation. However, the value of $\oint_C \mathbf{F} \cdot ds$ changes sign when the orientation of C is reversed.

EXERCISES for Section 16.3

For exercises 1-6: Consider the straight line segment C connecting the point P_1 to the point P_2. Find a path **r** starting at P_1 and traveling along C to P_2, and use it to find

$$\oint_C f\,ds \text{ and } \oint_C \mathbf{F} \cdot ds.$$

	P_1	P_2	$f(x, y, z)$	$\mathbf{F}(x, y, z)$
1.	$(1, 1, 1)$	$(2, 2, 2)$	$xy + yz$	$\langle x, y^2, z^2 \rangle$
2.	$(0, 1, 0)$	$(1, 0, 0)$	$xz + yz$	$\langle x^3, y, x^2 \rangle$
3.	$(0, -1, 1)$	$(0, 1, -1)$	$xy + z^2$	$\langle x^2, y^3, z \rangle$
4.	$(-1, -1, -1)$	$(1, 1, 1)$	$x^2 + z^2$	$\langle xy, y^2, yz \rangle$
5.	$(1, 2, 3)$	$(3, 2, 1)$	$y^2 + z^2$	$\langle x^2, xy, xz \rangle$
6.	$(1, -1, 0)$	$(0, 0, 1)$	$x^2 + y^2$	$\langle xz, yz, z^2 \rangle$

For exercises 7-10: Calculate

$$\oint_C \mathbf{F} \cdot \mathbf{T}\, ds,$$

where C is the curve parametrized by the given position function $\mathbf{r}$, and $\mathbf{F}$ is the given vector field.

7. $\mathbf{r} : [1,2] \longrightarrow \mathbb{R}^3$, where $\mathbf{r}(t) = \langle \dfrac{t^3}{3}, \dfrac{t^2}{\sqrt{2}}, t \rangle$

   $\mathbf{F}(x,y,z) = (\ln(xyz))\,\mathbf{i} \;+\; xyz\,\mathbf{j} \;+\; (x^2+y^2+z^2)\,\mathbf{k}$

8. $\mathbf{r} : [1,2] \longrightarrow \mathbb{R}^3$, where $\mathbf{r}(t) = \langle \dfrac{t^3}{3}, t^2, 4t \rangle$

   $\mathbf{F}(x,y,z) = (y^2 z^2)\,\mathbf{i} \;+\; \cos(x^2+z^2)\,\mathbf{j} \;+\; \ln(xyz)\,\mathbf{k}$

9. $\mathbf{r} : [0,\pi] \longrightarrow \mathbb{R}^3$, where $\mathbf{r}(t) = \langle \cos 2t,\, \sin 2t,\, t \rangle$

   $\mathbf{F}(x,y,z) = (x^2+y^2)\,\mathbf{i} \;+\; z\,\mathbf{j} \;+\; \arctan(y/x)\,\mathbf{k}$

10. $\mathbf{r} : [1,e] \longrightarrow \mathbb{R}^3$, where $\mathbf{r}(t) = \langle t^2, 2t, \ln t \rangle$

    $\mathbf{F}(x,y,z) = \tan(y^2+z^2)\,\mathbf{i} \;+\; xy^2 z^3\,\mathbf{j} \;+\; e^{xyz}\,\mathbf{k}$

For exercises 11-22: Consider the scalar field

$$f(x,y) = x^2 y^3,$$

the vector field $\quad \mathbf{F}(x,y) = y\mathbf{i} - x\mathbf{j},$

and the unit quarter circle C parametrized in three different ways:

$$\mathbf{r}(t) = \langle \cos t, \sin t \rangle \quad \text{where } 0 \le t \le \pi/2,$$

$$\mathbf{p}(t) = \langle \cos \pi t, \sin \pi t \rangle \quad \text{where } 0 \le t \le 1/2,$$

$$\mathbf{q}(t) = \langle \sin(t/2), \cos(t/2) \rangle \quad \text{where } 0 \le t \le \pi.$$

11. Which of the parametrizations have the same orientation? Which have opposite orientation?

12. If t represents time in seconds and each of the parametrizations describes the path of a different particle, which of the particles moves the slowest? Which moves the fastest? Which particle would collide with the other two, and when?

13. Find $\oint_C f\, ds$ using the parametrization $\mathbf{r}$.

14. Find $\oint_C f\, ds$ using the parametrization $\mathbf{p}$.

16.3 LINE INTEGRALS

15. Find $\oint_C f\,ds$ using the parametrization q.

16. Does the orientation of the curve affect the value of $\oint_C f\,ds$?

17. Find $\oint_C \mathbf{F}\cdot d\mathbf{s}$ using the parametrization r.

18. Find $\oint_C \mathbf{F}\cdot d\mathbf{s}$ using the parametrization p.

19. Find $\oint_C \mathbf{F}\cdot d\mathbf{s}$ using the parametrization q.

20. Does the orientation of the curve affect the value of $\oint_C \mathbf{F}\cdot d\mathbf{s}$?

21. The quarter circle C has endpoints at $(1,0)$ and $(0,1)$. Find $\oint_{C'} f\,ds$, where C' is the straight line segment connecting $(1,0)$ to $(0,1)$. Is the value different than over the quarter circle C?

22. Find $\oint_{C'} \mathbf{F}\cdot d\mathbf{s}$, where C' is the straight line segment connecting $(1,0)$ to $(0,1)$. Is the value different than over the quarter circle C?

---

23. Thinking of a scalar field f as measuring the mass density along a wire in the shape of the curve C, explain why it makes sense that $\oint_C f\,ds$ does not depend on the orientation of C.

24. Thinking of a vector field $\mathbf{F}$ as measuring gravitational force at each point along the curve C, explain why it makes sense that $\oint_C \mathbf{F}\,d\mathbf{s}$ does depend on the orientation of C.

25. Suppose that $\mathbf{F}(\mathbf{r}(t))$ is orthogonal to $\mathbf{r}'(t)$ at every point $\mathbf{r}(t)$ along the curve C. Show that

$$\oint_C \mathbf{F}\cdot d\mathbf{s} = 0.$$

26. Suppose that $\mathbf{F}(\mathbf{r}(t))$ points in the same direction as $\mathbf{r}'(t)$ at every point $\mathbf{r}(t)$ along the curve C. Show that

$$\oint_C \mathbf{F}\cdot d\mathbf{s} = \oint_C \|\mathbf{F}\|\,ds.$$

16.4 PARAMETRIZED SURFACES

A position function $\mathbf{r}(t) = \langle x(t), y(t), z(t) \rangle$ parametrizes a *curve* in space ($\mathbb{R}^3$). We can think of a particle constrained to move along this curve as having "one degree of freedom," represented by the single parameter t.

In contrast, we think of a particle constrained to move on a *surface* as having *two* degrees of freedom of movement. For example, when we move on the surface of the earth, both our *latitude* and *longitude* can change. Accordingly, a parametrization of a surface requires two parameters, say u and v.

Definition 8

> A surface S in space $\mathbb{R}^3$ is the image of a position function
> $$\mathbf{r}: D \longrightarrow \mathbb{R}^3$$
> $$\mathbf{r}(u,v) = \langle x(u,v), y(u,v), z(u,v) \rangle$$
> where D is some domain in the uv-plane.

The parameters u and v provide a sort of "coordinate system" for the surface. You have seen maps of regions of the earth where the latitude and longitude lines appear to be horizontal and vertical, respectively. This flat map corresponds to the uv-plane where u is latitude and v is longitude.

The position function $\mathbf{r}$ in this case associates or *maps* each ordered pair (u,v) to a corresponding point on the earth's surface. This correspondence allows us to use the coordinates (u,v) to chart our position and navigate about the surface.

To calculate the effects of a scalar or vector field on a surface, we need the navigational system provided by a parametrization. In this section we illustrate some techniques for parametrizing surfaces.

Parametrizing a function graph

To parametrize a surface S in space, we need to describe two things explicitly:

1. the coordinate functions x, y, z of the position function $\mathbf{r}$

$$\mathbf{r}(u,v) = \langle x(u,v), y(u,v), z(u,v) \rangle$$

2. the domain D of points (u,v) corresponding to points $\mathbf{r}(u,v)$ on the surface S.

16.4 PARAMETRIZED SURFACES

If our surface S is that of a function graph

$$z = f(x, y)$$

over some domain in the xy-plane, then it is an easy matter to parametrize the surface in terms of u and v. Simply let

$$x = u$$
$$y = v$$
$$z = f(u, v)$$

and let D be the domain of the function f in the uv-plane.

EXAMPLE 23 Parametrize the upper hemisphere of the unit sphere: $\{(x,y,z) : x^2+y^2+z^2 = 1, \quad z \geq 0\}$.

Solution The surface can be described as the graph of the function $z = \sqrt{1 - x^2 - y^2}$ over the unit disk (the unit circle and its interior) in the xy-plane. Hence, our parametrization is

$$\mathbf{r}(u, v) = \langle u, v, \sqrt{1 - u^2 - v^2} \rangle$$

for $D = \{(u,v) : u^2 + v^2 \leq 1\}$. Figure 16.13 illustrates the parametrization $\mathbf{r}$ and its domain D. Any surface having an equation where one of the coordinates x, y, z is expressed in terms of the other two can be parametrized in a similar way. ∎

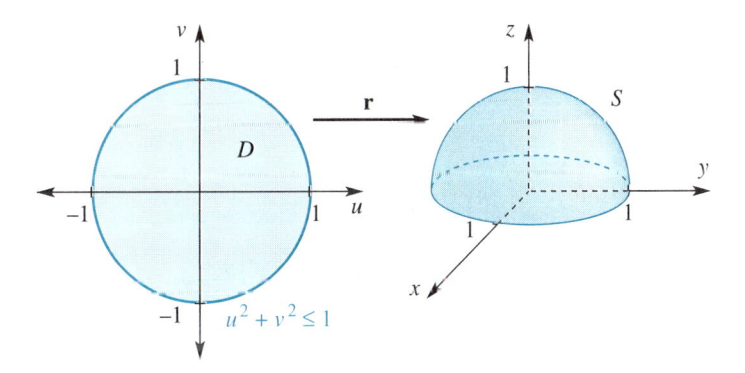

Figure 16.13 $\mathbf{r}(u,v) = \langle u, v, \sqrt{1 - u^2 - v^2} \rangle$ for D in the uv-plane.

Using cylindrical or spherical coordinates to parametrize a surface

Cylindrical or spherical coordinates can also prove useful in parametrizing a surface, if the equation of the surface allows us to express each coordinate x, y, and z in terms of two cylindrical or two spherical parameters.

We'll illustrate by parametrizing the surface of the previous example in two more ways.

EXAMPLE 24 Parametrize the upper hemisphere of the unit sphere using cylindrical coordinates.

Solution Since $x = r\cos\theta$ and $y = r\sin\theta$ and $z = \sqrt{1-x^2-y^2} = \sqrt{1-r^2}$, we have all three coordinates in terms of the parameters r and θ:

$$\mathbf{p}(r,\theta) = \langle r\cos\theta, r\sin\theta, \sqrt{1-r^2}\rangle,$$

where $0 \le r \le 1$ and $0 \le \theta \le 2\pi$. The domain of our parametrization is actually a rectangle in the $r\theta$-plane (see Figure 16.14). ∎

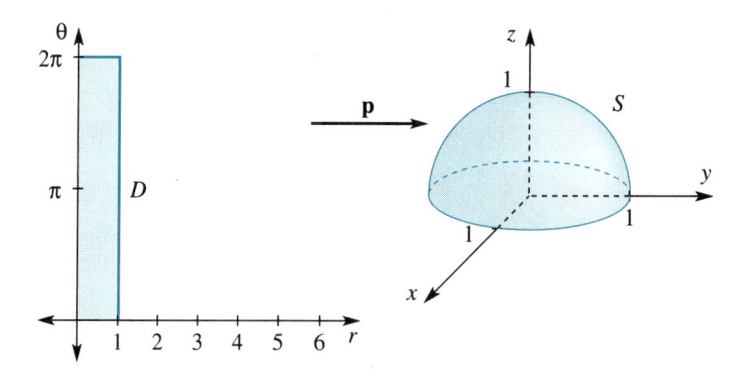

Figure 16.14 $\mathbf{p}(r,\theta) = \langle r\cos\theta, r\sin\theta, \sqrt{1-r^2}\rangle$ for D in the $r\theta$-plane.

EXAMPLE 25 Parametrize the upper hemisphere of the unit sphere using spherical coordinates.

Solution Since $\rho = 1$ on the unit sphere, we have

$$x = \rho\sin\varphi\cos\theta = \sin\varphi\cos\theta$$
$$y = \rho\sin\varphi\sin\theta = \sin\varphi\sin\theta$$
$$z = \rho\cos\varphi = \cos\varphi.$$

Hence all three coordinates are in terms of θ and φ,

$$\mathbf{q}(\theta,\varphi) = \langle \sin\varphi\cos\theta, \sin\varphi\sin\theta, \cos\varphi\rangle,$$

where $0 \le \theta \le 2\pi$ and $0 \le \varphi \le \frac{\pi}{2}$. The domain of this parametrization is a rectangle in the $\theta\phi$-plane (see Figure 16.15). ∎

16.4 PARAMETRIZED SURFACES

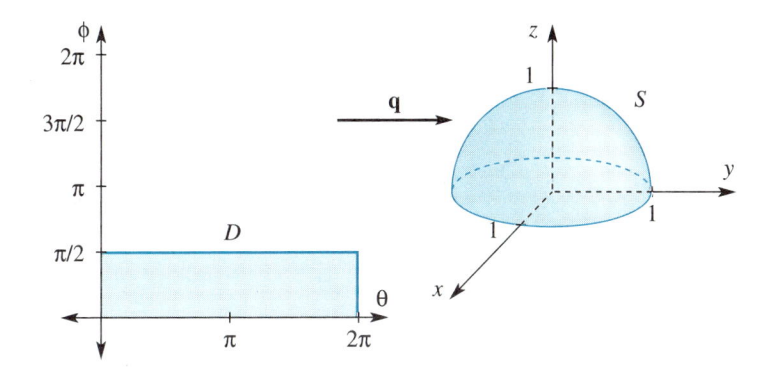

Figure 16.15 $\mathbf{q}(\theta, \varphi) = \langle \sin\varphi\cos\theta, \sin\varphi\sin\theta, \cos\varphi \rangle$ for D in the $\theta\phi$-plane.

Graphing parametrized surfaces

A parametrized surface can be plotted using a technique similar to the slicing method discussed in Chapter 13. Suppose that S is parametrized by

$$\mathbf{r}(u, v) = \langle x(u, v), y(u, v), z(u, v) \rangle$$

over some domain D in the uv-plane. If we set u or v equal to some constant value c, we obtain a parametrized *curve* in space. For example, if we set $u = 3$, then

$$\mathbf{r}(3, v) = \langle x(3, v), y(3, v), z(3, v) \rangle$$

will trace out a curve in space for values of the single remaining parameter v. Similarly, if we set $v = -2$, then

$$\mathbf{r}(u, -2) = \langle x(u, -2), y(u, -2), z(u, -2) \rangle$$

traces out another curve in space for values of the single parameter u. These curves are simply the images of vertical ($u = c$) and horizontal ($v = c$) grid lines in the uv-plane (see Figure 16.16).

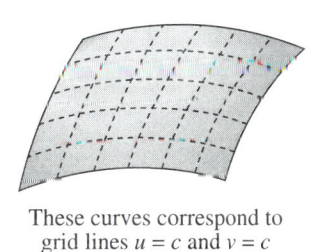

These curves correspond to grid lines $u = c$ and $v = c$ (c a constant)

Figure 16.16 Image curves of grid lines on a parametrized surface S.

If a machine plots several image curves (or their piece-wise linear approximations) of uv-grid lines, the result resembles a wireframe model of the surface. Indeed, for a function graph $z = f(x, y)$ parametrized by

$$\mathbf{r}(u, v) = \langle u, v, f(u, v) \rangle,$$

the image curves of the grid lines of the domain D correspond exactly to x-slices and y-slices of the function graph. For other parametrizations, these image curves of the grid lines of the domain D need not correspond to cross-sections of the surface S.

EXAMPLE 26   Describe the image curves of horizontal and vertical grid lines of the domain D in each of the parametrizations of the upper unit hemisphere of the three previous examples.

Solution   For the parametrization

$$\mathbf{r}(u, v) = \langle u, v, \sqrt{1 - u^2 - v^2} \rangle,$$

the image curve of $u = c$ for $-1 \leq c \leq 1$ is a vertical cross-section of the hemisphere cut parallel to the yz-plane. The image curve of $v = c$ for $-1 \leq c \leq 1$ is a vertical cross-section of the hemisphere cut parallel to the xz-plane. Each of these curves is in the shape of the upper half of a circle.

For the parametrization in cylindrical coordinates

$$\mathbf{p}(r, \theta) = \langle r \cos \theta, r \sin \theta, \sqrt{1 - r^2} \rangle,$$

the image curve of $r = c$ for $0 \leq r \leq 1$ is a circular horizontal cross-section of the hemisphere cut parallel to the xy-plane. The image curve of $\theta = c$ for $0 \leq \theta \leq 2\pi$ is a unit quarter circle extending from the base of the hemisphere (the unit circle in the xy-plane) up to the "North Pole" $(0, 0, 1)$.

For the parametrization in spherical coordinates

$$\mathbf{q}(\theta, \varphi) = \langle \sin \varphi \cos \theta, \sin \varphi \sin \theta, \cos \varphi \rangle,$$

the image curve of $\theta = c$ is the same as for cylindrical coordinates. The image curve of $\varphi = c$ for $0 \leq \varphi \leq \pi/2$ is a circular horizontal cross-section of the hemisphere cut parallel to the xy-plane. The values of θ and ϕ are the equivalent of longitude and latitude lines. ∎

Smooth surfaces, normal vectors, and tangent planes

If S is a surface parametrized by a C^1-differentiable position function

$$\mathbf{r}(u, v) = \langle x(u, v), y(u, v), z(u, v) \rangle,$$

16.4 PARAMETRIZED SURFACES

then two vectors can be computed at each point $\mathbf{r}(u,v)$:

$$\mathbf{T}_u = \frac{\partial \mathbf{r}}{\partial u} = \left\langle \frac{\partial x}{\partial u}, \frac{\partial y}{\partial u}, \frac{\partial z}{\partial u} \right\rangle$$

and

$$\mathbf{T}_v = \frac{\partial \mathbf{r}}{\partial v} = \left\langle \frac{\partial x}{\partial v}, \frac{\partial y}{\partial v}, \frac{\partial z}{\partial v} \right\rangle.$$

If these vectors are nonzero, they are tangent to the surface S at the point $\mathbf{r}(u,v)$ (see Figure 16.17).

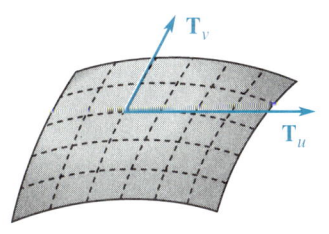

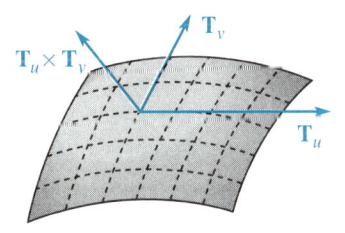

Figure 16.17 $\mathbf{T}_u = \partial \mathbf{r}/\partial u$ and $\mathbf{T}_v = \partial \mathbf{r}/\partial v$ are tangent vectors to S.

The vector
$$\mathbf{T}_u \times \mathbf{T}_v$$
is called the **fundamental vector product** of the parametrization $\mathbf{r}$. Every point at which $\mathbf{T}_u \times \mathbf{T}_v \ne 0$ is called a **smooth point**. A point at which $\mathbf{r}$ is not C^1-differentiable or where $\mathbf{T}_u \times \mathbf{T}_v = 0$ is called a **singular point**.

At a smooth point, the vector $\mathbf{T}_u \times \mathbf{T}_v$ is normal to the surface S (see Figure 16.18).

Figure 16.18 $\mathbf{T}_u = \partial \mathbf{r}/\partial u$ and $\mathbf{T}_v = \partial \mathbf{r}/\partial v$ are tangent vectors to S.

We can use the fundamental vector product to find the equation of the tangent plane to the surface at a smooth point. Recall, if (x_0, y_0, z_0) is a point in a plane and $\langle a, b, c \rangle$ is a vector normal to the plane, then for any point (x, y, z) in the plane we have $\langle a, b, c \rangle \cdot \langle x - x_0, y - y_0, z - z_0 \rangle = 0$, so

$$a(x - x_0) + b(y - y_0) + c(z - z_0) = 0.$$

EXAMPLE 27 Find $\mathbf{T}_u$, $\mathbf{T}_v$, the fundamental vector product $\mathbf{T}_u \times \mathbf{T}_v$, and the equation of the tangent plane to the surface parametrized by

$$\mathbf{r}(u,v) = \langle uv, u-v, \frac{u}{v} \rangle$$

at the point $\mathbf{r}(2,-1)$.

Solution First, we calculate $\mathbf{T}_u = \dfrac{\partial \mathbf{r}}{\partial u} = \langle v, 1, \dfrac{1}{v} \rangle$ and $\mathbf{T}_v = \dfrac{\partial \mathbf{r}}{\partial v} = \langle u, -1, \dfrac{-u}{v^2} \rangle$.

The fundamental vector product is

$$\mathbf{T}_u \times \mathbf{T}_v = \begin{vmatrix} \mathbf{i} & \mathbf{j} & \mathbf{k} \\ v & 1 & 1/v \\ u & -1 & -u/v^2 \end{vmatrix}$$

$$= \langle \frac{-u}{v^2} + \frac{1}{v}, \frac{u}{v} + \frac{u}{v}, -v - u \rangle$$

$$= \langle \frac{v-u}{v^2}, \frac{2u}{v}, -v-u \rangle.$$

At $(u,v) = (2,-1)$, we have

$$\mathbf{T}_u = \langle -1, 1, -1 \rangle \qquad \mathbf{T}_v = \langle 2, -1, -2 \rangle \qquad \mathbf{T}_u \times \mathbf{T}_v = \langle -3, -4, -1 \rangle.$$

The equation of the tangent plane to the surface at $\mathbf{r}(2,-1) = (-2, 3, -2)$ is

$$-3(x+2) - 4(y-3) - (z+2) = 0,$$

or, after expanding and collecting constant terms,

$$-3x - 4y - z + 4 = 0. \qquad \blacksquare$$

If the surface S is the graph of a function $z = f(x,y)$ then we can parametrize the surface as $\mathbf{r}(u,v) = \langle u, v, f(u,v) \rangle$. If f is C^1-differentiable, we have

$$\mathbf{T}_u = \frac{\partial \mathbf{r}}{\partial u} = \langle 1, 0, \frac{\partial f}{\partial u} \rangle$$

$$\mathbf{T}_v = \frac{\partial \mathbf{r}}{\partial v} = \langle 0, 1, \frac{\partial f}{\partial v} \rangle$$

$$\mathbf{T}_u \times \mathbf{T}_v = \frac{\partial \mathbf{r}}{\partial u} \times \frac{\partial \mathbf{r}}{\partial v} = \langle -\frac{\partial f}{\partial u}, -\frac{\partial f}{\partial v}, 1 \rangle.$$

EXAMPLE 28 Find a *unit* normal vector and the equation of the tangent plane to the graph of $z = x^2 y^2$ at the point $(-1, 2, 4)$.

16.4 PARAMETRIZED SURFACES

Solution We can parametrize the surface with

$$\mathbf{r}(u,v) = \langle u, v, u^2v^2 \rangle,$$

so
$$\mathbf{T}_u \times \mathbf{T}_v = \langle -2uv^2, -2u^2v, 1 \rangle.$$

At the point $(-1, 2, 4)$ we have $u = -1$ and $v = 2$, so

$$\mathbf{T}_u \times \mathbf{T}_v \Big|_{(u,v)=(-1,2)} = \langle 8, -4, 1 \rangle,$$

and a unit normal vector is given by

$$\frac{\langle 8, -4, 1 \rangle}{\sqrt{8^2 + (-4)^2 + 1^2}} = \langle \frac{8}{9}, -\frac{4}{9}, \frac{1}{9} \rangle.$$

An equation of the tangent plane is

$$8(x+1) - 4(y-2) + (z-4) = 0.$$

EXAMPLE 29 A surface S is parametrized by

$$\mathbf{r}(u,v) = (u^2 - v^2, u + v, u^2 + 4v)$$

over $D = \{(u,v) : 0 \leq u \leq 1, 0 \leq v \leq 1\}$. Find $\mathbf{T}_u$, $\mathbf{T}_v$, the fundamental vector product $\mathbf{T}_u \times \mathbf{T}_v$, and the equation of the tangent plane to the surface at the point $\mathbf{r}(0, 1/2) = (-1/4, 1/2, 2)$.

Solution We have
$$\mathbf{T}_u = \langle 2u, 1, 2u \rangle, \qquad \mathbf{T}_v = \langle -2v, 1, 4 \rangle,$$
so that

$$\mathbf{T}_u \Big|_{(0,1/2)} = \langle 0, 1, 0 \rangle, \qquad \mathbf{T}_v \Big|_{(0,1/2)} = \langle -1, 1, 4 \rangle.$$

The fundamental vector product at this point is

$$(\mathbf{T}_u \times \mathbf{T}_v) \Big|_{(0,1/2)} = \begin{vmatrix} \mathbf{i} & \mathbf{j} & \mathbf{k} \\ 0 & 1 & 0 \\ -1 & 1 & 4 \end{vmatrix} = 4\mathbf{i} - 0\mathbf{j} + 1\mathbf{k} = \langle 4, 0, 1 \rangle.$$

The equation of tangent plane is

$$4(x + 1/4) + 0(y - 1/2) + 1(z - 2) = 0,$$

or, after collecting terms:

$$4x + z - 1 = 0.$$

EXAMPLE 30 Find the equation of the tangent plane to $z = 3x^2 + 8xy$ at the point $(1, 0, 3)$.

Solution First, we need to parametrize the surface. If we let $x = u$, $y = v$, then $z = 3u^2 + 8uv$, so

$$\mathbf{r}(u, v) = \langle u, v, 3u^2 + 8uv \rangle.$$

We can compute

$$\mathbf{T}_u = \langle 1, 0, 6u + 8v \rangle, \qquad \mathbf{T}_v = \langle 0, 1, 8u \rangle.$$

At $(u, v) = (1, 0)$, we have $\mathbf{r}(1, 0) = \langle 1, 0, 3 \rangle$, and

$$\mathbf{T}_u = \langle 1, 0, 6 \rangle, \qquad \mathbf{T}_v = \langle 0, 1, 8 \rangle.$$

The fundamental vector product is

$$\mathbf{T}_u \times \mathbf{T}_v = \begin{vmatrix} \mathbf{i} & \mathbf{j} & \mathbf{k} \\ 1 & 0 & 6 \\ 0 & 1 & 8 \end{vmatrix} = -6\mathbf{i} - 8\mathbf{j} + 1\mathbf{k}.$$

The equation of the tangent plane is $-6x - 8y + z + 3 = 0$. ∎

EXAMPLE 31 Find the equation of the tangent plane to $x^3 + 3xy + z^2 = 2$ at the point $(1, 1/3, 0)$.

Solution First, we need to parametrize the surface. Let $x = u$ and $z = v$. Then

$$3xy = 2 - z^2 - x^3$$

and

$$y = \frac{2}{3x} - \frac{z^2}{3x} - \frac{x^2}{3} = \frac{2}{3u} - \frac{v^2}{3u} - \frac{u^2}{3}.$$

Now we can write the parametrization as

$$\mathbf{r}(u, v) = \langle u, \frac{2}{3u} - \frac{v^2}{3u} - \frac{u^2}{3}, v \rangle.$$

We can compute

$$\mathbf{T}_u = \left\langle 1, -\frac{2}{3u^2} + \frac{v^2}{3u^2} - \frac{2u}{3}, 0 \right\rangle$$

$$\mathbf{T}_v = \left\langle 0, \frac{-2v}{3u}, 1 \right\rangle.$$

At $(u, v) = (1, 0)$, we have

$$\mathbf{r}(1, 0) = \langle 1, 1/3, 0 \rangle, \quad \mathbf{T}_u = \langle 1, -4/3, 0 \rangle, \quad \mathbf{T}_v = \langle 0, 0, 1 \rangle.$$

16.4 PARAMETRIZED SURFACES

The fundamental vector product is

$$T_u \times T_v = \begin{vmatrix} i & j & k \\ 1 & -4/3 & 0 \\ 0 & 0 & 1 \end{vmatrix}$$

$$= -4/3\,i - 1\,j + 0\,k = \langle -4/3, -1, 0 \rangle.$$

An equation for the tangent plane at $(1, 1/3, 0)$ is

$$-4x/3 - y + 5/3 = 0 \quad \text{or} \quad -4x - 3y + 5 = 0. \qquad \blacksquare$$

EXERCISES for Section 16.4

For exercises 1-6: Parametrize the surfaces described as the image of

$$r : D \longrightarrow \mathbb{R}^3,$$

where D is a subset of $\mathbb{R}^2$ (the uv-plane). Sketch the region D.

1. $z = x^3 + y^2$; $x^2 + y^2 \leq 1$.
2. $x = z^2 y^2$; $0 \leq z \leq 1$; $0 \leq y \leq z$.
3. $y = \sqrt{x+z}$; $0 \leq x \leq z$; $0 \leq z \leq 1$.
4. $x^3 + y^3 + z = 0$; $x^2 + y^2 \leq 4$.
5. $x + 2y^2 - z^2 = 3$; $y \geq 0$; $z \geq 0$; $y^2 + z^2 \leq 1$.
6. $x^2 + y + z^3 = 10$; $x \geq 0$; $z \leq 0$; $x^2 + z^2 \leq 4$.

For exercises 7-13: Suppose that a surface S is parametrized by r over the domain D as indicated. Find the following:

$$T_u, \quad T_v, \quad T_u \times T_v,$$

and the equation of the tangent plane to the surface at the indicated point.

7. $r(u,v) = \langle v \cos u, v \sin u, v \rangle$; $\quad r(\pi, -\pi)$.
8. $r(u,v) = \langle u \cos v, u \sin v, u \rangle$; $\quad r(0, \pi/2)$.
9. $r(u,v) = \langle \cos u, \sin u, v \rangle$; $\quad r(0, 1)$.
10. $r(u,v) = \langle \cos v, \sin v, u \rangle$; $\quad r(1, \pi/4)$.
11. $r(u,v) = \langle u+v, u-v, v-u \rangle$; $\quad r(-1, 2)$.
12. $r(u,v) = \langle u-v, v-u, u+v \rangle$; $\quad r(1, -1)$.
13. $r(u,v) = \langle u^2, v^2, u^2+v^2 \rangle$; $\quad r(1, 1)$.

14. Find a unit normal vector to the surface in exercise 1 at the origin $(0,0,0)$.

15. Find a unit normal vector to the surface in exercise 2 at the point $(1/16, 1/2, 1/2)$.

16. Find a unit normal vector to the surface in exercise 3 at the point $(1/8, 1/2, 1/8)$.

17. Parametrize the surface in exercise 1 using cylindrical coordinates.

18. Parametrize the surface in exercise 4 using cylindrical coordinates.

19. Find the equation of the tangent plane to the surface in exercise 5 at the point $(5/2, 1/2, 0)$.

20. Find the equation of the tangent plane to the surface in exercise 6 at the point $(-1, 8, 1)$.

16.5 SURFACE INTEGRALS

A *surface integral* is to a parametrized surface S what a *line integral* is to a parametrized curve C. In this section we define surface integrals, explain some of their physical interpretations, and illustrate how they are computed.

Finding the surface area of a parametrized surface

In the last chapter, we noted how the double integral

$$\iint_D 1 \, dA$$

simply measures the area of the region D.

Suppose that $\mathbf{r}: D \longrightarrow \mathbb{R}^3$, where

$$\mathbf{r}(u, v) = \langle x(u, v), y(u, v), z(u, v) \rangle$$

is a parametrization of a surface S. How might we go about measuring the *surface area* of S?

If we consider a fine rectangular grid partition of the domain D in the uv-plane, we could "follow" its image to the surface S.

If the vertical grid lines of D are spaced Δu apart and the horizontal lines are spaced Δv apart, then each tiny rectangle of the original grid has area

$$\Delta A = \Delta u \Delta v.$$

Now, we want to examine the image of one of these grid rectangles under the parametrization $\mathbf{r}$ (see Figure 16.19).

16.5 SURFACE INTEGRALS

Close-up: image of a grid rectangle is approximately a parallelogram.

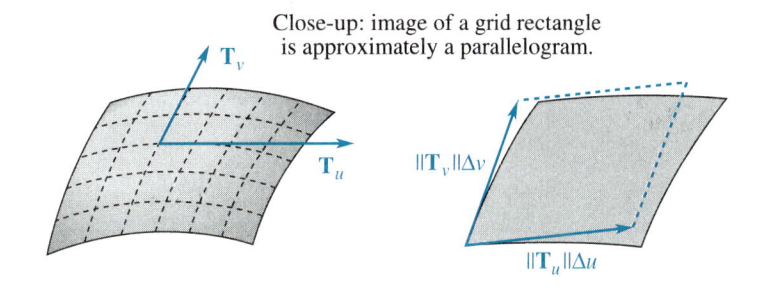

Figure 16.19 Approximating surface area using the image grid.

Provided the parametrization **r** is smooth and the partition grid of D is fine enough, the images of the sides of the rectangle will be approximately straight line segments (but not necessarily at right angles to each other). This suggests that we could approximate the image of the rectangle R with a *parallelogram*. Now, the grid lines correspond to holding either u constant (the vertical lines) or v constant (the horizontal lines). Hence the sides of the parallelogram are approximated by

$$\|\frac{\partial \mathbf{r}}{\partial u}\Delta u\| = \|\mathbf{T}_u\|\Delta u \quad \text{and} \quad \|\frac{\partial \mathbf{r}}{\partial v}\Delta v\| = \|\mathbf{T}_v\|\Delta v.$$

The area of this tiny parallelogram is

$$\Delta S \approx \|\mathbf{T}_u\|\,\|\mathbf{T}_v\|\,|\sin\theta|\,\Delta u \Delta v,$$

where θ is the angle between $\mathbf{T}_u$ and $\mathbf{T}_v$.

But the length of the fundamental vector product is

$$\|\mathbf{T}_u \times \mathbf{T}_v\| = \|\mathbf{T}_u\|\,\|\mathbf{T}_v\|\,|\sin\theta|,$$

so we can write $\qquad \Delta S \approx \|\mathbf{T}_u \times \mathbf{T}_v\|\Delta u \Delta v.$

If we add up these parallelogram surface area approximations we obtain an approximation of the entire surface area of S. The finer the partition, the better the approximation, and we are led to defining the surface area of S as follows:

Definition 9

> If S is a surface with smooth parametrization $\mathbf{r} : D \longrightarrow \mathbb{R}^3$, then the **surface area** of S is
> $$\iint_S dS = \iint_D \|\mathbf{T}_u \times \mathbf{T}_v\|\,du\,dv.$$

EXAMPLE 32 Suppose that the surface S is parametrized by

$$\mathbf{r}(u, v) = \langle u \cos v, u \sin v, u^2 \rangle; \qquad 0 \le u \le 1, \quad 0 \le v \le 2\pi.$$

Find the surface area of S.

Solution First, we need to find the length of the fundamental vector product:

$$\mathbf{T}_u = \langle \cos v, \sin v, 2u \rangle \qquad \mathbf{T}_v = \langle -u \sin v, u \cos v, 0 \rangle$$

and

$$\mathbf{T}_u \times \mathbf{T}_v = \begin{vmatrix} \mathbf{i} & \mathbf{j} & \mathbf{k} \\ \cos v & \sin v & 2u \\ -u \sin v & u \cos v & 0 \end{vmatrix}$$

$$= (-2u^2 \cos v)\mathbf{i} - (2u^2 \sin v)\mathbf{j} + (u \cos^2 v + u \sin^2 v)\mathbf{k}$$

$$= \langle -2u^2 \cos v, -2u^2 \sin v, u \rangle.$$

$$\|\mathbf{T}_u \times \mathbf{T}_v\| = \sqrt{4u^4 \cos^2 v + 4u^4 \sin^2 v + u^2}$$

$$= \sqrt{4u^4 + u^2} = u\sqrt{4u^2 + 1} \qquad (\text{since } u \ge 0).$$

The surface area of S is

$$\iint_D \|\mathbf{T}_u \times \mathbf{T}_v\| \, du \, dv = \int_0^{2\pi} \int_0^1 u\sqrt{4u^2 + 1} \, du \, dv.$$

If we substitute $t = 4u^2 + 1$, so that $dt = 8u \, du$, we have

$$\int u\sqrt{4u^2 + 1} \, du = \int \frac{1}{8}\sqrt{t} \, dt = \frac{1}{12}t^{3/2} = \frac{1}{12}(4u^2 + 1)^{3/2}.$$

Therefore,

$$\int_0^{2\pi} \int_0^1 u\sqrt{4u^2 + 1} \, du \, dv = \int_0^{2\pi} \frac{1}{12}(4u^2 + 1)^{3/2} \bigg]_0^1 dv$$

$$= \frac{5^{3/2} - 1}{12} \int_0^{2\pi} dv = \frac{5\sqrt{5}\pi - \pi}{6}. \quad \blacksquare$$

16.5 SURFACE INTEGRALS

EXAMPLE 33 Verify that the surface area of a sphere of radius R is $4\pi R^2$ by means of a surface integral.

Solution We'll use spherical coordinates to parametrize the surface of the sphere:

$$\mathbf{r} : D \longrightarrow \mathbb{R}^3,$$

where $D = \{(\theta, \varphi) : 0 \leq \theta \leq 2\pi; \ -\frac{\pi}{2} \leq \varphi \leq \frac{\pi}{2}\}$, and

$$\mathbf{r}(\theta, \varphi) = \langle R\cos\theta\sin\varphi, \ R\sin\theta\sin\varphi, \ R\cos\varphi \rangle.$$

We compute $\quad \mathbf{T}_\theta = \langle -R\sin\theta\sin\varphi, \ R\cos\theta\sin\varphi, \ 0 \rangle$

and $\quad \mathbf{T}_\varphi = \langle R\cos\theta\cos\varphi, \ R\sin\theta\cos\varphi, \ -R\sin\varphi \rangle.$

Hence,

$$\mathbf{T}_\theta \times \mathbf{T}_\varphi = \begin{vmatrix} \mathbf{i} & \mathbf{j} & \mathbf{k} \\ -R\sin\theta\sin\varphi & R\cos\theta\sin\varphi & 0 \\ R\cos\theta\cos\varphi & R\sin\theta\cos\varphi & -R\sin\varphi \end{vmatrix}$$

$$= (-R^2\cos\theta\sin^2\varphi)\mathbf{i} - (R^2\sin\theta\sin^2\varphi)\mathbf{j}$$
$$\quad + R^2(-\sin^2\theta\sin\varphi\cos\varphi - \cos^2\theta\sin\varphi\cos\varphi)\mathbf{k}$$
$$= R^2\langle -\cos\theta\sin^2\varphi, \ -\sin\theta\sin^2\varphi, \ -\sin\varphi\cos\varphi \rangle.$$

Therefore,

$$\|\mathbf{T}_\theta \times \mathbf{T}_\varphi\| = R^2\sqrt{\cos^2\theta\sin^4\varphi + \sin^2\theta\sin^4\varphi + \sin^2\varphi\cos^2\varphi}$$

$$= R^2\sqrt{\sin^4\varphi + \sin^2\varphi\cos^2\varphi}$$

$$= R^2\sqrt{\sin^2\varphi(\sin^2\varphi + \cos^2\varphi)}$$

$$= R^2\sqrt{\sin^2\varphi} = R^2|\sin\varphi|.$$

(Note the absolute value signs.) So, the surface area of the sphere is

$$\iint_D \|\mathbf{T}_\theta \times \mathbf{T}_\varphi\| \, d\theta \, d\varphi = \int_0^{2\pi} \int_{-\frac{\pi}{2}}^{\frac{\pi}{2}} R^2|\sin\varphi| \, d\varphi \, d\theta.$$

Since R^2 is a constant, we can factor it outside the integral signs. To handle the absolute value $|\sin\varphi|$, we split the inner integral over two intervals:

$$R^2 \left(\int_0^{2\pi} \int_{-\frac{\pi}{2}}^0 (-\sin\varphi)\, d\varphi\, d\theta + \int_0^{2\pi} \int_0^{\frac{\pi}{2}} (\sin\varphi)\, d\varphi\, d\theta \right)$$

$$= R^2 \left(\int_0^{2\pi} \cos\varphi \Big]_{-\frac{\pi}{2}}^0 d\theta + \int_0^{2\pi} (-\cos\varphi) \Big]_0^{\frac{\pi}{2}} d\theta \right)$$

$$= R^2 \left(\int_0^{2\pi} 1\, d\theta + \int_0^{2\pi} 1\, d\theta \right)$$

$$= R^2 (2\pi + 2\pi) = 4\pi R^2. \qquad \blacksquare$$

Surface integral of a scalar field

The surface integral of a scalar field $f(x,y,z)$ over a surface S is written

$$\iint_S f\, dS = \iint_D f(\mathbf{r}(u,v)) \|\mathbf{T}_u \times \mathbf{T}_v\|\, du\, dv.$$

For a physical interpretation of the surface integral $\iint_S f\, dS$, consider f as measuring the mass per unit of area of a thin sheet of material. If this density function is integrated over the surface, we obtain

$$\text{the total mass of the surface} = \iint_S f\, dS,$$

while

$$\text{the total area of the surface} = \iint_S 1\, dS.$$

EXAMPLE 34 A sheet of material S corresponds to the part of the plane $x = z$ inside the cylinder $x^2 + y^2 = 1$ (units in centimeters). The mass per square centimeter of the material is given by the density function $f(x,y,z) = x^2$ grams per square centimeter. Find the total mass of the sheet.

Solution First, we'll need to parametrize the surface S. If we let $x = u$ and $y = v$, then $z = x = u$ and we can parametrize S as

$$\mathbf{r}: D \longrightarrow \mathbb{R}^3$$

$$\mathbf{r}(u,v) = (u, v, u)$$

$$D = \{(u,v):\quad u^2 + v^2 \leq 1\}.$$

Now we compute $\quad \mathbf{T}_u = \langle 1, 0, 1 \rangle \quad \mathbf{T}_v = \langle 0, 1, 0 \rangle$

16.5 SURFACE INTEGRALS

and $\mathbf{T}_u \times \mathbf{T}_v = \langle -1, 0, 1 \rangle \qquad \|\mathbf{T}_u \times \mathbf{T}_v\| = \sqrt{2}.$

The mass of the sheet is given by

$$\iint_S f \, dS = \sqrt{2} \iint_D u^2 \, du \, dv.$$

Now, D is a unit disk in the uv-plane, so we'll make a change to polar coordinates by

$$u = r\cos\theta, \qquad v = r\sin\theta, \qquad dA = r\, dr\, d\theta.$$

We have

$$\sqrt{2} \iint_D u^2 \, du \, dv = \sqrt{2} \int_0^{2\pi} \int_0^1 r^3 \cos^2\theta \, dr \, d\theta$$

$$= \sqrt{2} \int_0^{2\pi} \frac{r^4}{4} \cos^2\theta \Big]_0^1 d\theta$$

$$= \frac{\sqrt{2}}{4} \int_0^{2\pi} \cos^2\theta \, d\theta$$

$$= \frac{\sqrt{2}}{4} \int_0^{2\pi} \frac{1+\cos 2\theta}{2} d\theta$$

$$= \frac{\sqrt{2}}{8}\theta + \frac{\sqrt{2}\sin 2\theta}{16}\Big]_0^{2\pi}$$

$$= \left(\frac{\sqrt{2}\pi}{4} + 0\right) - (0+0) = \frac{\sqrt{2}\pi}{4} \approx 1.11.$$

Hence, the sheet has a mass of approximately 1.11 grams. ∎

EXAMPLE 35 Find $\iint_S x \, dS$ where S is the part of the plane $x+y+z = 1$ in the first octant ($x \geq 0$, $y \geq 0$, $z \geq 0$).

Solution The surface S is a triangle with vertices $(1,0,0)$, $(0,0,1)$ and $(0,1,0)$. If we solve for z, then

$$z = 1 - x - y$$

and the parametrization we use is:

$$\mathbf{r} : D \longrightarrow \mathbb{R}^3, \qquad \mathbf{r}(u,v) = \langle u, v, 1-u-v \rangle.$$

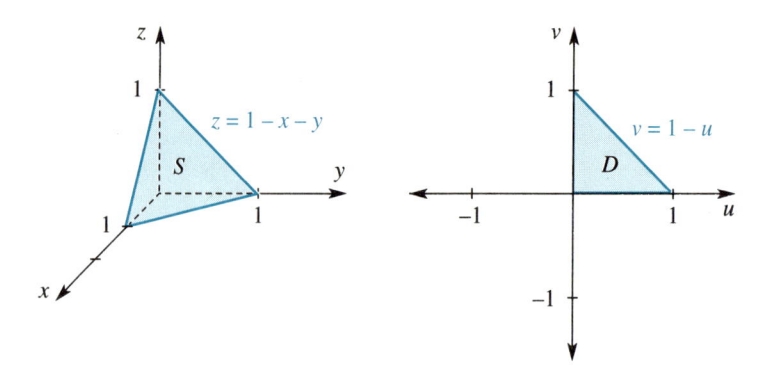

Figure 16.20 Surface S and its domain of parametrization D for $r(u,v) = \langle u, v, 1-u-v \rangle$.

The surface S and the domain D are shown in Figure 16.20.

We compute $\quad \mathbf{T}_u = \langle 1, 0, -1 \rangle \quad \mathbf{T}_v = \langle 0, 1, -1 \rangle$

$$\mathbf{T}_u \times \mathbf{T}_v = \langle 1, 1, 1 \rangle \quad \|\mathbf{T}_u \times \mathbf{T}_v\| = \sqrt{3}.$$

We have
$$\iint_S x \, dS = \iint_D u \|\mathbf{T}_u \times \mathbf{T}_v\| \, du \, dv$$

$$= \int_0^1 \int_0^{1-u} u\sqrt{3} \, dv \, du$$

$$= \int_0^1 uv\sqrt{3} \Big]_0^{1-u} du$$

$$= \sqrt{3} \int_0^1 (u - u^2) \, du$$

$$= \sqrt{3} \left(\frac{u^2}{2} - \frac{u^3}{3} \right) \Big]_0^1$$

$$= \sqrt{3} \left(\frac{1}{2} - \frac{1}{3} \right) = \frac{\sqrt{3}}{6}. \quad \blacksquare$$

Oriented surfaces

The fundamental vector product $\mathbf{T}_u \times \mathbf{T}_v$ provides a normal vector to a surface S parametrized by $\mathbf{r}(u,v)$. If we divide the fundamental vector product by its length, we obtain what is called the **principal unit normal vector**

$$\mathbf{n}(u,v) = \frac{\mathbf{T}_u \times \mathbf{T}_v}{\|\mathbf{T}_u \times \mathbf{T}_v\|}.$$

The direction of $\mathbf{n}(u,v)$ establishes an orientation to the surface, much like the direction of a tangent vector establishes an orientation to a curve.

16.5 SURFACE INTEGRALS

We generally think of a surface as having two sides, like a sheet of paper. Choosing one of the two sides distinguishes a particular orientation to the surface, like designating one side of a piece of paper as the "front" and the other as the "back." The surface shown in Figure 16.21 has an orientation with the principal unit normal pointing outward at each point. A different parametrization of the same surface might *reverse* the orientation, with the principal unit normal pointing inward.

Remarkably, there exist surfaces that are not orientable. The Möbius band is perhaps the most famous example of such a surface—it has only one side! You can make your own Möbius band by taking a rectangular strip of paper, twisting it once, and then gluing the opposite ends together (see Figure 16.22).

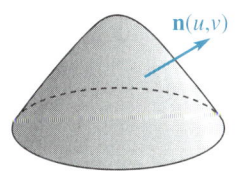

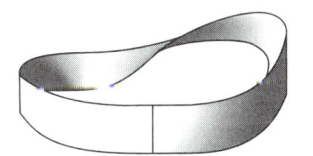

Figure 16.21 The principal unit normal $\mathbf{n}(u,v)$ orients a surface.

Figure 16.22 The Möbius band is a nonorientable surface.

▶▶▶ **Try this: make a Möbius band and draw a continuous line lengthwise along its center. What happens? Now try cutting the Möbius band "in half" along this line.**

Surface integral of a vector field

If we think of a vector field $\mathbf{F}$ as representing the velocity field of a fluid and a surface S as a porous membrane permitting the flow of fluid through it, then one might ask about the total volume flow rate through the surface.

If r parametrizes the surface, then at any smooth point $\mathbf{r}(u,v)$ on the surface we can calculate the velocity vector of the fluid flow

$$\mathbf{F}(\mathbf{r}(u,v))$$

as well as a normal vector to the surface by means of the **fundamental vector product**

$$\mathbf{T}_u \times \mathbf{T}_v.$$

We form the **principal unit normal vector** to the surface at this point as

$$\mathbf{n}(u,v) = \frac{\mathbf{T}_u \times \mathbf{T}_v}{\|\mathbf{T}_u \times \mathbf{T}_v\|}.$$

Hence, the component of fluid flow perpendicular to the surface at $\mathbf{r}(u,v)$ is

$$\mathbf{F} \cdot \mathbf{n} = \mathbf{F}(\mathbf{r}(u,v)) \cdot \mathbf{n}(u,v).$$

This dot product is a scalar field that we can integrate over the surface to obtain the total volume rate of fluid flow through the surface in the direction of the principal unit normal $\mathbf{n}$.

Definition 10

> If $\mathbf{F}$ is a C^1-differentiable vector field and S is a surface with a smooth parametrization $\mathbf{r}: D \longrightarrow \mathbb{R}^3$, then the surface integral of the normal component of $\mathbf{F}$ over S with respect to this parametrization is denoted
>
> $$\iint_S \mathbf{F} \cdot \mathbf{n}\, dS \quad \text{or} \quad \iint_S \mathbf{F} \cdot d\mathbf{S},$$
>
> where either notation indicates the double integral to be computed:
>
> $$\iint_D \mathbf{F}(\mathbf{r}(u,v)) \cdot \mathbf{n}(u,v) \|\mathbf{T}_u \times \mathbf{T}_v\|\, du\, dv = \iint_D \mathbf{F}(\mathbf{r}(u,v)) \cdot (\mathbf{T}_u \times \mathbf{T}_v)\, du\, dv.$$
>
> This surface integral value is called the **flux** of $\mathbf{F}$ over the surface S.

The last equation in the definition follows from noting that

$$\mathbf{n}(u,v)\|\mathbf{T}_u \times \mathbf{T}_v\| = \frac{\mathbf{T}_u \times \mathbf{T}_v}{\|\mathbf{T}_u \times \mathbf{T}_v\|} \|\mathbf{T}_u \times \mathbf{T}_v\| = \mathbf{T}_u \times \mathbf{T}_v.$$

▶▶▶ **The value of the flux of a vector field depends on the orientation of the surface, indicated by the direction of the principal unit normal vector. If q is a smooth parametrization of S reversing the orientation, then**

$$\iint_S \mathbf{F} \cdot \mathbf{n}\, dS$$

changes sign (since the principal unit normal vector now points in the opposite direction).

If you parametrize a surface S for the purpose of calculating the flux of a vector field, but find that your parametrization has the opposite of the desired orientation, then simply remember to reverse the sign at the end of the calculation.

16.5 SURFACE INTEGRALS

EXAMPLE 36 Calculate the *outward* flux of $\mathbf{F}(x,y,z) = \langle x, y, z \rangle$ over the upper unit hemisphere

$$S = \{(x,y,z) : x^2 + y^2 + z^2 = 1; \; z \geq 0\}.$$

Solution We can express S as the graph of $z = \sqrt{1 - x^2 - y^2}$ (positive square root, since this is the upper hemisphere), so one parametrization of the surface, using cylindrical coordinates, is

$$\mathbf{r} : D \longrightarrow \mathbb{R}^3$$

$$\mathbf{r}(r, \theta) = \langle r \cos \theta, \; r \sin \theta, \; \sqrt{1 - r^2} \rangle$$

$$D = \{(r, \theta) : 0 \leq r \leq 1; \; 0 \leq \theta \leq 2\pi\}.$$

We compute

$$\mathbf{T}_r = \langle \cos \theta, \; \sin \theta, \; \frac{-r}{\sqrt{1-r^2}} \rangle \qquad \mathbf{T}_\theta = \langle -r \sin \theta, \; r \cos \theta, \; 0 \rangle$$

and

$$\mathbf{T}_r \times \mathbf{T}_\theta = \langle \frac{r^2 \cos \theta}{\sqrt{1-r^2}}, \; \frac{r^2 \sin \theta}{\sqrt{1-r^2}}, \; r \rangle.$$

Does this normal vector have the desired direction? Since $0 \leq r \leq 1$, we can see that the z-component of $\mathbf{T}_r \times \mathbf{T}_\theta$ is nonnegative. Thus, the normal vector must point up and out of the hemisphere, as desired. Now, we can compute

$$\iint_S \mathbf{F} \cdot d\mathbf{S} = \iint_D \mathbf{F}(\mathbf{r}(r,\theta)) \cdot (\mathbf{T}_r \times \mathbf{T}_\theta) \, dr \, d\theta$$

$$= \int_0^{2\pi} \int_0^1 \langle r \cos \theta, \; r \sin \theta, \; \sqrt{1-r^2} \rangle \cdot \langle \frac{r^2 \cos \theta}{\sqrt{1-r^2}}, \; \frac{r^2 \sin \theta}{\sqrt{1-r^2}}, \; r \rangle \, dr \, d\theta$$

$$= \int_0^{2\pi} \int_0^1 \left(\frac{r^3 \cos^2 \theta}{\sqrt{1-r^2}} + \frac{r^3 \sin^3 \theta}{\sqrt{1-r^2}} + r\sqrt{1-r^2} \right) dr \, d\theta$$

$$= \int_0^{2\pi} \int_0^1 \left(\frac{r^3(\cos^2 \theta + \sin^2 \theta) + r(1-r^2)}{\sqrt{1-r^2}} \right) dr \, d\theta$$

$$= \int_0^{2\pi} \int_0^1 \frac{r}{\sqrt{1-r^2}} \, dr \, d\theta$$

$$= \int_0^{2\pi} \left(-\sqrt{1-r^2} \right) \bigg|_0^1 d\theta = \int_0^{2\pi} [(0) - (-1)] \, d\theta$$

$$= \int_0^{2\pi} 1 \, d\theta = 2\pi.$$

∎

EXAMPLE 37 A fluid is flowing at a constant rate straight upwards, so that its velocity vector field is

$$\mathbf{F}(x, y, z) = \mathbf{k}.$$

Calculate the outward flux of $\mathbf{F}$ over the cone $z = \sqrt{x^2 + y^2}$, for $x^2 + y^2 \leq 1$.

Solution We can parametrize the cone with

$$\mathbf{r} : D \longrightarrow \mathbb{R}^3,$$

where $D = \{(u, v) : u^2 + v^2 \leq 1\}$ and

$$\mathbf{r}(u, v) = \langle u, v, \sqrt{u^2 + v^2} \rangle.$$

We have $\quad \mathbf{T}_u = \langle 1, 0, \dfrac{u}{\sqrt{u^2 + v^2}} \rangle, \quad \mathbf{T}_v = \langle 0, 1, \dfrac{v}{\sqrt{u^2 + v^2}} \rangle,$

so $\quad \mathbf{T}_u \times \mathbf{T}_v = \begin{vmatrix} \mathbf{i} & \mathbf{j} & \mathbf{k} \\ 1 & 0 & \dfrac{u}{\sqrt{u^2 + v^2}} \\ 0 & 1 & \dfrac{v}{\sqrt{u^2 + v^2}} \end{vmatrix} = \dfrac{-u}{\sqrt{u^2 + v^2}} \mathbf{i} + \dfrac{-v}{\sqrt{u^2 + v^2}} \mathbf{j} + 1\mathbf{k}.$

Since the z-component is positive, this normal vector points up and into the cone, rather than outward. Hence, the outward flux is

$$-\iint_S \mathbf{F} \cdot d\mathbf{S} = -\iint_D \mathbf{k} \cdot (\mathbf{T}_u \times \mathbf{T}_v) \, du \, dv$$

$$= -\iint_D \langle 0, 0, 1 \rangle \cdot \langle \dfrac{-u}{\sqrt{u^2 + v^2}}, \dfrac{-v}{\sqrt{u^2 + v^2}}, 1 \rangle \, du \, dv$$

$$= -\iint_D du \, dv = -\pi.$$

(Since D is a unit circle, it has area π.) ∎

EXERCISES for Section 16.5

For exercises 1-6: Suppose a surface S is parametrized by $\mathbf{r}$ over the domain $\mathbf{D}$ as indicated. Find $\|\mathbf{T}_u \times \mathbf{T}_v\|$, the surface area of S, and the value of $\iint_S f \, dS$ where $f(x, y, z)$ is the scalar field indicated.

1. $\mathbf{r}(u, v) = \langle v \cos u, v \sin u, v \rangle; \quad D = [0, \frac{\pi}{2}] \times [0, 1];$
$f(x, y, z) = z.$

2. $\mathbf{r}(u, v) = \langle u \cos v, u \sin v, u \rangle; \quad D = [0, 1] \times [0, \frac{\pi}{2}];$
$f(x, y, z) = z^2.$

16.5 SURFACE INTEGRALS

3. $\mathbf{r}(u, v) = \langle \cos u, \sin u, v \rangle;$ $\quad D = [0, \pi] \times [0, 1];$
   $f(x, y, z) = x + y + z.$

4. $\mathbf{r}(u, v) = \langle \cos v, \sin v, u \rangle;$ $\quad D = [0, 1] \times [0, \pi];$
   $f(x, y, z) = x^2 + y^2 + z^2.$

5. $\mathbf{r}(u, v) = \langle u + v, u - v, v - u \rangle;$ $\quad D = [1, 2] \times [0, 1];$
   $f(x, y, z) = x^2 + y^2.$

6. $\mathbf{r}(u, v) = \langle u - v, v - u, u + v \rangle;$ $\quad D = [0, 1] \times [1, 2];$
   $f(x, y, z) = y^2 + z^2.$

---

7. Imagine a unit circle in the xz-plane whose center is at a distance $R > 1$ from the z axis. If this circle is rotated about the z-axis, it generates a surface S in the shape of a torus or "doughnut." This surface has a parametrization given by

$$\mathbf{r} : D \longrightarrow \mathbb{R}^3,$$

where $D = [0, 2\pi] \times [0, 2\pi]$ and

$$\mathbf{r}(\theta, \varphi) = \langle (R + \cos \varphi) \cos \theta, (R + \cos \varphi) \sin \theta, \sin \varphi \rangle.$$

Show that the surface area of the torus is $4\pi^2 R$.

---

8. Consider the left hemisphere of the unit sphere; in other words,

$$x^2 + y^2 + z^2 = 1, \quad y \leq 0.$$

Parametrize this hemisphere in terms of suitable variables u and v. Be sure to state the domain D of the parametrization and sketch it in the uv-plane.

9. Use your parametrization to verify that the surface area of the hemisphere is 2π.

---

10. Explain, without performing the calculation, why

$$\iint_S (x + y + z) \, dS = 0,$$

If $S = \{(x, y, z) : x^2 + y^2 + z^2 = 1\}$ is the unit sphere.

11. A fluid is flowing at a constant rate given by the constant velocity vector field

$$\mathbf{F}(x, y, z) = -\langle -\frac{\sqrt{2}}{2}, 0, \frac{\sqrt{2}}{2} \rangle.$$

Find the flux of $\mathbf{F}$ over the cone $z = \sqrt{x^2 + y^2}$, for $x^2 + y^2 \leq 1$.

17

Fundamental Theorems of Vector Calculus

In the last chapter, we saw how differential and integral calculus can be used to analyze *vector fields*, and we examined some generalizations of integral calculus to curves and surfaces in space. In this chapter, we see that each of these generalizations, in turn, gives rise to what could be called the *fundamental theorems of vector calculus*.

In the calculus of scalar-valued functions of a single variable, we have two fundamental theorems that state the intimate relationship between derivatives and integrals. Let's recall what those important theorems say.

First fundamental theorem of calculus: If f is continuous on $[a,b]$, then f has an antiderivative given by

$$A(x) = \int_a^x f(t)\,dt, \qquad a \leq x \leq b.$$

In other words, we are guaranteed that $A'(x) = f(x)$.

Second fundamental theorem of calculus: If F is *any* antiderivative of f (in other words, $F'(x) = f(x)$), then

$$\int_a^b f(x)\,dx = F(b) - F(a).$$

This is really quite amazing: the value of the definite integral of f over the entire interval $[a,b]$ depends only on the value of its antiderivative at the two endpoints of the interval!

17.1 CONSERVATIVE FIELDS AND POTENTIALS

In this section, we are going to discuss what could be called the fundamental theorem of calculus for line integrals. First, we need a little terminology.

Conservative fields–independence of path

Definition 1
> A vector field **F** is said to be **conservative** in a region if, for any piece-wise smooth curve C in that region, the value of line integral
> $$\oint_C \mathbf{F} \cdot d\mathbf{s}$$
> depends only on the endpoints of the curve C.

If a vector field **F** is conservative, then for two different piece-wise smooth curves C_1 and C_2, starting at the same point A and ending at the same point B, we have

$$\oint_{C_1} \mathbf{F} \cdot d\mathbf{s} = \oint_{C_2} \mathbf{F} \cdot d\mathbf{s}$$

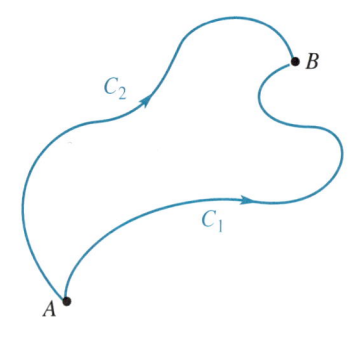

(see Figure 17.1).

Figure 17.1 $\oint_{C_1} \mathbf{F} \cdot d\mathbf{s} = \oint_{C_2} \mathbf{F} \cdot d\mathbf{s}$ for a conservative vector field **F**.

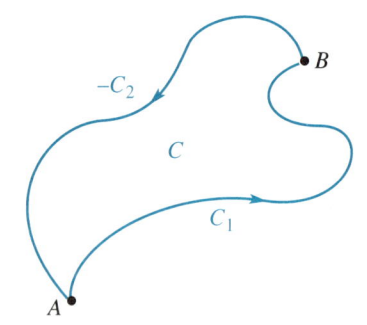

Figure 17.2 C_1 and $-C_2$ form a closed curve.

Let's examine this illustration a bit. Suppose that we reverse the orientation of the parametrization of C_2 (so that the curve runs from B to A). If we call this curve $-C_2$ for convenience, then for a conservative vector field **F**, we must have

$$\oint_{C_1} \mathbf{F} \cdot d\mathbf{s} = -\oint_{-C_2} \mathbf{F} \cdot d\mathbf{s}.$$

The two curves C_1 and $-C_2$ together form a piece-wise smooth closed curve C (see Figure 17.2), and we must have

$$\oint_C \mathbf{F} \cdot d\mathbf{s} = \oint_{C_1} \mathbf{F} \cdot d\mathbf{s} + \oint_{-C_2} \mathbf{F} \cdot d\mathbf{s} = 0.$$

17.1 CONSERVATIVE FIELDS AND POTENTIALS

Conversely, starting with any piece-wise smooth closed curve C, we could always break it into two pieces like this. We can conclude the following:

▶▶▶ **The vector field F is conservative in a region if and only if**

$$\oint_C \mathbf{F} \cdot d\mathbf{s} = 0$$

for every closed curve C in that region.

Let's put these remarks in a physical context. If $\mathbf{F}$ is a *conservative force field*, then the work performed in moving from point A to a point B depends only on the two points and *not* on the particular path we travel. Moreover, no matter what distance we travel along the way, if we arrive at our original starting point (a closed curve), then *zero* work is performed.

EXAMPLE 1 Show that the vector field $\mathbf{F}(x,y) = y\,\mathbf{i} - x\,\mathbf{j}$ is not conservative.

Solution To show that $\mathbf{F}$ is not conservative, we need find only one example of a pair of curves C_1 and C_2 with the same initial point A and terminal point B, but with

$$\oint_{C_1} \mathbf{F} \cdot d\mathbf{s} \neq \oint_{C_2} \mathbf{F} \cdot d\mathbf{s}.$$

Suppose that we connect $A = (1,0)$ and $B = (0,1)$ with the two curves illustrated in Figure 17.3.

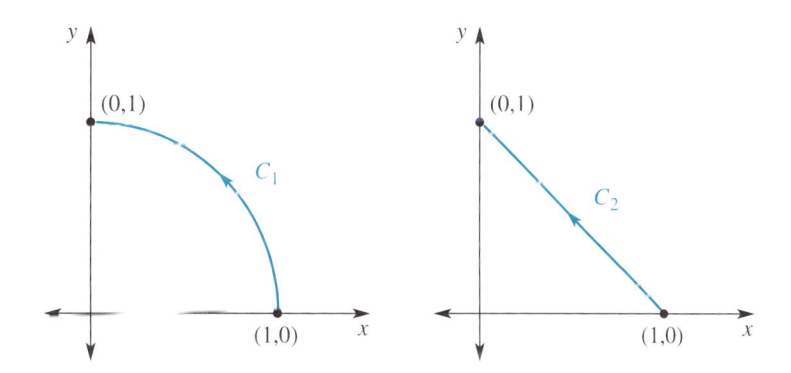

Figure 17.3 Two curves connecting $A = (1,0)$ to $B = (0,1)$.

1. The curve C_1 is parametrized by the path

$$\mathbf{r}_1(t) = \langle \cos t, \sin t \rangle \qquad \text{for } 0 \leq t \leq \pi/2.$$

This path traces out the unit quarter circle in the first quadrant.

2. The curve C_2 is parametrized by the path

$$\mathbf{r}_2(t) = \langle 1 - t, \, t \rangle \qquad \text{for } 0 \le t \le 1.$$

This path traces out the straight line segment connecting $(1, 0)$ to $(0, 1)$.

Now, let's calculate the two line integrals:

1. For C_1, we have $\mathbf{r}_1'(t) = \langle -\sin t, \, \cos t \rangle$, so

$$\oint_{C_1} \mathbf{F} \cdot d\mathbf{s} = \int_0^{\pi/2} \langle \sin t, \, -\cos t \rangle \cdot \langle -\sin t, \, \cos t \rangle \, dt$$

$$= \int_0^{\pi/2} (-\sin^2 t - \cos^2 t) \, dt$$

$$= \int_0^{\pi/2} (-1) \, dt = -t \Big]_0^{\pi/2} = -\pi/2.$$

2. For C_2, we have $\mathbf{r}_2'(t) = \langle -1, \, 1 \rangle$, so

$$\oint_{C_2} \mathbf{F} \cdot d\mathbf{s} = \int_0^1 \langle t, \, t - 1 \rangle \cdot \langle -1, \, 1 \rangle \, dt$$

$$= \int_0^1 (-t + (t - 1)) \, dt$$

$$= \int_0^1 (-1) \, dt = -t \Big]_0^1 = -1.$$

Since $-\pi/2 \ne -1$, we can see that the value of the line integrals differ, although C_1 and C_2 have the same initial and terminal points. We conclude that $\mathbf{F}$ is *not* conservative. ∎

This example illustrates a way one can tell that a vector field $\mathbf{F}$ is not conservative:

▶▶▶ **If we can find a single example of a pair of curves C_1 and C_2 having the same initial and terminal endpoints but**

$$\oint_{C_1} \mathbf{F} \cdot d\mathbf{s} \ne \oint_{C_2} \mathbf{F} \cdot d\mathbf{s},$$

then F is *not* conservative.

However, even if one chooses two such curves and finds that

$$\oint_{C_1} \mathbf{F} \cdot d\mathbf{s} = \oint_{C_2} \mathbf{F} \cdot d\mathbf{s},$$

17.1 CONSERVATIVE FIELDS AND POTENTIALS

we cannot conclude that **F** is conservative. The two values may be equal by coincidence, and we have no way of checking all of the infinitely many different pairs of curves connecting the same two points. There are simpler ways of detecting that a field **F** is conservative, as we will see.

Potentials

If we start with a differentiable scalar field f, we can obtain its gradient vector field ∇f. Now, let's reverse the situation: suppose that we start with a vector field **F**. Can we find a scalar field having **F** as its gradient? If so, we call this scalar field a **potential** for **F**.

It's traditional to use the Greek letter phi (φ) to denote a potential scalar field. Using this notation, we have the following definition.

Definition 2

> A continuous vector field **F** has a **potential** φ if and only if
> $$\nabla \varphi = \mathbf{F}.$$

If **F** is a 2-dimensional vector field with

$$\mathbf{F}(x, y) = \langle P(x, y), Q(x, y) \rangle,$$

then $\varphi(x, y)$ is a potential for **F** if and only if

$$\frac{\partial \varphi}{\partial x} = P(x, y) \quad \text{and} \quad \frac{\partial \varphi}{\partial y} = Q(x, y).$$

Similarly, if **F** is a 3-dimensional vector field,

$$\mathbf{F}(x, y, z) = P(x, y, z)\mathbf{i} + Q(x, y, z)\mathbf{j} + R(x, y, z)\mathbf{k},$$

then $\varphi(x, y, z)$ is a potential for **F** if and only if

$$\frac{\partial \varphi}{\partial x} = P(x, y, z), \quad \frac{\partial \varphi}{\partial y} = Q(x, y, z), \quad \frac{\partial \varphi}{\partial z} = R(x, y, z).$$

The potential φ is a kind of "antiderivative" for the vector field **F**. Just as a continuous function of one variable has infinitely many antiderivatives (remember that an arbitrary constant can be added), if a vector field **F** has one potential φ, then it has infinitely many (again by simply adding an arbitrary constant).

EXAMPLE 2 Determine a potential, if one exists, for the vector field

$$\mathbf{F}(x, y, z) = \langle 3x^2 y, x^3 + y^3, 2z \rangle.$$

Solution   We need a function $\varphi(x,y,z)$ such that

$$\nabla\varphi = \langle 3x^2y, x^3+y^3, 2z\rangle,$$

so we must have

$$\frac{\partial\varphi}{\partial x} = 3x^2y, \qquad \frac{\partial\varphi}{\partial y} = x^3+y^3, \qquad \frac{\partial\varphi}{\partial z} = 2z.$$

The first requirement $\dfrac{\partial\varphi}{\partial x} = 3x^2y$ forces the function φ to be of the form

$$\varphi(x,y,z) = x^3y + (\text{ expression in terms of } y \text{ and } z).$$

The second requirement $\dfrac{\partial\varphi}{\partial y} = x^3+y^3$ forces the function φ to be of the form

$$\varphi(x,y,z) = x^3y + \frac{y^4}{4} + (\text{ expression in terms of } x \text{ and } z).$$

The third requirement $\dfrac{\partial\varphi}{\partial z} = 2z$ forces $\varphi(x,y,z)$ to be of the form

$$\varphi(x,y,z) = z^2 + (\text{ expression in terms of } x \text{ and } y).$$

We note that any scalar field of the form

$$\varphi(x,y,z) = x^3y + \frac{y^4}{4} + z^2 + C \qquad (C \text{ a constant})$$

satisfies all three requirements, and we can check that

$$\frac{\partial\varphi}{\partial x} = 3x^2y, \qquad \frac{\partial\varphi}{\partial y} = x^3+y^3, \qquad \frac{\partial\varphi}{\partial z} = 2z. \qquad\blacksquare$$

Potentials and line integrals of conservative vector fields enjoy a relationship very similar to that between antiderivatives and definite integrals of continuous single-variable functions. If **F** has potential function in a region, then **F** will be conservative. Here's why: Suppose that $\mathbf{F} = \nabla\varphi$ and C is a smooth curve connecting points A and B. If **r** is a parametrization of the curve with $\mathbf{r}(a) = A$ and $\mathbf{r}(b) = B$, then

$$\oint_C \mathbf{F}\cdot d\mathbf{s} = \int_a^b \mathbf{F}(\mathbf{r}(t))\cdot\mathbf{r}'(t)\,dt$$

$$= \int_a^b \nabla\varphi(\mathbf{r}(t))\cdot\mathbf{r}'(t)\,dt.$$

Examine the integrand closely. This is simply the derivative of

$$\varphi(\mathbf{r}(t))$$

17.1 CONSERVATIVE FIELDS AND POTENTIALS

since

$$\frac{d}{dt}\Big[\varphi(\mathbf{r}(t))\Big] = \nabla\varphi(\mathbf{r}(t)) \cdot \mathbf{r}'(t).$$

Since $\varphi(\mathbf{r}(t))$ is an antiderivative of $\nabla\varphi(\mathbf{r}(t)) \cdot \mathbf{r}'(t)$, we have

$$\int_a^b \nabla\varphi(\mathbf{r}(t)) \cdot \mathbf{r}'(t) = \varphi(\mathbf{r}(b)) - \varphi(\mathbf{r}(a)) = \varphi(B) - \varphi(A).$$

So the integral's value depends only on the two endpoints $\mathbf{r}(a) = A$ and $\mathbf{r}(b) = B$ and not on any of the other points $\mathbf{r}(t)$ $(a < t < b)$.

EXAMPLE 3 Find $\oint_C \mathbf{F} \cdot \mathbf{T}\, ds$, where

$$\mathbf{F}(x, y, z) = \langle 3x^2 y, x^3 + y^3, 2z \rangle$$

and C is parametrized by

$$\mathbf{r}(t) = \langle \cos t, \sin t, \tan t \rangle \qquad \text{for } 0 \leq t \leq \pi/4.$$

Solution In the last example, we saw that $\mathbf{F}$ has a potential function

$$\varphi(x, y, z) = x^3 y + \frac{y^4}{4} + z^2.$$

The initial point and terminal points of the curve C are:

$$\mathbf{r}(0) = \langle 1, 0, 0 \rangle \qquad \text{and} \qquad \mathbf{r}(\pi/4) = \langle \sqrt{2}/2, \sqrt{2}/2, 1 \rangle.$$

Therefore, $\oint_C \mathbf{F} \cdot \mathbf{T}\, ds = \varphi(\sqrt{2}/2, \sqrt{2}/2, 1) - \varphi(1, 0, 0)$

$$= (\frac{1}{4} + \frac{1}{16} + 1) - (0 + 0 + 0) = \frac{21}{16}.$$

Knowing that $\mathbf{F}$ had a potential φ greatly simplified our task. ∎

The fundamental theorem for line integrals

In fact, the *only* conservative vector fields are those that have potentials:

Theorem 17.1   A vector field $\mathbf{F}$ is conservative if and only if $\mathbf{F} = \nabla\varphi$ for a scalar field φ.

Paraphrased, the theorem states that the line integral of $\mathbf{F}$ will depend only on the endpoints of the path if and only if $\mathbf{F}$ is the *gradient* of a scalar field φ.

When can we tell that a vector field is conservative? The fundamental theorem for line integrals tells us that the following statements are all equivalent:

$$\oint_C \mathbf{F} \cdot d\mathbf{s} \text{ depends only on the endpoints of } C.$$

$$\oint_C \mathbf{F} \cdot d\mathbf{s} = 0 \text{ for every closed curve } C.$$

$$\mathbf{F} = \nabla \varphi \text{ for some scalar potential } \varphi.$$

To determine that $\mathbf{F}$ is conservative, we need to demonstrate a potential φ. Here are some clues that can be used to tell whether or not to bother looking.

If $\mathbf{F}$ is a C^1-differentiable vector field

$$\mathbf{F}(x,y) = \langle P(x,y), Q(x,y) \rangle,$$

then P and Q have continuous partial derivatives. Now, if $\mathbf{F}$ has a potential φ, then

$$\frac{d\varphi}{dx} = P(x,y) \quad \frac{d\varphi}{dy} = Q(x,y)$$

and φ is C^2-differentiable. Let's look at the mixed second-order partials of φ:

$$\frac{\partial^2 \varphi}{\partial y\, \partial x} = \frac{\partial P}{\partial y} \quad \frac{\partial^2 \varphi}{\partial x\, \partial y} = \frac{\partial Q}{\partial x}.$$

These must match.

▶▶▶ **For a 2-dimensional vector field F to have a potential, we must have**

$$\frac{\partial P}{\partial y} = \frac{\partial Q}{\partial x}$$

or, equivalently,

$$\frac{\partial Q}{\partial x} - \frac{\partial P}{\partial y} = 0.$$

This won't guarantee that F has a potential, but if $\frac{\partial Q}{\partial x} - \frac{\partial P}{\partial y} \neq 0$**, then F is definitely *not* conservative.**

We can use the same type of reasoning for a 3-dimensional vector field $\mathbf{F}$. If $\mathbf{F}(x,y,z) = P(x,y,z)\mathbf{i} + Q(x,y,z)\mathbf{j} + R(x,y,z)\mathbf{k}$ is C^1-differentiable, then

17.1 CONSERVATIVE FIELDS AND POTENTIALS

we'll need

$$\frac{\partial P}{\partial y} = \frac{\partial Q}{\partial x} \quad \frac{\partial P}{\partial z} = \frac{\partial R}{\partial x} \quad \frac{\partial Q}{\partial z} = \frac{\partial R}{\partial y}$$

in order for all the second-order partials of φ to match. But this also means:

▶▶▶ **For a 3-dimensional vector field to have a potential, we must have:**

$$\operatorname{curl} \mathbf{F} = \left(\frac{\partial R}{\partial y} - \frac{\partial Q}{\partial z}\right)\mathbf{i} + \left(\frac{\partial P}{\partial z} - \frac{\partial R}{\partial x}\right)\mathbf{j} + \left(\frac{\partial Q}{\partial x} - \frac{\partial P}{\partial y}\right)\mathbf{k} = 0.$$

Again, this doesn't guarantee that F is conservative, but if curl F ≠ 0, then F is definitely *not* conservative.

EXAMPLE 4 Show that the vector field $\mathbf{F}(x,y) = y\mathbf{i} - x\mathbf{j}$ is not conservative.

Solution We have $P(x,y) = y$ and $Q(x,y) = -x$. We simply note that

$$\frac{\partial Q}{\partial x} = -1 \neq 1 = \frac{\partial P}{\partial y},$$

so **F** cannot have a potential function and therefore is not conservative. We noted this in an earlier example by directly showing that **F** was not path-independent for the purposes of line integrals. ■

EXAMPLE 5 Show that the vector field defined by $\mathbf{F}(x,y,z) = \langle 3x^2y, x^3+y^3, 2z\rangle$ is conservative.

Solution We note that for $\mathbf{F}(x,y,z) = \langle 3x^2y, x^3+y^3, 2z\rangle$, we have

$$\operatorname{curl} \mathbf{F} = \nabla \times \mathbf{F}$$

$$= \begin{vmatrix} \mathbf{i} & \mathbf{j} & \mathbf{k} \\ \partial/\partial x & \partial/\partial y & \partial/\partial z \\ 3x^2y & x^3+y^3 & 2z \end{vmatrix}$$

$$= (0-0)\mathbf{i} - (0-0)\mathbf{j} + (3x^2 - 3x^2)\mathbf{k}$$

$$= \langle 0, 0, 0\rangle.$$

This does not, by itself, prove that **F** is conservative. However, recall that earlier (Example 2) we successfully found a potential function for **F**:

$$\varphi(x,y,z) = x^3y + \frac{y^4}{4} + z^2.$$

■

Relationship between curl and line integrals

A closed oriented curve lying in a plane determines a particular normal direction by use of a "right-hand rule." If you place your right hand so that your fingers point in the direction of the curve, with your palm facing the interior of the plane region enclosed by the curve, then your thumb will extend in the normal direction determined by the curve.

Figure 17.4 illustrates this right-hand rule for three curves in planes parallel with the coordinate planes. For each of the curves C_1, C_2, C_3, we have indicated the unit normal vector n pointing in the direction determined by the right-hand rule.

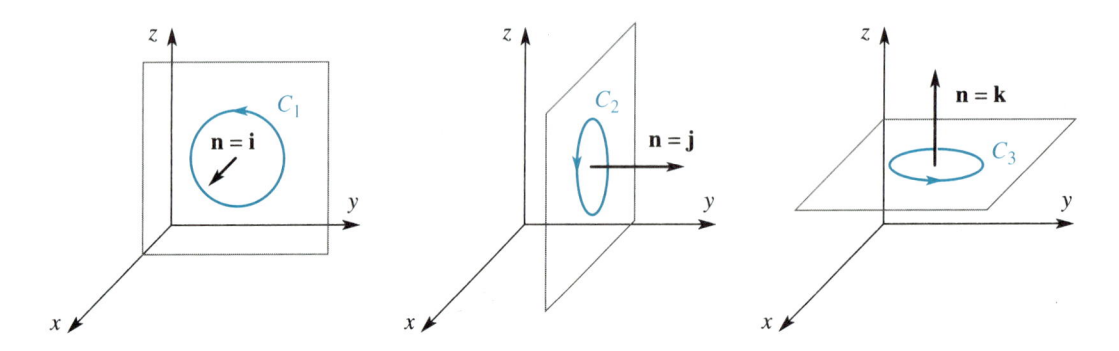

Figure 17.4 Right-hand rule for closed plane curves.

Given a vector field

$$\mathbf{F}(x, y, z) = P(x, y, z)\mathbf{i} + Q(x, y, z)\mathbf{j} + R(x, y, z)\mathbf{k},$$

we can express curl F in terms of line integrals along closed curves. Recall that if F represents a velocity vector field, we can think of

$$\oint_C \mathbf{F} \cdot \mathbf{T}\, ds$$

as giving us the circulation along the curve C. Now, if C is a very tiny closed plane curve around a point (x_0, y_0, z_0) and we measure the area ΔS enclosed by the curve, then the component of curl $\mathbf{F}(x_0, y_0, z_0)$ in the direction determined by the curve is approximately the circulation divided by the area enclosed. In other words, if n is a unit normal vector determined by the curve C, then

$$\text{curl } \mathbf{F}(x_0, y_0, z_0) \cdot \mathbf{n}\Delta S \approx \oint_C \mathbf{F} \cdot \mathbf{T}\, ds.$$

The approximation becomes better and better as ΔS approaches 0. In fact,

$$\text{curl } \mathbf{F}(x_0, y_0, z_0) \cdot \mathbf{n} = \lim_{\Delta S \to 0} \frac{1}{\Delta S} \oint_C \mathbf{F} \cdot \mathbf{T}\, ds.$$

17.1 CONSERVATIVE FIELDS AND POTENTIALS

Suppose that (x_0, y_0, z_0) is a point enclosed by the curves C_1, C_2, C_3 in the planes $x = x_0$, $y = y_0$, and $z = z_0$, respectively. If the curves are oriented as shown in Figure 17.4, the unit normal vectors are **i**, **j**, and **k**. The discussion above implies that

$$\frac{1}{\Delta S} \oint_{C_1} \mathbf{F} \cdot \mathbf{T} \, ds \approx \frac{\partial R}{\partial y} - \frac{\partial Q}{\partial z},$$

$$\frac{1}{\Delta S} \oint_{C_2} \mathbf{F} \cdot \mathbf{T} \, ds \approx \frac{\partial P}{\partial z} - \frac{\partial R}{\partial x},$$

$$\frac{1}{\Delta S} \oint_{C_3} \mathbf{F} \cdot \mathbf{T} \, ds \approx \frac{\partial Q}{\partial x} - \frac{\partial P}{\partial y}.$$

Indeed, the components of curl **F** are sometimes defined as the limits of these approximations as ΔS approaches 0.

EXERCISES for Section 17.1

For exercises 1-8: Consider the vector field

$$\mathbf{F}(x, y, z) = y e^{xy} \mathbf{i} + \left(x e^{xy} + \frac{1}{y}\right) \mathbf{j}.$$

Which of the statements are true, and which are false? Explain your answer in each case.

1. **F** is a conservative vector field.
2. div **F** = 0.
3. curl **F** = 0.
4. $\oint_C y e^{xy} \, dx + (x e^{xy} + \frac{1}{y}) \, dy = 0$ for any simple, closed, smooth curve C.
5. **F** = grad f for some function f.
6. **F** is irrotational at every point.
7. div (curl **F**) = 0.
8. curl (div **F**) = 0.

For exercises 9-13: Suppose that $\mathbf{F}(x, y, z) = \langle yz, xz, xy \rangle$. Which of the statements are true and which are false? Explain your reasons.

9. $\mathbf{F} = \nabla \varphi$ for some scalar field φ.
10. $\nabla \times \mathbf{F} = 0$.
11. **F** is conservative.
12. **F** is irrotational at every point (x, y, z).

13. $\oint_C \mathbf{F} \cdot \mathbf{T}\, ds = 0$ for every simple, smooth, closed curve C.

For exercises 14-17: Determine whether or not the following vector fields are conservative (explain your reasoning). If so, find a potential function for the vector field.

14. $\mathbf{F}(x, y) = (e^x \sin y + 8x)\mathbf{i} + (e^x \cos y + 1)\mathbf{j}$.

15. $\mathbf{G}(x, y) = (\ln y \cos x)\mathbf{i} + (y \sin x)\mathbf{j}$.

16. $\mathbf{H}(x, y) = \langle e^x - \cos y, e^y - x \sin y \rangle$.

17. $\mathbf{L}(x, y, z) = \frac{y^4}{4x}\mathbf{i} + y^3 \ln x\, \mathbf{j} + \cos z\, \mathbf{k}$.

18. Find the work done in moving a particle directly from $(-2, 0)$ to $(2, 0)$ along the x-axis by the force $\mathbf{F}(x, y) = \langle x^2, -y \rangle$.

19. Find the work done in moving a particle along the semicircle
$$x^2 + y^2 = 4, \quad y \geq 0,$$
from $(-2, 0)$ to $(2, 0)$, by the same force $\mathbf{F}(x, y) = \langle x^2, -y \rangle$.

20. Consider
$$\mathbf{F}(x, y) = P(x, y)\mathbf{i} + Q(x, y)\mathbf{j} = \frac{-y}{x^2 + y^2}\mathbf{i} + \frac{x}{x^2 + y^2}\mathbf{j}.$$

Show that
$$\frac{\partial Q}{\partial x} = \frac{\partial P}{\partial y},$$

but that the integral
$$\oint_C P(x, y)\, dx + Q(x, y)\, dy \neq 0$$

over the closed curve C, where C is the unit circle parametrized by $\mathbf{r}(t) = \langle \cos t, \sin t \rangle$ for $0 \leq t \leq 2\pi$. Is $\mathbf{F}$ conservative?

21. Show that a gravitational force field described by
$$\mathbf{F}(x, y, z) = \frac{-GMm}{(x^2 + y^2 + z^2)^{3/2}} \langle x, y, z \rangle$$

has a potential
$$\varphi(x, y, z) = \frac{GMm}{\sqrt{x^2 + y^2 + z^2}}.$$

(Note: This is called the *Newtonian potential*.)

17.2 STOKES' THEOREM

In the last section, we saw a fundamental theorem of calculus for line integrals of *conservative* vector fields **F**:

$$\text{If } \nabla \varphi(x) = \mathbf{F}(x), \text{ then } \int_C \mathbf{F} \cdot \mathbf{T}\, ds = \varphi(B) - \varphi(A),$$

where A and B are the two endpoints of the curve C. Remember that we called φ a *potential* for **F**.

In the remaining sections of the chapter we'll see some other of the fundamental theorems of vector calculus applied to vector fields in $\mathbb{R}^2$ and $\mathbb{R}^3$. To set the stage for these theorems, we need to discuss some terminology regarding curves and surfaces.

A parametrized surface S is said to be **simple** if its parametrization **r** is *one-to-one*. Like a simple curve, a simple surface never "crosses" itself, in the sense that no point on the surface can be the image of two different points in the domain D. In terms of its parametrization **r**, if $(u_1, v_1) \neq (u_2, v_2)$ are two points in the domain D, then we must have $\mathbf{r}(u_1, v_1) \neq \mathbf{r}(u_2, v_2)$.

A **closed** surface is a surface that has no boundary. For example, a sphere is a closed surface. Note that for an *oriented* closed surface, it makes sense to refer to the principal unit normal vector as being either *inward* or *outward*, depending on the specific parametrization.

Suppose that a simple surface S has a parametrization $\mathbf{r}(u, v)$ with domain D that is enclosed by a simple closed curve C^* in the uv-plane (as shown in Figure 17.5).

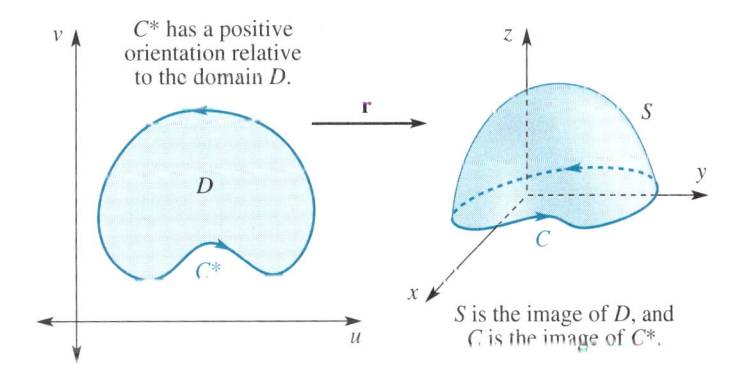

S is the image of *D*, and *C* is the image of *C**.

Figure 17.5 Domain D enclosed by a simple closed curve C^*.

If the closed curve C^* is oriented counterclockwise (so D is always to the left of the curve as we travel around it), then we say that C^* is **positively oriented** with respect to D. The image of the curve C^* forms the boundary C of the surface S. If C is given the orientation "inherited" from C^*, then

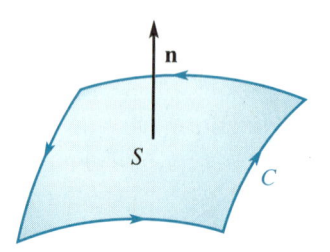

Figure 17.6 Surface S with positively oriented boundary C.

we say that C is **positively oriented** with respect to the orientation of S. Figure 17.6 illustrates a surface S with its principal unit normal vector n and a positively oriented boundary C.

What is Stokes' theorem?

Stokes' theorem relates a surface integral involving the curl of a vector field over a surface S, with the line integral of its tangential component around the boundary of S. It is named in honor of George G. Stokes (1819–1903), an Irish mathematician. In terms of the notation for surface and line integrals, the statement of Stokes' theorem can be stated precisely as follows:

Theorem 17.2

Stokes' theorem
Hypothesis 1: $\mathbf{F}$ is a C^1-differentiable vector field in some region of $\mathbb{R}^3$ containing a simple smooth surface S.
Hypothesis 2: C is the piece-wise smooth boundary of S with a positive orientation.
Conclusion: We have
$$\iint_S (\nabla \times \mathbf{F}) \cdot \mathbf{n} \, dS = \oint_C \mathbf{F} \cdot \mathbf{T} \, ds.$$

Reasoning Let's try to understand why Stokes' theorem makes sense. First, let's consider the line integral

$$\oint \mathbf{F} \cdot \mathbf{T} \, ds$$

around the positively oriented boundary C of S, as shown in Figure 17.7.

17.2 STOKES' THEOREM

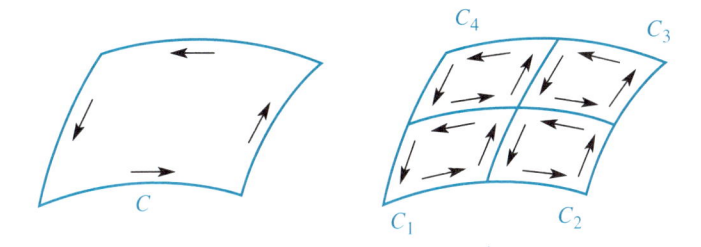

Figure 17.7 Subdividing S into 4 subregions.

Now, suppose that we subdivide S into 4 subregions as shown in Figure 17.7, with boundaries C_1, C_2, C_3, and C_4, each positively oriented.

If we traverse each of the four resulting curves C_1, C_2, C_3, and C_4, note that we end up traversing each of the interior curve segments twice, once in each direction. Those segments along the original boundary, however, get traversed only once, and always in the original direction of the boundary C. This means that if we sum up the line integrals of $\mathbf{F} \cdot \mathbf{T}$ along the four curves, the parts along the interior curve segments will cancel each other out and we'll find

$$\oint_C \mathbf{F} \cdot \mathbf{T}\, ds = \oint_{C_1} \mathbf{F} \cdot \mathbf{T}\, ds + \oint_{C_2} \mathbf{F} \cdot \mathbf{T}\, ds + \oint_{C_3} \mathbf{F} \cdot \mathbf{T}\, ds + \oint_{C_4} \mathbf{F} \cdot \mathbf{T}\, ds.$$

The same is true if we subdivide S into many small subregions as shown in Figure 17.8. All of the interior curve segments are still traversed twice, once in each direction, but those along the original boundary are traversed only once.

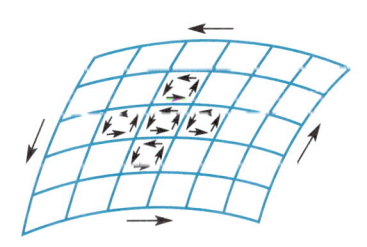

Figure 17.8 All the interior curve segments are traversed in both directions.

If we sum up all the line integrals over all these tiny closed curves C_i, the result is equal to our original line integral around the surface boundary C:

$$\oint_C \mathbf{F} \cdot \mathbf{T}\, ds = \sum_{i=1}^{N} \oint_{C_i} \mathbf{F} \cdot \mathbf{T}\, ds.$$

We noted earlier (in Section 16.4) the relationship between the curl of a vector field at a point and the line integral about a tiny closed planar curve C surrounding that point. Precisely, we have

$$\operatorname{curl} \mathbf{F} \cdot \mathbf{n} = \lim_{\Delta S \to 0} \frac{1}{\Delta S} \oint_C \mathbf{F} \cdot \mathbf{T}\, ds,$$

where C is a curve lying in a plane surrounding the point, ΔS is the area enclosed by curve C, and $\mathbf{n}$ is the unit normal to the plane determined by C.

Examine Figure 17.8 again. As we subdivide the surface S finer and finer, the individual closed curves become approximately planar. If ΔS is the approximate surface area enclosed by each C_i, and we choose a point on the surface within each grid curve C_i, then each of the line integrals $\oint_{C_i} \mathbf{F} \cdot \mathbf{T}\, ds$ is approximated by

$$\oint_{C_i} \mathbf{F} \cdot \mathbf{T}\, ds \approx \operatorname{curl} \mathbf{F} \cdot \mathbf{n}\, \Delta S.$$

Summing these, we obtain

$$\oint_C \mathbf{F} \cdot \mathbf{T}\, ds \approx \sum_{i=1}^N \operatorname{curl} \mathbf{F} \cdot \mathbf{n}\, \Delta S.$$

The hypotheses of the theorem guarantee that the approximation becomes better and better as $N \to \infty$. Hence

$$\oint_C \mathbf{F} \cdot \mathbf{T}\, ds = \lim_{N \to \infty} \sum_{i=1}^N \operatorname{curl} \mathbf{F} \cdot \mathbf{n}\, \Delta S = \iint_S (\nabla \times \mathbf{F}) \cdot \mathbf{n}\, dS,$$

and this is exactly the conclusion of Stokes' theorem. □

Let's illustrate Stokes' theorem with an example.

EXAMPLE 6 Suppose that $\mathbf{F}$ is the vector field defined by

$$\mathbf{F}(x, y, z) = (x^2 + y - 4)\mathbf{i} + 3xy\mathbf{j} + (2xz + z^2)\mathbf{k}$$

and that S is the hemisphere of radius 4 defined as the set of points

$$S = \{(x, y, z) : x^2 + y^2 + z^2 = 16, z \geq 0\}.$$

Verify that Stokes' theorem holds:

$$\iint_S \nabla \times \mathbf{F} \cdot \mathbf{n}\, dS = \int_C \mathbf{F} \cdot \mathbf{T}\, ds,$$

17.2 STOKES' THEOREM

where C is the positively oriented boundary of S.

Solution To verify Stokes' theorem, we essentially need to calculate both sides of the equation

$$\iint_S (\nabla \times \mathbf{F}) \cdot \mathbf{n}\, dS = \oint_C \mathbf{F} \cdot \mathbf{T}\, ds$$

and check to see that they are equal.

First we'll calculate the left-hand side of the equation. We have

$$\nabla \times \mathbf{F}(x,y,z) = \begin{vmatrix} \mathbf{i} & \mathbf{j} & \mathbf{k} \\ \dfrac{\partial}{\partial x} & \dfrac{\partial}{\partial y} & \dfrac{\partial}{\partial z} \\ x^2+y-4 & 3xy & 2xz+z^2 \end{vmatrix}$$

$$= 0\mathbf{i} - (2z - 0)\mathbf{j} + (3y - 1)\mathbf{k}$$

$$= \langle 0,\ -2z,\ 3y - 1\rangle.$$

The hemisphere is the graph of $z = \sqrt{16 - x^2 - y^2}$ (positive square root, since $z \geq 0$) for $x^2 + y^2 \leq 16$, so we can use the parametrization

$$\mathbf{r}: D \longrightarrow \mathbb{R}^3$$

$$\mathbf{r}(u,v) = \langle u,\ v,\ \sqrt{16 - u^2 - v^2}\rangle,$$

where $D = \{(u,v) : u^2 + v^2 \leq 16\}$. Figure 17.9 illustrates the surface S, with its boundary C, and the domain D of the parametrization.

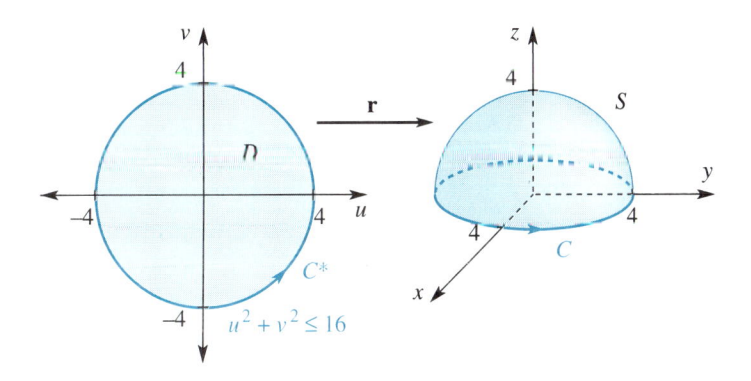

Figure 17.9 Parametrizing a sphere of radius 4.

We compute

$$\mathbf{T}_u = \langle 1, 0, \frac{-u}{\sqrt{16-u^2-v^2}} \rangle, \qquad \mathbf{T}_v = \langle 0, 1, \frac{-v}{\sqrt{16-u^2-v^2}} \rangle,$$

$$\mathbf{T}_u \times \mathbf{T}_v = \begin{vmatrix} \mathbf{i} & \mathbf{j} & \mathbf{k} \\ 1 & 0 & \frac{-u}{\sqrt{16-u^2-v^2}} \\ 0 & 1 & \frac{-v}{\sqrt{16-u^2-v^2}} \end{vmatrix}$$

$$= \langle \frac{u}{\sqrt{16-u^2-v^2}}, \frac{v}{\sqrt{16-u^2-v^2}}, 1 \rangle.$$

This means that we have

$$\nabla \times \mathbf{F}(\mathbf{r}(u,v)) \cdot (\mathbf{T}_u \times \mathbf{T}_v)$$

$$= \langle 0, -2\sqrt{16-u^2-v^2}, 3v-1 \rangle \cdot \langle \frac{u}{\sqrt{16-u^2-v^2}}, \frac{v}{\sqrt{16-u^2-v^2}}, 1 \rangle$$

$$= 0 - 2v + (3v-1) = v - 1.$$

Hence, the surface integral is computed as follows:

$$\iint_S (\nabla \times \mathbf{F}) \cdot \mathbf{n} \, dS = \int_{-4}^{4} \int_{-\sqrt{16-v^2}}^{\sqrt{16-v^2}} (\nabla \times \mathbf{F})(\mathbf{r}(u,v)) \cdot (\mathbf{T}_u \times \mathbf{T}_v) \, du \, dv$$

$$= \int_{-4}^{4} \int_{-\sqrt{16-v^2}}^{\sqrt{16-v^2}} (v-1) \, du \, dv$$

$$= \int_{-4}^{4} (uv - u) \Big]_{-\sqrt{16-v^2}}^{\sqrt{16-v^2}} dv$$

$$= \int_{-4}^{4} (2v\sqrt{16-v^2} - 2\sqrt{16-v^2}) \, dv$$

$$= \frac{-2(16-v^2)^{\frac{3}{2}}}{3} \Big]_{-4}^{4} - 2 \int_{-4}^{4} \sqrt{16-v^2} \, dv$$

$$= 0 - (\text{area of circle of radius } 4)$$

$$= -16\pi.$$

Now, let's calculate the right-hand side of the equation. We can parametrize C with

$$\mathbf{r} : [0, 2\pi] \longrightarrow \mathbb{R}^3 \qquad \mathbf{r}(t) = \langle 4\cos t, 4\sin t, 0 \rangle.$$

The velocity vector is $\qquad \mathbf{r}'(t) = \langle -4\sin t, 4\cos t, 0 \rangle,$

17.2 STOKES' THEOREM

so

$$\oint_C \mathbf{F} \cdot \mathbf{T}\, ds = \int_0^{2\pi} \mathbf{F}(\mathbf{r}(t)) \cdot \mathbf{r}'(t)\, dt$$

$$= \int_0^{2\pi} \langle 16\cos^2 t + 4\sin t - 4,\ 48\cos t \sin t,\ 0\rangle \cdot \langle -4\sin t,\ 4\cos t,\ 0\rangle\, dt$$

$$= \int_0^{2\pi} (-64\cos^2 t \sin t - 16\sin^2 t + 16\sin t + 192\cos^2 t \sin t)\, dt$$

$$= 128\int_0^{2\pi} \cos^2 t \sin t\, dt - 16\int_0^{2\pi} \sin^2 t\, dt + 16\int_0^{2\pi} \sin t\, dt$$

$$= 128\left(\frac{-\cos^3 t}{3}\right)\bigg]_0^{2\pi} - 16\left(\frac{t}{2} - \frac{\sin 2t}{4}\right)\bigg]_0^{2\pi} - 16\cos t\bigg]_0^{2\pi}$$

$$= 0 - 16(\pi) - 0 = -16\pi.$$

This matches exactly our original computation. ∎

Stokes' theorem tells us that the curl of a C^1-differentiable vector field has the property that, for all surfaces "capping" the same closed boundary curve C, the surface integral has the same value:

$$\oint_C \mathbf{F} \cdot \mathbf{T}\, ds.$$

(See Figure 17.10.)

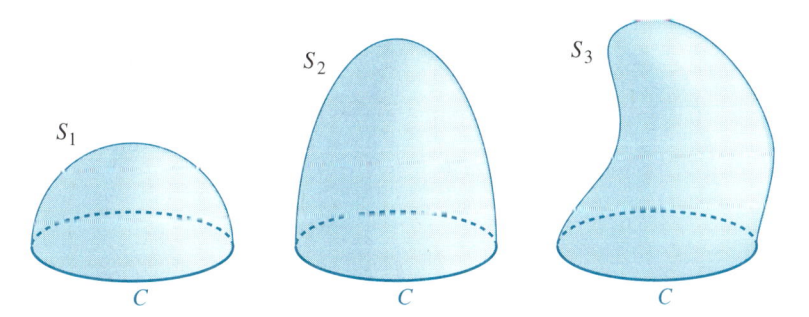

Figure 17.10 Stokes' Theorem: $\displaystyle\iint_S \nabla \mathbf{F} \times \mathbf{n}\, dS$ is the same for all three surfaces capping C.

This result is entirely consistent with our previous observation regarding conservative vector fields. Recall that, if $\mathbf{F}$ is a conservative vector field, then

$$\oint_C \mathbf{F} \cdot \mathbf{T}\, ds = 0$$

for every closed curve C. If $\mathbf{F}$ is C^1-differentiable, this also means that **curl** $\mathbf{F} = \mathbf{0}$. But then

$$\iint_S \text{curl } \mathbf{F} \cdot \mathbf{n}\, dS = \iint_S \mathbf{0} \cdot \mathbf{n}\, dS = 0$$

for any surface S capping the closed curve C.

EXAMPLE 7 Suppose that $\mathbf{F}(x, y, z) = \langle -2xz, x, y^2 \rangle$.

Find the value of the surface integral

$$\iint_S \nabla \times \mathbf{F} \cdot \mathbf{n}\, dS$$

for each of the surfaces shown in Figure 17.11, where the principal unit normal in each case points outward.

The boundary of each surface below is the circle of radius 2 in the xy-plane, centered at the origin.

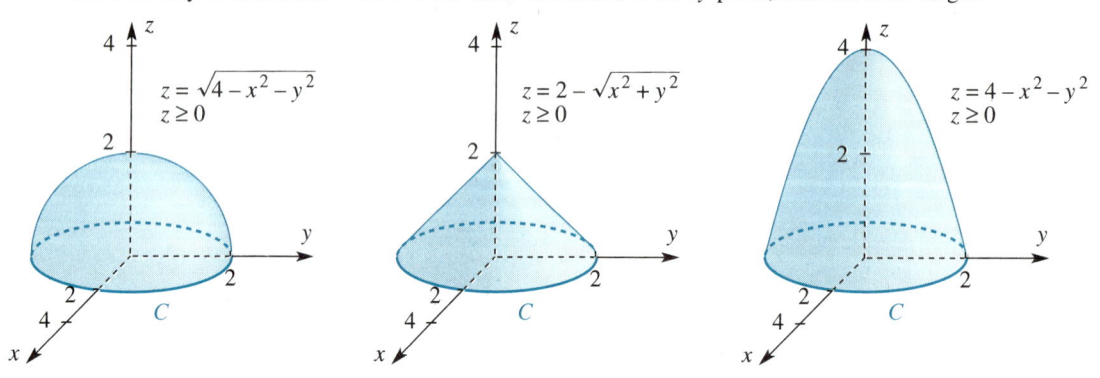

Figure 17.11 Three surfaces capping the same curve.

Solution   Each of the surfaces shown has the same boundary, namely the circle of radius 2 in the xy-plane. This circle is parametrized by

$$\mathbf{r}(t) = \langle 2\cos t,\ 2\sin t,\ 0 \rangle \quad \text{for } 0 \leq t \leq 2\pi.$$

The velocity vector is $\quad \mathbf{r}'(t) = \langle -2\sin t,\ 2\cos t,\ 0 \rangle.$

Hence, if S is any surface capping the positively oriented boundary C, we must have

17.2 STOKES' THEOREM

$$\iint_S \nabla \times \mathbf{F} \cdot \mathbf{n}\, dS = \oint_C \mathbf{F} \cdot \mathbf{T}\, ds$$

$$= \int_0^{2\pi} \langle 0,\ 2\cos t,\ 4\sin^2 t\rangle \cdot \langle -2\sin t, 2\cos t, 0\rangle\, dt$$

$$= \int_0^{2\pi} 4\cos^2 t\, dt = \int_0^{2\pi} 2 + 2\cos 2t\, dt$$

$$= 2t + \sin 2t \Big]_0^{2\pi}$$

$$= 4\pi.$$

This is the value of the surface integral for each of the surfaces shown. ∎

EXERCISES for Section 17.2

For exercises 1-5: Consider
$$\mathbf{F}(x,y,z) = \langle y, -x, z - x^2 - y^2 \rangle \quad \text{and} \quad \mathbf{G}(x,y,z) = \langle -2y, 2x, -2 \rangle.$$

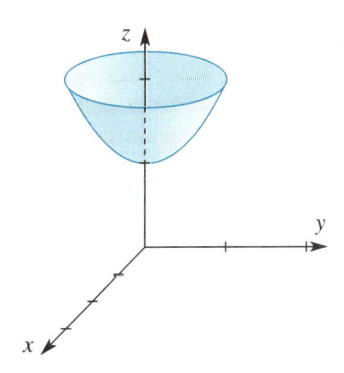

For Exercises 1-5

The surface S pictured above has parametrization $\mathbf{r} : D \to \mathbb{R}^3$, where

$$\mathbf{r}(u,v) = (u, v, u^2 + v^2 + 1) \quad \text{and} \quad D = \{(u,v)\ :\ u^2 + v^2 \le 1\}.$$

(Note: D is the unit disk in the uv plane.) C is the boundary of S, and has positively oriented parametrization)

$$\mathbf{g} : [0, 2\pi] \to \mathbb{R}^3 \quad \text{and} \quad \mathbf{g}(t) = \langle \cos t, \sin t, 2 \rangle.$$

1. Express the surface area of S in terms of a double integral.

2. Find $\displaystyle\iint_S \mathbf{G} \cdot \mathbf{n}\, dS$.

3. Find $\displaystyle\oint_C \mathbf{F} \cdot \mathbf{T}\, ds$.

4. Find $\nabla \times \mathbf{F}$.

5. Does Stokes' theorem hold for $\mathbf{F}$, S, and C? Explain why or why not.

For exercises 6-9: Verify that the conclusion of Stokes' theorem holds by computing both $\iint_S (\nabla \times \mathbf{F}) \cdot \mathbf{n}\, dS$ and $\oint_C \mathbf{F} \cdot \mathbf{T}\, ds$ directly for the given vector fields and surfaces.

6. S is the first octant portion of the plane $2x + 2y + z = 2$ and $\mathbf{F}(x, y, z) = \langle -2y, 3x, -4z \rangle$.

7. S is the portion of the upper half cone $z = \sqrt{x^2 + y^2}$ lying below the plane $z = 4$ and $\mathbf{F}(x, y, z) = \langle z, 2x, -y^2 \rangle$.

8. S is the portion of the sphere $x^2 + y^2 + z^2 = 4$ lying above the plane $z = 1$ and $\mathbf{F}(x, y, z) = \langle x, z^2, yz \rangle$.

9. S is the portion of the sphere $x^2 + y^2 + z^2 = 4$ lying above the plane $z = 1$ and $\mathbf{F}(x, y, z) = \langle yz^2, x^2, xyz \rangle$.

For exercises 10-13: The *circulation* of a vector field $\mathbf{F}$ along a curve C is given by $\oint_C \mathbf{F} \cdot \mathbf{T}\, ds$. Use Stokes' theorem to calculate the circulation of the given vector field around the curve C described.

10. $\mathbf{F}(x, y, z) = \langle x, 3xz, y \rangle$ and C is the curve of intersection of the plane $z = x + 2$ and the cylinder $x^2 + y^2 = 4$.

11. $\mathbf{F}(x, y, z) = \langle yz^2, x^2, xyz \rangle$ and C is the curve of intersection of the upper half of the ellipsoid $9x^2 + 9y^2 + z^2 = 1$ and the cylinder $x^2 + y^2 = 1$.

12. $\mathbf{F}(x, y, z) = \langle x^4, xy, z^4 \rangle$ and C is the boundary of the portion of the paraboloid $y = x^2 + z^2$ for $z \geq 0$.

13. $\mathbf{F}(x, y, z) = \langle x, z^2, yz \rangle$ and C is the boundary of the first octant portion of the sphere $x^2 + y^2 + z^2 = 1$.

14. Calculate the flux of **curl F** out through the portion of the sphere $x^2 + y^2 + (z-2)^2 = 4$ that lies above the plane $z = 3$ if $\mathbf{F} = \langle -y, x, \sqrt{z} \rangle$.

15. Referring to Figure 17.11, verify that

$$\iint_S (\nabla \times \mathbf{F}) \cdot \mathbf{n}\, dS = 4\pi$$

for each of the three surfaces, where $\mathbf{F}(x, y, z) = \langle -2xz, x, y^2 \rangle$.

17.3 GREEN'S THEOREM

It's also possible to apply Stokes' theorem to 2-dimensional vector fields. This special case of Stokes' theorem is usually called **Green's theorem**.

If $\mathbf{F}(x,y) = \langle P(x,y), Q(x,y) \rangle$ is a 2-dimensional vector field, we can consider it as a 3-dimensional vector field

$$\mathbf{F}(x,y,z) = \langle P(x,y), Q(x,y), 0 \rangle$$

just by restricting our attention to the xy-plane. In this case,

$$\text{curl } \mathbf{F} = \left\langle 0, 0, \frac{\partial Q}{\partial x} - \frac{\partial P}{\partial y} \right\rangle.$$

A smooth simple curve that lies entirely in the xy-plane encloses a plane region D that can be parametrized simply by

$$\mathbf{r}(x,y) = \langle x, y, 0 \rangle.$$

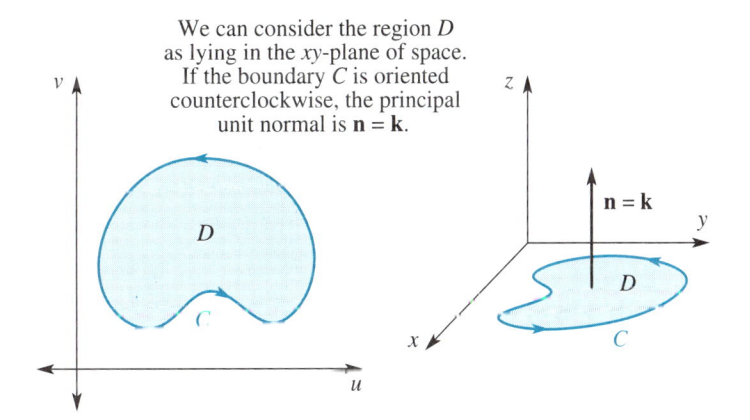

(See Figure 17.12.)

Figure 17.12 Considering a plane region in space.

If we traverse the boundary C of D in the counterclockwise direction, the principal unit normal vector to D will just be $\mathbf{k} = \langle 0, 0, 1 \rangle$ at every point. Now,

$$\text{curl } \mathbf{F} \cdot \mathbf{k} = \left\langle 0, 0, \frac{\partial Q}{\partial x} - \frac{\partial P}{\partial y} \right\rangle \cdot \langle 0, 0, 1 \rangle = \frac{\partial Q}{\partial x} - \frac{\partial P}{\partial y},$$

and the surface integral in this case is simply the double integral

$$\iint_D \frac{\partial Q}{\partial x} - \frac{\partial P}{\partial y}\,dx\,dy.$$

Hence, Stokes' theorem tells us that

$$\iint_D \frac{\partial Q}{\partial x} - \frac{\partial P}{\partial y}\,dx\,dy = \oint_C P\,dx + Q\,dy.$$

This particular application of Stokes' theorem is known as **Green's theorem** (as are several other theorems in vector calculus). It is stated more precisely as Theorem 17.3:

Theorem 17.3

Green's theorem

Hypothesis 1: C is a simple piece-wise smooth curve in the xy-plane, oriented counterclockwise, and D is the region enclosed by C.
Hypothesis 2: $\mathbf{F}(x,y) = \langle P(x,y), Q(x,y) \rangle$ is a C^1-differentiable vector field defined on D.
Conclusion:

$$\iint_D \left(\frac{\partial Q}{\partial x} - \frac{\partial P}{\partial y}\right)dx\,dy = \oint_C P\,dx + Q\,dy.$$

EXAMPLE 8 Verify Green's theorem for $\mathbf{F}(x,y) = \langle xy,\ x+y \rangle$, where C is the boundary of the unit square $[0,1] \times [0,1]$ (oriented counterclockwise).

Solution To calculate $\oint_C P\,dx + Q\,dy$ will require splitting the curve C into 4 smooth parts, as shown in Figure 17.13.

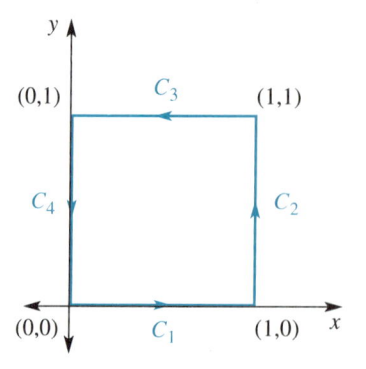

Figure 17.13 The boundary of the square can be split into four smooth curves.

17.3 GREEN'S THEOREM

We have the following parametrizations and line integral computations:

$$C_1: \quad \mathbf{r}_1(t) = \langle t, 0 \rangle \quad \text{for } 0 \leq t \leq 1, \quad \text{so } \mathbf{r}_1'(t) = \langle 1, 0 \rangle,$$

and

$$\oint_{C_1} P\,dx + Q\,dy = \int_0^1 \langle 0, 0+t \rangle \cdot \langle 1, 0 \rangle\,dt = \int_0^1 0\,dt = 0.$$

$$C_2: \quad \mathbf{r}_2(t) = \langle 1, t \rangle \quad \text{for } 0 \leq t \leq 1, \quad \text{so } \mathbf{r}_2'(t) = \langle 0, 1 \rangle,$$

and

$$\oint_{C_2} P\,dx + Q\,dy = \int_0^1 \langle t, 1+t \rangle \cdot \langle 0, 1 \rangle\,dt = \int_0^1 (1+t)\,dt = t + \frac{t^2}{2}\Big]_0^1 = \frac{3}{2}.$$

$$C_3: \quad \mathbf{r}_3(t) = \langle 1-t, 1 \rangle \quad \text{for } 0 \leq t \leq 1, \quad \text{so } \mathbf{r}_3'(t) = \langle -1, 0 \rangle,$$

and

$$\oint_{C_3} P\,dx + Q\,dy = \int_0^1 \langle (1-t)1, 1-t+1 \rangle \langle -1, 0 \rangle\,dt = \int_0^1 (t-1)\,dt = \frac{t^2}{2} - t\Big]_0^1 = -\frac{1}{2}.$$

$$C_4: \quad \mathbf{r}_4(t) = \langle 0, 1-t \rangle \quad \text{for } 0 \leq t \leq 1, \quad \text{so } \mathbf{r}_1'(t) = \langle 0, -1 \rangle,$$

and

$$\oint_{C_4} P\,dx + Q\,dy = \int_0^1 \langle 0(1-t), 0+(1-t) \rangle \cdot \langle 0, -1 \rangle\,dt = \int_0^1 (t-1)\,dt$$

$$= \frac{t^2}{2} - t\Big]_0^1 = -\frac{1}{2}.$$

Summing up the four pieces, we have

$$\oint_C P\,dx + Q\,dy = 0 + \frac{3}{2} - \frac{1}{2} - \frac{1}{2} = \frac{1}{2}.$$

Now, by Green's theorem we should achieve the same result by calculating

$$\iint_D \left(\frac{\partial Q}{\partial x} - \frac{\partial P}{\partial y} \right) dx\,dy,$$

where D is the unit square $[0, 1] \times [0, 1]$. We have

$$\frac{\partial Q}{\partial x} = \frac{\partial}{\partial x}(x+y) = 1 \qquad \frac{\partial P}{\partial y} = \frac{\partial}{\partial y}(xy) = x.$$

Hence,

$$\iint_D \left(\frac{\partial Q}{\partial x} - \frac{\partial P}{\partial y}\right) dx\, dy = \int_0^1 \int_0^1 (1-x)\, dx\, dy$$

$$= \int_0^1 \left[x - \frac{x^2}{2}\right]_0^1 dy$$

$$= \int_0^1 \frac{1}{2}\, dy = \frac{y}{2}\Big]_0^1 = \frac{1}{2}.$$

We see that the results agree. ∎

Calculating area using Green's theorem

Green's theorem provides an alternative for calculating the area of a region D enclosed by a closed piece-wise simple curve C (oriented counterclockwise). If $P(x,y)$ and $Q(x,y)$ are any two C^1-differentiable functions such that $\frac{\partial Q}{\partial x} - \frac{\partial P}{\partial y} = 1$, then the area of D is

$$\iint_D 1\, dx\, dy = \iint_D \left(\frac{\partial Q}{\partial x} - \frac{\partial P}{\partial y}\right) dx\, dy = \oint_C P\, dx + Q\, dy.$$

Thus, we can calculate the area of a region by means of a line integral around its boundary, provided that we make suitable choices for $P(x,y)$ and $Q(x,y)$. Here are three possibilities:

▶▶▶ **If D is a region enclosed by a closed piece-wise simple curve C (oriented counterclockwise), then each of the following line integrals has a value equal to the area of D.**

$$\textbf{area of } D = \oint_C x\, dy = -\oint_C y\, dx = \frac{1}{2}\oint_C x\, dy - y\, dx.$$

EXAMPLE 9 Find the area of the ellipse whose boundary has equation

$$\frac{x^2}{a^2} + \frac{y^2}{b^2} = 1$$

using a line integral.

Solution The boundary curve C can be parametrized with counterclockwise orientation with

$$\mathbf{r}(t) = \langle a\cos t, b\sin t\rangle \qquad 0 \leq t \leq 2\pi.$$

17.3 GREEN'S THEOREM

We have $\mathbf{r}'(t) = \langle -a\sin t,\ b\cos t \rangle$. Using a line integral to measure the area, we have

$$\text{area of } D = \frac{1}{2}\oint_C x\,dy - y\,dx = \frac{1}{2}\oint \langle -y, x\rangle \cdot \langle dx, dy\rangle$$

$$= \frac{1}{2}\int_0^{2\pi} \langle -b\sin t,\ a\cos t\rangle \cdot \langle -a\sin t,\ b\cos t\rangle\,dt$$

$$= \frac{1}{2}\int_0^{2\pi} (ab\sin^2 t + ab\cos^2 t)\,dt$$

$$= \frac{ab}{2}\int_0^{2\pi} 1\,dt$$

$$= \frac{ab}{2}(2\pi) = \pi ab.\quad\blacksquare$$

EXERCISES for Section 17.3

For exercises 1-5: Keeping in mind the usefulness of Green's theorem, evaluate the following integrals.

1. Use Green's theorem to evaluate

$$\oint_C (4y + \sqrt{\sin^3 x})\,dx + (6x - \sec^3 y)\,dy,$$

where C is a circle centered at the origin of radius 4.

2. Find

$$\oint_C e^x\,dx + xy\,dy,$$

where C is the path that starts at $(0,0)$, follows the x-axis to $(2,0)$, and then returns to $(0,0)$ along the parabola $y = 2x - x^2$.

3. Use Green's theorem to evaluate the line integral

$$\oint_C (2^x - e^y)\,dx + (4x^2 - e^y)\,dy,$$

where C is the path consisting of the straight line segment from $(0,0)$ to $(1,0)$, the straight line segment from $(1,0)$ to $(1,1)$, the straight line segment from $(1,1)$ to $(0,1)$, and the straight line segment from $(0,1)$ to $(0,0)$.

4. Use Green's theorem to transform the integral

$$\int_0^1 \int_0^x y\,dy\,dx$$

into a line integral, then evaluate the line integral.

5. If C is the boundary of $[0,2] \times [1,2]$, then find $\oint_C y^2\,dx - x^2\,dy$.

For exercises 6-9: Verify that the conclusion of Green's theorem holds by calculating both $\oint_C P\,dx + Q\,dy$ and $\iint_D \dfrac{\partial Q}{\partial x} - \dfrac{\partial P}{\partial y}\,dx\,dy$ directly.

6. $\mathbf{F}(x,y) = \langle x^2, xy^2 \rangle$ and C is the triangle made by connecting $(0,0)$, $(1,1)$, and $(0,1)$.

7. $\mathbf{F}(x,y) = \langle x^2 + y^2, x^2 - y^2 \rangle$ and C is the curve that bounds the first quadrant portion of the region bounded by the y-axis, $y = 2 - x^2$ and $y = x$.

8. $\mathbf{F}(x,y) = \langle x+y, xy \rangle$ and C is the curve bounding the region between $y = x^2$ and $x = y^2$.

9. $\mathbf{F}(x,y) = \langle xy, y + 4x \rangle$ and C is the curve bounding the region below the curve $y = x^2 - 4x + 3$ and above the x-axis.

10. Use Green's theorem to evaluate $\oint_C (\cos^2 x - 2y)\,dx + (x^2 - \sin^2 y)\,dy$, where C is the circle of radius 1 centered at $(0,0)$.

11. Use Green's theorem to evaluate $\oint_C (e^x - \cos y)\,dx + (e^y + x^2)\,dy$, where C is the path given in exercise 7.

12. Use Green's theorem to evaluate $\oint_C (3y - x^3)\,dx + (2xy + y^3)\,dy$, where C is the path given in exercise 9.

13. The curve C is made up of C_1 parametrized by $\mathbf{r}_1(t) = \langle t - \sin t, 1 - \cos t \rangle$ for $0 \leq t \leq 2\pi$ and C_2, which can be parametrized by $\mathbf{r}_2(t) = \langle t, 0 \rangle$ for $0 \leq t \leq 2\pi$. Use the area of the region bounded by C. (Hint: be careful with the orientation of the given curves.)

14. Use the area formula derived from Green's theorem to calculate the area of the region bounded by the curves $y = x^2$ and $x = y^2$.

15. The curve C, given by $\mathbf{r}(t) = \langle t^2\sqrt{3}, 3t - t^3/3 \rangle$, forms a loop for the values $-3 \leq t \leq 3$. Use the area formula derived from Green's theorem to find the area of the region bounded by this loop.

16. Use Green's theorem to calculate $\iint_D (2x - 2y)\,dA$ for the region D bounded by the circle $(x-1)^2 + (y-2)^2 = 1$. (Hint: the circle may be parametrized by $\mathbf{r}(t) = \langle 1 + \cos t, 2 + \sin t \rangle$ for $0 \leq t \leq 2\pi$.)

17. Verify that each of the line integrals

$$\oint_C x\,dy \quad \text{and} \quad -\oint_C y\,dx$$

also yields the area πab enclosed by the ellipse C having equation

$$\frac{x^2}{a^2} + \frac{y^2}{b^2} = 1.$$

17.4 THE DIVERGENCE THEOREM—GAUSS' THEOREM

Stokes' theorem relates the integral of the normal component of **curl F** over a surface to the integral of the tangential component of **F** over the closed boundary curve of the surface. **Gauss' theorem**, also known as the **divergence theorem**, is analogous to Stokes' theorem: it relates the integral of div **F** over a 3-dimensional region R to the outward flux of **F** over the closed boundary surface S of the region. Following is the precise statement of the theorem:

Theorem 17.4

The divergence theorem (Gauss' theorem)
Hypothesis 1: R is a region of $\mathbb{R}^3$ bounded by a closed simple surface S, oriented so that the principal unit normal **n** points outward.
Hypothesis 2: **F** is a C^1-differentiable vector field defined on R.
Conclusion:
$$\iint_S \mathbf{F} \cdot \mathbf{n} \, dS = \iiint_R \nabla \cdot \mathbf{F} \, dV.$$

In other words, the integral of the divergence of **F** over the region R has the same value as the total outward flux of **F** over its boundary surface S.

Reasoning   The reasoning behind the divergence theorem is quite similar to that behind Stokes' theorem. Imagine that we subdivide the region R into several subregions and consider the surface integral of the outward normal component of **F** over the boundary surface of each subregion (see Figure 17.14).

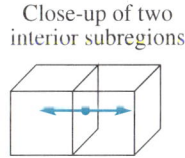

Close-up of two interior subregions

Within the region, the outward normals of the faces of the adjacent interior subregions point in exactly the opposite directions. The only normals that do not cancel out are those along the exterior surface of the region as shown.

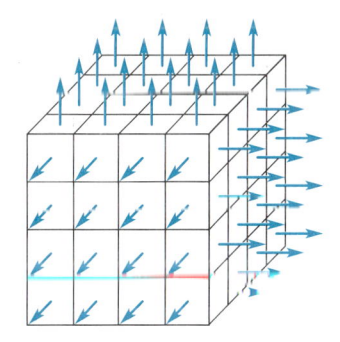

Figure 17.14 Subdividing a region R.

Two adjacent regions within the region will share a face, but their outward normals will point in opposite directions. Therefore, when we take the sum of all these surface integrals, the only parts that do not cancel belong to the faces that make up the original boundary of the region R. Now, as we partition our region R into finer and finer subregions, the surface integral of the outward flux of F over each tiny closed surface S_i has a value

$$\iint_{S_i} \mathbf{F} \cdot \mathbf{n}\, dS \approx \operatorname{div} \mathbf{F} \Delta V,$$

where div F is evaluated at a point within the subregion and ΔV is the volume of the subregion. This approximation becomes better and better as the partition is made finer and finer, and we obtain

$$\iint_S \mathbf{F} \cdot \mathbf{n}\, dS = \lim_{N \to \infty} \sum_{i=1}^N \iint_{S_i} \mathbf{F} \cdot \mathbf{n}\, dS = \lim_{N \to \infty} \sum_{i=1}^N \nabla \cdot \mathbf{F} \Delta V = \iiint_R \nabla \cdot \mathbf{F}\, dV.$$

This is simply the statement of the divergence theorem. □

EXAMPLE 10 Verify the divergence theorem for

$$\mathbf{F}(x, y, z) = x\mathbf{i} + y\mathbf{j} + z\mathbf{k}$$

for the cube $R = [-1, 1] \times [-1, 1] \times [-1, 1]$.

Solution First we calculate the volume integral. The divergence of $\mathbf{F}$ is

$$\nabla \cdot \mathbf{F} = 1 + 1 + 1 = 3,$$

and

$$\int_{-1}^1 \int_{-1}^1 \int_{-1}^1 3\, dx\, dy\, dz = 3(\text{volume of the cube}) = 3 \cdot 8 = 24.$$

Now we calculate the surface integral. Figure 17.15 illustrates the region R and the principal unit normals to each face of the cube.

17.4 THE DIVERGENCE THEOREM—GAUSS' THEOREM

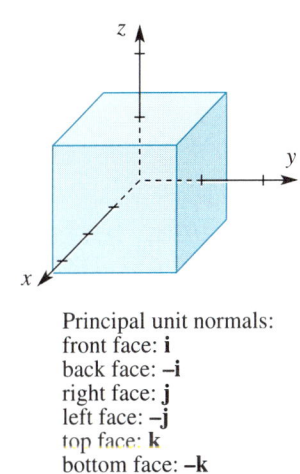

Principal unit normals:
front face: **i**
back face: **−i**
right face: **j**
left face: **−j**
top face: **k**
bottom face: **−k**

Figure 17.15 The region R and its principal unit normal vectors.

Let's calculate $\mathbf{F} \cdot \mathbf{n}$ for each face of the cube:

front face: $\langle x, y, z \rangle \cdot \langle 1, 0, 0 \rangle = x = 1$

(since $x = 1$ over the entire front face of the cube)

back face: $\langle x, y, z \rangle \cdot \langle -1, 0, 0 \rangle = -x = 1$

(since $x = -1$ over the entire back face of the cube)

left face: $\langle x, y, z \rangle \cdot \langle 0, -1, 0 \rangle = -y = 1$

right face: $\langle x, y, z \rangle \cdot \langle 0, 1, 0 \rangle = y = 1$

top face: $\langle x, y, z \rangle \cdot \langle 0, 0, 1 \rangle = z = 1$

bottom face: $\langle x, y, z \rangle \cdot \langle 0, 0, -1 \rangle = -z = 1.$

The front face has a surface area of 4, hence

$$\iint_{front} \mathbf{F} \cdot \mathbf{n} \, dS = \iint_{front} 1 \, dS = 4.$$

Similarly, the surface integral over each of the other faces has the same value 4. Thus,

$$\iint_{S} \mathbf{F} \cdot \mathbf{n} \, dS = 6 \cdot 4 = 24.$$

We can see that the surface integral of $\mathbf{F} \cdot \mathbf{n}$ over the surface S of the cube has the same value as the volume integral of $\nabla \cdot \mathbf{F}$ over the region R. ■

EXERCISES for Section 17.4

1. Find

$$\iint_S \mathbf{F} \cdot \mathbf{n}\, dS,$$

where S is the surface bounding the tetrahedron formed by the coordinate planes and the plane $x + y + z = 1$, $\mathbf{n}$ is the outward unit normal to S, and $\mathbf{F} = (2x + e^y)\mathbf{i} + (\ln x - 2y)\mathbf{j} + (\cos(xy) + z)\mathbf{k}$. You may use the fact that the volume of the tetrahedron is $\frac{1}{6}$.

2. Verify the divergence theorem for the *solid* hemisphere $\{(x, y, z) : x^2 + y^2 + z^2 \leq 1, z \geq 0\}$, where $\mathbf{F}(x, y, z) = \langle x, y, z \rangle$. (Note that the outward principal unit normal is $\mathbf{n} = \langle x, y, z \rangle$ along the spherical portion of the surface and $\mathbf{n} = -\mathbf{k}$ along the flat bottom of the surface.)

3. Use Gauss' theorem to calculate the outward flux of

$$\mathbf{F}(x, y, z) = \langle 2x, 3y, 4z \rangle$$

over the closed cylindrical surface

$$S = \{(x, y, z) : x^2 + y^2 = 1, \ -1 \leq z \leq 2\}.$$

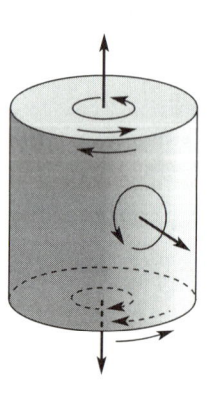

For Exercise 3

4. Justify the following statement: The outward flux of an incompressible C^1-differentiable vector field $\mathbf{F}$ is 0 over any simple, closed surface.

For exercises 5-8: Verify that the conclusion of the divergence theorem holds by computing both $\iint_S \mathbf{F} \cdot \mathbf{n}\, dS$ and $\iiint_R \nabla \cdot \mathbf{F}\, dV$ directly for the given vector fields $\mathbf{F}$ and regions R.

5. R is the region bounded by the coordinate planes and the plane $2x + 2y + z = 2$ and $\mathbf{F}(x, y, z) = \langle -2y, 3x, -4z \rangle$.

17.4 THE DIVERGENCE THEOREM—GAUSS' THEOREM

6. R is the region bounded by the upper half cone $z = \sqrt{x^2 + y^2}$ and the plane $z = 2$ and $\mathbf{F}(x, y, z) = \langle x, 3xz, y \rangle$.

7. R is the region bounded by the two paraboloids, $z = 2 - x^2 - y^2$ and $z = x^2 + y^2$, and $\mathbf{F}(x, y, z) = \langle x, y, z \rangle$

8. R is the region bounded by the paraboloid $z = x^2 + y^2$ and the plane $z = 1$, and $\mathbf{F}(x, y, z) = \langle y, x, z^2 \rangle$.

For exercises 9-12: Use the divergence theorem to calculate the outward flux of the given vector field $\mathbf{F}$ through the given surface S.

9. S is the boundary of the region that lies inside of the sphere $x^2 + y^2 + (z-2)^2 = 4$, above the paraboloid $x^2 + y^2 = z$, and above the plane $z = 1$. $\mathbf{F}(x, y, z) = \langle y, -x, z \rangle$.

10. S is the surface bounding a prism with bottom resting on the plane $z = 1$, top at $z = 5$, and $\mathbf{F}(x, y, z) = \langle \ln y, e^{x+z}, \ln z \rangle$.

11. S is the boundary of a "trough" with bottom formed by the surface $z = y^2$, ends at $x = 0$ and $x = 5$, and top formed by the plane $z = 4$, and $\mathbf{F}(x, y, z) = \langle x^2, y^2, x + y + z \rangle$.

12. S is the surface of the region interior to both of two spheres, $x^2 + y^2 + (z-2)^2 = 4$ and $x^2 + y^2 + z^2 = 4$, and $\mathbf{F}(x, y, z) = \langle y^2, x^2, z \rangle$.

A

Appendices

The appendices contain additional material for your review or reference.

Appendix 1 includes review material on right triangle trigonometry.

Appendix 2 explains techniques of integration using trigonometric identities and relationships.

Appendix 3 discusses the method of partial fractions, an algebraic technique useful for simplifying some integrals.

Appendix 4 consists of a brief introduction to polar coordinates and their use in differential and integral calculus.

Appendix 5 discusses complex numbers and their algebraic and geometric representations.

Appendix 6 extends the material on Taylor polynomials with a discussion of Taylor's formula.

Additional practice exercises on differentiation and integration are included.

The appendices conclude with a collection of useful facts and formulas from algebra, trigonometry, and analytic geometry. Tables of differentiation and integration formulas can be found in the end pages of the book, as well as formulas from plane and solid geometry.

A.1 TRIANGLE TRIGONOMETRY

In a right triangle with legs of lengths a and b and a hypotenuse of length c, we know from the Pythagorean theorem the familiar relationship

$$a^2 + b^2 = c^2.$$

Figure A.1 illustrates the definitions of the three principal trigonometric ratios associated with an angle θ (the Greek letter theta) in terms of the sides of the triangle.

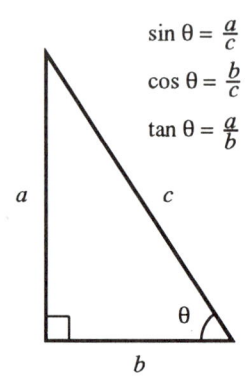

Figure A.1 Trigonometric ratios.

$$\text{sine } \theta = \frac{opposite}{hypotenuse} = \frac{a}{c}$$

$$\text{cosine } \theta = \frac{adjacent}{hypotenuse} = \frac{b}{c}$$

$$\text{tangent } \theta = \frac{opposite}{adjacent} = \frac{a}{b}$$

A.1 TRIANGLE TRIGONOMETRY

The nonsense word "SOHCAHTOA" may help you in memorizing these definitions. Note that every three letters of the word form the initials for one of the trigonometric ratio definitions.

The common abbreviations for sine θ, cosine θ, and tangent θ are

$$\sin \theta, \quad \cos \theta, \quad \tan \theta,$$

respectively. Three additional trigonometric ratios are obtained by taking the reciprocals of these ratios.

$$\operatorname{cosecant} \theta = \frac{1}{\operatorname{sine} \theta} = \frac{hypotenuse}{opposite} = \frac{c}{a}$$

$$\operatorname{secant} \theta = \frac{1}{\operatorname{cosine} \theta} = \frac{hypotenuse}{adjacent} = \frac{c}{b}$$

$$\operatorname{cotangent} \theta = \frac{1}{\operatorname{tangent} \theta} = \frac{adjacent}{opposite} = \frac{b}{a}.$$

The common abbreviations for cosecant θ, secant θ, and cotangent θ are

$$\csc \theta, \quad \sec \theta, \quad \cot \theta,$$

respectively.

EXAMPLE 1 A simple example of a right triangle is the "3–4–5" triangle, with $a = 3$, $b = 4$, and $c = 5$. If θ is the angle opposite side a, find the six trigonometric ratios associated with θ.

Solution Figure A.2 shows the right triangle and angle θ under consideration.

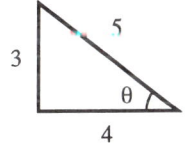

Figure A.2 A 3–4–5 right triangle.

Using the definitions of the trigonometric ratios, we have:

$$\text{sine } \theta = \frac{3}{5}, \quad \text{cosine } \theta = \frac{4}{5}, \quad \text{tangent } \theta = \frac{3}{4},$$

$$\text{cosecant } \theta = \frac{5}{3}, \quad \text{secant } \theta = \frac{5}{4}, \quad \text{cotangent } \theta = \frac{4}{3}.$$ ∎

Two triangles of different sizes are called *similar* if they have the same shape. More precisely, two triangles are similar if the three side lengths of one are all the same multiple of the corresponding side lengths of the other. Here's a key fact from geometry:

> If two triangles are similar, then corresponding angles must be equal in measure. Conversely, if corresponding angles of two triangles are equal in measure, then the two triangles must be similar.

Another useful fact from geometry is:

> The measures of the angles in any triangle sum to 180°.

For instance, if we double the lengths of all three sides of the triangle in the previous example, then we obtain a 6–8–10 right triangle having the same angle θ. Note that all the trigonometric ratios remain exactly the same (see Figure A.3).

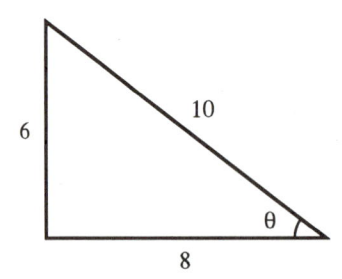

Figure A.3 A 6–8–10 right triangle.

The same observation holds for any triangle similar to our 3–4–5 triangle. On the other hand, any right triangle with this same angle θ must be similar to our 3–4–5 triangle. (Since the angle measures of a triangle sum up to 180°, the remaining angle in both triangles must measure $90° - \theta$.) This means that any trigonometric ratio like $\sin \theta$ really depends only on θ, and not on the size of any of the infinitely many similar right triangles that could be chosen to compute it.

A.1 TRIANGLE TRIGONOMETRY

This is why trigonometric ratios can be used to compute distances so effectively. Extensive trigonometry tables were indispensable aids before the relatively recent development of inexpensive calculators.

EXAMPLE 2 A rocket is launched straight up. From a ground tracking station exactly 50 miles away from the launch site, you record the angle of elevation of the rocket after five minutes to be $37.4°$. What is the altitude a of the rocket at that moment?

Solution Figure A.4 illustrates the problem situation.

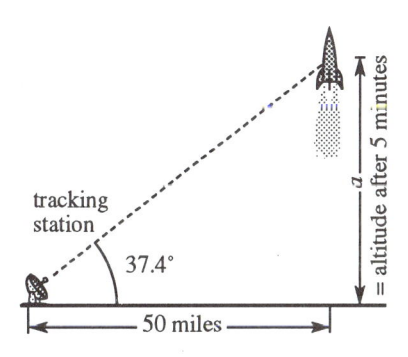

Figure A.4 Tracking the altitude of a rocket.

In the right triangle shown, we can see that the side adjacent to the angle of elevation is the distance between the tracking station and the launch site, and the side opposite is the altitude a. Hence,

$$\tan 37.4° = \frac{a}{50 \text{ miles}}.$$

Solving for a (and using a calculator to compute $\tan 37.4°$), we have

$$a = (\tan 37.4°)(50 \text{ miles}) \approx (.7646)(50 \text{ miles}) = 38.23 \text{ miles}.$$

Radian measure for angles

To extend the use of trigonometric functions beyond right-triangle computations, we will consider another system of angle measurement and broaden the range of values an angle measure can have.

While degree measure for angles is still commonly used in many applications, we will often find **radian** measure more convenient for calculus. If a circle is drawn having the vertex of an angle at its center, then an angle of one radian subtends an arc of length equal to the radius of the circle (see Figure A.5).

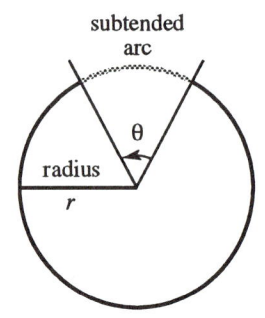

If θ = 1 radian, then the length of the subtended arc is equal to the radius.

Figure A.5 Radian measure of angles.

A circle of radius r has circumference $2\pi r$, so the full $360°$ angle of a circle is equivalent to 2π radians. This gives us the conversion formulas between degrees and radians:

$$1° = \frac{\pi}{180} \text{ radians} \approx 0.0174533 \text{ radians}$$

and

$$1 \text{ radian} = \frac{180°}{\pi} \approx 57.3°.$$

Now we can talk about positive and negative angle measures by agreeing that positive angles are measured counterclockwise and negative angles are measured clockwise. Angles larger than $360°$ or 2π radians in magnitude can also be considered as "wrapping around" the circle more than one revolution. Figure A.6 illustrates several angles measured in either degrees or radians.

A.1 TRIANGLE TRIGONOMETRY

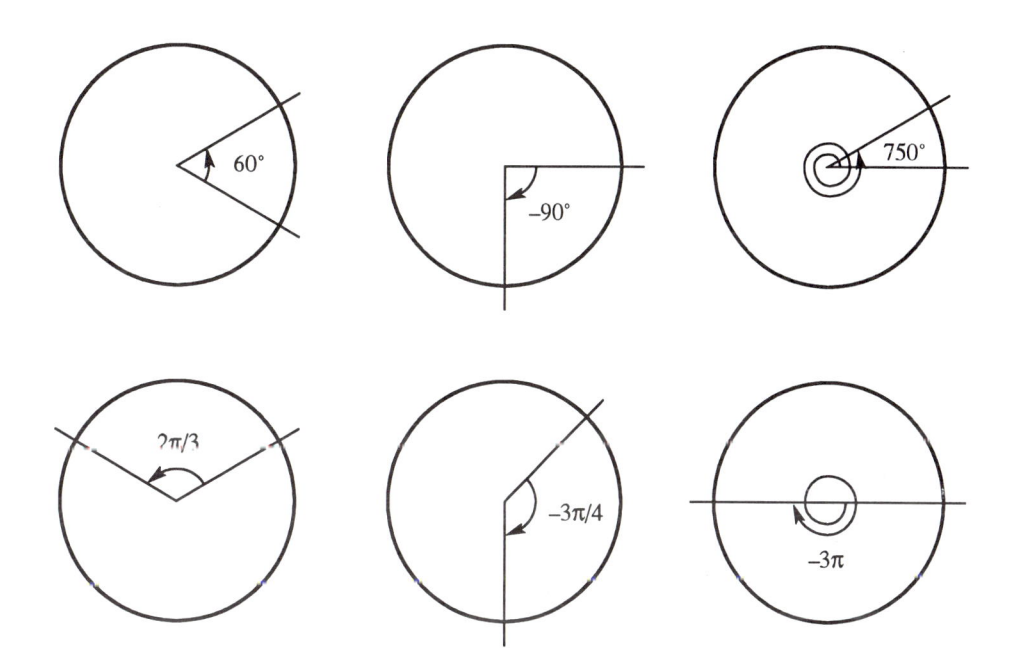

Figure A.6 Examples of angle measure.

EXAMPLE 3 Convert the following to radian measure:

$$45°, \quad 90°, \quad -120°, \quad 600°.$$

Solution $45° = 45\pi/180 = \pi/4$ radians.
$90° = 90\pi/180 = \pi/2$ radians.
$-120° = -120\pi/180 = -2\pi/3$ radians.
$600° = 600\pi/180 = 10\pi/3$ radians.

EXAMPLE 4 Convert the following to degree measure:

$$\pi/3 \text{ radians}, \quad -3\pi/2 \text{ radians}, \quad 25 \text{ radians}.$$

Solution $\pi/3$ radians $= \dfrac{\pi}{3}\left(\dfrac{180°}{\pi}\right) = 60°$.

$-3\pi/2$ radians $= \dfrac{-3\pi}{2}\left(\dfrac{180°}{\pi}\right) = -270°$.

25 radians $= (25)\left(\dfrac{180°}{\pi}\right) \approx 1432.4°$.

Law of sines and law of cosines

Relationships between the sides and angles of an arbitrary triangle (not just a right triangle) can be described using trigonometric functions. The two most important of these relationships are called the law of sines and the law of cosines.

Figure A.7 shows a triangle with sides a, b, c, and with angles α, β, γ opposite those sides, respectively.

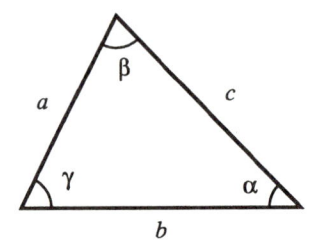

Figure A.7 Triangle with sides a, b, c, and opposite angles α, β, γ.

The law of sines states that the ratio of the sine of an angle to the length of the side opposite it is the same for each angle in the triangle.

Law of sines:
$$\frac{\sin \alpha}{a} = \frac{\sin \beta}{b} = \frac{\sin \gamma}{c}.$$

The law of cosines provides a generalization of the Pythagorean theorem, and allows us to compute the length of the third side of any triangle, provided we know the lengths of the other two sides and the angle between them.

Law of cosines:
$$a^2 = b^2 + c^2 - 2bc \cos \alpha$$
$$b^2 = a^2 + c^2 - 2ac \cos \beta$$
$$c^2 = a^2 + b^2 - 2ab \cos \gamma.$$

Notice that if γ is a right angle (90°), then $\cos \gamma = 0$, and the last equation is the usual statement of the Pythagorean theorem.

A.1 TRIANGLE TRIGONOMETRY

EXERCISES for Section A.1

For exercises 1-6: Draw a triangle with sides of length $a = 5$, $b = 12$, and $c = 13$, with angle θ opposite side a.
1. Find $\sin \theta$.
2. Find $\cos \theta$.
3. Find $\sec \theta$.
4. Find $\csc \theta$.
5. Find $\tan \theta$.
6. Find $\cot \theta$.

For exercises 7-12: Convert each angle measure θ in radians to degrees, and evaluate each of the six trigonometric functions at θ.
7. $\theta = \pi/5$ radians.
8. $\theta = 7\pi/2$ radians.
9. $\theta = -5\pi/4$ radians.
10. $\theta = 371\pi$ radians.
11. $\theta = 8\pi/3$ radians.
12. $\theta = -21\pi/6$ radians.

For exercises 13-18: Convert each angle measure θ in degrees to radians, and evaluate each of the six trigonometric functions at θ.
13. $\theta = 100°$.
14. $\theta = -45°$.
15. $\theta = 270°$.
16. $\theta = 1000°$.
17. $\theta = -270°$.
18. $\theta = \pi°$.

For exercises 19-23: Of the three angles and three sides of triangle ABC, you are given some of the measurements. Determine the three missing measurements.
19. angle $A = 35°$, angle $B = 45°$, side $c = 7$ cm.
20. side $b = 20$ cm, side $c = 60$ cm, angle $A = 40°$.
21. side $b = 130$ miles, side $c = 150$ miles, angle $B = 110°$.
22. side $a = 7$ meters, side $b = 8$ meters, side $c = 10$ meters.
23. side $a = 2$ inches, angle $C = 90°$, side $b = 3$ inches.

24. Suppose side $a = 7$ feet, and angle $B = 150°$. Find a value for the length of side b for which angle A has

a) no possible value, b) exactly one possible value, c) two possible values.

25. A rocket is launched straight up. From a ground tracking station exactly 40 miles away from the launch site, you record the angle of elevation of the rocket after ten minutes to be 62.5°. What is the altitude a of the rocket at that moment?

26. Radar station A is 120 miles due south of radar station B. Station A detects an unidentified flying object at a bearing of N 47° E (that is, at an angle 47° measured clockwise from due North) at exactly the same time that station B detects the UFO at a bearing of S 50° E (that is, at an angle 50° measured counterclockwise from due South). Find the distance from the UFO to both stations.

27. Two planes, one flying at 360 mph and the other at 540 mph, leave an airport at the same time. Three hours later they are exactly 1440 miles apart. What is the angle between their flight paths?

28. Two lighthouses are 74 miles apart on a beach and sight a ship in distress at angles of 59° and 40° with the beach. What is the nearest the ship could be to either lighthouse?

29. A surveyor at location C sights two points A and B on opposite sides of a reservoir. She knows that she is 1000 meters from A and 1500 meters from B, and measures angle $ACB = 30°$. How wide is the reservoir?

30. A television antenna tower is mounted at the top of a building. From a point 150 meters from the base of the building at ground level, the angle of elevation to the top of the building is 34°, while the angle of elevation to the top of the antenna is 50°. How tall is the antenna tower?

A.2 TECHNIQUES OF INTEGRATION

Numerical techniques of integration such as the trapezoidal or Simpson's rules provide powerful methods of approximating the value of a definite integral

$$\int_b^a f(x)\,dx.$$

The problem of finding an antiderivative (indefinite integral) of a function f requires finding another function F such that

$$F'(x) = f(x).$$

Once we have found one such antiderivative, any other antiderivative will differ from F only by a constant. Written in terms of integral notation

$$\int f(x)\,dx = F(x) + C,$$

where C represents an arbitrary constant.

There are two basic methods of antidifferentation–substitution and integration by parts. Essentially, the method of substitution requires one to recognize an antiderivative of the form

$$\int f(g(x))g'(x)\,dx.$$

By substituting $u = g(x)$ (so that $du = g'(x)\,dx$) we have

$$\int f(u)\,du$$

A.2 TECHNIQUES OF INTEGRATION

whose antiderivative may be easier for us to recognize.

Integration by parts requires two substitutions, and is based on the product rule of differentiation:

$$\int u\, dv = uv - \int v\, du.$$

Integration by parts is particularly useful when it is easier for us to identify the antiderivative $\int v\, du$ than the original $\int u\, dv$.

To facilitate the use of these basic methods of integration, there are several techniques involving algebraic and trigonometric manipulations. In this section, we discuss some of the most popular of these techniques, and provide examples of each. There are additional practice exercises at the end of the appendices.

Powers of trigonometric functions

To find antiderivatives of the form

$$\int \sin^m x \cos^n x\, dx \quad \text{and} \quad \int \tan^m x \sec^n x\, dx$$

you can take advantage of different trigonometric identities, depending on whether the powers m and n are even or odd. First, let's consider

$$\int \sin^m x \cos^n x\, dx.$$

Case 1. If n is odd, rewrite the integral as

$$\int (\sin^m x \cos^{n-1} x)(\cos x)\, dx$$

and make the substitution

$$u = \sin x \qquad du = (\cos x)\, dx$$

We can rewrite $\cos^{n-1} x$ (an even power of $\cos x$ because n is odd) in terms of $\sin x$ using the Pythagorean identity $\cos^2 x = 1 - \sin^2 x$.

EXAMPLE 5 Find $\int \cos^3 x \sin^2 x \, dx$.

Solution First, we rewrite the integral as

$$\int \cos^2 x \sin^2 x (\cos x \, dx).$$

Now using $\cos^2 x = 1 - \sin^2 x$, we have

$$\int (1 - \sin^2 x) \sin^2 x (\cos x \, dx) = \int (\sin^2 x - \sin^4 x)(\cos x \, dx).$$

Substituting $u = \sin x$ and $du = \cos x \, dx$ gives us

$$\int (u^2 - u^4) \, du = \frac{u^3}{3} - \frac{u^5}{5} + C.$$

Substituting $\sin x$ back for u yields the antiderivative in terms of x:

$$\int \cos^3 x \sin^2 x \, dx = \frac{\sin^3 x}{3} - \frac{\sin^5 x}{5} + C. \quad \blacksquare$$

Case 2. If m is odd, rewrite the integral as

$$-\int (\sin^{m-1} x \cos^n x)(-\sin x) \, dx$$

and make the substitution

$$u = \cos x \qquad du = (-\sin x) \, dx$$

We can rewrite $\sin^{m-1} x$ (an even power of $\sin x$ because m is odd) in terms of $\cos x$ using the Pythagorean identity $\sin^2 x = 1 - \cos^2 x$.

EXAMPLE 6 Find $\int \sin^5 x \, dx$.

Solution First, we rewrite the integral as

$$-\int \sin^4 x (-\sin x \, dx) = -\int (\sin^2 x)^2 (-\sin x \, dx).$$

Now, using $\sin^2 x = 1 - \cos^2 x$, we have

$$-\int (1 - \cos^2 x)^2 (-\sin x \, dx) = -\int (1 - 2\cos^2 x + \cos^4 x)(-\sin x \, dx).$$

A.2 TECHNIQUES OF INTEGRATION

Substituting $u = \cos x$ and $du = -\sin x\, dx$ gives us

$$-\int (1 - 2u^2 + u^4)\,du = -\left(u - \frac{2u^3}{3} + \frac{u^5}{5}\right) + C = -u + \frac{2u^3}{3} - \frac{u^5}{5} + C.$$

Substituting $\cos x$ back for u yields the antiderivative in terms of x:

$$\int \sin^5 x\, dx = -\cos x + \frac{2\cos^3 x}{3} - \frac{\cos^5 x}{5} + C.$$

∎

Note that if both m and n are odd, you can choose to treat it as either Case 1 or Case 2.

Case 3. If both m and n are even use the half-angle formulas,

$$\sin^2 x = \frac{1 - \cos 2x}{2} \quad \text{and} \quad \cos^2 x = \frac{1 + \cos 2x}{2},$$

repeating if necessary.

EXAMPLE 7 Find $\int \sin^4 x\, dx$.

Solution In this case we write

$$\int \sin^4 x\, dx = \int (\sin^2 x)(\sin^2 x)\, dx$$

and use the half-angle formula $\sin^2 x = \dfrac{1 - \cos 2x}{2}$ to obtain

$$\int \left(\frac{1 - \cos 2x}{2}\right)\left(\frac{1 - \cos 2x}{2}\right) dx$$

$$= \frac{1}{4} \int (1 - 2\cos 2x + \cos^2 2x)\, dx$$

$$= \frac{1}{4} \int 1\, dx - \frac{1}{2} \int \cos 2x\, dx + \frac{1}{4} \int \cos^2 2x\, dx.$$

Now, $\dfrac{1}{4}\int 1\, dx = \dfrac{x}{4} + C_1$, and $-\dfrac{1}{2}\int \cos 2x\, dx = -\dfrac{1}{4}\sin 2x + C_2$ (using the substitution $u = 2x$, $du = 2\, dx$). For the last term we apply the other half-angle formula

$$\cos^2 2x = \frac{1 + \cos 4x}{2}$$

to obtain

$$\frac{1}{4}\int\frac{1+\cos 4x}{2}\,dx = \frac{1}{4}\left(\frac{1}{2}\int 1\,dx + \frac{1}{2}\int\cos 4x\,dx\right) = \frac{x}{8} + \frac{1}{32}\sin 4x + C_3.$$

We add up the results and sum the three constants to a single arbitrary constant:

$$\int\sin^4 x\,dx = \frac{3x}{8} - \frac{\sin 2x}{4} + \frac{\sin 4x}{32} + C.$$

A similar set of strategies can be used to find antiderivatives of

$$\int\tan^m x\sec^n x\,dx.$$

Case 1. If n is even, write the integral as

$$\int(\tan^m x)(\sec^{n-2} x)(\sec^2 x\,dx),$$

make the substitutions $u = \tan x$ and $du = \sec^2 x\,dx$, and write $\sec^{n-2} x$ in terms of $\tan x$ using the Pythagorean identity

$$\sec^2 x = 1 + \tan^2 x.$$

EXAMPLE 8 Find $\int\tan x\sec^4 x\,dx$.

Solution First, we rewrite the integral as

$$\int\tan x\sec^4 x\,dx = \int\tan x\sec^2 x(\sec^2 x\,dx)$$

and use the Pythagorean identity $\sec^2 x = 1 + \tan^2 x$ to obtain

$$\int\tan x(1+\tan^2 x)(\sec^2 x\,dx) = \int(\tan x + \tan^3 x)(\sec^2 x\,dx).$$

Substituting $u = \tan x$ and $du = \sec^2 x\,dx$ gives us

$$\int(u+u^3)\,du = \frac{u^2}{2} + \frac{u^4}{4} + C.$$

Substituting $\tan x$ back for u yields the antiderivative in terms of x:

$$\int\tan x\sec^4 x\,dx = \frac{\tan^2 x}{2} + \frac{\tan^4 x}{4} + C.$$

A.2 TECHNIQUES OF INTEGRATION

Case 2. If m is odd, write the integral as

$$\int (\tan^{m-1} x)(\sec^{n-1} x)(\sec x \tan x \, dx),$$

make the substitutions $u = \sec x$ and $du = \sec x \tan x \, dx$, and write $\tan^{m-1} x$ in terms of $\sec x$ using the Pythagorean identity

$$\tan^2 x = \sec^2 x - 1.$$

EXAMPLE 9 Find $\int \tan^3 x \sec x \, dx$.

Solution First, we rewrite the integral as

$$\int \tan^3 x \sec x \, dx = \int \tan^2 x (\sec x \tan x \, dx)$$

and use the Pythagorean identity $\tan^2 x = \sec^2 x - 1$ to obtain

$$\int (\sec^2 x - 1)(\sec x \tan x \, dx).$$

Substituting $u = \sec x$ and $du = \sec x \tan x \, dx$ gives us

$$\int (u^2 - 1) du = \frac{u^3}{3} - u + C.$$

Substituting $\sec x$ back for u yields the antiderivative in terms of x:

$$\int \tan^3 x \sec x \, dx = \frac{\sec^3 x}{3} - \sec x + C.$$

∎

Note that if m is odd and n is even, you can choose to treat it as either Case 1 or Case 2.

Case 3. If m is even and n is odd, then one can use the Pythagorean identity $\tan^2 x = \sec^2 x - 1$ to rewrite $\tan^m x$ in terms of $\sec x$. This will produce a sum of integrals of odd powers of $\sec x$, and each such integral can be found using integration by parts, or by repeated use of the reduction formula

$$\int \sec^n x \, dx = \frac{1}{n-1} \tan x \sec^{n-2} x + \frac{n-2}{n-1} \int \sec^{n-2} x \, dx.$$

EXAMPLE 10 Find $\int \tan^2 x \sec^3 x \, dx$.

Solution Using the Pythagorean identity $\tan^2 x + 1 = \sec^2 x$, we can write

$$\int \tan^2 x \sec^3 x \, dx = \int (\sec^2 x - 1) \sec^3 x \, dx = \int \sec^5 x \, dx - \int \sec^3 x \, dx.$$

Applying the reduction formula to the first integral with $n = 5$, we have

$$\int \sec^5 x \, dx = \frac{1}{4} \tan x \sec^3 x + \frac{3}{4} \int \sec^3 x \, dx.$$

Hence,

$$\int \sec^5 x \, dx - \int \sec^3 x \, dx = \frac{1}{4} \tan x \sec^3 x - \frac{1}{4} \int \sec^3 x \, dx.$$

Now we apply the reduction formula again to find

$$\int \sec^3 x \, dx = \frac{1}{2} \sec x \tan x + \frac{1}{2} \int \sec x \, dx$$

$$= \frac{1}{2} \sec x \tan x + \frac{1}{2} \ln |\sec x + \tan x|.$$

Combining the results, we finally have

$$\int \tan^2 x \sec^3 x \, dx = \frac{1}{4} \tan x \sec^3 x - \frac{1}{4} \left(\frac{1}{2} \sec x \tan x + \frac{1}{2} \ln |\sec x + \tan x| \right) + C$$

$$= \frac{1}{4} \tan x \sec^3 x - \frac{1}{8} \sec x \tan x + \frac{1}{8} \ln |\sec x + \tan x| + C. \quad \blacksquare$$

Trigonometric substitutions

A trigonometric substitution is useful in many cases where the integrand involves the sum or difference of squares of a constant a and a differentiable function u. Here is a guide for making trigonometric substitutions:

Form	Substitution to make	Pythagorean identity used	Substitute back for
$a^2 - u^2$	$u = a \sin \theta$ $du = a \cos \theta \, d\theta$	$a^2 - a^2 \sin^2 \theta$ $= a^2 \cos^2 \theta$	$\theta = \arcsin(\frac{u}{a})$
$a^2 + u^2$	$u = a \tan \theta$ $du = a \sec^2 \theta \, d\theta$	$a^2 + a^2 \tan^2 \theta$ $= a^2 \sec^2 \theta$	$\theta = \arctan(\frac{u}{a})$
$u^2 - a^2$	$u = a \sec \theta$ $du = a \sec \theta \tan \theta \, d\theta$	$a^2 \sec^2 \theta - a^2$ $= a^2 \tan^2 \theta$	$\theta = \arctan(\frac{u}{a})$

A.2 TECHNIQUES OF INTEGRATION

EXAMPLE 11 Find $\int \dfrac{1}{\sqrt{9-x^2}}\,dx$.

Solution The expression $9 - x^2$ is of the form $a^2 - u^2$ where $a = 3$ and $u = x$. We use the substitution

$$x = 3\sin\theta \quad \text{and} \quad dx = 3\cos\theta\,d\theta$$

and rewrite the integral as

$$\int \frac{1}{\sqrt{9-x^2}}\,dx = \int \frac{1}{\sqrt{9-(3\sin\theta)^2}}(3\cos\theta\,d\theta) = \int \frac{3\cos\theta\,d\theta}{\sqrt{9-9\sin^2\theta}}$$

$$= \int \frac{3\cos\theta\,d\theta}{3\sqrt{1-\sin^2\theta}} = \int \frac{\cos\theta\,d\theta}{\sqrt{\cos^2\theta}}.$$

Note that we can write $\sqrt{\cos^2\theta} = \cos\theta$ since $\theta = \arcsin\dfrac{x}{3}$ will always lie between $-\pi/2$ and $\pi/2$ and $\cos\theta$ must be positive. Thus, we have

$$\int \frac{\cos\theta\,d\theta}{\sqrt{\cos^2\theta}} = \int \frac{\cos\theta}{\cos\theta}\,d\theta = \int d\theta = \theta + C.$$

Substituting back for θ yields $\int \dfrac{1}{\sqrt{9-x^2}}\,dx = \arcsin\dfrac{x}{3} + C.$ ■

EXAMPLE 12 Find $\int \dfrac{x^2\,dx}{(25-x^2)^{3/2}}.$

Solution The integrand involves $25 - x^2$ which is of the form

$$a^2 - u^2,$$

where $a = 5$ and $u = x$. We make the substitution

$$x = 5\sin\theta \quad \text{and} \quad dx = 5\cos\theta,$$

resulting in

$$\int \frac{x^2\,dx}{(25-x^2)^{3/2}} = \int \frac{(25\sin^2\theta)(5\cos\theta\,d\theta)}{(25-25\sin^2\theta)^{3/2}} = \int \frac{\sin^2\theta\cos\theta\,d\theta}{\cos^3\theta} = \int \tan^2\theta\,d\theta.$$

Now we carry out the integration:

$$\int \tan^2\theta\,d\theta = \int (\sec^2\theta - 1)\,d\theta = \tan\theta - \theta + C.$$

Finally, we substitute for θ in terms of x. Using $x = 5\sin\theta$, we have

$$\theta = \arcsin\left(\frac{x}{5}\right) \quad \text{and} \quad \tan\theta = \frac{x}{\sqrt{25-x^2}},$$

and we can write

$$\int \frac{x^2\,dx}{(25-x^2)^{3/2}} = \frac{x}{\sqrt{25-x^2}} - \arcsin\left(\frac{x}{5}\right) + C.$$

We can check by differentiating that

$$\frac{d}{dx}\left(\frac{x}{\sqrt{25-x^2}} - \arcsin\left(\frac{x}{5}\right) + C\right) = \frac{x^2}{(25-x^2)^{3/2}}.$$

■

EXAMPLE 13 Find $\displaystyle\int \frac{x^2\,dx}{4+25x^2}$.

Solution This integral expression involves the form $a^2 + u^2$, where $a = 2$ and $u = 5x$. We use the substitution $u = a\tan\theta$ or $5x = 2\tan\theta$. Thus,

$$x = \frac{2}{5}\tan\theta \quad \text{and} \quad dx = \frac{2}{5}\sec^2\theta\,d\theta,$$

and

$$\int \frac{x^2\,dx}{4+25x^2} = \int \frac{(\frac{4}{25}\tan^2\theta)(\frac{2}{5}\sec^2\theta\,d\theta)}{4+25(\frac{4}{25}\tan^2\theta)} = \frac{2}{125}\int \frac{\tan^2\theta\sec^2\theta\,d\theta}{1+\tan^2\theta}.$$

Using the identity $1 + \tan^2\theta = \sec^2\theta$ we can simplify the integral to

$$\frac{2}{125}\int \frac{\tan^2\theta\sec^2\theta\,d\theta}{\sec^2\theta} = \frac{2}{125}\int \tan^2\theta\,d\theta = \frac{2}{125}(\tan\theta - \theta) + C.$$

Substituting back into terms of x, we note that

$$\tan\theta = \frac{5}{2}x \quad \text{and} \quad \theta = \arctan\left(\frac{5}{2}x\right).$$

Hence,

$$\int \frac{x^2\,dx}{4+25x^2} = \frac{2}{125}\left(\frac{5}{2}x - \arctan\left(\frac{5}{2}x\right)\right) + C = \frac{x}{25} - \frac{2}{125}\arctan\left(\frac{5}{2}x\right) + C.$$

■

A.2 TECHNIQUES OF INTEGRATION

EXAMPLE 14 Find $\int \dfrac{dx}{x^2\sqrt{2x^2-10}}$ for $x > \sqrt{5}$.

Solution This expression involves the form $u^2 - a^2$ where $u = \sqrt{2}x$ and $a = \sqrt{10}$. We use the substitution $u = a\sec\theta$ or $\sqrt{2}x = \sqrt{10}\sec\theta$.

With $x = \dfrac{\sqrt{10}}{\sqrt{2}}\sec\theta = \sqrt{5}\sec\theta$ and $dx = \sqrt{5}\sec\theta\tan\theta\,d\theta$, we have

$$\int \dfrac{dx}{x^2\sqrt{2x^2-10}} = \int \dfrac{\sqrt{5}\sec\theta\tan\theta\,d\theta}{5\sec^2\theta\sqrt{2(5\sec^2\theta)-10}}$$

$$= \dfrac{1}{\sqrt{50}}\int \dfrac{\tan\theta\,d\theta}{\sec\theta\sqrt{\sec^2\theta-1}}.$$

Using the identity $\tan^2\theta = \sec^2\theta - 1$ we have $\dfrac{1}{\sqrt{50}}\int \dfrac{\tan\theta\,d\theta}{\sec\theta\sqrt{\tan^2\theta}}$.

Now with $x > \sqrt{5}$, we must have $0 < \theta < \dfrac{\pi}{2}$ and $\tan\theta > 0$. Thus $\sqrt{\tan^2\theta} = \tan\theta$ and the integral simplifies to

$$\dfrac{1}{\sqrt{50}}\int \dfrac{d\theta}{\sec\theta} = \dfrac{1}{\sqrt{50}}\int \cos\theta\,d\theta = \dfrac{1}{\sqrt{50}}\sin\theta + C.$$

To rewrite this in terms of x, note that

$$\sin\theta = \sqrt{1-\cos^2\theta} = \sqrt{1-\dfrac{1}{\sec^2\theta}} = \sqrt{1-\dfrac{5}{x^2}}.$$

Thus,

$$\int \dfrac{dx}{x^2\sqrt{2x^2-10}} = \dfrac{1}{\sqrt{50}}\sqrt{1-\dfrac{5}{x^2}} + C = \dfrac{\sqrt{2x^2-10}}{10x} + C. \qquad \blacksquare$$

EXAMPLE 15 Find $\int \dfrac{dx}{2x^2+6x+7}$.

Solution In this case, we need to write the denominator of the integrand as a sum of a squared term and a constant term. We do this by "completing the square," as illustrated here:

$$2x^2+6x+7 = 2(x^2+3x)+7 = 2(x^2+2\cdot\dfrac{3}{2}x+(\dfrac{3}{2})^2)-2(\dfrac{3}{2})^2+7.$$

We can factor $(x^2+2(\tfrac{3}{2}x)+(\tfrac{3}{2})^2) = (x+\tfrac{3}{2})^2$ and write the quadratic as

$$2(x+\dfrac{3}{2})^2 - \dfrac{9}{2} + 7 = 2(x+\dfrac{3}{2})^2 + \dfrac{5}{2}.$$

The original integral can now be written as

$$\int \frac{dx}{2x^2+6x+7} = \int \frac{dx}{2(x+\frac{3}{2})^2+\frac{5}{2}} = \frac{1}{2}\int \frac{dx}{(x+\frac{3}{2})^2+\frac{5}{4}}.$$

Now we can proceed with a trigonometric substitution for the form a^2+u^2, where

$$a = \frac{\sqrt{5}}{2} \quad \text{and} \quad u = (x+\frac{3}{2}).$$

Substituting $x+\frac{3}{2} = \frac{\sqrt{5}}{2}\tan\theta$ and $du = dx = \frac{\sqrt{5}}{2}\sec^2\theta\, d\theta$, we have

$$\frac{1}{2}\int \frac{dx}{(x+\frac{3}{2})^2+\frac{5}{4}} = \frac{1}{2}\int \frac{\frac{\sqrt{5}}{2}\sec^2\theta\, d\theta}{\frac{5}{4}\tan^2\theta+\frac{5}{4}} = \frac{\sqrt{5}/2}{5/2}\int \frac{\sec^2\theta\, d\theta}{\tan^2\theta+1}$$

$$= \frac{\sqrt{5}}{5}\int \frac{\sec^2\theta\, d\theta}{\sec^2\theta} = \frac{\sqrt{5}}{5}\int d\theta = \frac{\sqrt{5}\theta}{5}+C = \frac{\sqrt{5}\arctan\left(\frac{2}{\sqrt{5}}(x+\frac{3}{2})\right)}{5}+C. \blacksquare$$

EXERCISES for Section A.2

For exercises 1-16: Find each of the antiderivatives, making use of the Pythagorean identities of trigonometry.

1. $\int \cos^3 x\, dx$
2. $\int \sin^2 x \cos^2 x\, dx$
3. $\int \sin^3 x \cos^2 x\, dx$
4. $\int \sin^6 x\, dx$
5. $\int \tan^3 x \sec^4 x\, dx$
6. $\int \tan^3 x \sec^3 x\, dx$
7. $\int \tan^6 x\, dx$
8. $\int \sqrt{\sin x} \cos^3 x\, dx$
9. $\int (\tan x + \cot x)^2\, dx$
10. $\int \sin^3 x\, dx$
11. $\int \cos^4 x\, dx$
12. $\int \sec^4 x\, dx$
13. $\int \csc^4 x \cot^4 x\, dx$
14. $\int \frac{\cos x}{2-\sin x}\, dx$
15. $\int \frac{\sec^2 x}{(1+\tan x)^2}\, dx$
16. $\int \cos^5 x\, dx$

For exercises 17-26: Find each of the antiderivatives, using an appropriate trigonometric substitution.

17. $\int \frac{x^2}{\sqrt{4-x^2}}\, dx$
18. $\int \frac{1}{x\sqrt{9+x^2}}\, dx$
19. $\int \frac{1}{x^2\sqrt{x^2-25}}\, dx$
20. $\int \frac{x}{\sqrt{4-x^2}}\, dx$

21. $\int \dfrac{1}{(x^2 - 1)^{3/2}}\, dx$

22. $\int \dfrac{1}{(36 + x^2)^2}\, dx$

23. $\int \sqrt{9 - 4x^2}\, dx$

24. $\int \dfrac{x}{(16 - x^2)^2}\, dx$

25. $\int \dfrac{x^3}{\sqrt{9x^2 + 49}}\, dx$

26. $\int \dfrac{1}{x^4 \sqrt{x^2 - 3}}\, dx$

For exercises 27-34: Find each of the antiderivatives, completing the square as necessary.

27. $\int \dfrac{1}{x^2 - 4x + 8}\, dx$

28. $\int \dfrac{1}{\sqrt{4x - x^2}}\, dx$

29. $\int \dfrac{2x + 3}{\sqrt{9 - 8x - x^2}}\, dx$

30. $\int \dfrac{1}{x^3 - 1}\, dx$

31. $\int \dfrac{1}{(x^2 + 4x + 5)^2}\, dx$

32. $\int \dfrac{1}{(x^2 + 6x + 13)^{3/2}}\, dx$

33. $\int \dfrac{1}{2x^2 - 3x + 9}\, dx$

34. $\int \dfrac{(4 + x^2)^2}{x^3}\, dx$

A.3 METHOD OF PARTIAL FRACTIONS

Simplification is in the eye of the beholder. You are no doubt familiar with the algebra involved in "simplifying" the sum of rational expressions. For example, given

$$\dfrac{2x - 4}{x^2 + 1} + \dfrac{4}{5x - 3},$$

we can rewrite this by finding a common denominator for the two rational expressions:

$$\dfrac{2x - 4}{x^2 + 1} + \dfrac{4}{5x - 3} = \dfrac{(2x - 4)(5x - 3)}{(x^2 + 1)(5x - 3)} + \dfrac{4(x^2 + 1)}{(5x - 3)(x^2 + 1)}$$

$$= \dfrac{(2x - 4)(5x - 3) + 4(x^2 + 1)}{(5x - 3)(x^2 + 1)}$$

$$= \dfrac{(10x^2 - 26x + 12) + (4x^2 + 4)}{5x^3 - 3x^2 + 5x - 3}$$

$$= \dfrac{14x^2 - 26x + 16}{5x^3 - 3x^2 + 5x - 3}.$$

This final expression is simpler in the sense that it is a single rational expression instead of the sum of two. However, given the choice of finding an antiderivative of either form, we will have a much simpler time dealing with the original expression, for

$$\int \left(\frac{2x-4}{x^2+1} + \frac{4}{5x-3} \right) dx = \int \frac{2x}{x^2+1} dx - 4 \int \frac{1}{x^2+1} dx + \int \frac{4}{5x-3} dx$$

$$= \ln(x^2+1) - 4 \arctan x + \frac{4 \ln|5x-3|}{5} + C.$$

(In the first integral, we made use of the substitution $u = x^2 + 1$, $du = 2x\,dx$; in the third integral, we made use of the substitution $u = 5x - 3$, $du = 5\,dx$.)

Partial fractions provide a technique for breaking down a rational function into simpler parts (at least for integration purposes). In other words, starting with a rational expression such as

$$\frac{14x^2 - 26x + 16}{5x^3 - 3x^2 + 5x - 3}$$

we want to obtain its *partial fraction decomposition*

$$\frac{2x-4}{x^2+1} + \frac{4}{5x-3}.$$

Strictly speaking, the method of partial fractions is simply an algebraic procedure, and not a technique of integration. While it is used to simplify integrals, the method is also useful in other areas of mathematics. (If you go on in your study of differential equations, you will encounter the method of partial fractions in simplifying *Laplace transforms*.)

We will describe a step-by-step procedure for the method, and then apply it to several integration examples.

Procedure for the Method of Partial Fractions

Problem: Find $\int \frac{p(x)}{q(x)} dx$, where $p(x)$ and $q(x)$ are polynomials.

Step 1. Check that the degree of the numerator is less than the degree of the denominator. If not, do long division and work with the remainder.

Step 2. Factor the denominator completely into linear and quadratic terms. (This can always be done, even though the algebra might be complicated.)

Step 3. Examine the factors of the denominator.

If the factor $(x + a)$ appears n times, associate with it the sum of n terms

$$\frac{A_1}{(x+a)} + \frac{A_2}{(x+a)^2} + \cdots + \frac{A_n}{(x+a)^n}.$$

If $(x^2 + bx + c)$ appears m times, associate with it the sum of m terms

$$\frac{B_1 x + C_1}{(x+a)} + \frac{B_2 x + C_2}{(x^2+bx+c)^2} + \cdots + \frac{B_m x + c_m}{(x^2+bx+c)^m}.$$

A.3 METHOD OF PARTIAL FRACTIONS

Step 4. Add up these terms from Step 3 and set the sum equal to the original fraction.

Step 5. Clear the denominators, multiply out, and collect terms.

Step 6. Set the corresponding coefficients of the numerator of each side of the equation equal to each other. Use these equations to solve for the unknowns introduced in Step 3.

Step 7. Now each of these partial fractions can be antidifferentiated using either linear substitutions or trigonometric substitutions.

EXAMPLE 16 Find $\int \dfrac{2-x}{x^2 + x}\, dx$.

Solution To find the antiderivative, we use the method of partial fractions. The factors of the denominator are x and $(x+1)$. Now we find must find two real numbers, A and B, such that

$$\frac{2-x}{x^2+x} = \frac{A}{x} + \frac{B}{x+1}.$$

Multiply both sides of this equation by the common denominator $x(x+1)$, to obtain

$$2 - x = A(x+1) + B(x).$$

When $x = 0$, then $2 - 0 = A(0+1) + B(0)$, and we see that $A = 2$. When $x = -1$, then $2 - (-1) = A(0) + B(-1)$ and we see that $B = -3$. We can now write the integral as follows:

$$\int \frac{2-x}{x^2+x}\, dx = \int \left(\frac{2}{x} - \frac{3}{x+1}\right) dx$$

$$= 2 \int \frac{1}{x}\, dx - 3 \int \frac{1}{x+1}\, dx$$

$$= 2 \ln x - 3 \ln|x+1| + C.$$

∎

EXAMPLE 17 Find $\int \dfrac{3x+11}{(x+2)(x+3)} dx$

Solution To find the antiderivative, we use the method of partial fractions. First, we write

$$\frac{3x+11}{(x+2)(x+3)} = \frac{A}{x+2} + \frac{B}{x+3} = \frac{A(x+3) + B(x+2)}{(x+2)(x+3)}$$

The numerator of the right-hand side is

$$A(x+3) + B(x+2) = Ax + 3A + Bx + 2B = (A+B)x + (3A+2B).$$

Setting the numerators equal, we have

$$(A+B)x + (3A+2B) = 3x + 11.$$

Equating the coefficients of like powers of x, we have

$$A + B = 3 \quad \text{and} \quad 3A + 2B = 11.$$

Solving for A and B, we find that $A = 5$ and $B = -2$. Thus,

$$\frac{3x+11}{(x+2)(x+3)} = \frac{5}{x+2} - \frac{2}{x+3},$$

and

$$\int \frac{3x+11}{(x+2)(x+3)} dx = \int \left(\frac{5}{x+2} - \frac{2}{x+3} \right) dx = 5 \int \frac{dx}{x+2} - 2 \int \frac{dx}{x+3}.$$

Using the formula $\int \dfrac{du}{u} = \ln |u| + C$, where $u = (x+2)$ in the first integral, $u = (x+3)$ in the second, and $du = dx$ in both. We obtain

$$5 \ln|x+2| - 2 \ln|x+3| + C. \qquad \blacksquare$$

EXAMPLE 18 Find $\int \dfrac{x^3 + 6x^2 + 3x + 6}{x^3 + 2x^2} dx.$

Solution Since the denominator is of the same power as the numerator, we can divide the numerator by the denominator, and write the integral as :

$$\int \left(1 + \frac{4x^2 + 3x + 6}{x^3 + 2x^2} \right) dx = \int 1 \, dx + \int \frac{4x^2 + 3x + 6}{x^3 + 2x^2} dx$$

$$= x + \int \frac{4x^2 + 3x + 6}{x^3 + 2x^2} dx.$$

We now use partial fractions on the "remainder" integral.

A.3 METHOD OF PARTIAL FRACTIONS

Since $x^3 + 2x^2$ can be factored as $x^2(x+2)$, we have three terms:

$$\frac{A}{x}, \quad \frac{B}{x^2}, \quad \text{and} \quad \frac{C}{(x+2)}.$$

We now find three numbers A, B, and C such that

$$\frac{4x^2 + 3x + 6}{x^3 + 2x^2} = \frac{A}{x} + \frac{B}{x^2} + \frac{C}{x+2}.$$

Multiply both sides of the equation by $x^3 + 2x^2$ to obtain

$$4x^2 + 3x + 6 = Ax(x+2) + B(x+2) + Cx^2$$
$$= Ax^2 + 2Ax + Bx + 2B + Cx^2$$
$$= (A+C)x^2 + (2A+B)x + 2B.$$

Equate the coefficients of like powers of x to find

$$A + C = 4 \qquad 2A + B = 3 \qquad 2B = 6.$$

Therefore, $A = 0$, $B = 3$, and $C = 4$, and we substitute these values to finally obtain

$$\int \left(1 + \frac{4x^2 + 3x + 6}{x^3 + 2x^2}\right) dx = x + \int \frac{4x^2 + 3x + 6}{x^3 + 2x^2} dx$$
$$= x + \int \left(\frac{0}{x} + \frac{3}{x^2} + \frac{4}{x+2}\right) dx$$
$$= x - \frac{3}{x} + 4\ln|x+2| + C.$$

∎

EXAMPLE 19 Find $\displaystyle\int \frac{x^3 + 5x^2 + 2x - 4}{(x^4 - 1)} dx$.

Solution To antidifferentiate, we use the method of partial fractions. The denominator $(x^4 - 1)$ can be factored as

$$x^4 - 1 = (x^2 - 1)(x^2 + 1) = (x-1)(x+1)(x^2+1).$$

Thus,

$$\frac{x^3 + 5x^2 + 2x - 4}{(x^4 - 1)} = \frac{A}{x-1} + \frac{B}{x+1} + \frac{Cx+D}{x^2+1}$$

$$= \frac{A(x+1)(x^2+1) + B(x-1)(x^2+1) + (Cx+D)(x-1)(x+1)}{(x-1)(x+1)(x^2+1)}.$$

Multiplying out the terms in the numerator of the left-hand side and collecting like terms, we have

$$A(x+1)(x^2+1) + B(x-1)(x^2+1) + (Cx+D)(x-1)(x+1)$$
$$= Ax^3 + Ax^2 + Ax + A + Bx^3 - Bx^2 + Bx - B + Cx^3 + Dx^2 - Cx - D$$
$$= (A+B+C)x^3 + (A-B+D)x^2 + (A+B-C)x + (A-B-D).$$

Setting this numerator equal to $x^3 + 5x^2 + 2x - 4$, and equating coefficients of like powers of x, we obtain

$$A + B + C = 1 \qquad A - B + D = 5$$

$$A + B - C = 2 \qquad A - B - D = -4$$

The solutions to these equations are

$$A = 1, \quad B = \frac{1}{2}, \quad C = -\frac{1}{2}, \quad \text{and} \quad D = \frac{9}{2}.$$

Thus,

$$\frac{x^3 + 5x^2 + 2x - 4}{(x^4 - 1)} = \frac{1}{x-1} + \frac{\frac{1}{2}}{x+1} + \frac{-\frac{1}{2}x + \frac{9}{2}}{x^2+1}$$

$$= \frac{1}{x-1} + \frac{1}{2(x+1)} + \frac{9-x}{2(x^2+1)}$$

$$= \frac{1}{x-1} + \frac{1}{2(x+1)} + \frac{9}{2}\left(\frac{1}{x^2+1}\right) - \frac{1}{2}\left(\frac{x}{x^2+1}\right).$$

Then $\displaystyle\int \frac{x^3 + 5x^2 + 2x - 4}{x^4 - 1} \, dx$

$$= \int \frac{dx}{x-1} + \frac{1}{2}\int \frac{dx}{x+1} + \frac{9}{2}\int \frac{dx}{x^2+1} - \frac{1}{2}\int \frac{x\,dx}{x^2+1}$$

$$= \ln|x-1| + \frac{1}{2}\ln|x+1| + \frac{9}{2}\arctan x - \frac{1}{4}\ln(x^2+1) + C.$$

(Note, we used the substitution $\displaystyle\int \frac{du}{u} = \ln|u| + C$ three times: on the first integral with $u = x - 1$ and $du = dx$, on the second integral with $u = x + 1$ and $du = dx$, and on the fourth integral with $u = x^2 + 1$ and $du = 2x\,dx$.) ∎

A.3 METHOD OF PARTIAL FRACTIONS

EXAMPLE 20 Find $\int \dfrac{x^2 + 4x}{(x-2)^2(x^2+4)}\, dx$.

Solution To antidifferentiate, we use the method of partial fractions to write

$$\dfrac{x^2+4x}{(x-2)^2(x^2+4)} = \dfrac{A}{x-2} + \dfrac{B}{(x-2)^2} + \dfrac{Cx+D}{(x^2+4)}$$

$$= \dfrac{A(x-2)(x^2+4) + B(x^2+4) + (Cx+D)(x-2)^2}{(x-2)^2(x^2+4)}.$$

Expanding the numerator, we have

$$A(x-2)(x^2+4) + B(x^2+4) + (Cx+D)(x-2)^2$$
$$= (A+C)x^3 + (-2A+B-4C+D)x^2 + (4A+4C-4D)x + (-8A+4B+4D).$$

Equate the coefficients of like powers of x to obtain the equations

$$A + C = 0 \qquad\qquad -2A + B - 4C + D = 1$$

$$4A + 4C - 4D = 4 \qquad\qquad -8A + 4B + 4D = 0.$$

Solving these equations, we find

$$A = \dfrac{1}{4}, \qquad B = \dfrac{3}{2}, \qquad C = -\dfrac{1}{4} \qquad \text{and} \qquad D = -1.$$

Hence,

$$\int \dfrac{x^2+4x}{(x-2)^2(x^2+4)}\, dx = \int \dfrac{\frac{1}{4}}{x-2}\, dx + \int \dfrac{\frac{3}{2}}{(x-2)^2}\, dx - \int \dfrac{\frac{1}{4}x-1}{x^2+4}\, dx$$

$$= \dfrac{1}{4}\int \dfrac{dx}{x-2} + \dfrac{3}{2}\int \dfrac{dx}{(x-2)^2} - \dfrac{1}{4}\int \dfrac{x\, dx}{x^2+4} - \int \dfrac{dx}{x^2+4}.$$

We can use the substitution $u = x-2$ and $du = dx$ on the first two integrals, and $u = x^2 + 4$ and $du = 2x\, dx$ on the third to obtain

$$\int \dfrac{x^2+4x}{(x-2)^2(x^2+4)}\, dx$$

$$= \dfrac{1}{4}\ln|x-2| - \dfrac{3}{2(x-2)} - \dfrac{1}{8}\ln(x^2+4) - \dfrac{1}{2}\arctan\left(\dfrac{x}{2}\right) + C.$$

EXAMPLE 21 Find $\int \dfrac{x^4 - x^3 + 2x^2 - x + 2}{(x-1)(x^2+2)^2}\,dx$.

Solution To antidifferentiate, we use the method of partial fractions to write

$$\frac{x^4 - x^3 + 2x^2 - x + 2}{(x-1)(x^2+2)^2} = \frac{A}{x-1} + \frac{Bx+C}{x^2+2} + \frac{Dx+E}{(x^2+2)^2}$$

$$= \frac{A(x^2+2)^2 + (Bx+C)(x-1)(x^2+2) + (Dx+E)(x-1)}{(x-1)(x^2+2)^2}.$$

Expanding the numerator and collecting like terms, we have

$$A(x^2+2)^2 + (Bx+C)(x-1)(x^2+2) + (Dx+E)(x-1)$$
$$=(A+B)x^4 + (C-B)x^3 + (4A-C+2B+D)x^2$$
$$+ (2C-2B+E-D)x + (4A-2C-E).$$

Now equate the coefficients of like powers of x with $x^4 - x^3 + 2x^2 - x + 2$, the numerator of the left-hand side:

$$A+B=1, \qquad C-B=-1,$$

$$4A-C+2B+D=2, \qquad 2C-2B+E-D=-1, \qquad 4A-2C-E=2.$$

Solving for A, B, C, D, and E we find

$$A=\frac{1}{3}, \qquad B=\frac{2}{3}, \qquad C=-\frac{1}{3}, \qquad D=-1, \qquad E=0.$$

Substituting these values, we have $\int \dfrac{x^4 - x^3 + 2x^2 - x + 2}{(x-1)(x^2+2)^2}\,dx$

$$= \frac{1}{3}\int \frac{dx}{x-1} + \int \frac{\frac{2}{3}x - \frac{1}{3}}{x^2+2}\,dx - \int \frac{x\,dx}{(x^2+2)^2}$$

$$= \frac{1}{3}\int \frac{dx}{x-1} + \frac{1}{3}\int \frac{2x\,dx}{x^2+2} - \frac{1}{3}\int \frac{dx}{x^2+2} - \int \frac{x\,dx}{(x^2+2)^2}.$$

For the first integral, use the substitution $u = x - 1$ and $du = dx$.

For the second and fourth integrals, use the substitution $u = x^2 + 2$ and $du = 2x\,dx$. For the third integral, use the formula

$$\int \frac{du}{u^2+a^2} = \frac{1}{a}\arctan\frac{u}{a},$$

letting $u = x$, and $a = \sqrt{2}$. Finally, we obtain $\int \dfrac{x^4 - x^3 + 2x^2 - x + 2}{(x-1)(x^2+2)^2}\,dx$

$$= \frac{1}{3}\ln|x-1| + \frac{1}{3}\ln(x^2+2) - \frac{\sqrt{2}}{6}\arctan\frac{x}{\sqrt{2}} + \frac{1}{2(x^2+2)} + C.$$

■

EXERCISES for Section A.3

For exercises 1-16: Find each of the antiderivatives, using the method of partial fractions.

1. $\int \dfrac{5x-12}{x(x-4)}\,dx$

2. $\int \dfrac{37-11x}{(x+1)(x-2)(x-3)}\,dx$

3. $\int \dfrac{6x-11}{(x-1)^2}\,dx$

4. $\int \dfrac{x+16}{x^2+2x-8}\,dx$

5. $\int \dfrac{5x^2-10x-8}{x^3-4x}\,dx$

6. $\int \dfrac{2x^2-25x-33}{(x-1)^2(x-5)}\,dx$

7. $\int \dfrac{9x^4+17x^3+3x^2-8x+3}{x^5+3x^4}\,dx$

8. $\int \dfrac{x^3+3x^2+3x+63}{(x^2-9)^2}\,dx$

9. $\int \dfrac{5x^2+11x+17}{x^3+5x^2+4x+20}\,dx$

10. $\int \dfrac{x^2+3x+1}{x^4+5x^2+4}\,dx$

11. $\int \dfrac{2x^3+10x}{(x^2+1)^2}\,dx$

12. $\int \dfrac{x^3+3x-2}{x^2-x}\,dx$

13. $\int \dfrac{x^6-x^3+1}{x^4+9x^2}\,dx$

14. $\int \dfrac{2x^3-5x^2+46x+98}{(x^2+x-12)^2}\,dx$

15. $\int \dfrac{4x^3+2x^2-5x-18}{(x-4)(x+1)^3}\,dx$

16. $\int \dfrac{2x^4-2x^3+6x^2-5x+1}{x^3-x^2+x-1}\,dx$

A.4 POLAR COORDINATES

A very useful alternative coordinate system for the plane is the **polar coordinate system**. To locate a point in the (two-dimensional) plane requires specifying two real numbers or coordinates. In the rectangular or Cartesian coordinate system, the two coordinates make reference to two perpendicular real axes intersecting at the origin.

We can imagine a point's coordinates (x, y) as specifying directions to a person at the origin. The x-coordinate gives the person the precise distance and direction to walk in the horizontal direction, and the y-coordinate gives the precise distance and direction to walk in the vertical direction. For example, we can locate the point $(3, -2)$ by walking 3 units to the right and 2 units down.

An alternative method of communicating a point's location would be to specify the point's distance from the origin and the direction to walk straight out to the point. For example, if we were given the information that a point located exactly 6 units from the origin along a line making a 40° angle measured counterclockwise from the positive x-axis, we can find the point precisely (see Figure A.8).

Rectangular coordinates specify directions parallel to the coordinate axes.

(3,–2)

Polar coordinates specify a radial distance from the origin and angle direction from the positive x-axis.

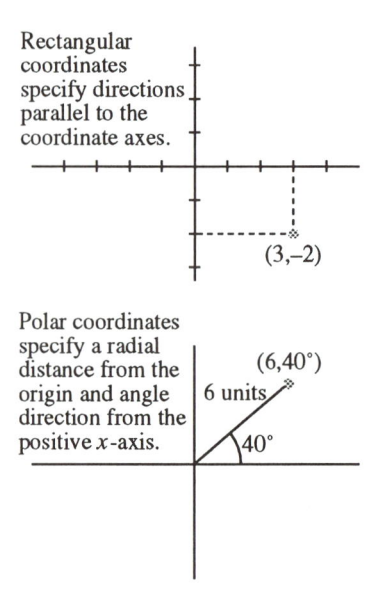

(6,40°)
6 units
40°

Figure A.8 Rectangular and polar coordinate systems.

The convention for specifying polar coordinates is to list the distance r and angle θ as an ordered pair (r, θ). The angle θ could be measured in either degrees or radians, but we will usually use radian measure.

EXAMPLE 22 Locate the following points given their polar coordinates:

$$P = (2, \frac{\pi}{4}), \qquad Q = (5, \frac{2\pi}{3}), \qquad R = (3, -\frac{5\pi}{6}) \qquad S = (2, \frac{9\pi}{4}).$$

Solution The points are illustrated in Figure A.9. Notice that the point $(2, \frac{9\pi}{4})$ represents the same point as $(2, \frac{\pi}{4})$. In other words, points P and S coincide.

A.4 POLAR COORDINATES

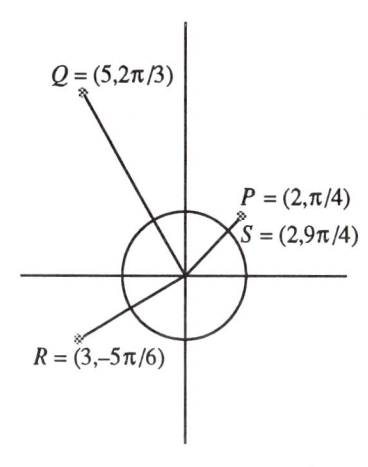

Figure A.9 Locating points with polar coordinates

Both the Cartesian and polar coordinate systems require an ordered pair of real numbers to specify a point's location on the plane. An important difference between the systems has to do with *uniqueness of representations*. In the Cartesian system, every point has exactly one ordered pair (x, y) associated with it. However, in the polar coordinate system, a point has infinitely many representations (r, θ) because

$$(r, \theta + 2\pi n)$$

gives the same location as (r, θ) for any integer n. For example, $(2, \frac{\pi}{4})$, $(2, \frac{9\pi}{4})$, $(2, \frac{17\pi}{4})$, and $(2, -\frac{15\pi}{4})$ all represent the same point.

A negative radius r can be specified as a polar coordinate by adopting the convention that $(-r, \theta)$ and $(r, \theta + \pi)$ represent the same point. Geometrically, the point $(-r, \theta)$ is the reflection of (r, θ) through the origin (see Figure A.10).

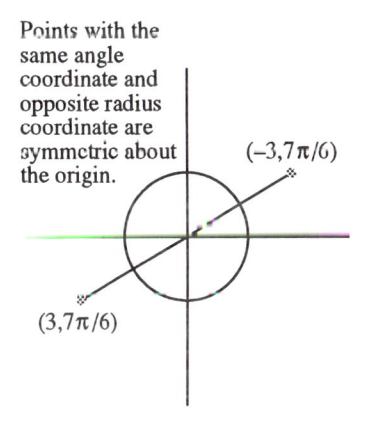

Figure A.10 Comparing (r, θ) and $(-r, \theta)$.

A special case occurs when $r = 0$. Notice that $(0, \theta)$ represents the origin, regardless of the value of θ.

For points other than the origin, we could gain a unique polar coordinate representation by arbitrarily restricting the radius r to be positive and the angle measure θ to lie in some predetermined interval of length 2π, say

$$0 \leq \theta < 2\pi$$

or

$$-\pi < \theta \leq \pi.$$

Such restrictions may be useful in certain applications, but we will generally allow both r and θ to take on any real number values.

Conversions between coordinate systems

Given a point whose location is specified by polar coordinates (r, θ), we can find the corresponding rectangular coordinates (x, y) through some simple trigonometry. Figure A.11 illustrates the situation for a point in the first quadrant.

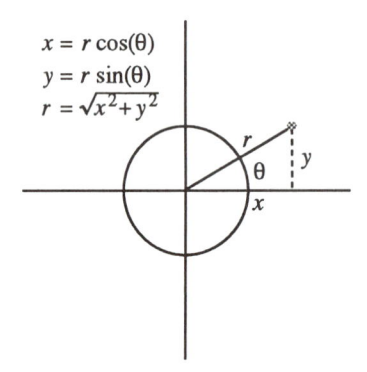

Figure A.11 Conversions between rectangular and polar coordinates.

Specifically,

$$x = r\cos(\theta) \quad \text{and} \quad y = r\sin(\theta).$$

A.4 POLAR COORDINATES

EXAMPLE 23 Convert the following from polar coordinates to rectangular coordinates:

$$(4, 3\pi/4) \quad \text{and} \quad (-2.5, -638°).$$

Solution For $(4, 3\pi/4)$, we have $r = 4$ and $\theta = 3\pi/4$, so

$$x = r\cos(\theta) = 4 \cdot \frac{-\sqrt{2}}{2} = -2\sqrt{2} \quad \text{and} \quad y = r\sin(\theta) = 4 \cdot \frac{\sqrt{2}}{2} = 2\sqrt{2}.$$

Hence, $(4, 3\pi/4)$ in polar coordinates corresponds to

$$(-2\sqrt{2}, 2\sqrt{2}) \approx (-2.8284, 2.8284)$$

in rectangular coordinates.

For $(-2.5, -638°)$, we have

$$x = (-2.5)\cos(-638°) \approx 0.3470 \quad \text{and} \quad y = (-2.5)\sin(-638°) \approx -2.4757,$$

so $(-2.5, -638°)$ in polar coordinates corresponds approximately to

$$(-0.3479, -2.4757)$$

in rectangular coordinates. ∎

Given a point whose location is specified with rectangular coordinates (x, y), we can find corresponding polar coordinates (r, θ) for the same point. First, we note that

$$r = \sqrt{x^2 + y^2}$$

by using the Pythagorean theorem.

If $r = 0$, then any value θ may be used. If $r \neq 0$, we can find a suitable value $0 \leq \theta < 2\pi$ by noting that

$$\sin(\theta) = y/r \quad \text{and} \quad \cos(\theta) = x/r.$$

More directly, if we compute $\arctan(y/x)$, then either

$$\theta = \arctan(y/x) \quad \text{or} \quad \theta = \arctan(y/x) + \pi,$$

depending on the quadrant (x, y) is in (the first value applies to Quadrants I and IV; the second value applies to Quadrants II and III).

EXAMPLE 24 Convert the following from rectangular coordinates to polar coordinates:

$$(3,4) \quad \text{and} \quad (-7.51, -6.28).$$

Solution   For $(3,4)$ we have $x = 3$ and $y = 4$, so

$$r = \sqrt{3^2 + 4^2} = \sqrt{25} = 5.$$

Since $(3,4)$ is in the first quadrant, we can use

$$\theta = \arctan(y/x) = \arctan(4/3) \approx 0.9273 \text{ radians}$$

or, using degree measure, $\theta \approx 53.13°$. Thus, $(3,4)$ in rectangular coordinates corresponds approximately to

$$(5, 0.9273) \quad \text{or} \quad (5, 53.13°)$$

in polar coordinates.

For $(-7.51, -6.28)$, we have

$$r = \sqrt{(-7.51)^2 + (-6.28)^2} \approx 9.79.$$

Since $(-7.51, -6.28)$ is in the third quadrant, we can use

$$\theta = \arctan(y/x) + \pi = \arctan\left(\frac{-6.28}{-7.51}\right) + \pi \approx 3.838 \text{ radians}$$

or, using degree measure, $\theta \approx 219.9°$. Thus, $(-7.51, -6.28)$ in rectangular coordinates corresponds approximately to

$$(9.79, 3.838) \quad \text{or} \quad (9.79, 219.9°)$$

in polar coordinates. ∎

Polar graphs

Sometimes it is more convenient to describe a curve using polar coordinates rather than Cartesian coordinates. This is particularly true when the curve is defined in terms of distance to a reference point. The simplest example of such a curve is a circle. For example, a circle of radius 3 centered at the origin has the equation

$$r = 3$$

in polar coordinates.

A.4 POLAR COORDINATES

The grid markings on graph paper correspond to constant values of the two coordinates. For Cartesian coordinates, the grid lines are horizontal and vertical, corresponding to equations of the form

$$x = a \quad \text{and} \quad y = b,$$

where a and b are constants.

For polar coordinates, the grid lines are concentric circles centered at the origin, corresponding to equations of the form

$$r = a,$$

and lines through the origin, corresponding to equations of the form

$$\theta = b,$$

where a and b are constants. Figure A.12 shows a polar grid that could be used as an aid in plotting polar graphs by hand.

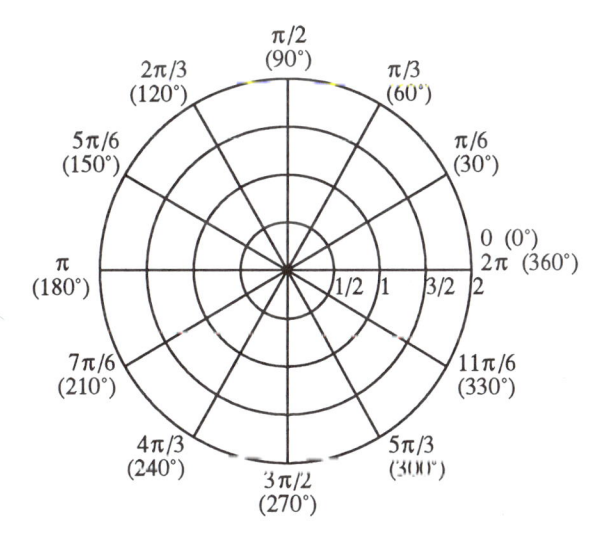

Figure A.12 Polar grid lines are concentric circles and lines through the origin.

A functional equation of the form

$$r = r(\theta)$$

describes the distance r from the origin as a function of the angle θ. You might think of a radar screen with a beam emanating from the origin sweeping around in a circle, recording the distance r out to a curve for each value θ.

EXAMPLE 25 Plot $r = 2\sin(\theta)$.

Solution Figure A.13 shows the plot of points for this polar equation. We can see that we have obtained what looks like a circle of radius 1 centered at the Cartesian point $(0, 1)$.

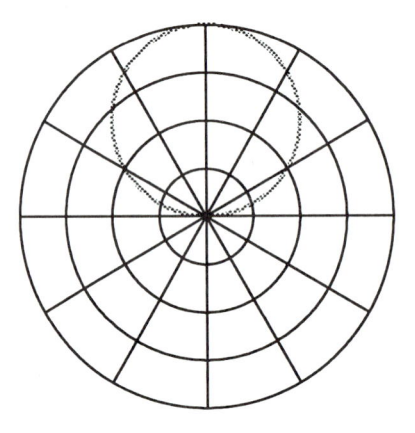

Figure A.13 Graph of $r = 2\sin(\theta)$.

Can we be sure this is a circle? If we multiply both sides of the equation by r, we obtain

$$r^2 = 2r\sin(\theta).$$

Converting to Cartesian coordinates, we find that this is equivalent to

$$x^2 + y^2 = 2y.$$

On the other hand, a circle of radius 1 centered at $(0, 1)$ will have the equation

$$x^2 + (y - 1)^2 = 1^2$$

or, after expanding,

$$x^2 + y^2 - 2y + 1 = 1.$$

We can see that this is equivalent to our polar equation. ∎

Some curves are much more easily described with polar coordinates than rectangular coordinates. Figure A.14 shows the graph of a "hyperbolic spiral" $r = 2\pi/\theta$ for $0 < \theta \leq 4\pi$. The radius r is undefined for $\theta = 0$, and the spiral approaches the horizontal asymptote $y = 2\pi$ as θ approaches 0 through positive values. In comparison, the rectangular description of this curve is much more complicated (try it!).

A.4 POLAR COORDINATES

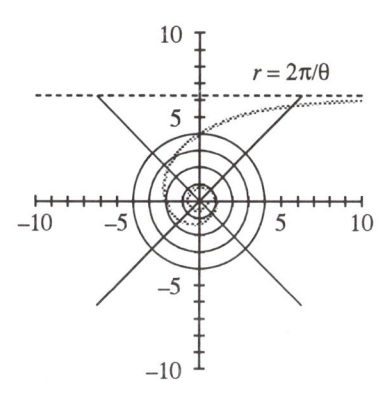

Figure A.14 The hyperbolic spiral $r = 2\pi/\theta$ for $0 < \theta \leq 4\pi$.

When r is a periodic function of θ (like many functions built up from trigonometric functions), then the graph of $r = r(\theta)$ will come back to its starting point over one period. In this case, the graph may determine one or more closed curves. The appearance of the resulting "loops" of such a curve provide the motivation for many colorful (and descriptive) names for particular polar graphs. For example, Figure A.15 shows an example of a heart-shaped polar graph, aptly named a "cardioid."

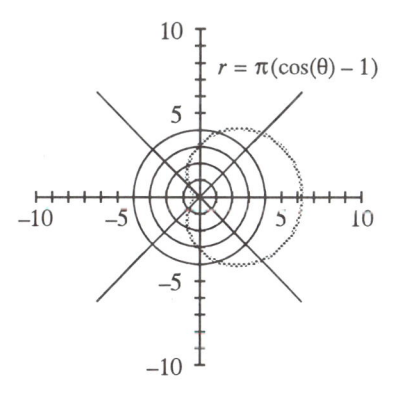

Figure A.15 The cardioid $r = \pi(\cos(\theta) - 1)$.

Polar graphs as parametric curves

We can think of polar plotting as a special case of parametric plotting. Recall that a curve can be described by expressing the coordinates as functions of a third variable (parameter) t:

$$(x(t), y(t)).$$

If we use t instead of θ as a parameter, then we can use the fact that $x = r \cos t$ and $y = r \sin t$ to express the curve

$$r = f(t)$$

parametrically as

$$x(t) = f(t)\cos(t)$$
$$y(t) = f(t)\sin(t).$$

EXAMPLE 26 Express the polar graph $r = 2\sin(t)$, $0 \le t \le 2\pi$ as a parametric curve $(x(t), y(t))$.

Solution Substituting into the usual conversion formulas for changing from polar to rectangular coordinates, we obtain

$$x(t) = 2\sin(t)\cos(t)$$
$$y(t) = 2\sin^2(t).$$

Try graphing this with a parametric plotter over the interval $0 \le t \le 2\pi$ to see that it does trace out the same circle of radius 1 with center at $(0, 1)$. ∎

Hence, any machine grapher capable of plotting parametric equations is automatically capable of plotting polar equations. Of course, some graphers have a special polar graphing feature.

Differentiation using polar coordinates

The slope $\frac{dy}{dx}$ of a polar graph $r = r(\theta)$ can be determined by using the chain rule and the two conversion formulas

$$x = r(\theta)\cos(\theta) \quad \text{and} \quad y = r(\theta)\sin(\theta)$$

to obtain

$$\frac{dy}{dx} = \frac{\frac{dy}{d\theta}}{\frac{dx}{d\theta}}.$$

Taking the indicated derivatives with respect to θ in the numerator and denominator yields

$$\frac{dy}{dx} = \frac{\frac{dr}{d\theta}\sin\theta + r(\theta)\cos\theta}{\frac{dr}{d\theta}\cos\theta - r(\theta)\sin\theta}.$$

A.4 POLAR COORDINATES

EXAMPLE 27 Find the slope of the polar graph of $r = 2\sin\theta$ at $\theta = \pi/3$.

Solution Using the derivative formula for $\dfrac{dy}{dx}$ we just developed, we have

$$\frac{dy}{dx} = \frac{2\cos\theta\sin\theta + 2\sin\theta\cos\theta}{2\cos\theta\cos\theta - 2\sin\theta\sin\theta} = \frac{2\cos\theta\sin\theta}{\cos^2\theta - \sin^2\theta} = \frac{\sin(2\theta)}{\cos(2\theta)} = \tan(2\theta).$$

For $\theta = \pi/3$, we have

$$\left.\frac{dy}{dx}\right|_{\theta=\pi/3} = \tan(2\pi/3) = -\sqrt{3}.$$

Examine the polar graph of $r = 2\sin\theta$ for $\theta = \pi/3$ to see that this slope value is reasonable.

Integration using polar coordinates

To see how we can extend definite integration to regions defined in terms of polar coordinates, let's first review the situation for rectangular coordinates. The area of a region bounded by the graph of $y = f(x)$ and the x-axis between $x = a$ and $x = b$ is given by the definite integral

$$\int_a^b |f(x)|\,dx.$$

The absolute value signs are necessary to guarantee that area both above and below the x-axis is counted as positive contributions to the total area of the region. The value $|f(x)|\,\Delta x$ is the area of a small rectangle of width Δx and height $|f(x)|$, and it represents the approximate area of the region swept out by the graph over a small change in the independent variable Δx. In the limiting process associated with definite integration, this value corresponds to the differential $|f(x)|\,dx$.

Now, let's use that same idea for polar graphs. Suppose $r = r(\theta)$ is given and we want to approximate the area swept out by the graph for a small change in the independent variable $\Delta\theta$. Figure A.16 illustrates the area swept out by such a graph.

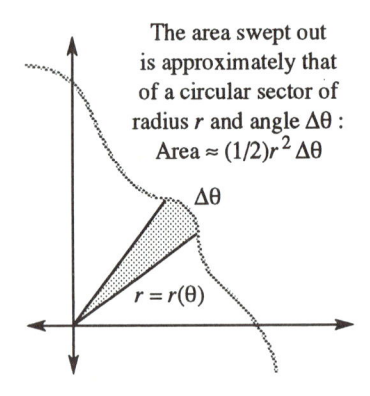

Figure A.16 Measuring area bounded by a polar graph.

The shaded region in the figure has approximately the area of a circular sector of angle $\Delta\theta$ and radius $r = f(\theta)$, for some θ in the tiny angle interval. Since a circle of radius r has area πr^2, the area of a circular sector of angle $\Delta\theta$ is given by

$$\frac{\Delta\theta}{2\pi}\pi r^2 = \frac{1}{2}r^2\Delta\theta.$$

To approximate the total area enclosed by the graph between $\theta = a$ and $\theta = b$, we can partition the angle interval into many such tiny subintervals and add the areas of the resulting circular sector area approximations:

$$\text{total area enclosed} \approx \sum_{i=1}^{n} \frac{1}{2}r^2(\theta_i)\Delta\theta,$$

where θ_i is the particular angle value chosen from each subinterval. The limiting value of these approximations as $\Delta\theta$ approaches 0 is given by a definite integral:

$$\text{total area enclosed} = \frac{1}{2}\int_a^b r^2(\theta)\, d\theta.$$

EXAMPLE 28 Using definite integration, find the total area bounded by the curve $r = 2\sin(\theta)$.

Solution   The graph of this polar function is shown in Figure A.17.

A.4 POLAR COORDINATES

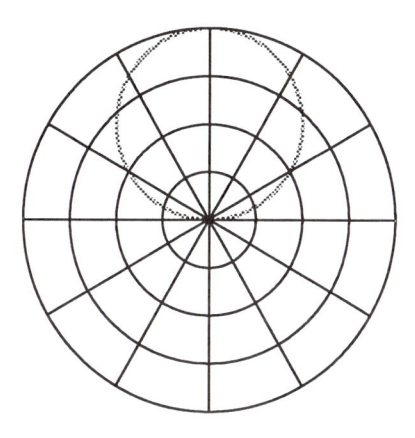

Figure A.17 Polar graph of $r = 2\sin(\theta)$.

In this case, we know that the graph is a circle of radius 1, so the area bounded by the graph certainly must be $\pi \cdot 1^2 = \pi$. To find this area by definite integration, we need to first determine the interval of integration. Although the period of $2\sin(\theta)$ is 2π, the entire graph is traced out over the interval $[0, \pi]$. Hence, the area is given by the integral

$$\frac{1}{2}\int_0^\pi r^2 \, d\theta = \frac{1}{2}\int_0^\pi 4\sin^2(\theta) \, d\theta = 2\int_0^\pi \sin^2(\theta) \, d\theta \approx 2 \cdot (1.5708) = 3.1416.$$

(The integral was computed by machine.) ∎

A polar graphing utility on a computer or calculator can be a great aid in checking that the interval of integration is the one you want. In the previous example, we would have obtained an erroneous area of 2π if we had used the interval $[0, 2\pi]$, because the circle is actually traced *twice* over this interval.

EXAMPLE 29 Find the total area of the cardioid $r = \pi(\cos(\theta) - 1)$.

Solution In this case, the entire cardioid is traced out over an interval $[0, 2\pi]$. Hence, the total area is given by

$$\frac{1}{2}\int_0^{2\pi} r^2 \, d\theta = \frac{1}{2}\int_0^{2\pi} \pi^2(\cos(\theta) - 1)^2 \, d\theta = \frac{\pi^2}{2}\int_0^{2\pi} (\cos^2(\theta) - 2\cos(\theta) + 1) \, d\theta = \frac{3\pi^3}{2}.$$

To judge the reasonableness of this answer, we can estimate the area from the graph (see Figure A.18).

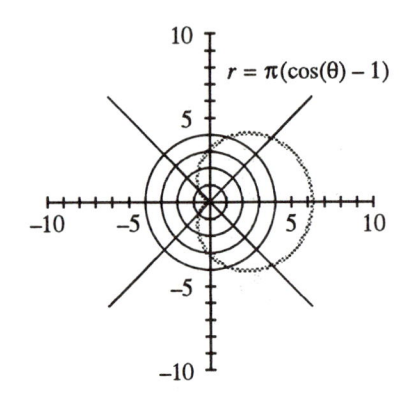

Figure A.18 Find the area of the cardioid.

The graph is roughly approximated by a circle of radius 4, whose area is $\pi \cdot 4^2 \approx 50$. On the other hand, the value we obtained for the area of the cardioid is

$$\frac{3\pi^3}{2} \approx 46.5.$$

More generally, the area bounded between the two graphs $y = f(x)$ and $y = g(x)$ between $x = a$ and $x = b$ is given by the definite integral

$$\int_a^b |f(x) - g(x)|\, dx.$$

For the region bounded between two polar graphs $r = r_1(\theta)$ and $r = r_2(\theta)$ between $\theta = a$ and $\theta = b$, we have the integral

$$\frac{1}{2}\int_a^b |r_1^2(\theta) - r_2^2(\theta)|\, d\theta.$$

The absolute value signs here simply indicate that we should always subtract the smaller squared radius from the larger in our integral.

EXERCISES for Section A.4

For exercises 1-8: Convert the indicated polar coordinates to rectangular coordinates.

1. $(3, 60°)$
2. $(1.414, -765°)$
3. $(3, -60°)$
4. $(-3.5, -600°)$
5. (π, π)
6. $(-5, -8\pi/3)$
7. $(0, 1000)$
8. $(1000, 0)$

For exercises 9-16: Convert the indicated rectangular coordinates to polar coordinates, both with θ measured in degrees and radians.

A.4 POLAR COORDINATES

9. $(3, 4)$
10. $(-5, -5)$
11. $(2.789, -3.254)$
12. $(-2.789, -3.254)$
13. (π, π)
14. $(-5, -8\pi/3)$
15. $(0, 1000)$
16. $(1000, 0)$

For exercises 17-26: Graph the given polar function over an appropriate interval (experiment with intervals having lengths that are multiples of π) for values of the parameter $a = 1/4$, $1/2$, 1, 2 and 4. On the basis of these graphs, determine what characteristic of the graph a controls. Predict the appearance of the graph for $a = 1/3$ and for $a = 3$, and then check your predictions by graphing the polar function for these parameter values. (When the given graph has a special name, it appears in parentheses.)

17. $r = 2a\cos(\theta)$ (Circle)
18. $r = a\sin(\theta)\cos^2(\theta)$ (Bifolium)
19. $r = a(\cos(\theta) + 1)$ (Cardioid)
20. $r = a\sin(\theta)\tan(\theta)$ (Cissoid of Diocles)
21. $r = a\sec(\theta) + 1$ (Conchoid of Nicomedes)
22. $r = \sec(\theta) + a$ (Conchoid of Nicomedes)
23. $r = 1 + a\cos(\theta)$ (Limaçon of Pascal)
24. $r = a + \cos(\theta)$ (Limaçon of Pascal)
25. $r = a\cos(2\theta)\sec(\theta)$ (Strophoid)
26. $r = 2a\theta/\pi$ (Spiral of Archimedes)

Conic sections (parabolas, hyperbolas, and ellipses) can all be described using polar coordinates. In fact, a single polar equation can generate all three types of conic sections:

$$r = \frac{ae}{1 - e\cos(\theta)}$$

where a and e are parameters (e is *not* the special transcendental number used in exponential functions). The value of e is called the **eccentricity** of the conic section.

For exercises 27-30: Let $a = 3$ and graph the polar function

$$r = \frac{3e}{1 - e\cos(\theta)}$$

over the interval $0 \leq \theta \leq 4\pi$ for values of the parameter $e = 1/4$, $1/2$, 1, 2 and 4.

27. On the basis of these graphs, guess which type of conic is generated for $e = 1/3$ and for $e = 3$, and then check your predictions by graphing the polar function for these values e.

28. Graph the polar function for $e = .1$, $e = .01$, and $e = .001$. What kind of curve is the limiting case as e approaches 0?

29. How will the answers to exercise 27 change if the value of a is changed from 3 to some other real value?

30. If $e = 1$, how does changing the value a change the appearance of the graph?

For exercises 31-34: A "rose" is obtained by graphing either of the polar equations
$$r = a\cos(k\theta) \quad \text{or} \quad r = a\sin(k\theta).$$
Graph both of these polar functions over the interval $0 \leq \theta \leq 12\pi$ for values of the parameter $a = 1/2$, 1, and 2 and for $k = 1, 2, 3,$ and 4 (twenty-four graphs in all).

31. What does the value a control? (Look at the size of the leaves.)

32. What does the value k control? (Count the number of the leaves.)

33. What does the choice of $\sin(\theta)$ or $\cos(\theta)$ control? (Look at the location of the leaves.)

34. On the basis of your answers to exercises 31-33, predict the appearance of the graphs of $r = 3\cos(5\theta)$ and $r = 1.5\sin(6\theta)$, including size, number, and location of the leaves of the rose. Then check your prediction by graphing the polar equations.

For exercises 35-41: Using as many examples from exercises 17-26 as you like, determine the graphical effect of each of the substitutions described. In other words, how is the new graph related to the old graph?

35. Replace θ with $-\theta$.

36. Replace θ with $(\pi + \theta)$.

37. Replace θ with $(\pi - \theta)$.

38. Replace r by $-r$.

39. Replace r by $-r$ and θ with $-\theta$.

40. Replace r by $-r$ and θ with $(\pi - \theta)$.

41. Replace θ by $(\theta + \pi/2)$.

For exercises 42-50: For each of the polar equations described, convert from the polar description of the curve to a rectangular description in terms of x and y. Then, using the original polar form, replace θ by t and express in terms of parametric equations $x(t)$ and $y(t)$. Graph both the polar and the parametric forms to check your answers. Find $\dfrac{dr}{d\theta}$ and the slope $\dfrac{dy}{dx}$ at $\theta = \pi/3$ and $\theta = 5\pi/4$. Judge the reasonableness of your slope values from the graph.

A.4 POLAR COORDINATES

42. $r = 6\cos(\theta)$
43. $r = 3\sin(\theta)\cos^2(\theta)$
44. $r = -2(\cos(\theta) + 1)$
45. $r = 5\sin(\theta)\tan(\theta)$
46. $r = 4\sec(\theta) + 1$
47. $r = \sec(\theta) - 7$
48. $r = 1 + \frac{1}{3}\cos(\theta)$
49. $r = -3 + \cos(\theta)$
50. $r = \cos(2\theta)\sec(\theta)$

For exercises 51-60: Find the area of region swept out by the graph over the interval $\pi/6 \leq \theta \leq \pi/4$. You will need to use a machine numerical integrator.

51. $r = 6\cos(\theta)$
52. $r = 3\sin(\theta)\cos^2(\theta)$
53. $r = -2(\cos(\theta) + 1)$
54. $r = 5\sin(\theta)\tan(\theta)$
55. $r = 4\sec(\theta) + 1$
56. $r = \sec(\theta) - 7$
57. $r = 1 + \frac{1}{3}\cos(\theta)$
58. $r = -3 + \cos(\theta)$
59. $r = \cos(2\theta)\sec(\theta)$
60. $r = 4\theta/\pi$

For exercises 61-68: Find the total area enclosed by the indicated polar graph.

61. $r = 1 - \sin(\theta)$
62. $r = 1 + \cos(\theta)$
63. $r = \sin(\theta) - 1$
64. $r = \cos(\theta) - 1$
65. $r = 1 + \cos(2\theta)$
66. $r = 1 + \sin(2\theta)$
67. $r = 1 - \sin(2\theta)$
68. $r = 1 - \cos(2\theta)$

For exercises 69-72: Find the area enclosed by the inner loop of the indicated polar graph.

69. $r = 1 + 2\cos(\theta)$
70. $r = 1 + 2\sin(\theta)$
71. $r = 1 - 2\sin(\theta)$
72. $r = 1 - 2\cos(\theta)$

For exercises 73-78: Find the total area enclosed by one "leaf" of the indicated polar graph.

73. $r = \cos(2\theta)$
74. $r = \sin(2\theta)$
75. $r = \cos(3\theta)$
76. $r = \sin(3\theta)$
77. $r = -\cos(3\theta)$
78. $r = -\sin(3\theta)$

For exercises 79-80: A "rose" is obtained by graphing either of the polar equations

$$r = a\cos(k\theta) \quad \text{or} \quad r = a\sin(k\theta),$$

where a is a positive real number and k is a positive integer.

79. Find the area of one leaf of the rose $r = a\cos(k\theta)$ in terms of a and k.
80. Find the area of one leaf of the rose $r = a\sin(k\theta)$ in terms of a and k.

A.5 COMPLEX NUMBERS

Points on a line can represent real numbers. We'll see that points in a plane can represent **complex numbers**. Usually, a complex number z is thought of as having the form $z = x + iy$, where x, y are real numbers and i formally satisfies the identity

$$i^2 = -1.$$

Since no *real* number satisfies this property, i has traditionally been called an **imaginary number**. This is rather unfortunate terminology, since i is just as "genuine" as any real number—it simply satisfies different properties.

If $z = x + iy$ is a complex number, we call x the **real part** and y the **imaginary part** of z and write

$$x = Re(z) \text{ and } y = Im(z).$$

Note that the imaginary part of a complex number is itself a *real number* **, because it represents the** *coefficient* **of i in** $z = x + yi$.

EXAMPLE 30 Identify $Re(z)$ and $Im(z)$ for each of the following complex numbers:

$$2 + 3i, \quad -7 + \frac{i}{2}, \quad -\pi i - \sqrt{2}, \quad \sqrt[3]{17}, \quad 13i, \quad 0$$

Solution

$Re(2 + 3i) = 2$ $\quad Im(2 + 3i) = 3$
$Re(-7 + \frac{i}{2}) = -7$ $\quad Im(-7 + \frac{i}{2}) = \frac{1}{2}$
$Re(-\pi i - \sqrt{2}) = -\sqrt{2}$ $\quad Im(-\pi i - \sqrt{2}) = -\pi$
$Re(\sqrt[3]{17}) = \sqrt[3]{17}$ $\quad Im(\sqrt[3]{17}) = 0$
$Re(13i) = 0$ $\quad Im(13i) = 13$
$Re(0) = 0$ $\quad Im(0) = 0.$ ∎

The real and/or the imaginary part of a complex number may be 0. If $z = x$ had no imaginary part ($Im(z) = 0$) then we call it a *pure real number*.

A.5 COMPLEX NUMBERS

If $z = iy$ has no real part ($Re(z) = 0$) then we call it a *pure imaginary number*. For example, $\sqrt[3]{17}$ is a pure real number; $13i$ is a pure imaginary number; 0 is *both* a pure real and pure imaginary number!

Arithmetic of complex numbers

Addition, subtraction and multiplication of complex numbers is accomplished by simply following the usual rules of algebra, and then using the identity $i^2 = -1$ to reduce the results if necessary.

EXAMPLE 31 Suppose $z = 2 + 3i$ and $w = -1 + 2i$. Find $z + w$, $z - w$, and zw.

Solution

$$z + w = (2 + 3i) + (-1 + 2i) = 1 + 5i$$

$$z - w = (2 + 3i) - (-1 + 2i) = 3 + i$$

$$zw = (2 + 3i)(-1 + 2i) = -2 - 3i + 4i + 6i^2 = -2 + i - 6 = -8 + i.$$

The **conjugate** $\bar{z}$ of a complex number z has the same real part, but the opposite imaginary part of z:

$$\text{if} \quad z = x + iy \quad \text{then} \quad \bar{z} = x - iy.$$

The product of any complex number and its conjugate is a pure real number, as we can verify in general. Writing $z = x + iy$ and $\bar{z} = x - iy$, note that

$$z\bar{z} = (x + iy)(x - iy) = x^2 - i^2 y^2 = x^2 + y^2.$$

The **modulus** or **absolute value** of a complex number z is *defined* as the positive square root of $z\bar{z}$:

$$|z| = \sqrt{z\bar{z}} = \sqrt{x^2 + y^2}.$$

Conjugates are useful in reducing the result of a quotient of two complex numbers as shown in the following example.

EXAMPLE 32   Suppose $z = 2 + 3i$ and $w = -1 + 2i$. Find $\dfrac{z}{w}$.

Solution   We can simplify $\dfrac{z}{w}$ by multiplying by $1 = \dfrac{\bar{w}}{\bar{w}}$.

$$\frac{z}{w} = \frac{2+3i}{-1+2i} = \frac{2+3i}{-1+2i} \cdot \frac{(-1-2i)}{(-1-2i)} = \frac{4-7i}{5} = \frac{4}{5} - \frac{7}{5}i.$$

∎

Ordered pair representations of complex numbers

If we use a Cartesian coordinate system for the plane, then we can associate each point to a complex number by identifying the real part with the x-coordinate and the imaginary part with the y-coordinate:

$$x + iy \quad \text{corresponds to the point} \quad (x, y).$$

In fact, we can refer to the x-axis as the *real axis* and the y-axis as the *imaginary axis*.

A.5 COMPLEX NUMBERS

EXAMPLE 33 Here are some correspondences between complex numbers and points in the Cartesian plane:

$$\begin{array}{lll} 2+3i & \text{corresponds to the point} & (2,3) \\ -1+2i & \text{corresponds to the point} & (-1,2) \\ 1 & \text{corresponds to the point} & (1,0) \\ i & \text{corresponds to the point} & (0,1) \\ 0 & \text{corresponds to the point} & (0,0). \end{array}$$

■

This identification allows us to drop any mention of the imaginary number i, if we wished. We could simply restrict ourselves to the ordered pair notation and define the arithmetic operations accordingly. If $z = (x_1, y_1)$ and $w = (x_2, y_2)$, then we have

$$z + w = (x_1, y_1) + (x_2, y_2) = (x_1 + x_2, y_1 + y_2)$$

$$z - w = (x_1, y_1) - (x_2, y_2) = (x_1 - x_2, y_1 - y_2)$$

$$zw = (x_1, y_1)(x_2, y_2) = (x_1 x_2 - y_1 y_2, x_1 y_2 + y_1 x_2)$$

$$\frac{z}{w} = \frac{(x_1, y_1)}{(x_2, y_2)} = \left(\frac{x_1 x_2 + y_1 y_2}{\sqrt{x_2^2 + y_2^2}}, \frac{-x_1 y_2 + y_1 x_2}{\sqrt{x_2^2 + y_2^2}} \right).$$

This is all well and good, but the fact of the matter is that it is sometimes much easier to maintain the "i" notation for computational purposes, because it only requires the usual algebraic rules and the single rule $i^2 = -1$, without the need for four more special definitions to handle the operations of $+$, $-$, $\cdot$ and $\div$ of ordered pairs.

Geometry of complex numbers

The advantage of using points to represent complex numbers becomes evident when we examine the geometrical interpretations they afford. In Figure A.19 we illustrate some of these interpretations.

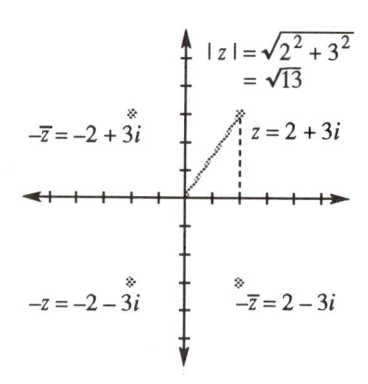

Figure A.19 Geometric interpretations of conjugates and absolute value.

The conjugate $\bar{z}$ is the reflection of z in the x-axis. The additive inverse $-z$ is the reflection of z in the origin. (The additive inverse of the conjugate $-\bar{z}$ is the reflection of z in the y-axis.) The absolute value $|z|$ is simply its distance from the origin (exactly the same geometrical meaning of absolute value for the real number line).

Addition of complex numbers also has a geometric interpretation, sometimes known as the parallelogram law: if we draw line segments from the origin to the points representing z and w, then the sum $z + w$ can be found by drawing the diagonal of the parallelogram determined by z and w (unless z and w are collinear with origin) as shown in Figure A.20.

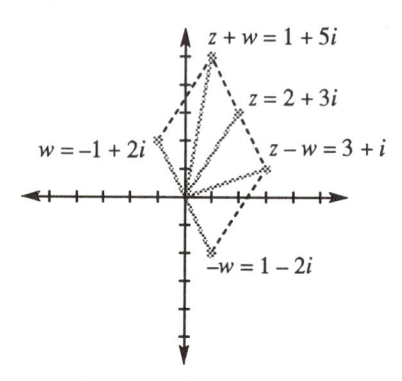

Figure A.20 Geometric interpretations of complex addition and subtraction: Parallelogram Law

If we write $z - w = z + (-w)$ then we can also find the difference of two complex numbers by the parallelogram law.

The geometric interpretation of complex number multiplication is best appreciated if we use polar coordinates.

For a complex number $z = x + iy$, we have

$$r = \sqrt{x^2 + y^2} = |z|,$$

so r is just the modulus (absolute value) of z.

A.5 COMPLEX NUMBERS

The **principal argument** of z is the angle $-\pi < \theta \leq \pi$ such that $x = r\cos\theta$ and $y = r\sin\theta$. Hence,

$$z = r(\cos(\theta) + i\sin(\theta)).$$

This polar form is used so often, there is a special shorthand notation for complex numbers written in this way. By definition,

$$\operatorname{cis}(\theta) = \cos\theta + i\sin\theta.$$

So

$$z = r\operatorname{cis}\theta.$$

We'll call this the polar form of z, in contrast to the rectangular form $z = x + iy$.

EXAMPLE 34 Express $4\operatorname{cis}(\frac{2\pi}{3})$ in rectangular form.

Solution

$$4\operatorname{cis}(\frac{2\pi}{3}) = 4(\cos(\frac{2\pi}{3}) + i\sin(\frac{2\pi}{3})) = 4[(-\frac{1}{2}) + i(\frac{\sqrt{3}}{2})] = -2 + 2\sqrt{3}i.$$

■

EXAMPLE 35 Express $-3 - 4i$ in polar form.

Solution If we convert $(-3, -4)$ to polar form, we obtain $(r, \theta) \approx (5, -2.2143)$. The polar form of the complex number is $5\operatorname{cis}(-2.2143)$

■

If we multiply two complex numbers written in polar form together, we notice a remarkable property. Suppose

$$z = r_1 \operatorname{cis}\theta_1 \quad \text{and} \quad w = r_2 \operatorname{cis}\theta_2$$

Then

$$zw = r_1 \operatorname{cis}\theta_1 \cdot r_2 \operatorname{cis}\theta_2 = r_1(\cos\theta_1 + i\sin\theta_1) \cdot r_2(\cos\theta_2 + i\sin\theta_2).$$

If we multiply the expression out using $i^2 = -1$, we find

$$zw = r_1 r_2[(\cos\theta_1 \cos\theta_2 - \sin\theta_1 \sin\theta_2) + i(\cos\theta_1 \sin\theta_2 + \sin\theta_1 \cos\theta_2)]$$

This looks rather messy, but if we examine the angle addition formulas for sine and cosine from trigonometry, we find

$$\cos(\theta_1 + \theta_2) = \cos\theta_1 \cos\theta_2 - \sin\theta_1 \sin\theta_2$$

$$\sin(\theta_1 + \theta_2) = \cos\theta_1 \sin\theta_2 + \sin\theta_1 \cos\theta_2$$

If we substitute these into our previous expression, we now have

$$zw = r_1 r_2 (\cos(\theta_1 + \theta_2) + i\sin(\theta_1 + \theta_2)).$$

Summarizing, we can see that if z has modulus r_1 and argument θ_1 and w has modulus r_2 and argument θ_2, then

$$zw \text{ has modulus } r_1 r_2 \text{ and argument } \theta_1 + \theta_2.$$

In other words, the modulus of the product is the product of the moduli, while the argument of the product is the *sum* of the arguments. (Note that we can't say *principal* argument here, since the sum of two arguments between $-\pi$ and π could fall outside that interval.)

A similar property holds for division of complex numbers:

$$\frac{z}{w} \text{ has modulus } \frac{r_1}{r_2} \text{ and argument } \theta_1 - \theta_2.$$

In other words, the modulus of the quotient is the quotient of the moduli (provided $r_2 \neq 0$), while the argument of the quotient is the *difference* of the arguments.

EXAMPLE 36 Suppose $z = 10\operatorname{cis}(\pi/2) = 10i$ and $w = \sqrt{2}\operatorname{cis}(-\pi/4) = 1 - i$. Calculate the product zw and quotient $\dfrac{z}{w}$ using both polar and rectangular forms, and then check that the results are equivalent between forms.

Solution Using the polar forms, we have

$$zw = 10\operatorname{cis}(\pi/2) \cdot \sqrt{2}\operatorname{cis}(-\pi/4) = 10\sqrt{2}\operatorname{cis}(\pi/2 + (-\pi/4)) = 10\sqrt{2}\operatorname{cis}(\pi/4),$$

and

$$\frac{z}{w} = \frac{10\operatorname{cis}(\pi/2)}{\sqrt{2}\operatorname{cis}(-\pi/4)} = \frac{10}{\sqrt{2}}\operatorname{cis}(\pi/2 - (-\pi/4)) = 5\sqrt{2}\operatorname{cis}(3\pi/4).$$

Using the rectangular forms, we have

$$zw = (10i)(1 - i) = 10i - 10i^2 = 10 + 10i,$$

and

$$\frac{z}{w} = \frac{10i}{1 - 1i} = \frac{10i}{1 - 1i}\frac{1 + 1i}{1 + 1i} = \frac{10i + 10i^2}{2} = -5 + 5i.$$

A.5 COMPLEX NUMBERS

We can check that these are equivalent by converting our polar results to rectangular form:

$$zw = 10\sqrt{2}\operatorname{cis}(\pi/4) = 10\sqrt{2}(\cos(\pi/4) + i\sin(\pi/4)) = 10\sqrt{2} \cdot \frac{1+i}{\sqrt{2}} = 10 + 10i,$$

and

$$\frac{z}{w} = 5\sqrt{2}\operatorname{cis}(3\pi/4) = 5\sqrt{2}(\cos(3\pi/4) + i\sin(3\pi/4)) = 5\sqrt{2} \cdot \frac{-1+i}{\sqrt{2}} = -5 + 5i.$$

We can see that the results match exactly. ∎

Powers and roots of complex numbers

The observation just made regarding the product of two complex numbers leads inductively to a result known as **de Moivre's Theorem**:

> If $z = \operatorname{cis}(\theta)$, then $z^n = \operatorname{cis}(n\theta)$ for any positive integer n.

This result follows simply by considering the power z^n as a product of n factors z. De Moivre's Theorem is a special case of a more general power rule for complex numbers:

> If $z = r\operatorname{cis}(\theta)$, then $z^n = r^n \operatorname{cis}(n\theta)$.

If we work backwards, then we can use this to find the nth root of a complex number $z = r\operatorname{cis}(\theta)$. If $w_0 = r^{1/n}\operatorname{cis}(\frac{\theta}{n})$, then

$$(w_0)^n = (r^{1/n})^n \operatorname{cis}(n \cdot (\frac{\theta}{n})) = r\operatorname{cis}(\theta) = z.$$

This is not the only nth root of z. In fact, every nonzero complex number z has exactly n distinct nth roots. To see why, note that $z = r\operatorname{cis}(\theta)$ can also be written as

$$z = r\operatorname{cis}(\theta + 2\pi k)$$

for any integer k. If we let

$$w_k = r^{1/n}\operatorname{cis}(\frac{\theta}{n} + \frac{2\pi k}{n}),$$

then we also have

$$(w_k)^n = r\operatorname{cis}(\theta + 2\pi k) = z.$$

Each w_k is a distinct complex number for $k = 0, 1, 2, \ldots, n-1$. When $k = n$ we have come "full circle," since

$$w_n = r^{1/n} \operatorname{cis}\left(\frac{\theta}{n} + \frac{2\pi n}{n}\right) = r^{1/n} \operatorname{cis}\left(\frac{\theta}{n} + 2\pi\right) = r^{1/n} \operatorname{cis}\left(\frac{theta}{n}\right) = w_0.$$

For any other integers $k > n$ or $k < 0$ we also obtain duplicates of the $w_0, w_1, w_2, \ldots, w_{n-1}$ that we have already accounted for.

All n of these roots have the same modulus $r^{1/n}$, so geometrically they are spaced at equal angles around a circle of radius $r^{1/n}$.

EXAMPLE 37 Find all four fourth roots of $16i = 16 \operatorname{cis} \frac{\pi}{2}$.

Solution Here $n = 4$, $r = 16$ and $\theta = \frac{\pi}{2}$. So,

$$r^{1/4} = 2 \quad \text{and} \quad \theta/n = \pi/8.$$

The four roots w_0, w_1, w_3, and w_4 will be spaced at angle intervals $2\pi/4 = \pi/2$ apart on a circle of radius 2 centered at the origin:

$$w_0 = 2\operatorname{cis}(\pi/8), \quad w_1 = 2\operatorname{cis}(5\pi/8), \quad w_2 = 2\operatorname{cis}(9\pi/8), \quad w_3 = 2\operatorname{cis}(13\pi/8).$$

The four roots are pictured in Figure A.21.

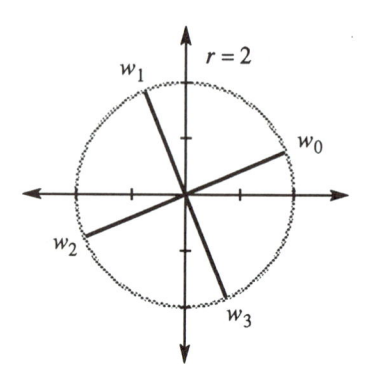

Figure A.21 The four fourth roots of $16i$.

EXERCISES for Section A.5

1. Convert $z = 4 - 7i$ to polar form $r \operatorname{cis}(\theta)$ and plot z.
2. Convert $z = 3.6 \operatorname{cis}(3.69)$ to rectangular form $x + yi$ and plot z.

For exercises 3-6: Evaluate the expressions, writing your answers in rectangular $(a + bi)$ form.

3. $(6 - 5i) + (-4 - 2i)$
4. $(-6 - 3i) - (2 + 8i)$

A.5 COMPLEX NUMBERS

5. $(6 - 5i)(-4 - 2i)$

6. $\dfrac{6 - 5i}{2 - 4i}$

7. Convert the complex numbers in exercises 5 and 6 to polar form and recompute their values in polar form. Then compute your results back to rectangular form and check with your original answers in rectangular form.

For exercises 8-10: Evaluate the expressions, writing your answers in polar $(r\,\text{cis}(\theta))$ form, where $-180° < \theta \leq 180°$.

8. $(25\,\text{cis}(-70°))(4\,\text{cis}(120°)$

9. $\dfrac{20\,\text{cis}(-40°)}{5\,\text{cis}(-10°)}$

10. $(3\,\text{cis}(20°))^{15}$

11. Shown in the illustration is one 9th root of a complex number. Graphically display the other 8 roots. Be as precise as possible. Are there any real roots? If so, what are they? Which pairs of roots are conjugates?

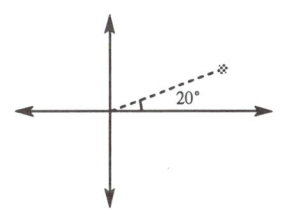

12. Suppose one of the six 6th roots of a complex number is the positive real number 3. What are the other five 6th roots?

13. For a given positive integer n, what is the maximum number of distinct real nth roots that a complex number z can have? Justify your answer.

14. Graph $f(x) = x^5 + 32$. What does the information from the sketch tell you about the type and number of roots (real vs. complex non-real) you can expect to find when solving $x^5 + 32 = 0$? Solve $x^5 + 32 = 0$ for all 5 roots.

15. Find all square roots of $z = -4 + 3i$. Plot these roots. Check your answers by squaring each.

16. Find all tenth roots of $z = 25$. Plot these roots. Check your answers by raising each to the tenth power.

17. Find all fourth roots of $z = 16$. Plot these roots. Check your answers by raising each to the fourth power.

18. Find all sixth roots of $z = 1$. Plot these roots. Check your answers by raising each to the sixth power. (These roots are called the sixth roots of unity.)

19. Solve $x^3 + 10 = 0$ for all 3 roots.

20. The transcendental number e can be raised to an imaginary power θi by using the definition

$$e^{\theta i} = \text{cis}(\theta).$$

Using this definition, verify that the following equation holds:

$$e^{\pi i} + 1 = 0.$$

This equation involves all five of the most important numbers in mathematics!

A.6 TAYLOR'S THEOREM

While it is true that a Taylor polynomial is not guaranteed to be a good approximation to a function other than very near the specified point $x = a$, we can get a handle on the error by means of a theorem which relates the error to the first "missing" derivative of the function.

Theorem A.1

Taylor's Theorem
Hypothesis 1: f is continuous and $(n+1)$-times differentiable on $[a, b]$.
Hypothesis 2: $p_n(x)$ is the nth degree Taylor polynomial approximation at $x = a$ to $f(x)$.
Conclusion: There is at least one point c, with $a < c < b$, such that

$$\frac{f^{(n+1)}(c)}{(n+1)!} = \frac{f(b) - p_n(b)}{(b-a)^{n+1}}.$$

Similarly, the theorem holds when the order of a and b are reversed ($b < c < a$).

The conclusion of Taylor's Theorem is often written as follows (after multiplying both sides by $(b-a)^{n+1}$ and adding $p_n(b)$ to both sides:

A.6 TAYLOR'S THEOREM

$$f(b) = p_n(b) + \frac{f^{(n+1)}(c)(b-a)^{n+1}}{(n+1)!}.$$

The quantity $\frac{f^{(n+1)}(c)}{(n+1)!}(b-a)^{n+1}$ is often called the **Lagrange form** of the remainder (error term) $R_n(b)$ of the Taylor polynomial. Lagrange (1736-1813) was an Italian-French mathematician (who played a leading role in the introduction of the metric system in France). Using this notation, the conclusion of Taylor's theorem can be written

$$f(b) = p_n(b) + R_n(b) = \sum_{k=0}^{n} \frac{f^{(k)}(a)}{k!}(b-a)^k + R_n(b).$$

In the special case $n = 0$, we can consider the Taylor polynomial to be the constant function $p_0(x) = f(a)$. The conclusion of Taylor's Theorem in this case becomes

$$f'(c) = \frac{f(b) - f(a)}{b - a},$$

which is simply a restatement of the Mean Value Theorem, and we can think of Taylor's Theorem as a generalization of the Mean Value Theorem.

Taylor's theorem can be used to turn an estimate for the $(n+1)$st derivative of a function into an estimate of the error in the n degree Taylor polynomial approximation:

$$|f(b) - p_n(b)| = \left| \frac{f^{(n+1)}(c)}{(n+1)!}(b-a)^{n+1} \right|.$$

EXAMPLE 38 Suppose f is continuous and five times differentiable on $[1,3]$, and that

$$\left| f^{(5)}(x) \right| < 0.1$$

for $x \in [1,3]$. What is the biggest possible error if we use fourth Taylor polynomial approximation at $x = 1$ to approximate $f(3)$?

Solution
$$|f(3) - p_4(3)| = \left| \frac{f^{(5)}(c)}{(5)!}(3-1)^5 \right| \quad \text{for some } 1 < c < 3 \text{ so that}$$

$$|f(3) - p_4(3)| < \left| \frac{0.1}{120} 2^5 \right| \approx 0.026667.$$

EXAMPLE 39 Use Taylor's Theorem to estimate the error at 0.5 for the fourth degree Maclaurin polynomial approximation to $y = \sin(x)$.

Solution Since $|\sin(x)| \leq 1$ and $|\cos(x)| \leq 1$ for all x. Together these mean that we have a simple estimate for the nth derivative of f, namely $|f^{(n)}(x)| \leq 1$ for all x. Using this estimate in the remainder, we have

$$|\sin(0.5) - p_4(0.5)| \leq \left|\frac{1}{120}(0.5)^5\right| \approx 0.00026.$$

■

EXAMPLE 40 For which values of x does the Maclaurin series for $f(x) = e^x$ converge to e^x?

Solution The remainder for the nth degree Maclaurin polynomial in this case is

$$R_n(x) = \frac{e^c}{(n+1)!} x^{n+1}$$

where c is between 0 and x (note that c depends on n). Since e^x is positive and increasing, we can estimate the remainder by noting that $e^c \leq e^{|x|}$ for any c between 0 and x. Using the ratio test, we find that the series converges for all x, so its terms must approach 0.

This means $\lim_{k \to \infty} \frac{x^k}{k!} = 0$, and since $|R_n(x)| \leq e^{|x|} \left|\frac{x^k}{k!}\right|$, we can conclude that $\lim_{k \to \infty} R_k(x) = 0$ *for any* x. ■

The remainder term $R_n(b)$ is sometimes expressed in its **integral form** as

$$R_n(b) = \frac{1}{n!} \int_a^b (b-x)^n f^{(n+1)}(x)\, dx.$$

EXERCISES for Section A.6

For exercises 1-10: Graphically determine the point c between the given a and b that determines the remainder $R_3(b)$ of the third degree Taylor polynomial. In other words, find where the graph of $y = f^{(4)}(x)/24$ crosses the graph of the horizontal line

$$y = \frac{f(b) - p_3(b)}{(b-a)^4}$$

A.6 TAYLOR'S THEOREM

over the interval $[a, b]$.

1. $y = \sin(2x + 1)$ $a = -1, \ b = -0.5$
2. $y = \cos(\frac{2\pi}{x^2+1})$ $a = 0, \ b = 1$
3. $y = \tan(x)$ $a = \pi/4, \ b = \pi/6$
4. $y = \sec(x)$ $a = 0, \ b = 0.5$
5. $y = \csc(-x)$ $a = \pi/2, \ b = 5\pi/6$
6. $y = -\cot(x)$ $a = 3\pi/4, \ b = 7\pi/8$
7. $y = \exp(2x + 1)$ $a = -1, \ b = 0$
8. $y = \frac{2}{x^2+1}$ $a = 0, \ b = 1$
9. $y = \sqrt{x^2 + 1}$ $a = 0, \ b = 1$
10. $y = \frac{1}{x}$ $a = 4, \ b = 4.5$

11. Calculate $\ln(1.2)$ with an error of at most 0.001 using a Taylor polynomial at 1 of $y = \ln(x)$.

12. Calculate $\sqrt{(1.2)}$ with an error of at most 0.001 using a Taylor polynomial at 1 of $y = \sqrt{(x)}$.

13. Calculate $e = \exp(1)$ with an error of at most 0.001 using a Taylor polynomial at 0 of $y = \exp(x)$.

14. Calculate $\sin(0.5)$ with an error of at most 0.0001 using a Taylor polynomial at 0 of $y = \sin(x)$.

DIFFERENTIATION PRACTICE

If you want lots of practice in finding derivatives "by the rules," here are 100 exercises to try. For each, find the derivative of the function indicated.

1. $\dfrac{ds}{dt}$, if $s = (2t+1)^2$
2. $\dfrac{dy}{dx}$, if $y = \dfrac{x^{-2} + 2x^{1/3}}{3x^3 - x + 1}$
3. $\dfrac{dM}{dw}$, if $M = \dfrac{w^2 - 4w + 3}{w^{3/2}}$
4. $\dfrac{dy}{dx}$, if $y = 10x^{3/2} + 3x^{-1/2}$
5. $\dfrac{dy}{dx}$, if $y = 6x^2 - 5x^{-1} + 2x^{-2/3}$
6. $\dfrac{dy}{dx}$, if $y = x^{-2} + 2x^{1/3}$
7. $\dfrac{dN}{dt}$, if $N = t^{2/3} - t^{-1/3}$
8. $\dfrac{dy}{dx}$, if $y = 6x^{4/3} + 4x^{-1/2}$
9. $\dfrac{dy}{dr}$, if $y = (5r - 4)^2$
10. $\dfrac{dx}{dw}$, if $x = (2w + 1)^3$
11. $\dfrac{dy}{dx}$, if $y = \sqrt{2x}$
12. $\dfrac{dy}{dx}$, if $y = \sqrt{4x^2 - 7x + 4}$
13. $\dfrac{dy}{dt}$, if $y = \sqrt{6t + 5}$
14. $\dfrac{dy}{ds}$, if $y = \sqrt[3]{5s - 8}$
15. $\dfrac{dy}{dz}$, if $y = \sqrt[3]{2z^2 + 4z + 8}$
16. $\dfrac{dy}{dv}$, if $y = \dfrac{1}{\sqrt{(v^4 + 7v^2)^3}}$
17. $\dfrac{dy}{dx}$, if $y = 6\sqrt[3]{x^4} + \dfrac{4}{\sqrt{x}}$
18. $\dfrac{dy}{dx}$, if $y = 10\sqrt{x^3} + \dfrac{3}{\sqrt[3]{x}}$
19. $\dfrac{dy}{dt}$, if $y = \sqrt[3]{t^2} - \dfrac{1}{\sqrt{t^3}}$
20. $\dfrac{dy}{dx}$, if $y = 6x^2 - \dfrac{5}{x} + \dfrac{2}{\sqrt[3]{x^2}}$
21. $\dfrac{dy}{dx}$, if $y = (x^5 - 4x + 8)^7$
22. $\dfrac{dy}{dx}$, if $y = (8x^3 - 2x^2 + x - 7)^5$
23. $\dfrac{dL}{dw}$, if $L = (w^4 - 8w^2 + 15)^4$
24. $\dfrac{dy}{dv}$, if $y = (17v - 5)^{1000}$
25. $\dfrac{dy}{dx}$, if $y = \dfrac{(3x^2 - 1)^4}{6}$
26. $\dfrac{dy}{ds}$, if $y = (3s)^{-4}$
27. $\dfrac{dT}{ds}$, if $T = (2s^{-4} + 3s^{-2} + 2)^{-6}$
28. $\dfrac{dP}{dv}$, if $P = (v^{-1} - 2v^{-2})^{-3}$
29. $\dfrac{dy}{dx}$, if $y = (3 + 2x^{-2} + 4x^{-3})^{1/3}$
30. $\dfrac{dy}{dt}$, if $y = \left[\left(1 + \dfrac{1}{t}\right)^{-1} + 1\right]^{-1}$
31. $\dfrac{dy}{dx}$, if $y = [7x + \sqrt{x^2 + 6}]^4$
32. $\dfrac{dy}{dx}$, if $y = [7x + \sqrt{x^3 + 3}]^6$
33. $\dfrac{dA}{dz}$, if $A = [1 + (1 + 2z)^{1/2}]^{1/2}$
34. $\dfrac{ds}{dt}$, if $s = \sqrt{t^2 + t + 1}\sqrt[3]{4t - 9}$
35. $\dfrac{dA}{ds}$, if $A = \sqrt[4]{s^2 + 9}(4s + 5)^4$
36. $\dfrac{dV}{dw}$, if $V = w^3(9w + 1)^5$
37. $\dfrac{dy}{dx}$, if $y = 7x(x^2 + 1)^2$
38. $\dfrac{dy}{dz}$, if $y = (2z + 5)^3(3z - 1)^4$
39. $\dfrac{dy}{dx}$, if $y = \sqrt{2x - 5}$
40. $\dfrac{dA}{ds}$, if $A = \sqrt[4]{s^2 + 9}$
41. $\dfrac{ds}{dt}$, if $s = \sqrt{t^2 + t + 1}$
42. $\dfrac{dy}{dx}$, if $y = \sqrt[3]{3x^3 - x + 1}$
43. $\dfrac{dy}{dx}$, if $y = 7x + \sqrt{x^2 + 6}$
44. $\dfrac{dy}{dz}$, if $y = [(z^2 - 1)^5 - 1]^5$
45. $\dfrac{dx}{dz}$, if $x = [z^2 + (z^2 + 9)^{1/2}]^{1/2}$
46. $\dfrac{dy}{dx}$, if $y = [(x - 6)^{-1} + 1]^{-1}$
47. $\dfrac{dy}{dx}$, if $y = [(x^2 - 6)^3 + 1]^3$
48. $\dfrac{dy}{dx}$, if $y = \left[\sqrt[4]{3x^2 - x + 1} + 1\right]^4$
49. $\dfrac{dy}{dx}$, if $y = \left(\dfrac{x}{x + 1}\right)^{5/2}$
50. $\dfrac{ds}{dt}$, if $s = \left(\dfrac{3t + 4}{6t - 7}\right)^3$

A APPENDICES

51. $\dfrac{dy}{dx}$, if $y = \left(\dfrac{3x^2 - 5}{2x^2 + 7}\right)^2$

52. $\dfrac{dy}{ds}$, if $y = \left(\dfrac{8s^2 - 4}{1 - 9s^3}\right)^4$

53. $\dfrac{dy}{dx}$, if $y = (6x - 7)^3(8x^2 + 9)^2$

54. $\dfrac{dy}{dw}$, if $y = (2w^2 - 3w + 1)(3w + 2)^4$

55. $\dfrac{dx}{dy}$, if $x = (15y + 2)(y^2 - 2)^{3/4}$

56. $\dfrac{dy}{dx}$, if $y = (x^2 + 4)^{5/3}(x^3 + 1)^{3/5}$

57. $\dfrac{dy}{dx}$, if $y = (x^6 + 1)^5(3x + 2)^3$

58. $\dfrac{dy}{dx}$, if $y = (3x - 8)^{-2}(7x^2 + 4)^{-3}$

59. $\dfrac{dx}{dy}$, if $x = (7y - 2)^{-2}(2y + 1)^{2/3}$

60. $\dfrac{dy}{dx}$, if $y = (3x + 1)^6 \sqrt{2x - 5}$

61. $\dfrac{dA}{dr}$, if $A = \sqrt{r}\sqrt{r + 1}\sqrt{r + 2}$

62. $\dfrac{ds}{dt}$, if $s = 2t(2t + 1)^2(2t + 3)^3$

63. $\dfrac{dy}{dx}$, if $y = \sqrt[4]{3x^3 - x + 1}(x^2 - 6)^3$

64. $\dfrac{dy}{dx}$, if $y = \sqrt[4]{3x^3 - x + 1}\left(x^{-2} + 2x^{1/3}\right)^3$

65. $\dfrac{dy}{dx}$, if $y = 8x^2\sqrt{x} + 3x^3\sqrt{x}$

66. $\dfrac{dy}{dw}$, if $y = \sqrt{w^3(9w + 1)^5}$

67. $\dfrac{dy}{dx}$, if $y = \sqrt[5]{(3x + 2)^4}$

68. $\dfrac{dy}{dv}$, if $y = \sqrt{(v^4 + 7v^2)^3}$

69. $\dfrac{dy}{dx}$, if $y = \sqrt{\dfrac{x^{-2} + 2x^{1/3}}{3x^3 - x + 1}}$

70. $\dfrac{dy}{dx}$, if $y = \sqrt{\dfrac{x^2 - 6}{3x^3 - x + 1}}$

71. $\dfrac{dy}{du}$, if $y = \sqrt{\dfrac{2u + 5}{7u - 9}}$

72. $\dfrac{dx}{dw}$, if $x = \dfrac{(w^2 + 1)^3}{(4w - 5)^5}$

73. $\dfrac{dy}{dz}$, if $y = \dfrac{9z^3 + 5z}{(6z + 1)^3}$

74. $\dfrac{dy}{dx}$, if $y = \dfrac{(x^2 - 6)^3}{\sqrt[4]{3x^2 - x + 1}}$

75. $\dfrac{dR}{dw}$, if $R = \dfrac{(w - 1)(w - 3)}{(w + 1)(w + 3)}$

76. $\dfrac{dA}{ds}$, if $A = \dfrac{1}{(8 - 5s + 7s^2)^{10}}$

77. $\dfrac{ds}{dt}$, if $s = \dfrac{4}{(9t^2 + 16)^{2/3}}$

78. $\dfrac{dy}{dx}$, if $y = \dfrac{6}{(3x^2 - 1)^4}$

79. $\dfrac{dy}{dx}$, if $y = \dfrac{7x(x^2 - 1)^2}{(3x + 10)^4}$

80. $\dfrac{dy}{dx}$, if $y = \dfrac{(x^{-2} + 2x^{1/3})^3}{\sqrt[4]{(3x^3 - x + 1)}}$

81. $f'(x)$, if $f(x) = \dfrac{3x + 1}{e^{x^2}}$

82. $f'(x)$, if $f(x) = \sin^{1/3}(4x)$

83. $f'(x)$, if $f(x) = \tan^{5/3}(x^2)$

84. $f'(x)$, if $f(x) = \cos^3(\sqrt{x})$

85. $f'(x)$, if $f(x) = \sec^3(4x^2 - 8)$

86. $f'(x)$, if $f(x) = \sin(x^3 + 1)^{1/3}$

87. $f'(x)$, if $f(x) = \tan^2(3x^2 - 5)^3$

88. $f'(x)$, if $f(x) = \sec(\sqrt[3]{x})$

89. $f'(x)$, if $f(x) = \sin^4(\csc(x))$

90. $f'(x)$, if $f(x) = e^{\cos(x)}$

91. $f'(x)$, if $f(x) = [\tan(e^x)]^3$

92. $f'(x)$, if $f(x) = e^{\sec(x^2)}$

93. $f'(x)$, if $f(x) = \tan(\sqrt{x})$

94. $f'(x)$, if $f(x) = \sin(e^{2x} + e^{-2x})$

95. $f'(x)$, if $f(x) = (e^x + e^{3x})^7$

96. $f'(x)$, if $f(x) = (x\cos(3x))^{1/2}$

97. $f'(x)$, if $f(x) = x\cos\left(\dfrac{1}{x}\right)$

98. $f'(x)$, if $f(x) = e^{-1/x}\tan(x^2)$

99. $f'(x)$, if $f(x) = \sin^{3/2}\left(\dfrac{1}{x}\right)e^{\cos(x)}$

100. $f'(x)$, if $f(x) = \tan\left(\dfrac{x^2 - 1}{x^3 + 1}\right)$

INTEGRATION PRACTICE

1. $\int x \arcsin x \, dx$
2. $\int \ln(1+x) \, dx$
3. $\int \cos^3 2x \sin^2 2x \, dx$
4. $\int \tan x \sec^5 x \, dx$
5. $\int \frac{1}{(x^2+25)^{3/2}} \, dx$
6. $\int \frac{\sqrt{4-x^2}}{x} \, dx$
7. $\int \frac{x^3+1}{x(x-1)^3} \, dx$
8. $\int \frac{x^3 - 20x^2 - 63x - 198}{x^4 - 81} \, dx$
9. $\int \frac{x}{\sqrt{4+4x-x^2}} \, dx$
10. $\int \frac{\sqrt[3]{x+8}}{x} \, dx$
11. $\int e^{2x} \sin 3x \, dx$
12. $\int \sin^3 x \cos^3 x \, dx$
13. $\int \frac{x}{\sqrt{4-x^2}} \, dx$
14. $\int \frac{x^5 - x^3 + 1}{x^3 + 2x^2} \, dx$
15. $\int \frac{1}{x^{3/2} + x^{1/2}} \, dx$
16. $\int e^x \sec e^x \, dx$
17. $\int x^2 \sin 5x \, dx$
18. $\int \sin^3 x \cos^{1/2} x \, dx$
19. $\int e^x \sqrt{1+e^x} \, dx$
20. $\int \frac{x^2}{\sqrt{4x^2+25}} \, dx$
21. $\int \sec^2 x \tan^2 x \, dx$
22. $\int x \cot x \csc x \, dx$
23. $\int x^2 (8-x^3)^{1/3} \, dx$
24. $\int \cos \sqrt{x} \, dx$
25. $\int \frac{e^{3x}}{1+e^x} \, dx$
26. $\int \frac{x^2 - 4x + 3}{\sqrt{x}} \, dx$
27. $\int \frac{x^3}{\sqrt{16-x^2}} \, dx$
28. $\int \frac{1-2x}{x^2+12x+35} \, dx$
29. $\int \arctan 5x \, dx$
30. $\int \frac{e^{\tan x}}{\cos^2 x} \, dx$
31. $\int \frac{1}{\sqrt{7+5x^2}} \, dx$
32. $\int \cot^6 x \, dx$
33. $\int x^3 \sqrt{x^2 - 25} \, dx$
34. $\int (x^2 - \text{sech}^2 4x) \, dx$
35. $\int x^2 e^{-4x} \, dx$
36. $\int \frac{3}{\sqrt{11 - 10x - x^2}} \, dx$
37. $\int \tan 7x \cos 7x \, dx$
38. $\int \frac{4x^2 - 12x - 10}{(x-2)(x^2-4x+3)} \, dx$
39. $\int (x^3+1) \cos x \, dx$
40. $\int \frac{\sqrt{9-4x^2}}{x^2} \, dx$
41. $\int (x - \cot 3x)^2 \, dx$
42. $\int \frac{1}{x(\sqrt{x} + \sqrt[4]{x})} \, dx$
43. $\int \frac{\sin x}{\sqrt{1+\cos x}} \, dx$
44. $\int \frac{x^2}{(25+x^2)^2} \, dx$
45. $\int \tan^3 x \sec x \, dx$
46. $\int \frac{2x^3 + 4x^2 + 10x + 13}{x^4 + 9x^2 + 20} \, dx$
47. $\int \frac{(x^2-2)^2}{x} \, dx$
48. $\int x^{3/2} \ln x \, dx$
49. $\int \frac{x^2}{\sqrt[3]{2x+3}} \, dx$
50. $\int x^3 e^{x^2} \, dx$

USEFUL FACTS AND FORMULAS

English Weights and Measures

LENGTH

12 inches = 1 foot
3 feet = 1 yard
$16\frac{1}{2}$ feet = 1 rod

63,360 inches = 1 mile
5,280 feet = 1 mile
1760 yards = 1 mile

AREA

144 square inches = 1 square foot
9 square feet = 1 square yard
160 square rods = 1 acre

$30\frac{1}{4}$ square yards = 1 square rod
4840 square yards = 1 acre
640 acres = 1 square mile

VOLUME

1728 cubic inches = 1 cubic foot   27 cubic feet = 1 cubic yard

DRY MEASURE

2 pints = 1 quart
8 quarts = 1 peck

4 pecks = 1 bushel
1 bushel = 2150.42 cubic inches

LIQUID MEASURE

8 ounces = 1 cup
2 cups = 1 pint
2 pints = 1 quart

4 quarts = 1 gallon
$31\frac{1}{2}$ gallons = 1 barrel
1 gallon = 231 cubic inches

WEIGHT

16 ounces = 1 pound   2000 pounds = 1 ton

TEMPERATURE

Freezing point of water occurs at 0° Celsius or 32° Fahrenheit.
Boiling point of water occurs at 100° Celsius or 212° Fahrenheit.
If C represents degrees Celsius and F represents degrees Fahrenheit, then

$$C = \frac{5}{9}(F - 32) \quad \text{and} \quad F = \frac{9}{5}C + 32.$$

Metric System

The metric weights and measures are derived by combining the following six numerical prefixes with the words *meter*, *gram*, and *liter*:

milli = one-thousandth   *deka* = ten
centi = one-hundredth   *hecto* = one hundred
deci = one-tenth   *kilo* = one thousand

For example,

10 millimeters = 1 centimeter	10 meters = 1 dekameter
10 centimeters = 1 decimeter	10 dekameters = 1 hectometer
10 decimeters = 1 meter	10 hectometers = 1 kilometer

APPROXIMATE METRIC EQUIVALENTS

1 inch	= 2.54001 centimeters
1 foot	= 30.48006 centimeters
1 yard	= 91.4402 centimeters
1 centimeter	= 0.032808 feet
1 mile	= 1.70935 kilometers
1 square mile	= 2,589,998 square meters
1 acre	= 4046.873 square meters
1 meter	= 39.37 inches

1 kilometer = 0.62137 mile = 1093.61 yards = 3,280.83 feet
1 kilogram = 2.20462 pounds = 35.27396 ounces
1 liter = 1.05671 liquid quarts = 0.908102 dry quart

1 pound	= 0.453592 kilograms
1 gallon	= 3.78533 liters
1 liquid quart	= 0.946333 liter
1 liter	= 61.0250 cubic inches
1 cubic inch	= 16.3872 cubic centimeters
1 cubic meter	= 35.314 cubic feet
1 dry quart	= 1.10120 liters
1 kilogram per square meter	= 0.204817 pound per square foot
1 pound per square fool	= 4.88241 kilograms per square meter
1 fluid ounce	= 29.573 cubic centimeters

Miscellaneous physical constants

g (average value) = 32.16 ft/sec^2 = 980 cm/sec^2

g (sea level, lat. 45°) = 32.172 ft/sec^2 = 980.616 cm./sec^2

1 hp (horsepower) = 550 ft-lb/min = 76.0404 kg-m/sec = 745.70 watts

Weight of 1 cubic foot of water = 62.425 lb (mass density)

Velocity of sound in dry air at 0° C = 33,136 cm/sec = 1,087 ft/sec

1 mph (miles per hour) = 88 ft/min = 1.467 ft/sec = 0.8684 knot

1 knot = 101.3 ft/min = 1.689 ft/sec = 1.152 mph

1 micron = 10^{-4} cm   1 angstrom unit = 10^{-8} cm

Mean radius of earth = 3,959 miles = 6,371 km

Equatorial diameter of earth = 7,926.68 miles = 12,756.78 km

Polar diameter of earth = 7,926.68 miles = 12,756.78 km

Constant of gravitation = 6.670 × 10^{-8} dyne· cm^2/gram^2

Electronic charge = 4.803 × 10^{-10} e.s.u.

Mass of electron = 9.107 × 10^{-28} gram

Mass of hydrogen atom = 1.673 × 10^{-24} gram

Avogadro's number = 6.023 × 10^{23} mole^{-1}

Planck's constant = 6.624 × 10^{-27} erg sec

Variation

If y varies directly as x, then $y = kx$.   (k = constant)

If y varies inversely as x, then $y = \dfrac{k}{x}$.

If y varies jointly as x and z, then $y = kxz$.

If y varies directly as x and inversely as z, then $y = \dfrac{kx}{z}$.

FORMULAS FROM ALGEBRA

Exponents

If p is a positive integer, $a^p = a \cdot a \cdot a \cdots$ to p factors.

$a^0 = 1 \quad (a \neq 0)$ \qquad $a^{-n} = \dfrac{1}{a^n}$

$a^m \cdot a^n = a^{m+n}$ \qquad $(ab)^n = a^n b^n$

$a^m \div a^n = a^{m-n}$ \qquad $\left(\dfrac{a}{b}\right)^n = \dfrac{a^n}{b^n}$

$(a^m)^n = a^{mn}$ \qquad $a^{p/q} = \sqrt[q]{a^p} = (\sqrt[q]{a})^p$

Radicals

If $r^q = A$, then r is a qth root of A. The **principal** qth **root** is denoted $\sqrt[q]{A}$.

$$\sqrt[q]{A} \text{ is } \begin{cases} \text{positive if } A \text{ is positive} \\ \text{negative if } A \text{ is negative and } q \text{ odd} \\ \text{imaginary if } A \text{ is negative and } q \text{ even} \end{cases}$$

$$\sqrt[q]{a^q} = \begin{cases} a \text{ if } a \geq 0 \\ a \text{ if } a < 0 \text{ and } q \text{ odd} \\ -a \text{ if } a < 0 \text{ and } q \text{ even} \end{cases}$$

$\sqrt[q]{a} \cdot \sqrt[q]{b} = \sqrt[q]{ab}$ except for $a < 0$, $b < 0$, and q even. For this case, use the rules given below for complex numbers.

$$\sqrt[q]{\dfrac{a}{b}} = \dfrac{\sqrt[q]{a}}{\sqrt[q]{b}}$$

Logarithms

Let M, N, b be positive and $b \neq 1$. Then

$\log_b MN = \log_b M + \log_b N$ \qquad $\log_b \dfrac{M}{N} = \log_b M - \log_b N$

$\log_b M^k = k \log_b M$ \qquad $\log_b \sqrt[q]{M} = \dfrac{1}{q} \log_b M$

$\log_b b = 1$ \qquad $\log_b 1 = 0$ \qquad $b^{\log_b M} = M$ \qquad $\log_a M = \dfrac{\log_b M}{\log_b a}$

$\ln M = \log_e M \approx 2.3026 \log_{10} M$ \qquad $\log_{10} M \approx (0.43429) \log_e M$

Complex numbers

$$i = \sqrt{-1}, \quad i^2 = -1, \quad i^3 = -i, \quad i^4 = 1, \quad i^5 = i, \quad \text{etc.}$$
$$i^{4p} = 1, \quad i^{4p+1} = i, \quad i^{4p+2} = -1, \quad i^{4p+3} = -i. \quad (p \text{ an integer})$$

$a + bi = c + di$ if and only if $a = c$ and $b = d$, and $a + bi = 0$ if and only if $a = b = 0$.

$$(a + bi) + (c + di) = (a + c) + (b + d)i$$
$$(a + bi) - (c + di) = (a - c) + (b - d)i$$
$$(a + bi)(c + di) = (ac - bd) + (ad + bc)i$$
$$\frac{a + bi}{c + di} = \frac{(a + bi)(c - di)}{(c + di)(c - di)} = \frac{ac + bd}{c^2 + d^2} + \frac{bc - ad}{c^2 + d^2}i$$

Polar form of complex numbers:

$a + bi = r(\cos\theta + i\sin\theta)$, where $r = \sqrt{a^2 + b^2}$ and $\tan\theta = \frac{b}{a}$.

Suppose $z = r(\cos\alpha + i\sin\alpha)$ and $w = R(\cos\beta + i\sin\beta)$. Then

$$zw = rR[\cos(\alpha + \beta) + i\sin(\alpha + \beta)]$$
$$\frac{z}{w} = \frac{r}{R}[\cos(\alpha - \beta) + i\sin(\alpha - \beta)]$$
$$z^n = r^n[\cos n\alpha + i\sin n\alpha]$$
$$z^{1/n} = r^{1/n}\left[\cos\frac{\alpha + k360°}{n} + i\sin\frac{\alpha + k360°}{n}\right], \quad k = 0, 1, 2, \ldots, n - 1.$$

Quadratic equations

Let r_1, r_2 be the roots of $ax^2 + bx + c = 0$ $(a \neq 0)$. Then

$$r_1, r_2 = \frac{-b \pm \sqrt{b^2 - 4ac}}{2a} \qquad r_1 + r_2 = -\frac{b}{a} \qquad r_1 r_2 = \frac{c}{a}.$$

If a, b, c are real and if

$b^2 - 4ac > 0$, then the roots are real and unequal;

$b^2 - 4ac = 0$, then the roots are real and equal;

$b^2 - 4ac < 0$, then the roots are imaginary and unequal.

If a, b, c are rational and if

$b^2 - 4ac$ is a perfect square, then the roots are rational;

$b^2 - 4ac$ is not a perfect square, then the roots are irrational.

TRIGONOMETRIC IDENTITIES

The following identities hold for all values of the angles *for which both sides of the equation are defined.* Where a double sign (±) occurs, the choice depends upon the quadrant in which the angle is located.

Basic identities

$$\csc\theta = \frac{1}{\sin\theta} \qquad \sec\theta = \frac{1}{\cos\theta} \qquad \tan\theta = \frac{\sin\theta}{\cos\theta} = \frac{1}{\cot\theta}$$

Pythagorean identities

$$\sin^2\theta + \cos^2\theta = 1 \qquad 1 + \tan^2\theta = \sec^2\theta \qquad 1 + \cot^2\theta = \csc^2\theta$$

Power identities

$$\sin^2\theta = \tfrac{1}{2}(1 - \cos 2\theta) \qquad \cos^2\theta = \tfrac{1}{2}(1 + \cos 2\theta)$$

$$\sin^3\theta = \tfrac{1}{4}(3\sin\theta - \sin 3\theta) \qquad \cos^3\theta = \tfrac{1}{4}(3\cos\theta + \cos 3\theta)$$

$$\sin^4\theta = \tfrac{1}{8}(3 - 4\cos 2\theta + \cos 4\theta) \qquad \cos^4\theta = \tfrac{1}{8}(3 + 4\cos 2\theta + \cos 4\theta)$$

Double and multiple angle identities

$$\sin 2\theta = 2\sin\theta\cos\theta = \frac{2\tan\theta}{1 + \tan^2\theta} \qquad \tan^2\theta = \frac{1 - \cos 2\theta}{1 + \cos 2\theta}$$

$$\cos 2\theta = \cos^2\theta - \sin^2\theta = 2\cos^2\theta - 1 = 1 - 2\sin^2\theta = \frac{1 - \tan^2\theta}{1 + \tan^2\theta}$$

$$\tan 2\theta = \frac{2\tan\theta}{1 - \tan^2\theta} \qquad \cot 2\theta = \frac{\cot^2\theta - 1}{2\cot\theta}$$

$$\sin 3\theta = 3\sin\theta - 4\sin^3\theta \qquad \cos 3\theta = 4\cos^3\theta - 3\cos\theta$$

$$\sin 4\theta = 4\sin\theta\cos\theta - 8\sin^3\theta\cos\theta \qquad \cos 4\theta = 8\cos^4\theta - 8\cos^2\theta + 1$$

Half-angle identities

$$\sin\tfrac{1}{2}\theta = \pm\sqrt{\frac{1 - \cos\theta}{2}} \qquad \cos\tfrac{1}{2}\theta = \pm\sqrt{\frac{1 + \cos\theta}{2}}$$

$$\tan\tfrac{1}{2}\theta = \pm\sqrt{\frac{1 - \cos\theta}{1 + \cos\theta}} = \frac{1 - \cos\theta}{\sin\theta} = \frac{\sin\theta}{1 + \cos\theta}$$

A APPENDICES

Angle sum and difference identities

$$\sin(\alpha + \beta) = \sin\alpha\cos\beta + \cos\alpha\sin\beta \qquad \cos(\alpha + \beta) = \cos\alpha\cos\beta - \sin\alpha\sin\beta$$

$$\tan(\alpha + \beta) = \frac{\tan\alpha + \tan\beta}{1 - \tan\alpha\tan\beta} \qquad \cot(\alpha + \beta) = \frac{\cot\beta\cot\alpha - 1}{\cot\beta + \cot\alpha}$$

$$\sin(\alpha - \beta) = \sin\alpha\cos\beta - \cos\alpha\sin\beta \qquad \cos(\alpha - \beta) = \cos\alpha\cos\beta + \sin\alpha\sin\beta$$

$$\tan(\alpha - \beta) = \frac{\tan\alpha - \tan\beta}{1 + \tan\alpha\tan\beta} \qquad \cot(\alpha - \beta) = \frac{\cot\beta\cot\alpha + 1}{\cot\beta - \cot\alpha}$$

$$\sin\alpha + \sin\beta = 2\sin\frac{\alpha+\beta}{2}\cos\frac{\alpha-\beta}{2} \qquad \sin\alpha - \sin\beta = 2\cos\frac{\alpha+\beta}{2}\sin\frac{\alpha-\beta}{2}$$

$$\cos\alpha + \cos\beta = 2\cos\frac{\alpha+\beta}{2}\cos\frac{\alpha-\beta}{2} \qquad \cos\alpha - \cos\beta = -2\sin\frac{\alpha+\beta}{2}\sin\frac{\alpha-\beta}{2}$$

$$\sin(\alpha+\beta) + \sin(\alpha-\beta) = 2\sin\alpha\cos\beta \qquad \sin(\alpha+\beta) - \sin(\alpha-\beta) = 2\cos\alpha\sin\beta$$

$$\cos(\alpha+\beta) + \cos(\alpha-\beta) = 2\cos\alpha\cos\beta \qquad \cos(\alpha+\beta) - \cos(\alpha-\beta) = -2\sin\alpha\sin\beta$$

$$\sin(\alpha+\beta)\sin(\alpha-\beta) = \sin^2\alpha - \sin^2\beta = \cos^2\beta - \cos^2\alpha$$

$$\cos(\alpha+\beta)\cos(\alpha-\beta) = \cos^2\alpha - \sin^2\beta = \cos^2\beta - \sin^2\alpha$$

$$\tan\alpha + \tan\beta = \frac{\sin(\alpha+\beta)}{\cos\alpha\cos\beta} \qquad \tan\alpha - \tan\beta = \frac{\sin(\alpha-\beta)}{\cos\alpha\cos\beta}$$

$$\cot\alpha + \cot\beta = \frac{\sin(\alpha+\beta)}{\sin\alpha\sin\beta} \qquad \cot\alpha - \cot\beta = \frac{\sin(\beta-\alpha)}{\sin\alpha\sin\beta}$$

$$\frac{\sin\alpha + \sin\beta}{\sin\alpha - \sin\beta} = \frac{\tan\frac{1}{2}(\alpha+\beta)}{\tan\frac{1}{2}(\alpha-\beta)} \qquad \frac{\sin\alpha + \sin\beta}{\cos\alpha - \cos\beta} = \cot\tfrac{1}{2}(\beta - \alpha)$$

$$\frac{\sin\alpha + \sin\beta}{\cos\alpha + \cos\beta} = \tan\tfrac{1}{2}(\alpha+\beta) \qquad \frac{\sin\alpha - \sin\beta}{\cos\alpha + \cos\beta} = \tan\tfrac{1}{2}(\alpha-\beta)$$

Equivalent expressions for $\sin\theta$ and $\cos\theta$

$$\sin\theta = \pm\sqrt{1 - \cos^2\theta} = \frac{\tan\theta}{\pm\sqrt{1+\tan^2\theta}} = \frac{1}{\pm\sqrt{1+\cot^2\theta}} = \frac{\pm\sqrt{\sec^2\theta - 1}}{\sec\theta}$$

$$= \frac{1}{\csc\theta} = \cos\theta\tan\theta = \pm\sqrt{\frac{1-\cos 2\theta}{2}} = 2\sin\frac{1}{2}\theta\cos\frac{1}{2}\theta$$

$$\cos\theta = \pm\sqrt{1 - \sin^2\theta} = \frac{1}{\pm\sqrt{1+\tan^2\theta}} = \frac{\cot\theta}{\pm\sqrt{1+\cot^2\theta}} = \frac{1}{\sec\theta}$$

$$= \frac{\pm\sqrt{\csc^2\theta - 1}}{\csc\theta} = \sin\theta\cot\theta = \pm\sqrt{\frac{1+\cos 2\theta}{2}} = \cos^2\frac{1}{2}\theta - \sin^2\frac{1}{2}\theta.$$

INVERSE TRIGONOMETRIC FUNCTIONS

Principal values

$$-\frac{\pi}{2} \leq \arcsin x \leq \frac{\pi}{2}, \qquad -1 \leq x \leq 1$$

$$-\frac{\pi}{2} < \arctan x \leq \frac{\pi}{2}, \qquad -\infty < x < \infty$$

$$0 \leq \arccos x \leq \pi, \qquad -1 \leq x \leq 1$$

$$0 < \text{arccot}\, x < \pi, \qquad -\infty < x < \infty$$

$$0 \leq \text{arcsec}\, x < \frac{\pi}{2}, \qquad x \geq 1$$

$$-x \leq \text{arcsec}\, x < -\frac{\pi}{2}, \qquad x \leq -1$$

$$0 < \text{arccsc}\, x \leq \frac{\pi}{2}, \qquad x \geq 1$$

$$-\pi < \text{arccsc}\, x \leq -\frac{\pi}{2}, \qquad x \leq -1$$

Fundamental identities involving principal values

$$\arcsin x + \arccos x = \frac{\pi}{2} \qquad\qquad \arctan x + \text{arccot}\, x = \frac{\pi}{2}$$

If $\theta = \arcsin x$, then

$$\sin \theta = x \qquad \tan \theta = \frac{x}{\sqrt{1-x^2}} \qquad \sec \theta = \frac{1}{\sqrt{1-x^2}}$$

$$\cos \theta = \sqrt{1-x^2} \qquad \cot \theta = \frac{\sqrt{1-x^2}}{x} \qquad \csc \theta = \frac{1}{x}$$

If $\theta = \arccos x$, then

$$\sin \theta = \sqrt{1-x^2} \qquad \tan \theta = \frac{\sqrt{1-x^2}}{x} \qquad \sec \theta = \frac{1}{x}$$

$$\cos \theta = x \qquad \cot \theta = \frac{x}{\sqrt{1-x^2}} \qquad \csc \theta = \frac{1}{\sqrt{1-x^2}}$$

If $\theta = \arctan x$, then

$$\sin \theta = \frac{x}{\sqrt{1+x^2}} \qquad \tan \theta = x \qquad \sec \theta = \sqrt{1+x^2}$$

$$\cos \theta = \frac{1}{\sqrt{1+x^2}} \qquad \cot \theta = \frac{1}{x} \qquad \csc \theta = \frac{\sqrt{1+x^2}}{x}$$

FORMULAS FROM ANALYTIC GEOMETRY

Lines

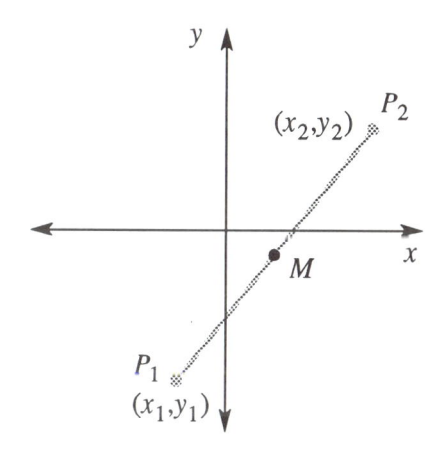

length of $\overline{P_1P_2}$: $d = \sqrt{(x_2-x_1)^2 + (y_2-y_1)^2}$

midpoint M : $\left(\dfrac{x_1+x_2}{2}, \dfrac{y_1+y_2}{2}\right)$

slope of $\overleftrightarrow{P_1P_2}$: $m = \dfrac{y_2-y_1}{x_2-x_1}$

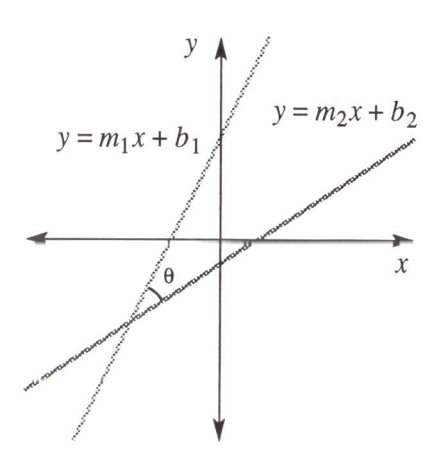

angle between two lines :

$\tan\theta = \dfrac{m_2-m_1}{1+m_1m_2}$

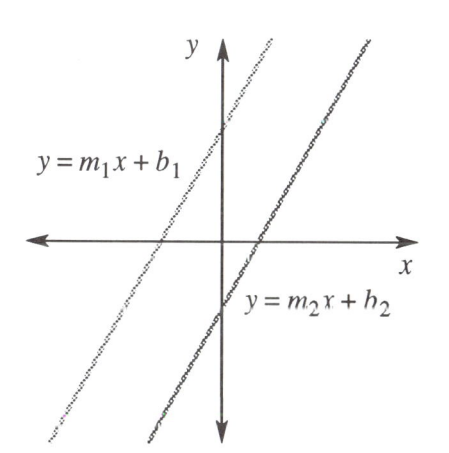

parallel lines : $m_1 = m_2$

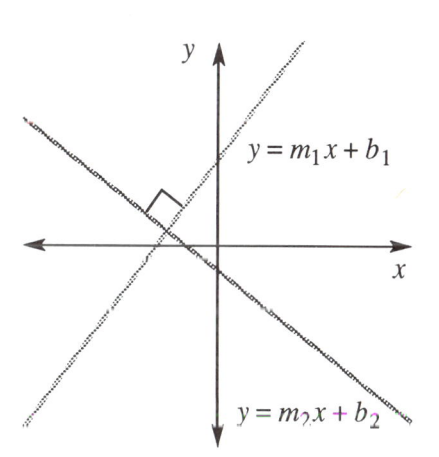

perpendicular lines : $m_1m_2 = -1$

Circles

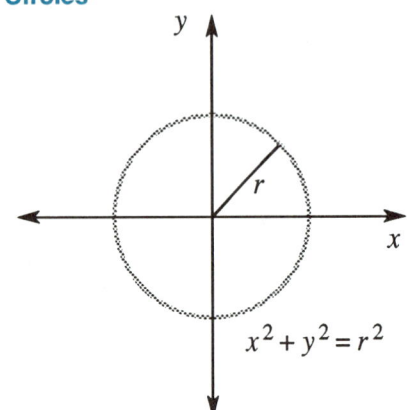

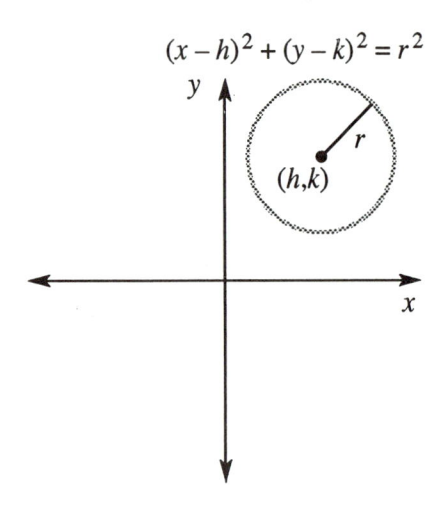

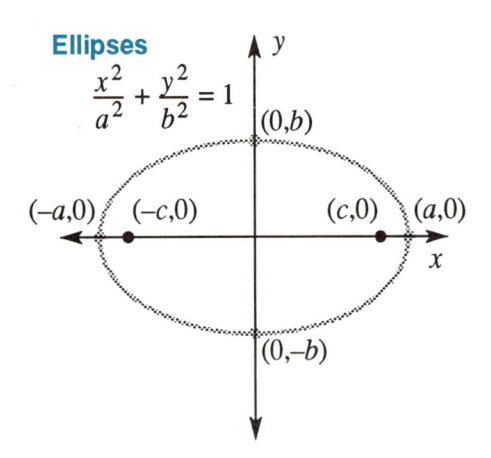

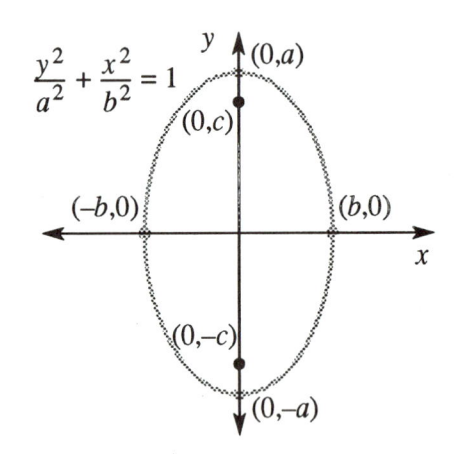

$$x^2 + y^2 = r^2$$

$$(x - h)^2 + (y - k)^2 = r^2$$

Ellipses

$$\frac{x^2}{a^2} + \frac{y^2}{b^2} = 1$$

Center at (0,0) with horizontal major axis
Foci at (±c,0)
where $c = \sqrt{a^2 - b^2}$

$$\frac{y^2}{a^2} + \frac{x^2}{b^2} = 1$$

Center at (0,0) with vertical major axis
Foci at (0,±c)
where $c = \sqrt{a^2 - b^2}$

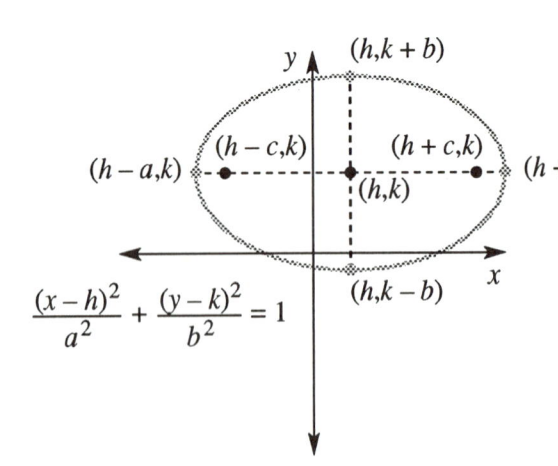

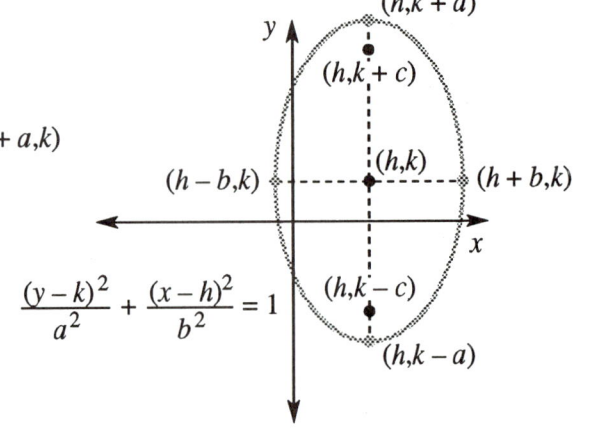

$$\frac{(x - h)^2}{a^2} + \frac{(y - k)^2}{b^2} = 1$$

Center at (h,k)
Foci at (h ± c, k)

$$\frac{(y - k)^2}{a^2} + \frac{(x - h)^2}{b^2} = 1$$

Center at (h,k)
Focus at (h, k ± c)

A. APPENDICES

Parabolas

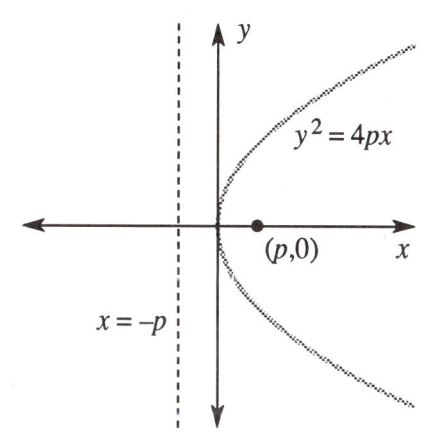

$y^2 = 4px$

Vertex at $(0,0)$
Focus at $(p,0)$
Directrix: $x = -p$

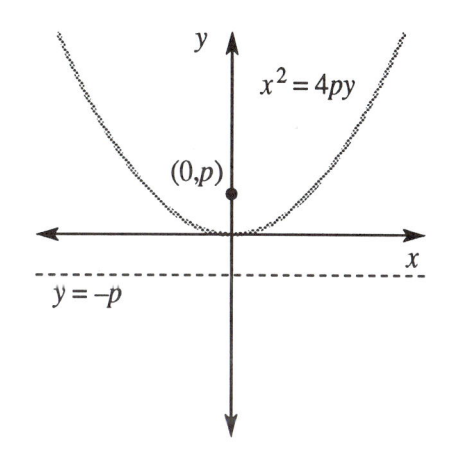

$x^2 = 4py$

Vertex at $(0,0)$
Focus at $(0,p)$
Directrix: $y = -p$

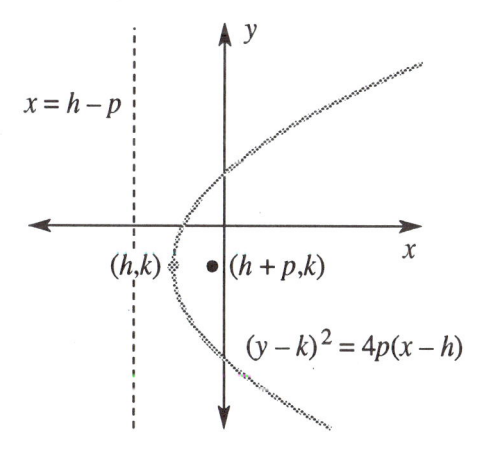

$(y-k)^2 = 4p(x-h)$

Vertex at (h,k)
Focus at $(h+p,k)$
Directrix: $x = h-p$

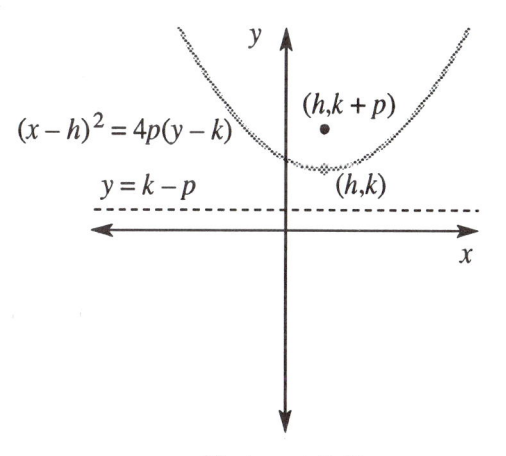

$(x-h)^2 = 4p(y-k)$

Vertex at (h,k)
Focus at $(h,k+p)$
Directrix: $y = k-p$

Hyperbolas

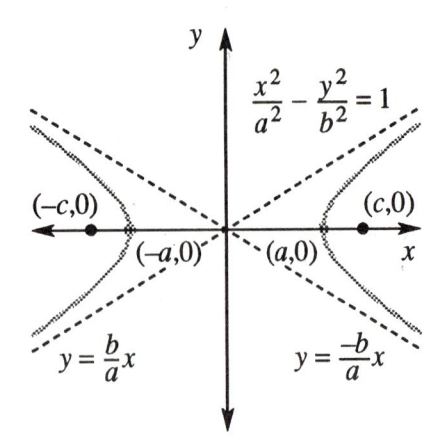

$$\frac{x^2}{a^2} - \frac{y^2}{b^2} = 1$$

Center at $(0,0)$
Foci : $(\pm c, 0)$
Asymptotes : $y = \pm \frac{b}{a} x$
Vertices at $(\pm a, 0)$
Eccentricity $e = \frac{\sqrt{a^2 + b^2}}{a}$

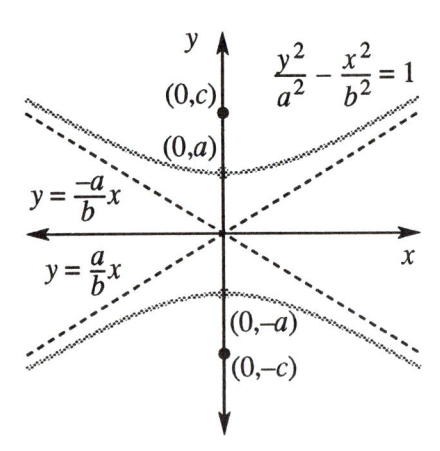

$$\frac{y^2}{a^2} - \frac{x^2}{b^2} = 1$$

Center at $(0,0)$
Foci : $(0, \pm c)$
Asymptotes : $y = \pm \frac{a}{b} x$
Vertices at $(0, \pm a)$
Eccentricity $e = \frac{\sqrt{a^2 + b^2}}{a}$

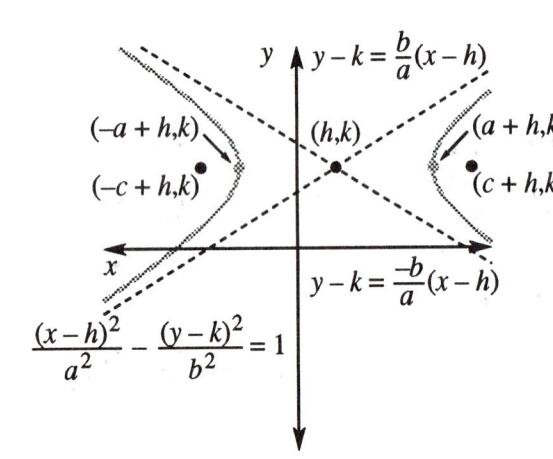

$$\frac{(x-h)^2}{a^2} - \frac{(y-k)^2}{b^2} = 1$$

Center at (h, k)
Foci : $(\pm c + h, k)$
Asymptotes : $y - k = \pm \frac{b}{a}(x - h)$
Vertices at $(\pm a + h, k)$
Eccentricity $e = \frac{\sqrt{a^2 + b^2}}{a}$

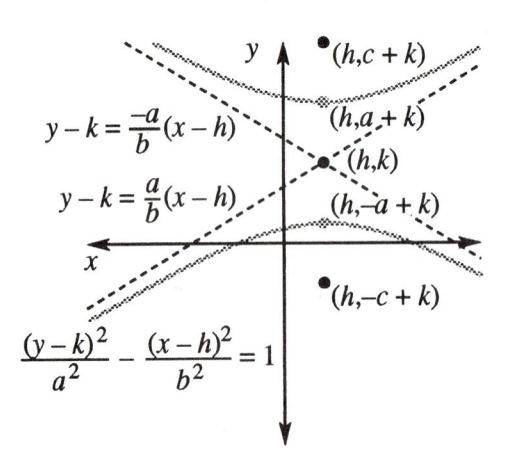

$$\frac{(y-k)^2}{a^2} - \frac{(x-h)^2}{b^2} = 1$$

Center at (h, k)
Foci : $(h, \pm c + k)$
Asymptotes : $y - k = \pm \frac{a}{b}(x - h)$
Vertices at $(h, \pm a + k)$
Eccentricity $e = \frac{\sqrt{a^2 + b^2}}{a}$

Answers to Selected Exercises

Chapter 10

Section 10.1

1. $\{3 - 5n\}_{n=0}^{\infty}$. **3.** $\{\pi\}_{n=0}^{\infty}$.
5. $\{3(-5)^n\}_{n=0}^{\infty}$. **7.** $\{\pi(0)^n\}_{n=0}^{\infty}$.
9. True.
11. False. (Example: $-1, 1, -1, 1, \ldots$.)
13. $\frac{1}{2}, 2, \frac{1}{2}, 2, \ldots$;
bounded, non-monotonic, divergent.
15. $-2, -3.1, -4.01, -5.001, -6.0001$;
unbounded, monotonic, divergent.
17. $0, \frac{3}{2}, \frac{2}{3}, \frac{5}{4}, \frac{4}{5}, \frac{7}{6}, \frac{6}{7}, \ldots$
19. a_{12}.
21. $\{1/n\}_{n=1}^{\infty}$; $\{(1/2)^n\}_{n=0}^{\infty}$; $\{3\}_{n=1}^{\infty}$.
23. $\{n\}_{n=1}^{\infty}$, $\{(3/2)^n\}_{n=0}^{\infty}$, $\{1 - 2n\}_{n=0}^{\infty}$.
25. $\{(-1)^n\}_{n=1}^{\infty}$, $\{\cos(\frac{n\pi}{4})\}_{n=1}^{\infty}$, $\{\sin(n)\}_{n=1}^{\infty}$.
27. $L = 0, N = 6$.
29. $\left\{\frac{1}{n!}\right\}_{n=0}^{\infty}$ converges faster than $\left\{\frac{1}{2^n}\right\}_{n=0}^{\infty}$.

Section 10.2

1. Converges to $-\frac{1}{4}$. **3.** Converges to $\frac{1}{e-1}$.
5. Diverges. **7.** Converges to $\frac{1}{21}$.
9. Diverges. **11.** Converges to $\frac{3e^2}{3-e}$.
13. Converges to $\frac{5}{6}$. **15.** Converges to $-\frac{1}{3}$.
17. $\sum_{n=0}^{\infty} \left(\frac{14}{17}\right)^n$; $\sum_{n=0}^{\infty} \left(\frac{1}{3}\right)\left(\frac{16}{17}\right)^n$.

19. The first few terms in the sequence are
1, 2, 2.5, $2.\overline{6}$, $2.708\overline{3}$, $2.71\overline{6}$, $2.7180\overline{5}$, 2.71825..., 2.71827..., 2.71828..., 2.71828...
and the sequence converges to e.

21. 42 ft.

23. The total length is 1. $0, \frac{2}{9}$, and $\frac{1}{3}$ are examples of points never erased.

Section 10.3

1. Converges. **3.** Converges.
5. $-.4515, .1598, -.6795, -.8179$.
7. Converges. **9.** Converges.
11. $f(x) = e^{-x}\sin(x)$ is not always positive.
13. $N = 61$; $S_{61} \approx 1.017$;
$$1.01734306175 \leq \sum_{n=1}^{\infty} \frac{1}{n^6} \leq 1.01734306179.$$

15. $\int_1^b \ln(1 + \frac{1}{x})\,dx$
$= x\ln(1 + \frac{1}{x})\big]_1^b - \int_1^b \frac{-1}{1+x}\,dx$
$= (x\ln(1 + \frac{1}{x}) + \ln(1 + x))\big]_1^b$
$= b\ln(1 + \frac{1}{b}) + \ln(1 + b) - \ln(2) - \ln(2)$.
$\int_1^{\infty} \ln(1 + \frac{1}{x})\,dx$ diverges since
$\lim_{b\to\infty}(b\ln(1 + \frac{1}{b}) + \ln(1 + b)) = \infty$.

(The second term in the limit expression approaches infinity.)

17. $\sum_{n=1}^{\infty} \frac{1}{n}$ diverges but $\left\{\frac{1}{n}\right\}_{n=1}^{\infty}$ converges.

19. $\sum_{n=1}^{\infty} \frac{1}{n}$ and $\sum_{n=1}^{\infty} \left(-\frac{1}{n}\right)$ diverge,

but $\sum_{n=1}^{\infty} \left(\frac{1}{n} + \left(-\frac{1}{n}\right)\right)$ converges.

Section 10.4

1. Converges. **3.** Diverges.
5. Converges. **7.** No information.
9. Converges absolutely.
11. $\lim_{n\to\infty} \left|\frac{1}{(n+1)^p}\left(\frac{n^p}{1}\right)\right|$
$= \lim_{n\to\infty} \left(\frac{n}{n+1}\right)^p = 1^p = 1$ for all p.
13. Converges absolutely.

15. Converges.

17. Diverges (p-series).

19. Diverges. **21.** Converges.

23. Diverges. **25.** Converges.

27. Diverges (p-series).

29. Diverges. **31.** Diverges.

33. Converges conditionally.

35. Converges.

37. Converges absolutely.

Section 10.5

1. $R = \infty$; $(-\infty, \infty)$.

3. $R = 2$; interval $(1, 5)$.

5. $h(x) = x - \dfrac{x^2}{2} + \dfrac{x^3}{3} - \dfrac{x^4}{4} + \cdots$
$= \sum_{n=1}^{\infty} \dfrac{(-1)^{n+1} x^n}{n}$; $h(0) = 0$; interval $(-1, 1]$.

7. $|R_{101}| < 0.0098$;

$\ln(2) \approx \sum_{n=1}^{101} \dfrac{(-1)^{n+1}}{n} \approx 0.698073$;

$\ln(2) \approx 0.693147$, by machine.

9. $\sum_{n=0}^{\infty} \dfrac{(-1)^n x^n}{n!}$.   **11.** $\sum_{n=0}^{\infty} \dfrac{(-1)^n x^n}{(n+1)!}$.

13. $S_9 \approx 0.4438$.

15. $\dfrac{\dfrac{x^3}{3!} - \dfrac{x^5}{5!} + \dfrac{x^7}{7!} - \dfrac{x^9}{9!} + \cdots}{x^3 - \dfrac{x^5}{2!} + \dfrac{x^7}{4!} - \dfrac{x^9}{6!} + \cdots}$

$= \dfrac{\sum_{n=1}^{\infty} \dfrac{(-1)^{n+1} x^{2n+1}}{(2n+1)!}}{\sum_{n=1}^{\infty} \dfrac{(-1)^{n+1} x^{2n+1}}{(2n-2)!}}$.

Section 10.6

1. $a_0 = 3$, $a_n = a_{n-1} - 5$, $n = 1, 2, 3, \ldots$

3. $a_0 = \pi$, $a_n = a_{n-1}$, $n = 1, 2, 3, \ldots$

5. $a_0 = 3$, $a_n = (-5)a_{n-1}$, $n = 1, 2, 3, \ldots$

7. $a_0 = \pi$, $a_n = 2 \cdot a_{n-1}$, $n = 1, 2, 3, \ldots$

9. $\sqrt{3}$, $\sqrt{3\sqrt{3}}$, $\sqrt{3\sqrt{3\sqrt{3}}}$,

$\sqrt{3\sqrt{3\sqrt{3\sqrt{3}}}}$, $\sqrt{3\sqrt{3\sqrt{3\sqrt{3\sqrt{3}}}}}$, $\ldots$.

11. 0, -1, -1.54, $\ldots$,

-1.57079, $\ldots$, -1.570796, $\ldots$

13. 0.5, 0.25, 0.0625,

0.00390625, 0.00001525, $\ldots$

15. The denominators form the Fibonacci sequence. The numerators form the Fibonacci sequence starting with the second term.

17. $89/55$.

19. $1, 1, 2, 3, 5, 8, 13, 21, 34, 55$;

The Fibonacci sequence.

21. $x_1 = 2.100$, $x_2 \approx 1.812$, $x_3 \approx 1.789$.

23. $x_1 \approx 0.773$, $x_2 \approx 0.775$, $x_3 \approx 0.775$.

25. $x_1 \approx 2.226$, $x_2 \approx 2.215$, $x_3 \approx 2.215$.

29. $x_1 \approx 1.699$, $x_2 \approx 1.522$, $x_3 \approx 1.496$.

ANSWERS TO SELECTED EXERCISES

Chapter 11

Section 11.1

1.

vector	initial point	terminal point
n	$(-1, 1)$	$(-1, -1)$
p	$(-1, 1)$	$(-1, 4)$
q	$(-1, 1)$	$(2, 1)$
u	$(2, 4)$	$(2, 1)$
v	$(2, 1)$	$(5, 1)$
w	$(-1, 1)$	$(5, -1)$

3. Using the Pythagorean Theorem, we find that the length of **m** is $\sqrt{13}$ and the length of **w** is $2\sqrt{10}$.

5. p and u, p and q, p and v, q and u, p and u, v and u.

7. Area $= (3)(2) = 6$.

9. $(9, -6)$.   **11.** $\overrightarrow{PQ} = \langle -3, -5, 1 \rangle$.

13. $\overrightarrow{PQ} = \langle -2.5, 3.4, 4.1 \rangle$.

15. $\overrightarrow{PQ} = \langle 3, 5, -1 \rangle$.

17. (11) $T = (5.5, 1.6, -5.1)$

(12) $T = (-7.5, -1.4, -4.1)$

(13) $T = (5, -6.8, -8.2)$

(14) $T = (2.5, -3.4, -4.1)$

(15) $T = (-0.5, -8.4, -3.1)$.

19. $\langle -2\sqrt{3}, 2 \rangle$.   **21.** $\langle 0, 0 \rangle$.

23. **m**: $\sqrt{13}$ units at $\approx 236.3°$;

n: 2 units due South $(180°)$;

p: 3 units due North $(0°)$;

q: 3 units due East $(90°)$;

u: 3 units due South $(180°)$;

v: 3 units due East $(90°)$;

w: $2\sqrt{10}$ units at $\approx 108.4°$.

25. In addition to the North Pole, if you start 1 mile north of any circle of latitude of circumference $1/n$ miles (n a positive integer), then moving 1 mile south places you on this circle, moving 1 mile east brings you around the circles a whole number of times, and moving 1 mile north brings you back to your starting point.

Section 11.2

1. $\mathbf{p} + \mathbf{q} + \mathbf{u} = \mathbf{q}$.

3. $\mathbf{q} - \mathbf{v} = \mathbf{0}$.

5. $\mathbf{w} - \mathbf{n} = 2\mathbf{q}$.

7. Listing clockwise: $\mathbf{w} - \mathbf{v}, -\mathbf{v}, -\mathbf{w}, \mathbf{v} - \mathbf{w}$.

9. $2\mathbf{v} = \langle -8, -6 \rangle$.

11. $\mathbf{v} + \mathbf{w} = \langle -1.28, 0.14 \rangle$.

13. $\mathbf{w} - \mathbf{v} = \langle 6.72, 6.14 \rangle$.

15. $\frac{\mathbf{w}}{5} - 4\mathbf{v} = \langle 16.544, 12.628 \rangle$.

17. $2\mathbf{v} = \langle 4, -6, 2 \rangle$.

19. $\mathbf{v} + \mathbf{w} = \langle 2.5, -2, -6 \rangle$.

21. $\mathbf{w} - \mathbf{v} = \langle -1.5, 4, -8 \rangle$.

23. $\frac{\mathbf{w}}{5} - 4\mathbf{v} = \langle -7.9, 12.2, -5.4 \rangle$.

25. (5) (using components) Let $\mathbf{u} = \langle u_1, u_2 \rangle$ be any vector in $\mathbb{R}^2$ and a, b any scalars. Show $a(b\mathbf{u}) = (ab)\mathbf{u}$.

$$a(b\mathbf{u}) = a\langle bu_1, bu_2 \rangle = \langle abu_1, abu_2 \rangle$$
$$= ab\langle u_1, u_2 \rangle = (ab)\mathbf{u}.$$

(geometric illustration) If we represent all the vectors as directed line segments, then $b\mathbf{u}$ is a vector whose length is $|b|$ times the length of $\mathbf{u}$ with the same direction if $b > 0$, opposite if $b < 0$. $a(b\mathbf{u})$ is a vector whose length is $|a|$ times the length of $b(\mathbf{u})$, so $|a|$ times $|b|$ times the length of $\mathbf{u}$ with same direction as $b\mathbf{u}$ if $a > 0$, opposite if $a < 0$. On the other hand $(ab)\mathbf{u}$ is a vector with the length $|ab|$ times the length of $\mathbf{u}$ and having the same direction as $\mathbf{u}$ if $ab > 0$, opposite if $ab < 0$. Since $|ab| = |a||b|$, the vectors are the same.

Section 11.3

1. $\mathbf{p} \cdot \mathbf{n} = -6$.

3. $\mathbf{u} \cdot \mathbf{p} = -9$.

5. $(\mathbf{m} \cdot \mathbf{u})\mathbf{u} = \langle 0, -2 \rangle$ where $\mathbf{u}$ is the unit vector in the direction $\mathbf{u}$.

7. $\mathbf{v} \cdot \mathbf{w} = \frac{1}{2}$.

8. $\mathbf{v} \cdot (\mathbf{w} - \mathbf{v}) = -\frac{1}{2}$; $\mathbf{v} \cdot (-\mathbf{v}) = -1$; $\mathbf{v} \cdot (-\mathbf{w}) = -\frac{1}{2}$; $\mathbf{v} \cdot (\mathbf{v} - \mathbf{w}) = \frac{1}{2}$.

9. $\mathbf{v} \cdot \mathbf{w} = -20.3$.   **11.** $\|\mathbf{v}\| = 5$.

13. $\|\mathbf{v} + \mathbf{w}\| \approx 1.28763$.

15. $\|\mathbf{v} - \mathbf{w}\| \approx 9.10264$.

17. $\langle -\frac{4}{5}, -\frac{3}{5} \rangle$.   **19.** $\theta \approx 167.77060°$.

21. $(\mathbf{v} \cdot \mathbf{u})\mathbf{u} \approx \langle 3.19044, 3.69348 \rangle$ where $\mathbf{u}$ is the unit vector in the direction $\mathbf{w}$.

23. $\mathbf{v} \cdot \mathbf{w} = -9$.

25. $\|\mathbf{v}\| \approx 3.74166$.

27. $\|\mathbf{v} + \mathbf{w}\| \approx 6.80074$.

29. $\|\mathbf{v} - \mathbf{w}\| \approx 9.06918$.

31. $\approx \langle 0.53452, -0.80178, 0.26726 \rangle$.

33. $\theta \approx 109.83550°$.

35. $(\mathbf{v} \cdot \mathbf{u})\mathbf{u} \approx \langle -8.95522, -0.17910, 1.25373 \rangle$.

37. No. For example, let $\mathbf{u} = \langle 3, 2 \rangle$ and $\mathbf{v} = \langle -2, 3 \rangle$.

39. Since $\mathbf{v} - \text{comp}_\mathbf{u}\mathbf{v} = \mathbf{v} - (\mathbf{v} \cdot \mathbf{u})\mathbf{u}$, we have

$$\mathbf{u} \cdot (\mathbf{v} - \text{comp}_\mathbf{u}\mathbf{v}) = \mathbf{u} \cdot (\mathbf{v} - (\mathbf{v} \cdot \mathbf{u})\mathbf{u})$$
$$= \mathbf{u} \cdot \mathbf{v} - \mathbf{u} \cdot (\mathbf{v} \cdot \mathbf{u})\mathbf{u}$$
$$= \mathbf{u} \cdot \mathbf{v} - (\mathbf{v} \cdot \mathbf{u})\mathbf{u} \cdot \mathbf{u}$$
$$= \mathbf{u} \cdot \mathbf{v} - (\mathbf{v} \cdot \mathbf{u})(1)$$
$$= 0 \Rightarrow \mathbf{u} \perp (\mathbf{v} - \text{comp}_\mathbf{u}\mathbf{v}).$$

41. Assuming θ is expressed such that $0 \leq \theta < 2\pi$, we have
$(r, \theta) = (\|\mathbf{v}\|, \arccos(\mathbf{v} \cdot \mathbf{i}/\|\mathbf{v}\|))$,
if $0 < \theta \leq \pi$;
$(r, \theta) = (\|\mathbf{v}\|, 2\pi - \arccos(\mathbf{v} \cdot \mathbf{i}/\|\mathbf{v}\|))$,
if $\pi < \theta < 2\pi$.

43. For $\mathbf{v}$: $\cos^2 \alpha + \cos^2 \beta = \frac{25}{25} = 1$;
for $\mathbf{w}$: $\cos^2 \alpha + \cos^2 \beta = 1$.

45. For $\mathbf{v}$:
$\alpha \approx 57.68897°$, $\beta \approx 143.30077°$, $\gamma \approx 74.49864°$;
for $\mathbf{w}$:
$\alpha \approx 85.95531°$, $\beta \approx 81.89028°$, $\gamma \approx 170.92541°$.

47. Let $\mathbf{u} = \langle u_1, u_2, u_3 \rangle$ be any vector in $\mathbb{R}^3$ and let α, β, and γ be the direction angles.

$$\cos^2 \alpha + \cos^2 \beta + \cos^2 \gamma$$
$$= \left(\frac{u_1}{\|\mathbf{u}\|}\right)^2 + \left(\frac{u_2}{\|\mathbf{u}\|}\right)^2 + \left(\frac{u_3}{\|\mathbf{u}\|}\right)^2$$
$$= \frac{u_1^2 + u_2^2 + u_3^2}{\|\mathbf{u}\|^2} = \frac{\|\mathbf{u}\|^2}{\|\mathbf{u}\|^2} = 1.$$

49. Let $\mathbf{v} = \langle v_1, v_2, v_3 \rangle$ be any vector in $\mathbb{R}^3$.

$$\frac{\mathbf{v}}{\|\mathbf{v}\|} = \frac{\langle v_1, v_2, v_3 \rangle}{\sqrt{v_1^2 + v_2^2 + v_3^2}};$$

$$\left\|\frac{\mathbf{v}}{\|\mathbf{v}\|}\right\| = \left(\left(\frac{v_1}{\sqrt{v_1^2 + v_2^2 + v_3^2}}\right)^2 + \left(\frac{v_2}{\sqrt{v_1^2 + v_2^2 + v_3^2}}\right)^2 + \left(\frac{v_3}{\sqrt{v_1^2 + v_2^2 + v_3^2}}\right)^2\right)^{1/2}$$

$$= \left(\frac{v_1^2 + v_2^2 + v_3^2}{v_1^2 + v_2^2 + v_3^2}\right)^{1/2}$$

$$= \sqrt{1} = 1.$$

Section 11.4

1. $U = \begin{bmatrix} 5 & -3 \\ 4 & 4 \\ 1 & -2 \end{bmatrix}$.

3. $X = \begin{bmatrix} 4 & 2 \\ 9 & 2 \\ -5 & -11 \end{bmatrix}$.

5. Associativity for addition.

7. $W = \begin{bmatrix} -10 & 6 \\ -8 & -8 \\ -2 & 4 \end{bmatrix}$.

9. Distributive property of scalar multiplication over matrix addition.

11. $F = \begin{bmatrix} 10 & 5 \\ 23 & 1 \end{bmatrix}$.

13. $I_3 C = C$, $CI_3 = C$.

15. $\det(E) = 0$.   **17.** $\det(I_3) = 1$.

19. $\det(0_n) = 0$ for any n.

21. $\det(P) = -11.$ **23.** $\det(P) = 33.$
25. $\det(P) = -11a.$ **27.** $\det(P) = 0.$
29. $\det(P) = 0.$

31. Let A be the original matrix. Assume we multiply row 2 of A by the scalar d. So, we have $B = \begin{bmatrix} a_1 & a_2 & a_3 \\ db_1 & db_2 & db_3 \\ c_1 & c_2 & c_3 \end{bmatrix}$ and

$\det(B) = a_1(db_2c_3 - db_3c_2) - a_2(db_1c_3 - db_3c_1)$
$\quad + a_3(db_1c_2 - db_2c_1)$
$= d(a_1(b_2c_3 - b_3c_2) - a_2(b_1c_3 - b_3c_1)$
$\quad + a_3(b_1c_2 - b_2c_1))$
$= d \det(A).$

Similar results occur if we multiply all the entries of any other row by the same scalar.

33. Let A be the original matrix.

Let $B = \begin{bmatrix} b_1 & b_2 & b_3 \\ a_1 & a_2 & a_3 \\ c_1 & c_2 & c_3 \end{bmatrix}.$

$\det(B) = b_1(a_2c_3 - a_3c_2) - b_2(a_1c_3 - a_3c_1)$
$\quad + b_3(a_1c_2 - c_1a_2)$
$= b_1a_2c_3 - b_1a_3c_2 - b_2a_1c_3 + a_3b_2c_1$
$\quad + b_3a_1c_2 - b_3c_1a_2$
$= -a_1(b_2c_3 - b_3c_2) + a_2(b_1c_3 - b_3c_1)$
$\quad - a_3(b_1c_2 - b_2c_1)$
$= -\det(A).$

Similar results occur if we interchange any 2 rows.

35. Let A be the original matrix.

Let $B = \begin{bmatrix} db_1 & db_2 & db_3 \\ b_1 & b_2 & b_3 \\ c_1 & c_2 & c_3 \end{bmatrix}$ where d is a scalar.

$\det(B) = db_1(b_2c_3 - b_3c_2) - db_2(b_1c_3 - b_3c_1)$
$\quad + db_3(b_1c_2 - b_2c_1)$
$= db_1b_2c_3 - db_1b_3c_2 - db_2b_1c_3 + db_2b_3c_1$
$\quad + db_3b_1c_2 - db_3b_2c_1$
$= 0.$

Section 11.5

1. $\mathbf{b} \times \mathbf{a} = 5\mathbf{j} + 5\mathbf{k}.$
3. $(\mathbf{b} - \mathbf{a}) \times (\mathbf{c} - \mathbf{a}) = -10\mathbf{i} - 5\mathbf{k}.$
5. $-3\mathbf{i} \times \mathbf{a} = -3\mathbf{j} - 3\mathbf{k}.$
7. $\frac{1}{2}\|\mathbf{a} \times \mathbf{c}\| = \frac{5}{2}\sqrt{6}.$
9. Area $\approx 8.66025.$
11. The volume of the parallelepiped determined by the three vectors is not zero. Therefore, all three vectors do not lie in the same plane.
13. $8x + 31y - 9z - 69 = 0.$
15. $\approx 70.89339°.$ **17.** False.
19. False. **21.** False.
23. True. **25.** False.

Section 11.6

1. $\langle x(t), y(t), z(t) \rangle = \langle 2+t, -1-3t, 1-2t \rangle,$ other answers are possible.
3. $\langle x(t), y(t), z(t) \rangle = \langle 3t, t, -t \rangle.$
5. $\langle x(t), y(t), z(t) \rangle$
$\quad = \langle 1 + 3s + 5t, 4 + 3s - t, 7 + 13s + t \rangle.$
7. $<x(t), y(t), z(t)> = <1 - 2t, 3 + t, 5 + 3t>.$
9. $<x(t), y(t), z(t)> = <1 + t, -5, 7 - t>.$
11. $<x(t), y(t), z(t)> = <-t, 2t, t>.$
13. $(1, 0, 1).$
15. $<x(s,t), y(s,t), z(s,t)>$
$\quad = <1 + s + 2t, 1 + 3s + 3t, 1 + 5s + 4t>.$
17. $-x + 3y - z + 4 = 0.$
19. No; yes (notice the vector is perpendicular to the plane).
21. $<x(s,t), y(s,t), z(s,t)>$
$\quad = <1 - s + 3t, 3 + 5s, 4 + 2s - 4t>.$
23. $<x(s,t), y(s,t), z(s,t)>$
$\quad = <1 + s - t, 2 + 3s - 3t, 4 - s + t>.$
25. $<x(t), y(t), z(t)> = <2 + 5t, -2t, 1 - t>.$

27. a) The lines must be distinct and either parallel or intersecting, i.e. not skew. That they are parallel can be determined by checking that their direction vectors are parallel. That they intersect can be determined by equating the vector components to determine a point of intersection.

b) The line must not pass through the point. This can be determined by checking that the position vector $< x(t), y(t), z(t) >$ does not yield the given point for any value of t.

c) The planes must not be parallel. This can be determined by checking that the normal vectors are not parallel.

d) The direction vector for the line must be parallel to the normal vector for the plane.

e) If the line is not in the plane then the line must not be parallel to any line in the plane.

Chapter 12

Section 12.1

1. and 3.

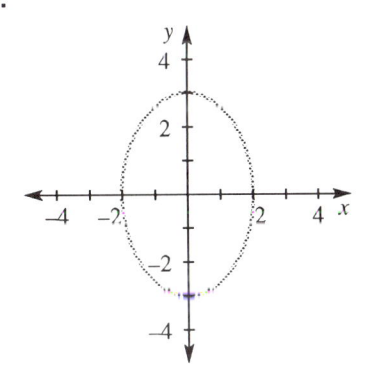

5. and 7.

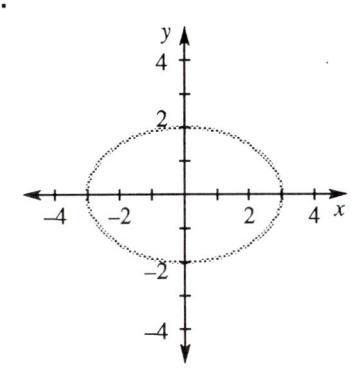

9. The orientation of the curves in problems 1,4,5, and 8 is counterclockwise. The orientation of the curves in problems 2,3,6, and 7 is clockwise. The conic section traced out by the position functions in problems 1-8 is an ellipse.

11.

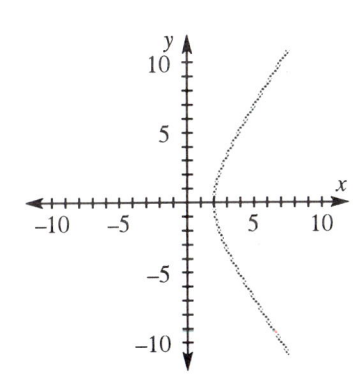

13.

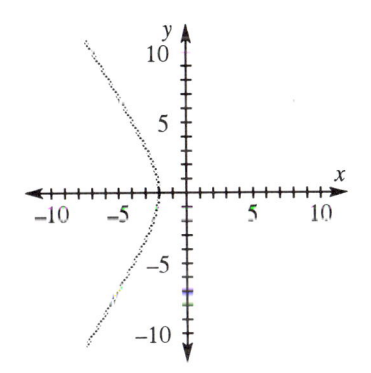

15.

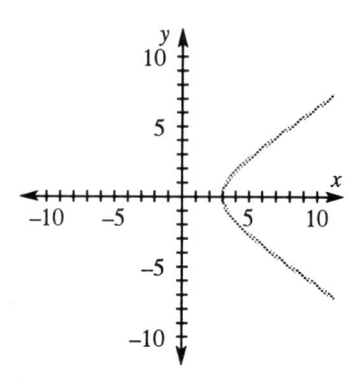

17.

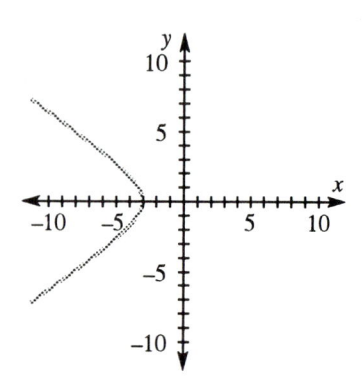

19. The orientation of the curves in problems 11, 14, 15, and 18 is clockwise. The orientation of the curves in problems 12, 13, 16, and 17 is counterclockwise. The conic section traced out by each position function is one branch of a hyperbola.

21.

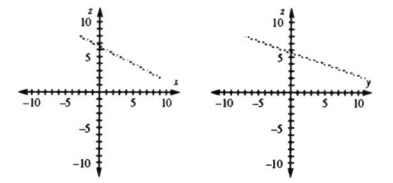

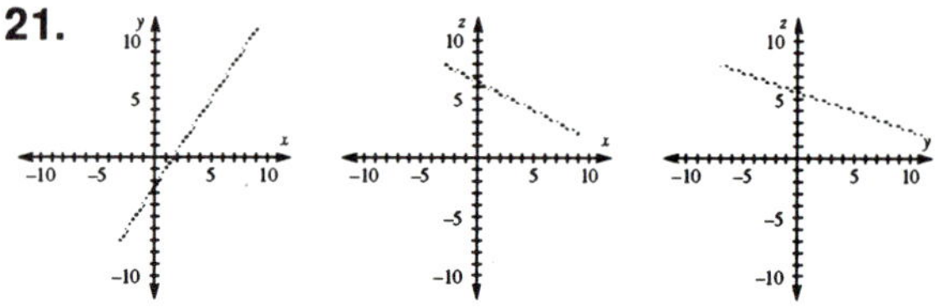

23.

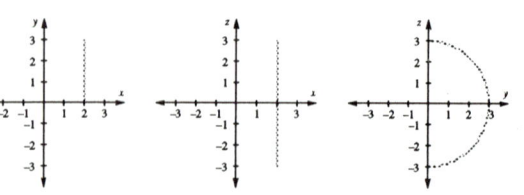

25.

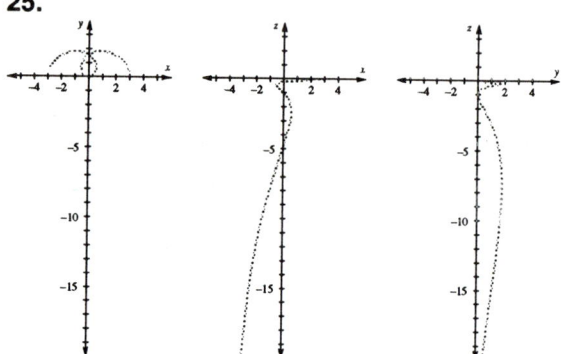

27.

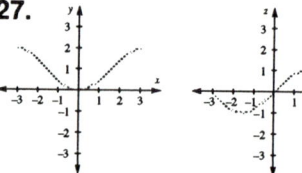

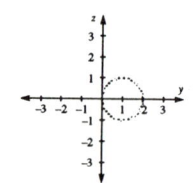

29.

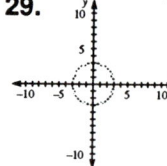

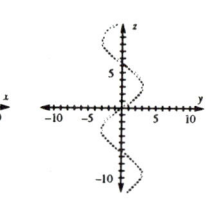

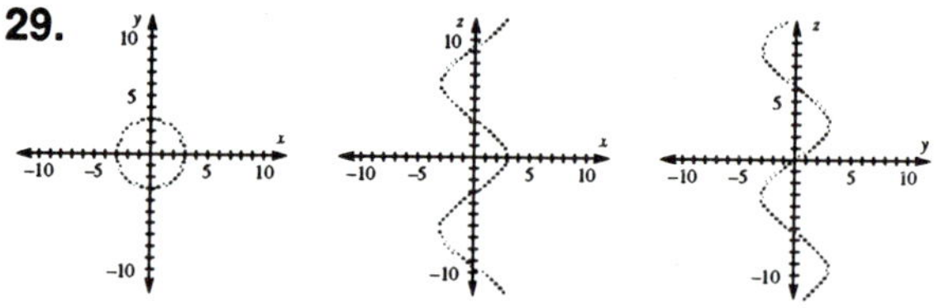

Section 12.2

1. $\mathbf{v}(t) = \langle 6t, \frac{1}{2\sqrt{t}} \rangle$; $\mathbf{v}(2) = \langle 12, \frac{1}{2\sqrt{2}} \rangle$.

3. $\mathbf{T}(t) = \frac{\langle 12t^{3/2}, 1 \rangle}{\sqrt{144t^3+1}}$; $\mathbf{T}(2) \approx \langle 0.99957, 0.02945 \rangle$.

5. $\langle x(t), y(t) \rangle = \langle 12 + 12t, \sqrt{2} + \frac{1}{2\sqrt{2}} t \rangle$.

7. $\mathbf{a}(t) = \langle -9\sin(3t), -9\cos(3t), \frac{3}{2} t^{-1/2} \rangle$;
$\mathbf{a}(1) \approx \langle -1.27008, 8.90993, 1.5 \rangle$.

9. Speed ≈ 4.24264.

11. $\mathbf{v}(t) = \langle t, 2t^{1/2}, 2 \rangle$; $\mathbf{v}(1) = \langle 1, 2, 2 \rangle$.

13. $\mathbf{T}(t) = \frac{\langle t, 2t^{1/2}, 2 \rangle}{t+2}$; $\mathbf{T}(1) = \langle \frac{1}{3}, \frac{2}{3}, \frac{2}{3} \rangle$.

15. $\langle x(t), y(t), z(t) \rangle = \langle \frac{1}{2} + t, \frac{4}{3} + 2t, 2 + 2t \rangle$.

17. $\mathbf{a}(t) = \langle -\pi^2 \sin(\pi t), -\frac{1}{t^2}, e^t \rangle$;
$\mathbf{a}(1) = \langle 0, -1, e \rangle$.

19. Speed $= \sqrt{\pi^2 + 1 + e^2} \approx 4.27302$.

21. $t = e$.

23. $t = 1$.

25. $\langle x(t), y(t), z(t) \rangle = \langle -\pi t, t, e + et \rangle$.

27. $\mathbf{a}(t) = \langle \frac{\pi^2}{4} \cos(\frac{\pi}{2}t), -\frac{\pi^2}{4} \sin(\frac{\pi}{2}t), 2 \rangle$;
$\mathbf{a}(3) = \langle 0, \frac{\pi^2}{4}, 2 \rangle$.

29. $t = 4$.

31. $\|\mathbf{a}(t)\|$ is constant on $(3,4)$.

33. The object crosses the xz-plane when $t = 4$. It does not ever cross either of the other coordinate planes.

35. $\mathbf{v}(t) = \langle -3\pi \sin(\frac{\pi}{2}t), \frac{\pi}{2}\cos(\frac{\pi}{2}t), 1 \rangle$;
$\mathbf{v}(1) = \langle -3\pi, 0, 1 \rangle$.

37. $\|\mathbf{v}(2)\| = \sqrt{\frac{\pi^2}{4} + 1} \approx 1.86210$.

39. Maximum speed, $t = 1$ and $t = 3$; minimum speed, $t = 2$.

41. The object never crosses any of the coordinate axes.

43. $\langle x(t), y(t), z(t) \rangle = \langle -6, -\frac{\pi}{2}t, 2 + t \rangle$.

Section 12.3

1. $\mathbf{r}'(t) = \langle 6t, \frac{1}{2\sqrt{t}} \rangle$; $\mathbf{r}'(\frac{1}{2}) = \langle 3, \frac{1}{\sqrt{2}} \rangle$.

3. $\theta \approx 19.98405°$.

5. $\mathbf{T}(t) = \frac{\langle 12t^{3/2}, 1 \rangle}{\sqrt{144t^3 + 1}}$;
$\mathbf{T}(\frac{1}{2}) = \frac{\langle 3\sqrt{2}, 1 \rangle}{\sqrt{19}} \approx \langle 0.97333, 0.22942 \rangle$.

7. $\|\mathbf{r}'(t)\| = \sqrt{\frac{144t^3 + 1}{4t}}$.

9. $f(t)\mathbf{r}(t) = \langle 3t^2 - 3t^4, t^{1/2} - t^{5/2} \rangle$.

Computing directly:
$\frac{d}{dt}(f(t)\mathbf{r}(t)) = \langle 6t - 12t^3, \frac{1}{2}t^{-1/2} - \frac{5}{2}t^{3/2} \rangle$.

Applying the product rule:
$f'(t)\mathbf{r}(t) + f(t)\mathbf{r}'(t)$
$= -2t\langle 3t^2, t^{1/2} \rangle + (1 - t^2)\langle 6t, \frac{1}{2}t^{-1/2} \rangle$
$= \langle -6t^3, -2t^{3/2} \rangle + \langle 6t - 6t^3, \frac{1}{2}t^{-1/2} - \frac{1}{2}t^{3/2} \rangle$
$= \langle 6t - 12t^3, \frac{1}{2}t^{-1/2} - \frac{5}{2}t^{3/2} \rangle$.

11. $\mathbf{v}_1(t) = \langle 3\cos(3t), -3\sin(3t), 3t^{1/2} \rangle$;
$\mathbf{v}_2(t) = \langle t, 2t^{1/2}, 2 \rangle$.

13. $\mathbf{T}_1(t) = \frac{\langle \cos(3t), -\sin(3t), t^{1/2} \rangle}{\sqrt{1+t}}$;
$\mathbf{T}_2(t) = \frac{\langle t, 2t^{1/2}, 2 \rangle}{t + 2}$.

15. $\mathbf{a}(t) = \langle 6\cos(3t^2) - 36t^2 \sin(3t^2),$
$-6\sin(3t^2) - 36t^2 \cos(3t^2), 12t \rangle$.

17. $\mathbf{r}_1 \cdot \mathbf{r}_2 = \frac{t^2 \sin(3t)}{2} + \frac{4t^{3/2}\cos(3t)}{3} + 4t^{5/2}$.

Computing directly:
$\frac{d}{dt}(\mathbf{r}_1 \cdot \mathbf{r}_2)$
$= \frac{2t\sin(3t) + 3t^2 \cos(3t)}{2} + 2t^{1/2}\cos(3t)$
$\quad - 4t^{3/2}\sin(3t) + 10t^{3/2}$
$= \sin(3t)(t - 4t^{3/2})$
$\quad + \cos(3t)(\frac{3}{2}t^2 + 2t^{1/2}) + 10t^{3/2}$.

Applying the dot product rule:
$\frac{d}{dt}(\mathbf{r}_1 \cdot \mathbf{r}_2)$
$= \frac{d\mathbf{r}_1}{dt} \cdot \mathbf{r}_2 + \mathbf{r}_1 \cdot \frac{d\mathbf{r}_2}{dt}$
$= \langle 3\cos(3t), -3\sin(3t), 3t^{1/2} \rangle \cdot \langle \frac{t^2}{2}, \frac{4}{3}t^{3/2}, 2t \rangle$
$\quad + \langle \sin(3t), \cos(3t), 2t^{3/2} \rangle \cdot \langle t, 2t^{1/2}, 2 \rangle$
$= \left(\frac{3t^2 \cos(3t)}{2} - 4t^{3/2}\sin(3t) + 6t^{3/2} \right)$
$\quad + (t\sin(3t) + 2t^{1/2}\cos(3t) + 4t^{3/2})$
$= \frac{3t^2 \cos(3t) + 2t\sin(3t)}{2} - 4t^{3/2}\sin(3t)$
$\quad + 2t^{1/2}\cos(3t) + 10t^{3/2}$
$= \sin(3t)(t - 4t^{3/2})$
$\quad + \cos(3t)(\frac{3}{2}t^2 + 2t^{1/2}) + 10t^{3/2}$.

19. $\frac{d}{dt}(\|\mathbf{r}_1\|) = \frac{6t^2}{\sqrt{1 + 4t^3}}$.

21. $\mathbf{r}_1 \times \mathbf{r}_2 = (y_1 z_2 - y_2 z_1)\mathbf{i}$
$\quad + (-x_1 z_2 + x_2 z_1)\mathbf{j} + (x_1 y_2 - x_2 y_1)\mathbf{k}$.

Computing directly:
$\frac{d}{dt}(\mathbf{r}_1 \times \mathbf{r}_2)$
$= (y_1' z_2 + y_1 z_2' - y_2' z_1 - y_2 z_1')\mathbf{i}$
$\quad + (-x_1' z_2 - x_1 z_2' + x_2' z_1 + x_2 z_1')\mathbf{j}$
$\quad + (x_1' y_2 + x_1 y_2' - x_2' y_1 - x_2 y_1')\mathbf{k}$.

(Continued on next page.)

21. (Continued from previous page.)

Applying the cross product rule:

$\frac{d}{dt}(\mathbf{r}_1 \times \mathbf{r}_2)$

$= \frac{d\mathbf{r}_1}{dt} \times \mathbf{r}_2 + \mathbf{r}_1 \times \frac{d\mathbf{r}_2}{dt}$

$= \langle x_1', y_1', z_1' \rangle \times \langle x_2, y_2, z_2 \rangle$
$\quad + \langle x_1, y_1, z_1 \rangle \times \langle x_2', y_2', z_2' \rangle$

$= \langle y_1'z_2 - y_2z_1', -x_1'z_2 + x_2z_1', x_1'y_2 - x_2y_1' \rangle$
$\quad + \langle y_1z_2' - y_2'z_1, -x_1z_2' + x_2'z_1, x_1y_2' - x_2'y_1 \rangle$

$= (y_1'z_2 + y_1z_2' - y_2'z_1 - y_2z_1')\mathbf{i}$
$\quad + (-x_1'z_2 - x_1z_2' + x_2'z_1 + x_2z_1')\mathbf{j}$
$\quad + (x_1'y_2 + x_1y_2' - x_2'y_1 - x_2y_1')\mathbf{k}$.

23. Let $\mathbf{r}(t) = \langle \sin(t), \cos(t) \rangle$.

25. Let $\mathbf{r}(t) = \langle x(t), y(t), z(t) \rangle$, then

$\frac{d\mathbf{T}}{dt} = \left(\frac{y'(y'x''-x'y'')+z'(z'x''-x'z'')}{((x')^2+(y')^2+(z')^2)^{3/2}} \right)\mathbf{i}$

$\quad + \left(\frac{x'(x'y''-y'x'')+z'(z'y''-y'z'')}{((x')^2+(y')^2+(z')^2)^{3/2}} \right)\mathbf{j}$

$\quad + \left(\frac{x'(x'z''-z'x'')+y'(y'z''-z'y'')}{((x')^2+(y')^2+(z')^2)^{3/2}} \right)\mathbf{k};$

$\frac{d\mathbf{T}}{dt} \times \mathbf{T}$

$= \left(\frac{[x'(x'y''-y'x'')+z'(z'y''-y'z'')]z'}{((x')^2+(y')^2+(z')^2)^2} \right.$

$\quad - \left. \frac{[x'(x'z''-z'x'')+y'(y'z''-z'y'')]y'}{((x')^2+(y')^2+(z')^2)^2} \right)\mathbf{i}$

$\quad - \left(\frac{[(y'(y'x''-x'y'')+z'(z'x''-x'z'')]z'}{((x')^2+(y')^2+(z')^2)^2} \right.$

$\quad + \left. \frac{[(x'(x'z''-z'x'')+y'(y'z''-z'y'')]x'}{((x')^2+(y')^2+(z')^2)^2} \right)\mathbf{j}$

$\quad + \left(\frac{[y'(y'x''-x'y'')+z'(z'x''-x'z'')]y'}{((x')^2+(y')^2+(z')^2)^2} \right.$

$\quad - \left. \frac{[(x'(x'y''-y'x'')+z'(z'y''-y'z'')]x'}{((x')^2+(y')^2+(z')^2)^2} \right)\mathbf{k}$

$= \frac{[(x')^2+(y')^2+(z')^2]}{[(x')^2+(y')^2+(z')^2]^2}$

$\quad \cdot \langle y''z' - z''y', z''x' - x''z', x''y' - y''x' \rangle$

$= \frac{\langle y''z'-z''y', z''x'-x''z', x''y'-y''x' \rangle}{(x')^2+(y')^2+(z')^2}$

$= \frac{\mathbf{r}''(t)}{\|\mathbf{r}'(t)\|} \times \mathbf{T}(t)$.

Section 12.4

1. $\mathbf{a}(t) = \langle 6t, \frac{1}{2\sqrt{t}} \rangle$.

3. $\|\langle 63, \frac{14}{3} \rangle\| \approx 63.17620$.

5. $\mathbf{r}(4) = \langle 61, \frac{23}{3} \rangle$.

7. $s(t) = \int_1^t (9u^4 + u)^{1/2}\, du$.

9. $\int \mathbf{r}(t)\, dt = \langle -\frac{\cos(3t)}{3}, \frac{\sin(3t)}{3}, \frac{4}{5}t^{5/2} \rangle + \mathbf{C}$.

11. $\|\langle \sin(6), \cos(6) - 1, 2^{5/2} \rangle\| \approx 5.66389$.

13. $s(t) = 2[(1+t)^{3/2} - 1]$.

15. $\mathbf{v}(t) = \langle \frac{t^3}{3} + 1, \frac{t^4}{4} + 2, t^2 + 3 \rangle$;
terminal velocity $= \langle \frac{11}{3}, 6, 7 \rangle$.

17. $\|\langle \frac{10}{3}, \frac{28}{5}, \frac{26}{3} \rangle\| \approx 10.84353$.

19. Total distance $= 10\pi$;
$\langle 3\sin(\frac{s}{5}), 3\cos(\frac{s}{5}), \frac{4s}{5} \rangle$ with domain $[0, 10\pi]$.

Section 12.5

1. $\mathbf{T}(t) = \frac{\langle 12t^{3/2}, 1 \rangle}{\sqrt{144t^3+1}}$.

3. $\mathbf{N}(t) = \frac{\langle 1, -12t^{3/2} \rangle}{\sqrt{144t^3+1}}$.

7. $\mathbf{B} \cdot \mathbf{B} = \|\mathbf{B}\|^2 = \|1\|^2 = 1$.

9. Let $\mathbf{B} \cdot \mathbf{B} = 1$.

Differentiate with respect to s to get

$\frac{d\mathbf{B}}{ds} \cdot \mathbf{B} + \mathbf{B} \cdot \frac{d\mathbf{B}}{ds} = 0 \Rightarrow 2\mathbf{B} \cdot \frac{d\mathbf{B}}{ds} = 0$

$\Rightarrow \mathbf{B} \cdot \frac{d\mathbf{B}}{ds} = 0 \Rightarrow \frac{d\mathbf{B}}{ds}$ is orthogonal to $\mathbf{B}$.

11. $\mathbf{T}(t) = \frac{\langle \cos(3t), -\sin(3t), t^{1/2} \rangle}{\sqrt{1+t}}$;

$\mathbf{T}(1) \approx \langle -0.70003, -0.09979, 0.70711 \rangle$.

13. Approximate equation:
$0.71174x - 0.01677y + 0.70225z - 1.52154 = 0$.

15. $\tau \approx -0.479$.

17. $\mathbf{N}(1) = \langle \frac{2}{3}, \frac{1}{3}, -\frac{2}{3} \rangle$.

19. $\mathbf{B}(1) = \langle \frac{2}{3}, -\frac{2}{3}, \frac{1}{3} \rangle$.

21. $\mathbf{T}(1) \approx \langle 0.56821, 0.28411, 0.77228 \rangle$.

23. Approximate equation:
$0.80547x - 0.59263z - 0.21284 = 0$.

25. $\tau = -\frac{2}{3}$.

Section 12.6

1. $\kappa(2) \approx 0.00184$; $\rho(2) \approx 543.76527$.

3. $\kappa(1) = \frac{1}{9}$; $\rho(1) = 9$.

5. $\mathbf{v}(t) = \langle x'(t), y'(t) \rangle$;

$$\mathbf{T} = \frac{x'(t)}{\sqrt{(x'(t))^2+(y'(t))^2}}\mathbf{i} + \frac{y'(t)}{\sqrt{(x'(t))^2+(y'(t))^2}}\mathbf{j};$$

$$\frac{d\mathbf{T}}{dt} = \frac{\sqrt{(x')^2+(y')^2}\,x'' - \frac{x'(2x'x''+2y'y'')}{2\sqrt{(x')^2+(y')^2}}}{(x')^2+(y')^2}\mathbf{i}$$

$$+ \frac{\sqrt{(x')^2+(y')^2}\,y'' - \frac{y'(2x'x''+2y'y'')}{2\sqrt{(x')^2+(y')^2}}}{(x')^2+(y')^2}\mathbf{j}$$

$$= \frac{((x')^2+(y')^2)x'' - (x')^2 x'' - x'y'y''}{((x')^2+(y')^2)^{3/2}}\mathbf{i}$$

$$+ \frac{((x')^2+(y')^2)y'' - x'y'x'' - (y')^2 y''}{((x')^2+(y')^2)^{3/2}}\mathbf{j}$$

$$= \frac{(y')^2 x'' - x'y'y''}{((x')^2+(y')^2)^{3/2}}\mathbf{i} + \frac{(x')^2 y'' - x'y'x''}{((x')^2+(y')^2)^{3/2}}\mathbf{j}$$

$$= \frac{y'(y'x''-x'y'')}{((x')^2+(y')^2)^{3/2}}\mathbf{i} + \frac{x'(x'y''-y'x'')}{((x')^2+(y')^2)^{3/2}}\mathbf{j};$$

$$\left\|\frac{d\mathbf{T}}{dt}\right\| = \left[\frac{(y')^2(y'x''-x'y'')^2 + (x')^2(x'y''-y'x'')^2}{((x')^2+(y')^2)^3}\right]^{1/2}$$

$$= \left[\frac{((y')^2+(x')^2)(x'y''-y'x'')^2}{((x')^2+(y')^2)^3}\right]^{1/2}$$

$$= \frac{|x'y''-y'x''|((x')^2+(y')^2)^{1/2}}{((x')^2+(y')^2)^{3/2}};$$

$$\kappa = \frac{\|d\mathbf{T}/dt\|}{\|\mathbf{r}'(t)\|} = \frac{|x'y''-y'x''|((x')^2+(y')^2)^{1/2}}{((x')^2+(y')^2)^{3/2}((x')^2+(y')^2)^{1/2}}$$

$$= \frac{|x'y''-y'x''|}{((x')^2+(y')^2)^{3/2}}.$$

7. $\mathbf{v}(t) = v(t)\mathbf{T}(t)$; $\mathbf{a}(t) = \frac{dv}{dt}\mathbf{T}(t) + v(t)\frac{d\mathbf{T}}{dt}$ (differentiating both sides with respect to t).

9. $\mathbf{a}(t) = \frac{dv}{dt}\mathbf{T}(t) + v(t)[\kappa(t)v(t)\mathbf{N}(t)]$

$$= \frac{dv}{dt}\mathbf{T}(t) + \kappa(t)v^2(t)\mathbf{N}(t).$$

11. $\|\mathbf{a}(t) \times \mathbf{v}(t)\| = \|\kappa(t)v^3(t)\mathbf{N}(t) \times \mathbf{T}(t)\|$

$= \kappa(t)v^3(t)$ since $\|\mathbf{N}(t) \times \mathbf{T}(t)\| = 1$.

13. $\kappa(t) = \frac{\|d\mathbf{T}/dt\|}{\|\mathbf{v}(t)\|} = \frac{\left\|\frac{d}{dt}\left(\frac{\mathbf{v}(t)}{\|\mathbf{v}(t)\|}\right)\right\|}{\|\mathbf{v}(t)\|} = \frac{\left\|\frac{d}{dt}\left(\frac{\mathbf{v}(t)}{b}\right)\right\|}{b}$

$$= \frac{\left\|\frac{\mathbf{v}'(t)}{b}\right\|}{b} = \frac{\|\mathbf{a}(t)\|}{b^2} = \frac{a}{b^2}$$

$\Rightarrow \kappa(t)$ is constant.

15. $\kappa(1) \approx 2.12132$; $\rho(1) \approx 0.47140$.

Chapter 13

Section 13.1

1. $\langle x(t), y(t), z(t) \rangle = \langle 0, 0, 0 \rangle + t\langle 1, 0, 1 \rangle$.

3. Domain $= \{(x, y) : y \neq 0\}$.

5. Hyperbolic cylinder running parallel to the y-axis. The cross-section in the xz plane is shown.

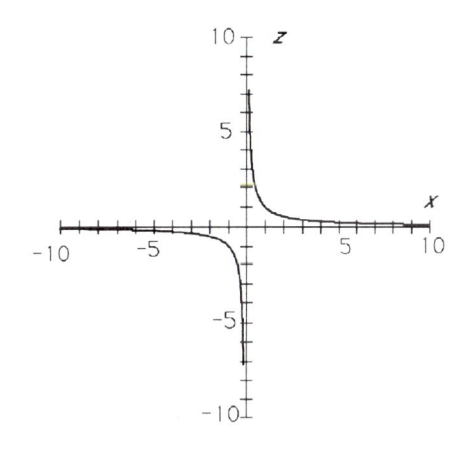

7. Circular cylinder with the y-axis running down its center. The cross section in the xz plane is shown.

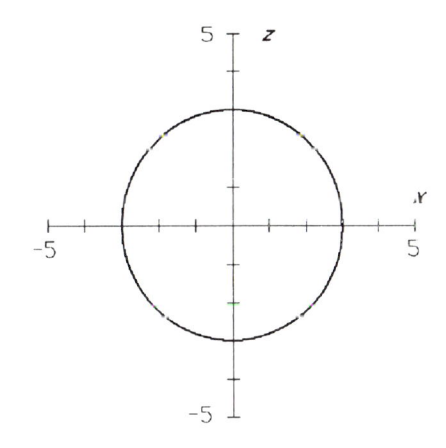

9. Square cylinder with the z-axis running down its center. The cross-section in the xy plane is shown.

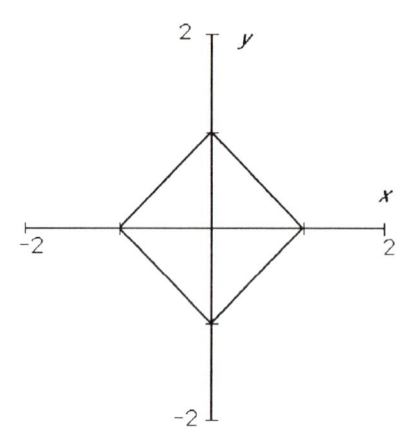

11. $(x+2)^2 + (y-4)^2 + (z+7)^2 = 49$.

13. When $a = b = c$.

15. It is not the graph of a function.

17. It is not the graph of a function.

19. It is the graph of a function.

21. $\frac{y^2}{a^2} + \frac{z^2}{b^2} = 1$; $\quad \frac{x^2}{a^2} + \frac{z^2}{b^2} = 1$.

23. $\frac{x^2}{c^2} = \frac{y^2}{a^2} + \frac{z^2}{b^2} - 1$; $\quad \frac{y^2}{c^2} = \frac{x^2}{a^2} + \frac{z^2}{b^2} - 1$.

25. $cx = \frac{y^2}{a^2} + \frac{z^2}{b^2}$; $\quad cy = \frac{x^2}{a^2} + \frac{z^2}{b^2}$.

Section 13.2

1. Minima at $\approx (1, 0.6), (1.35, 0), (0, 1.65)$; maxima at $(0,0), (3,3)$.

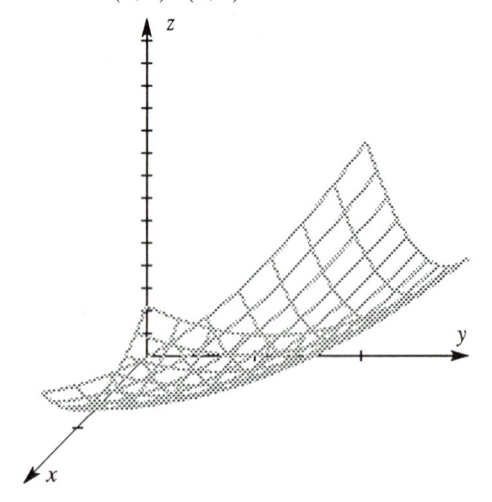

3. Minima at $(0, x)$ for $0 \leq x \leq 3$; maximum at $(3, 3)$.

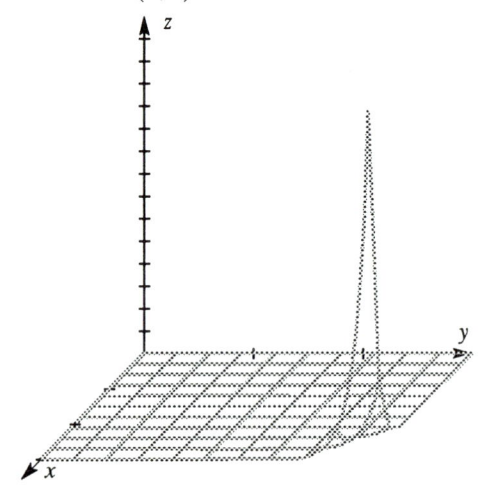

5. Minimum at $(0, 3)$; maximum at $(3, 0)$.

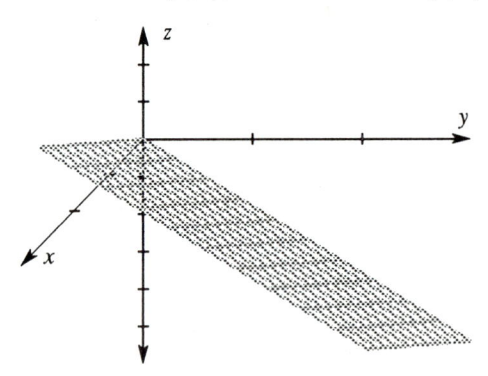

7. Saddle.

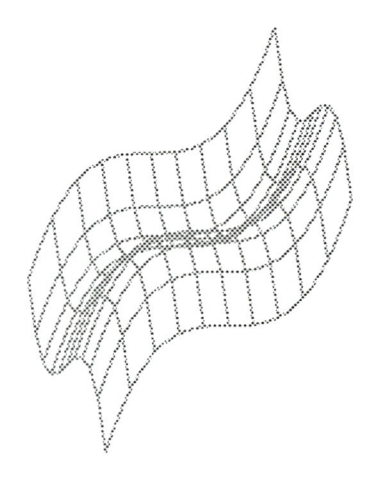

A127

9. Saddle.

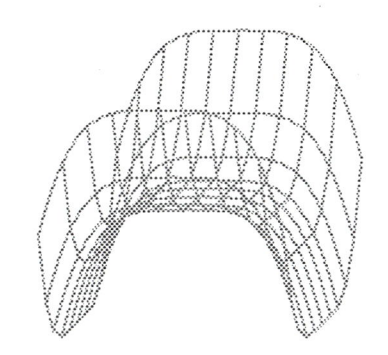

11. Saddle.

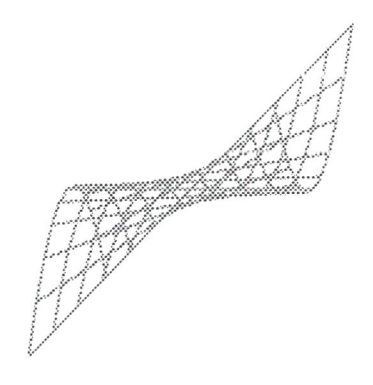

13. Bowl up.

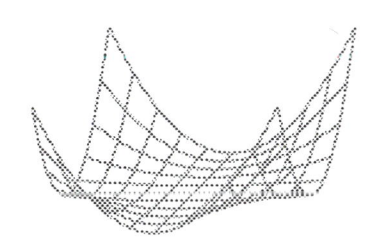

15. Uphill. **17.** Uphill.
19. Downhill. **21.** Downhill.
23. N, S.

25.

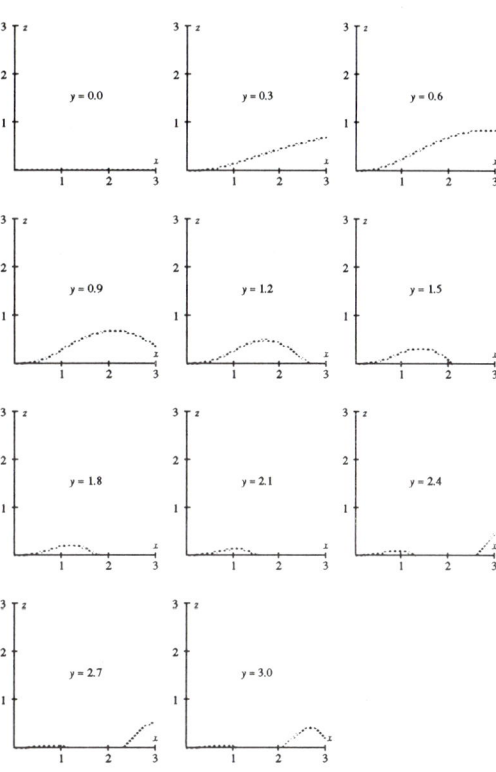

By looking at the graphs, it appears that a maximum occurs near $(3, 0.6)$ and a minimum occurs near $(3, 1.5)$.

Section 13.3

1. $m_1 = 2$.

3. $z = 2x + \frac{1}{2}y - \frac{3}{2}$.

5. $z = 2(x+3) + \frac{1}{2}(y-2) - \frac{13}{2}$.

7. $\dfrac{\partial z}{\partial x} = 0$; $\dfrac{\partial z}{\partial y} = 1$;

z-intercept $= (0, 0, 0)$;

$z = (y - 4) + 4$.

9. Slopes undefined, plane contains entire z-axis, plane is not the graph of a function.

11. A saddle point occurs at $(0, 0)$.

A128

13. No strict maximum, minimum, nor saddle point occurs at $(0,0)$. Graph is a trough whose bottom edge lies along the line $x - 2y = 0$.

15. Graph is the xy plane.

17. $z = -3 - 2(x+1) - 6(y-3) + (x+1)^2 - (y-3)^2$.

19. $z = -12 + (x+1)^2 + 2(y-3)^2$.

21. A saddle point occurs at $(-1, 3, 17)$.

23. 6 points.

(Answers may vary for exercises 24-30.)

25. $z = -2x^2 - 3(y-1)^2 - 5$.

27. $z = (2x - y + 3)^2 + 5$.

29. $2y + 3z = 6$.

Chapter 14

Section 14.1

1. $\frac{\partial f}{\partial x} = \frac{-2y}{(x-y)^2}$; $\left.\frac{\partial f}{\partial x}\right|_{(1,2)} = -4$;

$\frac{\partial f}{\partial y} = \frac{2x}{(x-y)^2}$; $\left.\frac{\partial f}{\partial y}\right|_{(1,2)} = 2$.

3. $\frac{\partial f}{\partial x} = e^{x^2+y^2}(2x)$; $\left.\frac{\partial f}{\partial x}\right|_{(1,-2)} = 2e^5$;

$\frac{\partial f}{\partial y} = e^{x^2+y^2}(2y)$; $\left.\frac{\partial f}{\partial y}\right|_{(1,-2)} = -4e^5$.

5. $\frac{\partial f}{\partial x} = 2xy\cos(x^2y)$; $\left.\frac{\partial f}{\partial x}\right|_{(2,0)} = 0$;

$\frac{\partial f}{\partial y} = x^2 \cos(x^2 y)$; $\left.\frac{\partial f}{\partial y}\right|_{(2,0)} = 4$.

7. $\frac{\partial f}{\partial x} = -(\frac{y}{x^2})\sec^2(\frac{y}{x})$; $\left.\frac{\partial f}{\partial x}\right|_{(2,\frac{\pi}{2})} = -\frac{\pi}{4}$;

$\frac{\partial f}{\partial y} = (\frac{1}{x})\sec^2(\frac{y}{x}) + \csc^2(y)$; $\left.\frac{\partial f}{\partial y}\right|_{(2,\frac{\pi}{2})} = 2$.

9. $\frac{\partial f}{\partial x} = \frac{x}{\sqrt{x^2+y^2}}$; $\left.\frac{\partial f}{\partial x}\right|_{(-3,4)} = -\frac{2}{3}$;

$\frac{\partial f}{\partial y} = \frac{y}{\sqrt{x^2+y^2}}$; $\left.\frac{\partial f}{\partial y}\right|_{(-3,4)} = 0.8$;

11. $f_x = \frac{1}{y}e^{x/y}$; $f_y = -\frac{x}{y^2}e^{x/y}$;

$f_x(1,2) = \frac{1}{2}e^{1/2}$; $f_y(1,2) = -\frac{1}{4}e^{1/2}$.

13. $f_x = -2y\cos(xy)\sin(xy)$; $f_x\left(\frac{1}{2}, \frac{\pi}{2}\right) = -\frac{\pi}{2}$;

$f_y = -2x\cos(xy)\sin(xy)$; $f_y\left(\frac{1}{2}, \frac{\pi}{2}\right) = -\frac{1}{2}$.

15. $f_x = \frac{1}{x}$; $f_y = -e^{-y}$;

$f_x(1,1) = 1$; $f_y(1,1) = -\frac{1}{e}$.

17. $f_x = \frac{x^2+2xy}{(x+y)^2}$; $f_y = -\frac{x^2}{(x+y)^2}$;

$f_x(-1,-1) = -\frac{1}{4}$; $f_y(-1,-1) = -\frac{1}{4}$.

19. $f_x = -\frac{z}{(x+y)^2}$; $f_x(1,2,3) = -\frac{1}{3}$;

$f_y = -\frac{z}{(x+y)^2}$; $f_y(1,2,3) = -\frac{1}{3}$;

$f_z = \frac{1}{x+y}$; $f_z(1,2,3) = \frac{1}{3}$.

21. $f_x = 5$; $f_x(-3,2,4) = 5$;

$f_y = -4$; $f_y(-3,2,4) = -4$;

$f_z = 3$; $f_z(-3,2,4) = 3$.

23. $f_x = \arcsin(z)$; $f_x(0.5, 0.5, 0.5) = \frac{\pi}{6}$;

$f_y = -\arctan(z)$; $f_y(0.5, 0.5, 0.5) \approx -0.46365$;

$f_z = \frac{x}{\sqrt{1-z^2}} - \frac{y}{1+z^2}$; $f_z(0.5, 0.5, 0.5) \approx 0.17735$.

25. $f_x = yz\cos(xyz)$; $f_x(1, \pi, 0) = 0$;

$f_y = xz\cos(xyz)$; $f_y(1, -\pi, 0) = 0$;

$f_z = xy\cos(xyz)$; $f_z(1, -\pi, 0) = -\pi$.

27. $f_x = \frac{yz(-x^2+y^2-z^2)}{(x^2+y^2-z^2)^2}$; $f_x(1,1,-1) = 1$;

$f_y = \frac{xz(x^2-y^2-z^2)}{(x^2+y^2-z^2)^2}$; $f_y(1,1,-1) = 1$;

$f_z = \frac{xy(x^2+y^2+z^2)}{(x^2+y^2-z^2)^2}$; $f_z(1,1,-1) = 3$.

29. $f_x = \frac{y^2+z^3}{(x+y^2+z^3)^2}$; $f_x(3,2,1) = \frac{5}{64}$;

$f_y = -\frac{2xy}{(x+y^2+z^3)^2}$; $f_y(3,2,1) = -\frac{3}{16}$;

$f_z = -\frac{3xz^2}{(x+y^2+z^3)^2}$; $f_z(3,2,1) = -\frac{9}{64}$.

31. $f_x = \frac{-2x^3+y^2+z-3x^2y}{(x^3+y^2+z)^2}$; $f_x(1,-1,1) = \frac{1}{3}$;

$f_y = \frac{x^3-y^2+z-2xy}{(x^3+y^2+z)^2}$; $f_y(1,-1,1) = \frac{1}{3}$;

$f_z = -\frac{(x+y)}{(x^3+y^2+z)^2}$; $f_z(1,-1,1) = 0$.

33. Downhill. **35.** Uphill.
37. Downhill. **39.** Uphill.

41. $\left.\frac{\partial P}{\partial y}\right|_{(0,0)} \approx 2.47113$.

43. $\left.\frac{\partial P}{\partial y}\right|_{(0.5,-0.12)} \approx 3.71943$.

45. $\left.\frac{\partial P}{\partial y}\right|_{(0,0)} \approx 2.51018$.

47. $\left.\frac{\partial P}{\partial y}\right|_{(-0.35,-0.42)} \approx 2.75972$.

49. $\left.\frac{\partial T}{\partial y}\right|_{(0,-3,0)} = 18$.

51. At any point in the (x,y) plane, moving straight up or down will result in a temperature decrease.

Section 14.2

1. $Df = \left[-\frac{2y}{(x-y)^2} \quad \frac{2x}{(x-y)^2}\right]$;
$f(x,y) \approx -3 - 4(x-1) + 2(y-2)$.

3. $Df = \left[2xe^{x^2+y^2} \quad 2ye^{x^2+y^2}\right]$;
$f(x,y) \approx e^5 + 2e^5(x-1) - 4e^5(y+2)$.

5. $Df = \left[2xy\cos(x^2y) \quad x^2\cos(x^2y)\right]$;
$f(x,y) \approx 4y$.

7. $Df = \left[-\frac{y}{x^2}\sec^2(\frac{y}{x}) \quad (\frac{1}{x})\sec^2(\frac{y}{x}) + \csc^2(y)\right]$;
$f(x,y) \approx 1 - \frac{\pi}{4}(x-2) + 2(y-\frac{\pi}{2})$.

9. $Df = \left[\frac{x}{\sqrt{x^2+y^2}} \quad \frac{y}{\sqrt{x^2+y^2}}\right]$;
$f(x,y) \approx 5 - 0.6(x+3) + 0.8(y-4)$.

11. $Df = \left[-\frac{z}{(x+y)^2} \quad -\frac{z}{(x+y)^2} \quad \frac{1}{x+y}\right]$;
$f(x,y,z) \approx 1 - \frac{1}{3}(x-1) - \frac{1}{3}(y-2) + \frac{1}{3}(z-3)$.

13. $Df = [5 \quad -4 \quad 3]$;
$f(x,y,z) \approx -11 + 5(x+3) - 4(y-2) + 3(z-4)$.

15. $Df = \left[\arcsin(z) \quad -\arctan(z) \quad \frac{x}{\sqrt{1-z^2}} - \frac{y}{1+z^2}\right]$;
$f(x,y,z) \approx 0.02998 + 0.52360(x-0.5)$
$\qquad -0.46365(y-0.5) + 0.17735(z-0.5)$.

17. $DH(2,3) = [-112 \quad 12]$;
$H(2.1,3.1) - (-47) \approx -10$.

19. $DH(-1,2) = [28 \quad 16]$;
$H(-1.3,1.8) - 1 \approx -11.6$.

21. $DH(3,-2) = [0.5 \quad -0.75]$;
$H(3.2,-2.1) - (-1.5) \approx 0.175$.

23. $DH(-2,-1) = [-1 \quad 2]$;
$H(-2.2,-0.9) - 2 \approx 0.4$.

25. $DT = \left[-\frac{600x}{(1+x^2+y^2+z^2)^2} \quad -\frac{600y}{(1+x^2+y^2+z^2)^2} \right.$
$\left. \qquad -\frac{600z}{(1+x^2+y^2+z^2)^2}\right]$;
$T(1.1,1.9,3.2) - 20 \approx -\frac{4}{3}$.

27. (a) $z = 2.5 + 2.5(x-1) + 0.75(x-2)$.

29. (b) $z = \sqrt{5} + \frac{1}{\sqrt{5}}(x-1) + \frac{2}{\sqrt{5}}(x-2)$.

Section 14.3

1. $D_{\mathbf{u}}f(1,2) = -\sqrt{2}$.

3. $D_{\mathbf{u}}f(2,0) = 4$.

5. $D_{\mathbf{u}}f(0,0) = -0.8$.

7. $\mathbf{u} \approx \langle -0.57735, -0.57735, 0.57735 \rangle$.

9. $\mathbf{u} \approx \langle 0.70711, -0.56569, 0.42426 \rangle$.

11. $\mathbf{u} \approx \langle 0.72570, -0.64261, 0.24580 \rangle$.

13. $D_{\mathbf{u}}T(-1,-3,-2) \approx -9.08688$.

15. $\mathbf{u} \approx \langle 0.97333, 0.16222, -0.16222 \rangle$.

17. $\left.D_{\mathbf{r}}T\right|_{t=4} \approx -3.52801$.

A130

19. $D_{\mathbf{i}}f$
$= \left\langle \frac{2x}{1+(x^2+y^2+z^2)^2}, \frac{2y}{1+(x^2+y^2+z^2)^2}, \frac{2z}{1+(x^2+y^2+z^2)^2} \right\rangle$
$\cdot \langle 1, 0, 0 \rangle$
$= \frac{2x}{1+(x^2+y^2+z^2)^2} = f_x;$

$D_{\mathbf{j}}f$
$= \left\langle \frac{2x}{1+(x^2+y^2+z^2)^2}, \frac{2y}{1+(x^2+y^2+z^2)^2}, \frac{2z}{1+(x^2+y^2+z^2)^2} \right\rangle$
$\cdot \langle 0, 1, 0 \rangle$
$= \frac{2y}{1+(x^2+y^2+z^2)^2} = f_y;$

$D_{\mathbf{k}}f$
$= \left\langle \frac{2x}{1+(x^2+y^2+z^2)^2}, \frac{2y}{1+(x^2+y^2+z^2)^2}, \frac{2z}{1+(x^2+y^2+z^2)^2} \right\rangle$
$\cdot \langle 0, 0, 1 \rangle$
$= \frac{2z}{1+(x^2+y^2+z^2)^2} = f_z.$

21. $f(x, y, z) \approx \frac{\pi}{4} + x - 1.$

23. $D_{\mathbf{u}}H(-1, 2) \approx -31.11270;$
$\mathbf{u} \approx -\langle 0.86824, 0.49614 \rangle.$

25. $D_{\mathbf{u}}H(-2, -1) \approx -31.11270;$
$\mathbf{u} \approx \langle 0.49614, -0.86824 \rangle.$

27. $D_{\mathbf{u}}H(-1, 2) \approx 0.17677;$
$\mathbf{u} \approx \langle 0.89443, 0.44721 \rangle.$

29. $D_{\mathbf{u}}H(-2, -1) = -\frac{1}{\sqrt{2}};$
$\mathbf{u} \approx \langle -0.44721, 0.89443 \rangle.$

Section 14.4

1. $f_{xx} = 2;\quad f_{xy} = 0;\quad f_{yy} = 2;$
$f(x, y) = x^2 + y^2;$ bowl up.

3. $f_{xx} = 6x;\quad f_{xy} = 0;\quad f_{yy} = -2;$
$f(x, y) \approx -y^2;$ trough down.

5. $f_{xx} = 0;\quad f_{xy} = 1;\quad f_{yy} = 0;$
$f(x, y) = xy;$ saddle.

7. $f_{xx} = 0;\quad f_{xy} = 3y^2;\quad f_{yy} = 6xy;$
$f(x, y) \approx 0;$ flat.

9. Fixing $x = a$, we have a curve of the form $f(a, y)$. Since $f_y(a, b) < 0$, this curve is decreasing at $y = b$ as y increases. Fixing $y = b$, we have a curve of the form $f(x, b)$. Since $f_x(x, b) > 0$, this curve is increasing at $x = a$ as x increases. Since $f_{xy}(a, b) < 0$, the rate at which the curves increase or decrease as we fix x or y at different values nearby is decreasing.

11. $f_{xy}(x, y) = \frac{2z}{(x+y)^3} = f_{yx}(x, y);$
$f_{xz}(x, y) = -\frac{1}{(x+y)^2} = f_{zx}(x, y);$
$f_{yz}(x, y) = -\frac{1}{(x+y)^2} = f_{zy}(x, y);$
$f_{xx}(x, y) = \frac{2z}{(x+y)^3} = f_{yy}(x, y);$
$f_{zz}(x, y) = 0.$

13. All second order partial derivatives are 0.

15. $f_{xy}(x, y) = 0 = f_{yx}(x, y);$
$f_{xz}(x, y) = \frac{1}{\sqrt{1-z^2}} = f_{zx}(x, y);$
$f_{yz}(x, y) = -\frac{1}{1+z^2} = f_{zy}(x, y);$
$f_{xx}(x, y) = f_{yy}(x, y) = 0;$
$f_{zz}(x, y) = \frac{xz}{(1-z^2)^{3/2}} + \frac{2zx}{1+z^2}.$

17. $f_{xx}(x, y) = 0;\quad f_{xy}(x, y) = 1 - \frac{1}{y^2};$
$f_{yy}(x, y) = \frac{2x}{y^3};$
$f(x, y) \approx -\frac{10}{3} + \frac{10}{3}(x+1) - \frac{8}{9}(y-3)$
$\qquad + \frac{8}{9}(x+1)(y-3) - \frac{1}{27}(y-3)^2.$

19. $f_{xx}(x, y) = 2y\cos(x^2y) - 4x^2y^2\sin(x^2y);$
$f_{xy}(x, y) = 2x\cos(x^2y) - 2x^3y\sin(x^2y);$
$f_{yy}(x, y) = -x^4\sin(x^2y);$
$f(x, y) \approx 4y + 4(x-2)y.$

21. $f_{yy} = \frac{x^2}{y^4}e^{x/y} + \frac{2x}{y^3}e^{x/y};$
$f_{xx} = \frac{1}{y^2}e^{x/y};\quad f_{xy} = -\frac{1}{y^2}e^{x/y}\left(\frac{x}{y}+1\right).$

23. $f_{xx} = -2y^2\cos^2(xy) + 2y^2\sin^2(xy);$
$f_{xy} = 2xy\sin^2(xy) - 2xy\cos^2(xy)$
$\qquad - 2\sin(xy)\cos(xy);$
$f_{yy} = -2x^2\cos^2(xy) + 2x^2\sin^2(xy).$

25. $f_{xx} = \frac{4y^2}{(x-y)^3};$
$f_{xy} = -\frac{4yx}{(x-y)^3};\quad f_{yy} = \frac{4x^2}{(x-y)^3}.$

27. $f_{xx} = (x^2 + y^2)(30x^2 + 6y^2)$;
$f_{xy} = 24xy(x^2 + y^2)$;
$f_{yy} = (x^2 + y^2)(30y^2 + 6x^2)$.

29. $f_{xx} = \frac{-2x^2+2y^2}{(x^2+y^2)^2}$; $f_{xy} = \frac{4xy}{(x^2+y^2)^2}$;
$f_{yy} = \frac{2x^2-2y^2}{(x^2+y^2)^2}$.

31. $f_{xx} = 2\sin y$; $f_{xy} = 2x\cos y$;
$f_{yy} = -(x^2+1)\sin y$.

33. $f_{xx} = \frac{2(1-x^2+y^2-z^2)}{(1+x^2+y^2-z^2)^2}$; $f_{xz} = \frac{4xz}{(1+x^2+y^2-z^2)^2}$;
$f_{xy} = \frac{-4xy}{(1+x^2+y^2-z^2)^2}$; $f_{yy} = \frac{2(1+x^2-y^2-z^2)}{(1+x^2+y^2-z^2)^2}$;
$f_{yz} = \frac{4yz}{(1+x^2+y^2-z^2)^2}$; $f_{zz} = \frac{-2(1+x^2+y^2+z^2)}{(1+x^2+y^2-z^2)^2}$.

35. $f_{xx} = 4e^{2x+y^2+z}$; $f_{xy} = 4ye^{2x+y^2+z}$;
$f_{xz} = 2e^{2x+y^2+z}$; $f_{yy} = (4y^2+2)e^{2x+y^2+z}$;
$f_{yz} = 2ye^{2x+y^2+z}$; $f_{zz} = e^{2x+y^2+z}$.

37. $f_{xx} = -\sin x \sin y \sin z$; $f_{xy} = \cos x \cos y \sin z$;
$f_{xz} = \cos x \sin y \cos z$; $f_{yy} = -\sin x \sin y \sin z$;
$f_{yz} = \sin x \cos y \cos z$; $f_{zz} = -\sin x \sin y \sin z$.

39. $f_{xx} = \frac{2yz(y+z)}{(x+y+z)^3}$;
$f_{xy} = \frac{(2yz+z^2)(x+y+z)-2(y^2z+yz^2)}{(x+y+z)^3}$;
$f_{xz} = \frac{(2yz+y^2)(x+y+z)-2(y^2z+yz^2)}{(x+y+z)^3}$;
$f_{yz} = \frac{(2xz+x^2)(x+y+z)-2(x^2z+xz^2)}{(x+y+z)^3}$;
$f_{yy} = \frac{2xz(x+z)}{(x+y+z)^3}$; $f_{zz} = \frac{2xy(x+y)}{(x+y+z)^3}$.

Section 14.5

1. $(-0.5, -0.5)$: saddle point.

3. $(0,0)$: saddle point;
$(1,1)$: relative maximum.

5. $(0,0)$: saddle point.

7. $(0,1)$: saddle point;
$(0,-1)$: relative maximum;
$(2,1)$: relative minimum;
$(2,-1)$: saddle point;
$(-2,1)$: relative minimum;
$(-2,-1)$: saddle point.

9. $(0,0)$: saddle point.

11. $(0,0)$: saddle point.

13. $(\frac{\pi}{2} + n\pi, 0)$, n an integer: saddle points.

15. $(0,a)$: minima; $(b,0)$: minima.

17. Absolute minimum value: -2;
absolute maximum value: 1.

19. Absolute minimum value: 0;
absolute maximum value: 2.

21. Absolute minimum value: 0;
absolute maximum value: 4.

23. Absolute minimum value: -1;
absolute maximum value: 2.

25. Absolute minimum value: 0;
absolute maximum value: 1.

Section 14.6

1. $\left(\frac{9}{7}, \frac{6}{7}, \frac{3}{7}\right)$.

3. $(2,2,2)$, $(2,-2,-2)$,
$(-2,-2,2)$, $(-2,2,-2)$.

5. $(2,2,2)$, $(-2,-2,-2)$.

7. $\sqrt[3]{6000}$ cm $\times$ $\sqrt[3]{6000}$ cm $\times$ $\sqrt[3]{6000}$ cm;
surface area ≈ 1981.15635 cm^2.

9. $\sqrt[3]{6000}$ cm $\times$ $\sqrt[3]{6000}$ cm $\times$ $\sqrt[3]{6000}$ cm.

11. farthest: $(-2.30278, -2.30278, 10.60555)$;
closest: $(1.30278, 1.30278, 3.39445)$.

Chapter 15

Section 15.1

1. $\int_1^2 \int_1^4 x + y \, dx \, dy = 12.$

3. $\int_0^1 \int_0^1 x^2 + y^2 \, dx \, dy = \frac{2}{3}.$

5. $\int_0^1 \int_1^3 x - 3x^2 y + y^{1/2} \, dx \, dy = -\frac{23}{3}.$

7. $\int_{-1}^1 \int_0^{\pi/2} x \sin(y) - y e^x \, dy \, dx = \frac{\pi^2}{8}(e^{-1} - e).$

9. $\int_0^{\pi/4} \int_0^{\pi/2} \frac{\cos(x)}{\cos^2(y)} \, dx \, dy = 1.$

11. $f(0.5, -0.5) + f(1.5, -0.5) + \ldots + f(6.5, -0.5)$
    $+ f(0.5, 0.5) + f(1.5, 0.5) + \ldots + f(6.5, 0.5)$
    $+ \ldots + f(0.5, 5.5) + f(1.5, 5.5)$
    $+ \ldots + f(6.5, 5.5) = 980.$

13. $\int_{-1}^2 \int_1^4 (2x + 6x^2 y) \, dx \, dy = 234.$

15. $\int_1^3 \int_{\pi/6}^{2\pi} 2y \cos(x) \, dx \, dy = -4.$

17. $\int_0^3 \int_1^2 \int_1^4 x + y + z \, dx \, dy \, dz = \frac{99}{2}.$

19. $\int_{-1}^1 \int_{-1}^1 \int_{-1}^1 \frac{y^2 e^z}{1+x^2} \, dx \, dy \, dz = \frac{\pi}{3}(e - e^{-1}).$

21. $\int_p^q \int_c^d \int_a^b 1 \, dx \, dy \, dz = \int_p^q \int_c^d x \Big]_{x=a}^{x=b} dy \, dz$
    $= \int_p^q \int_c^d (b-a) \, dy \, dz = \int_p^q (b-a) y \Big]_{y=c}^{y=d} dz$
    $= \int_p^q (b-a)(d-c) \, dz$
    $= (b-a)(d-c) z \Big]_{z=p}^{z=q} = (b-a)(d-c)(q-p),$
    which is the volume of the cube.

23. $\int_1^2 \int_{-1}^0 \int_0^3 (x + 2y + 4z) \, dz \, dy \, dx = 19.5;$

$\int_{-1}^0 \int_1^2 \int_0^3 (x + 2y + 4z) \, dz \, dx \, dy;$

$\int_0^3 \int_1^2 \int_{-1}^0 (x + 2y + 4z) \, dy \, dx \, dz;$

$\int_1^2 \int_0^3 \int_{-1}^0 (x + 2y + 4z) \, dy \, dz \, dx;$

$\int_{-1}^0 \int_0^3 \int_1^2 (x + 2y + 4z) \, dx \, dz \, dy;$

$\int_0^3 \int_{-1}^0 \int_1^2 (x + 2y + 4z) \, dx \, dy \, dz.$

Section 15.2

1. $\int_{-3}^2 \int_0^{y^2} (x^2 + y) \, dx \, dy = \frac{7895}{84}.$

3. $\int_0^{\pi/2} \int_{\cos(x)}^0 y \sin(x) \, dy \, dx = -\frac{1}{6}.$

5. $\int_1^2 \int_{-y}^y (x + y) \, dx \, dy = \frac{14}{3}.$

7. $\int_0^1 \int_{-\sqrt{y}}^{\sqrt{y}} y x^2 \, dx \, dy = \frac{4}{21}.$

9. $\int_0^{\sqrt{5}} \int_{x^2}^{10-x^2} y^2 \sqrt{x} \, dy \, dx \approx 380.21.$

11. $\int_0^1 \int_{e^y}^e x e^y \, dx \, dy.$

13. $\int_{-1}^0 \int_x^{x^3} (x + y) \, dy \, dx = 0.$

15. $\int_1^2 \int_{1-x}^{\sqrt{x}} x^2 y \, dy \, dx = \frac{163}{120}.$

17. $\int_1^2 \int_{x^3}^x e^{y/x} \, dy \, dx = 2e - 0.5 e^4.$

19. $\int_{\pi/6}^{\pi/4} \int_{\tan(x)}^{\sec(x)} y + \sin(x) \, dy \, dx \approx 0.20867.$

21. $\int_0^4 \int_{\sqrt{x}}^2 f(x, y) \, dy \, dx.$

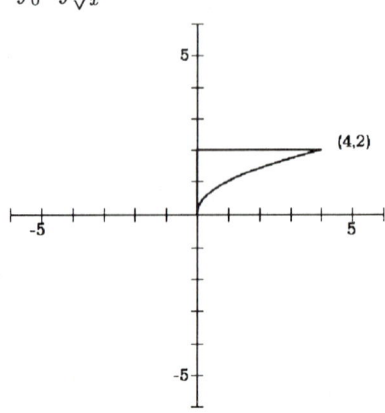

23. $\int_0^1 \int_{x^3}^{\sqrt{x}} f(x,y)\, dy\, dx.$

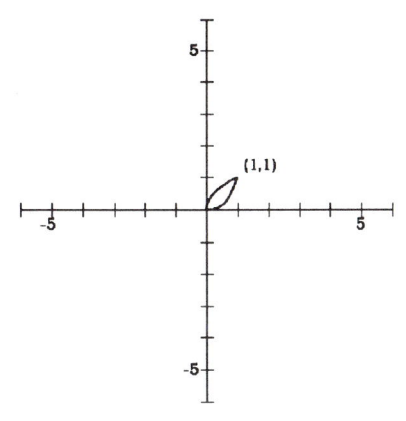

25. $\int_1^4 \int_{-x}^{\sqrt{x}} dy\, dx = \frac{73}{6}.$

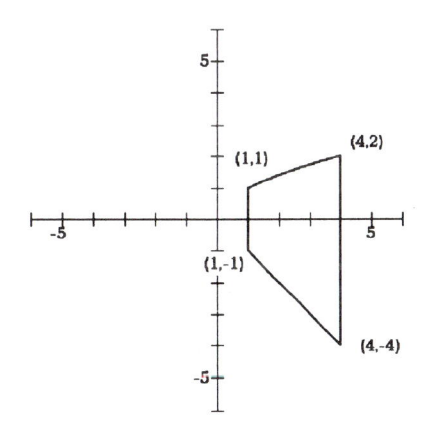

27. $\int_0^1 \int_x^{3x} dy\, dx + \int_1^2 \int_x^{4-x} dy\, dx = 2.$

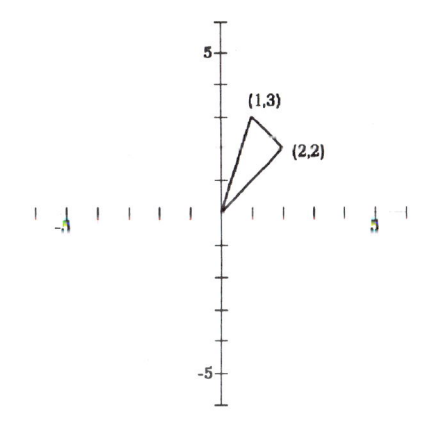

29. $2\int_0^1 \int_0^{e^y} dx\, dy = 2(e-1).$

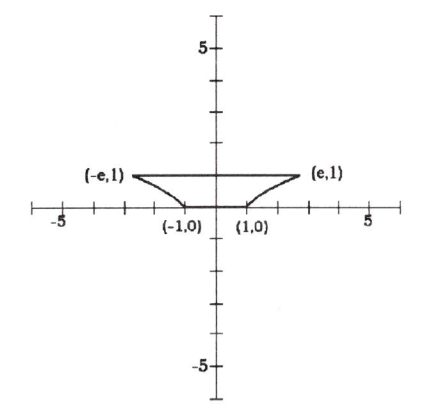

31. $\int_0^1 \int_0^2 4x^2 + y^2\, dx\, dy = \frac{34}{3}.$

33. $\int_{-1}^2 \int_{2-x}^{4-x^2} x^2 + 4\, dy\, dx = 21.15.$

Section 15.3

1. $\int_D \int x^2 + y^2\, dx\, dy = 2\pi.$

3. $\int_{-\sqrt{2}/2}^{\sqrt{2}/2} \int_{-\sqrt{1-x^2}}^{|x|} x^2 y^3\, dy\, dx$

$= 2\int_{3\pi/2}^{7\pi/4} \int_0^1 (r\cos(\theta))^2 (r\sin(\theta))^3 r\, dr\, d\theta$

$+ 2\int_{-\pi/4}^{\pi/4} \int_0^{1/(\sqrt{2}\cos(\theta))} (r\cos(\theta))^2 (r\sin(\theta))^3 r\, dr\, d\theta$

$\approx -0.02357.$

See sketch of the region of integration below.

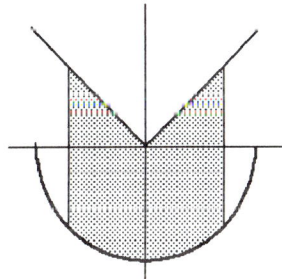

5. $\int_0^{2\pi} \int_0^{1+\sin(2\theta)} r\, dr\, d\theta = \frac{3\pi}{2}.$

7. $\int_{7\pi/6}^{11\pi/6} \int_0^{1+2\sin(\theta)} r\,dr\,d\theta = \pi - \frac{3\sqrt{3}}{2}$.

9. $\int_0^{\pi/3} \int_0^{\sin(3\theta)} r\,dr\,d\theta = \frac{\pi}{12}$.

11. $\int_0^{\pi/2} \int_0^{\sqrt{8\sin(2\theta)}} r\,dr\,d\theta = 4$.

13. $\int_0^{2\pi} \int_0^{3+2\sin\theta} r\,dr\,d\theta = 11\pi$.

15. $\int_{-5}^{5} \int_0^{\sqrt{25-x^2}} e^{-(x^2+y^2)}\,dy\,dx = \frac{\pi}{2}(1-e^{-25})$.

17. $\int_0^3 \int_0^{\sqrt{9-x^2}} (x^2+y^2)^{3/2}\,dy\,dx = \frac{243\pi}{10}$.

Section 15.4

1. $\int_0^1 \int_{x^2}^{x} \int_0^{xy} (x+2y+3z)\,dz\,dy\,dx = \frac{67}{630}$.

3. $\int_0^1 \int_0^{1-x} \int_0^{1-x-y} dz\,dy\,dx = \frac{1}{6}$.

5. $\int_0^{2\pi} \int_0^2 \int_{r^2}^4 r\,dz\,dr\,d\theta = 8\pi$.

7. $\int_0^1 \int_0^{\sqrt{1-x^2}} \int_0^{\sqrt{1-x^2-y^2}} dz\,dy\,dx = \frac{\pi}{6}$.

9. $(\frac{16}{35}, \frac{16}{35}, \frac{16}{35})$.

11. M
$$= \int_0^1 \int_0^{1-x} \int_0^{1-x-y} (x^2+y^2+z^2)\,dz\,dy\,dx$$
$$= \frac{1}{20}.$$

13. I_{xy}
$$= \int_0^1 \int_0^{1-z} \int_0^{1-y-z} x^2z^2+y^2z^2+z^4\,dx\,dy\,dz$$
$$= \frac{2}{315};$$
$I_{xz} = I_{yz} = \frac{2}{315}$
since all these integrals are essentially the same.

15. $\int_0^1 \int_{1+x}^{2x} \int_z^{x+z} x\,dy\,dz\,dx = -\frac{1}{12}$.

17. $\int_{-1}^2 \int_1^{x^2} \int_0^{x+y} 2x^2y\,dz\,dy\,dx = 64.125$.

19. $4\int_0^{3/2} \int_0^{\sqrt{9-4x^2}} \int_0^{9-4x^2-y^2} dz\,dy\,dx;$

$4\int_0^3 \int_0^{\sqrt{9-y^2}/2} \int_0^{9-4x^2-y^2} dz\,dx\,dy;$

$4\int_0^{3/2} \int_0^{9-4x^2} \int_0^{\sqrt{9-4x^2-z}} dy\,dz\,dx;$

$4\int_0^9 \int_0^{\sqrt{9-z}/2} \int_0^{\sqrt{9-4x^2-z}} dy\,dx\,dz;$

$4\int_0^3 \int_0^{9-y^2} \int_0^{\sqrt{9-y^2-z}/2} dx\,dz\,dy;$

$4\int_0^9 \int_0^{\sqrt{9-z}} \int_0^{\sqrt{9-y^2-z}/2} dx\,dy\,dz.$

All yield a value of $\frac{81\pi}{4}$.

21. The region is a horizontal slice of a cone.

Section 15.5

1. $(\sqrt{5}, \arctan(2), -2) \approx (2.23607, 1.10715, -2)$;
$(3, \arctan(2), \pi + \arctan(-\frac{\sqrt{5}}{2}))$
$\approx (3, 1.10715, 2.30052)$.

3. $(-\frac{5\sqrt{3}}{4}, \frac{15}{4}, -\frac{5}{2}) \approx (-2.16506, 3.75, -2.5)$;
$(5\frac{\sqrt{3}}{2}, \frac{2\pi}{3}, -\frac{5}{2}) \approx (4.33013, 2.09440, -2.5)$.

5. $(-\frac{7}{\sqrt{2}}, \frac{7}{\sqrt{2}}, 6) \approx (-4.94975, 4.94975, 6)$;
$(\sqrt{85}, \frac{3\pi}{4}, \arctan(\frac{7}{6})) \approx (9.21954, 2.35619, 0.86217)$.

7. $(0,0,-5);\quad (5,0,\pi)$.

9. $(0,4,0);\quad (4,\frac{\pi}{2},0)$.

11. $(\sqrt{3},-1,-1) \approx (1.73205,-1,-1)$;
$(\sqrt{5}, 330°, \pi + \arctan(-2))$
$\approx (2.23607, 330°, 116.56505°)$.

13. The half of the xz plane where $x \geq 0$.

15. $(0,0,0)$.

16. $(0,0,0)$.

17. Points on the z-axis have more than one representation in the cylindrical and spherical coordinate systems. Since $r = 0$, θ can be any angle.

19. $z = r^2 \cos 2\theta;\quad \rho\cos\phi = \rho^2\sin^2\phi\cos 2\theta$
or $1 = \rho\tan\phi\sin\phi\cos 2\theta$.

21. $z = 5;\quad \rho\cos\phi = 5$.

23. $r^2 = 9$ or $r = 3;\quad \rho^2\sin^2\phi = 9$ or $\rho\sin\phi = 3$.

25. $z = r^2 \sin^2 \theta$; $\rho \cos \phi = \rho^2 \sin^2 \phi \sin^2 \theta$
or $1 = \rho \tan \phi \sin \phi \sin^2 \theta$.

27. $z = \cot \theta$; $\rho \cos \phi = \cot \theta$.

29. $z^2 = r^2 + 1$; $\rho^2 \cos^2 \phi = \rho^2 \sin^2 \phi + 1$
or $\rho^2 \cos 2\phi = 1$.

31. $\int_0^{\pi/2} \int_0^{2\pi} \int_0^1 \rho^2 \sin \phi \, d\rho \, d\theta \, d\phi = \frac{2\pi}{3}$.

33. $\int_0^{\pi/2} \int_0^{\pi/4} \int_0^{3\cos\theta} \rho^2 \sin \phi \, d\rho \, d\phi \, d\theta$
$= 6(1 - \frac{1}{\sqrt{2}})$.

Section 15.6

1. $\int_0^1 \int_0^1 e^{x^2+y^2} \, dx \, dy \approx 1.7124$.

3. $\int_0^1 \int_0^1 e^{x^2+y^2} \, dx \, dy \approx 2.1144$.

5. $\iint_D x^2 + y^2 \, dA \approx 0.67$.

7. $\int_{0.2}^{0.6} \int_{0.4}^1 x^2 + y^2 \, dx \, dy \approx 0.1648$.

9. Answers vary.

Chapter 16

Section 16.1

11. $\nabla f = \langle 2xy^3z^4, 3x^2y^2z^4, 4x^2y^3z^3 \rangle$.

13. $\nabla h = \langle yze^{xyz}, xze^{xyz}, xye^{xyz} \rangle$.

15. $\nabla \cdot \mathbf{F} = 0$.

17. $\nabla \cdot \mathbf{H} = yz + 2x^2yz^2 + 2z$.

19. $\nabla \times \mathbf{F} = \langle 2y - 2z, 2z - 2x, 2x - 2y \rangle$.

21. $\nabla \times \mathbf{H} = \langle 2y - 2x^2y^2z, xy - 2x, 2xy^2z^2 - xz \rangle$.

23. $\nabla^2 f = 2y^3z^4 + 6x^2yz^4 + 12x^2y^3z^2$.

25. $\nabla^2 y = y^2z^2e^{xyz} + x^2z^2e^{xyz} + x^2y^2e^{xyz}$.

27. $\nabla^2 F = \langle 4, 4, 4 \rangle$.

29. $\nabla^2 H = \langle 0, 2y^2z^2 + 2x^2z^2 + 2x^2y^2, 6 \rangle$.

31. $\nabla f \times \nabla g$
$= \langle f_y g_z - f_z g_y, g_x f_z - f_x g_z, f_x g_y - f_y g_x \rangle$;
$\nabla \cdot (\nabla f \times \nabla g)$
$= \frac{\partial}{\partial x}(f_y g_z - f_z g_y)$
$\quad + \frac{\partial}{\partial y}(g_x f_z - f_x g_z) + \frac{\partial}{\partial z}(f_x g_y - f_y g_x)$
$= f_{yx} g_z + f_y g_{zx} - f_{zx} g_y - f_z g_{yx}$
$\quad + g_{xy} f_z + g_x f_{zy} - f_{xy} g_z - f_x g_{zy}$
$\quad + f_{xz} g_y + f_x g_{yz} - f_{yz} g_x - f_y g_{xz} = 0$.

33. Let $\mathbf{F} = \langle F_1, F_2, F_3 \rangle$.
curl $(f\mathbf{F})$
$= \left\langle \frac{\partial (fF_3)}{\partial y} - \frac{\partial (fF_2)}{\partial z}, \right.$
$\qquad \frac{\partial (fF_1)}{\partial z} - \frac{\partial (fF_3)}{\partial x},$
$\qquad \left. \frac{\partial (fF_2)}{\partial x} - \frac{\partial (fF_1)}{\partial y} \right\rangle$
$= \left\langle \frac{\partial f}{\partial y}F_3 + f\frac{\partial F_3}{\partial y} - \frac{\partial f}{\partial z}F_2 - f\frac{\partial F_2}{\partial z}, \right.$
$\qquad \frac{\partial f}{\partial z}F_1 + f\frac{\partial F_1}{\partial z} - \frac{\partial f}{\partial x}F_3 - f\frac{\partial F_3}{\partial x},$
$\qquad \left. \frac{\partial f}{\partial x}F_2 + f\frac{\partial F_2}{\partial x} - \frac{\partial f}{\partial y}F_1 - f\frac{\partial F_1}{\partial y} \right\rangle$
$= \left\langle f\left(\frac{\partial F_3}{\partial y} - \frac{\partial F_2}{\partial z}\right) + \left(\frac{\partial f}{\partial y}F_3 - \frac{\partial f}{\partial z}F_2\right), \right.$
$\qquad f\left(\frac{\partial F_1}{\partial z} - \frac{\partial F_3}{\partial x}\right) + \left(\frac{\partial f}{\partial z}F_1 - \frac{\partial f}{\partial x}F_3\right),$
$\qquad \left. f\left(\frac{\partial F_2}{\partial x} - \frac{\partial F_1}{\partial y}\right) + \left(\frac{\partial f}{\partial x}F_2 - \frac{\partial f}{\partial y}F_1\right) \right\rangle$
$= f\left\langle \frac{\partial F_3}{\partial y} - \frac{\partial F_2}{\partial z}, \frac{\partial F_1}{\partial z} - \frac{\partial F_3}{\partial x}, \frac{\partial F_2}{\partial x} - \frac{\partial F_1}{\partial y} \right\rangle$
$\quad + \left\langle \frac{\partial f}{\partial y}F_3 - \frac{\partial f}{\partial z}F_2, \frac{\partial f}{\partial z}F_1 - \frac{\partial f}{\partial x}F_3, \right.$
$\qquad \left. \frac{\partial f}{\partial x}F_2 - \frac{\partial f}{\partial y}F_1 \right\rangle$
$= f \text{ curl } \mathbf{F} + \nabla f \times \mathbf{F}$.

35. Let $\mathbf{F} = \langle F_1, F_2, F_3 \rangle$.

div $(f\mathbf{F})$

$= \frac{\partial}{\partial x}(fF_1) + \frac{\partial}{\partial y}(fF_2) + \frac{\partial}{\partial z}(fF_3)$

$= \frac{\partial f}{\partial x}F_1 + f\frac{\partial F_1}{\partial x} + \frac{\partial f}{\partial y}F_2$

$\quad + f\frac{\partial F_2}{\partial y} + \frac{\partial f}{\partial z}F_3 + f\frac{\partial F_3}{\partial z}$

$= f\left(\frac{\partial F_1}{\partial x} + \frac{\partial F_2}{\partial y} + \frac{\partial F_3}{\partial z}\right)$

$\quad + \left(F_1\frac{\partial f}{\partial x} + F_2\frac{\partial f}{\partial y} + F_3\frac{\partial f}{\partial z}\right)$

$= f \text{ div } \mathbf{F} + \mathbf{F} \cdot \nabla f$

37. p.

39. None.

41. $\nabla \times \mathbf{F} = \langle 0, 0, -0.25 \rangle$.

Section 16.2

1. $D\mathbf{F}(u, v) = \begin{bmatrix} 2 & -3 \\ 3 & 1 \end{bmatrix}$.

3. $D\mathbf{H}(x, y, z) = \begin{bmatrix} yz & xz & xy \\ y^2z^3 & 2xyz^3 & 3xy^2z^2 \end{bmatrix}$.

5. Does not make sense because $\mathbf{G} : \mathbb{R}^2 \to \mathbb{R}^3$ and $\mathbf{F} : \mathbb{R}^2 \to \mathbb{R}^2$.

7. $D(\mathbf{H} \circ \mathbf{P})(x, y, z) = \begin{bmatrix} a & b & c \\ d & e & f \end{bmatrix}$ where

$a = (x + y^2 + z^3)(x^3 - y^2 - z)$
$\quad + (x + y + z)(x^3 - y^2 - z)$
$\quad + 3x^2(x + y + z)(x + y^2 + z^3);$

$b = (x + y^2 + z^3)(x^3 - y^2 - z)$
$\quad + (x + y + z)(x^3 - y^2 - z)(2y)$
$\quad + (x + y + z)(x + y^2 + z^3)(-2y);$

$c = (x + y^2 + z^3)(x^3 - y^2 - z)$
$\quad + (x + y + z)(x^3 - y^2 - z)(3z^2)$
$\quad + (x + y + z)(x + y^2 + z^3)(-1);$

$d = (x + y^2 + z^3)^2(x^3 - y^2 - z)^3$
$\quad + 2(x + y + z)(x + y^2 + z^3)(x^3 - y^2 - z)^3$
$\quad + 3(x + y + z)(x + y^2 + z^3)^2$
$\quad \cdot (x^3 - y^2 - z)^2(3x^2);$

$e = (x + y^2 + z^3)^2(x^3 - y^2 - z)^3$
$\quad + 2(x + y + z)(x + y^2 + z^3)(x^3 - y^2 - z)^3(2y)$
$\quad + 3(x + y + z)(x + y^2 + z^3)^2$
$\quad \cdot (x^3 - y^2 - z)^2(-2y);$

$f = (x + y^2 + z^3)^2(x^3 - y^2 - z)^3$
$\quad + 2(x + y + z)(x + y^2 + z^3)(x^3 - y^2 - z)^3(3z^2)$
$\quad + 3(x + y + z)(x + y^2 + z^3)^2$
$\quad \cdot (x^3 - y^2 - z)^2(-1).$

9. Does not make sense because $\mathbf{F} : \mathbb{R}^2 \to \mathbb{R}^2$ and $\mathbf{P} : \mathbb{R}^3 \to \mathbb{R}^3$.

11. $D(\mathbf{H} \circ \mathbf{G})(u, v) = \begin{bmatrix} v & u \\ 0 & 2v \end{bmatrix}$.

13. Does not make sense because $\mathbf{G} : \mathbb{R}^2 \to \mathbb{R}^3$ and $\mathbf{P} : \mathbb{R}^3 \to \mathbb{R}^3$.

15. $D(\mathbf{P} \circ \mathbf{G})(u, v)$

$= \begin{bmatrix} v + \frac{1}{v} - \frac{v}{u^2} & u - \frac{u}{v^2} + \frac{1}{u} \\ v + \frac{2u}{v^2} - \frac{3v^3}{u^4} & u - \frac{2u^2}{v^3} + \frac{3v^2}{u^3} \\ 3u^2v^3 - \frac{2u}{v^2} + \frac{v}{u^2} & 3u^3v^2 + \frac{2u^2}{v^3} - \frac{1}{u} \end{bmatrix}$.

17. $\int_1^2 \int_1^2 \frac{1}{2v} \, du \, dv = \frac{1}{2}\ln(2)$.

19. $J(r, \theta, z) = \begin{vmatrix} \cos(\theta) & -r\sin(\theta) & 0 \\ \sin(\theta) & r\cos(\theta) & 0 \\ 0 & 0 & 1 \end{vmatrix}$

$= r\cos^2(\theta) + r\sin^2(\theta) = r.$

Section 16.3

1. $\mathbf{r}(t) = \langle 1+t, 1+t, 1+t \rangle$, $0 \le t \le 1$;

$\oint_C f \, ds = \frac{14\sqrt{3}}{3}$; $\oint_C \mathbf{F} \cdot d\mathbf{s} = \frac{37}{6}$.

3. $\mathbf{r}(t) = \langle 0, -1 + 2t, 1 - 2t \rangle$, $0 \le t \le 1$;

$\oint_C f \, ds = \frac{\sqrt{8}}{3}$; $\oint_C \mathbf{F} \cdot d\mathbf{s} = 0$.

5. $r(t) = \langle 1+2t, 2, 3-2t \rangle$, $0 \le t \le 1$;
$\oint_C f\, ds = \frac{50\sqrt{2}}{3}$; $\oint_C \mathbf{F} \cdot ds = \frac{4}{3}$.

7. $\oint_C \mathbf{F} \cdot \mathbf{T}\, ds \approx 21.12579$.

9. $\oint_C \mathbf{F} \cdot \mathbf{T}\, ds \approx 9.86960$.

11. $r(t)$ and $p(t)$ have the same orientation. $q(t)$ has orientation opposite to that of $r(t)$ and $p(t)$.

13. $\oint_C f\, ds = \frac{2}{15}$.

15. $\oint_C f\, ds = \frac{2}{15}$.

17. $\oint_C \mathbf{F} \cdot ds = -\frac{\pi}{2}$.

19. $\oint_C \mathbf{F} \cdot ds = \frac{\pi}{2}$.

21. $\oint_{C'} f\, ds = \frac{\sqrt{2}}{60}$. Yes, it is different.

23. Measuring the mass of a wire should not depend on which direction we move along the wire.

25. $\oint_C \mathbf{F} \cdot ds = \int_a^b \mathbf{F}(r(t)) \cdot r'(t)\, dt = \int_a^b 0\, dt = 0$.

Section 16.4

1. $r(u,v) = \langle u, v, u^3 + v^2 \rangle$;
D = circle with radius 1 and center at $(0,0)$.

3. $r(u,v) = \langle u, \sqrt{u+v}, v \rangle$;
D = triangle with vertices at the (u,v) coordinates $(0,0)$, $(1,1)$, and $(0,1)$.

5. $r(u,v) = \langle 3 - 2u^2 + v^2, u, v \rangle$;
D = 1st quadrant portion of the circle with radius 1 and center at $(0,0)$ in the uv plane.

7. $\mathbf{T}_u = \langle -v\sin u, v\cos u, 0 \rangle$;
$\mathbf{T}_v = \langle \cos u \sin u, 1 \rangle$;
$\mathbf{T}_u \times \mathbf{T}_v = \langle v\cos u, v\sin u, -v \rangle$;
$\pi(x - \pi) + \pi(z + \pi) = 0$.

9. $\mathbf{T}_u = \langle -\sin u, \cos u, 0 \rangle$; $\mathbf{T}_v = \langle 0, 0, 1 \rangle$;
$\mathbf{T}_u \times \mathbf{T}_v = \langle \cos u, \sin u, 0 \rangle$; $x - 1 = 0$.

11. $\mathbf{T}_u = \langle 1, 1, -1 \rangle$;
$\mathbf{T}_v = \langle 1, -1, 1 \rangle$; $\mathbf{T}_u \times \mathbf{T}_v = \langle 0, -2, -2 \rangle$;
$-2(y+3) - 2(z-3) = 0$.

13. $\mathbf{T}_u = \langle 2u, 0, 2u \rangle$; $\mathbf{T}_v = \langle 0, 2v, 2v \rangle$;
$\mathbf{T}_u \times \mathbf{T}_v = \langle -4u, -4u, 4u \rangle$;
$-x - y + z = 0$.

15. $\mathbf{u} = \frac{\langle 4, -1, -1 \rangle}{3\sqrt{2}}$.

17. $r(r,\theta) = \langle r\cos\theta, r\sin\theta, r^3\cos^3\theta + r^2\sin^2\theta \rangle$;
$D = \{(r,\theta) : 0 \le r \le 1, 0 \le \theta \le 2\pi\}$.

19. $(x - \frac{5}{2}) + 2(y - \frac{1}{2}) = 0$.

Section 16.5

1. $\|\mathbf{T}_u \times \mathbf{T}_v\| = v\sqrt{2}$;
Surface area: $\int_0^1 \int_0^{\pi/2} \sqrt{2}v\, du\, dv = \frac{\pi\sqrt{2}}{4}$;
$\iint_S f\, dS = \frac{\pi\sqrt{2}}{6}$.

3. $\|\mathbf{T}_u \times \mathbf{T}_v\| = 1$;
Surface area: $\int_0^1 \int_0^\pi du\, dv = \pi$;
$\iint_S f\, dS = 2 + \frac{\pi}{2}$.

5. $\|\mathbf{T}_u \times \mathbf{T}_v\| = \sqrt{8}$;
Surface area: $\int_0^1 \int_1^2 \sqrt{8}\, du\, dv = \sqrt{8}$.
$\iint_S f\, dS = \frac{16\sqrt{8}}{3}$.

7. $\mathbf{T}_\varphi = \langle -\sin\varphi\cos\theta, -\sin\varphi\sin\theta, \cos\varphi\rangle$;
$\mathbf{T}_\theta = \langle (R+\cos\varphi)(-\sin\theta), (R+\cos\varphi)(\cos\theta), 0\rangle$;
$\mathbf{T}_\theta \times \mathbf{T}_\varphi$
$= [(R+\cos\varphi)\cos\varphi\cos\theta]\mathbf{i}$
$\quad -[(R+\cos\varphi)\cos\varphi(-\sin\theta)]\mathbf{j}$
$\quad +[(R+\cos\varphi)(-\sin^2\theta)(-\sin\varphi)$
$\quad\quad -(R+\cos\varphi)\cos^2\theta(-\sin\varphi)]\mathbf{k}$
$= [(R+\cos\varphi)\cos\varphi\cos\theta]\mathbf{i}$
$\quad +[(R+\cos\varphi)\cos\varphi\sin\theta]\mathbf{j}$
$\quad +[(R+\cos\varphi)\sin\varphi]\mathbf{k}$.

$\|\mathbf{T}_\theta \times \mathbf{T}_\varphi\|$
$= ((R+\cos\varphi)^2(\cos^2\varphi)(\cos^2\theta)$
$\quad +(R+\cos\varphi)^2(\cos^2\varphi)(\sin^2\theta)$
$\quad +(R+\cos\varphi)^2\sin^2\varphi)^{1/2}$
$= ((R+\cos\varphi)^2\cos^2\varphi+(R+\cos\varphi)^2\sin^2\varphi)^{1/2}$
$= ((R+\cos\varphi)^2)^{1/2} = |R+\cos\varphi|$.

(Notice the absolute value signs!) Thus,

$$\iint_D \|\mathbf{T}_\theta \times \mathbf{T}_\varphi\|\,d\varphi\,d\theta = \int_0^{2\pi}\int_0^{2\pi} |R+\cos\varphi|\,d\varphi\,d\theta.$$

Since $R > 1$, $(R+\cos\varphi) > 0$, so we can safely remove the absolute value signs:

$$\int_0^{2\pi}\int_0^{2\pi}(R+\cos\varphi)\,d\varphi\,d\theta = \int_0^{2\pi}(R\varphi+\sin\varphi)\Big|_0^{2\pi}d\theta$$
$$= \int_0^{2\pi} 2\pi R\,d\theta = 2\pi R\theta\Big|_0^{2\pi} = 4\pi^2 R.$$

9. Using the parametrization from exercise 8,

$\mathbf{T}_u = \langle -\sin u\sin v,\ 0,\ \cos u\sin v\rangle$;
$\mathbf{T}_v = \langle \cos u\cos v,\ -\sin v,\ \sin u\cos v\rangle$;
$\mathbf{T}_u \times \mathbf{T}_v = \langle \cos u\sin^2 v,\ \sin v\cos v,\ \sin u\sin^2 v\rangle$;
$\|\mathbf{T}_u \times \mathbf{T}_v\| = |\sin v| = \sin v$.

(We can remove the absolute value signs since $\sin v \geq 0$ for $\frac{\pi}{2} \leq v \leq \pi$.)

The surface area of the hemisphere is
$$\int_{\pi/2}^{\pi}\int_0^{2\pi}\sin v\,du\,dv = \int_{\pi/2}^{\pi} u\sin v\Big]_{u=0}^{u=2\pi}dv$$
$$= \int_{\pi/2}^{\pi} 2\pi\sin v\,dv = -2\pi\cos v\Big]_{v=\pi/2}^{v=\pi} = 2\pi.$$

11. $\displaystyle\iint_S \mathbf{F}\cdot\mathbf{n}\,dS$

$$= \iint_D \mathbf{F}(\mathbf{r}(u,v))\cdot\mathbf{T}_u\times\mathbf{T}_v\,du\,dv = -\tfrac{\sqrt{2}}{2}\pi.$$

(Use polar coordinates.)

Chapter 17

Section 17.1

1. True; a potential is $\varphi(x,y) = e^{xy} + \ln(y)$.
3. True.
5. True; $f(x,y) = e^{xy} + \ln(y)$ is an example.
7. True; div(curl $\mathbf{F}$) = div(0) = 0.
9. True; $\varphi(x,y,z) = xyz + c$, where c is a constant.
11. True; a potential is $\varphi(x,y,z) = xyz$.
13. True; $\mathbf{F}$ is conservative.
15. No, $-\dfrac{\partial P}{\partial y} = -\dfrac{\cos(x)}{y} \neq y\cos(x) = \dfrac{\partial Q}{\partial x}$.
17. L is conservative;
a potential is $\varphi(x,y,z) = \dfrac{y^4}{4}\ln x + \sin z$.
19. Since $\mathbf{F}$ is conservative, we should have the same answer as in exercise 18: $\tfrac{16}{3}$.
21. $\nabla\phi(x,y,z) = \nabla\left(\dfrac{GMm}{(x^2+y^2+z^2)^{1/2}}\right)$
$= \mathbf{F}(x,y,z) = -\dfrac{GMm}{(x^2+y^2+z^2)^{3/2}}\langle x,\ y,\ z\rangle$.

Section 17.2

1. $\int_0^{2\pi}\int_0^1 \sqrt{4r^2+1}\, r\, dr\, d\theta = \frac{\pi}{6}(5^{3/2}-1)$.

3. $\int_0^{2\pi} \langle \sin t, -\cos t, 2-\cos^2 t - \sin^2 t\rangle$
   $\cdot \langle -\sin t, \cos t, 0\rangle\, dt = -2\pi$.

5. $\iint_S (\nabla \times \mathbf{F})\cdot \mathbf{n}\, dS = -2\pi$. Stokes' Theorem holds since all the hypotheses are satisfied.

7. $\iint_S (\nabla \times \mathbf{F})\cdot \mathbf{n}\, dS = \oint_C \mathbf{F}\cdot \mathbf{T}\, ds = 32\pi$.

9. $\iint_S (\nabla \times \mathbf{F})\cdot \mathbf{n}\, dS = \oint_C \mathbf{F}\cdot \mathbf{T}\, ds = -3\pi$.

11. $\iint_S (\nabla \times \mathbf{F})\cdot \mathbf{n}\, dS = \oint_C \mathbf{F}\cdot \mathbf{T}\, ds = 7\pi$.

13. $\iint_S (\nabla \times \mathbf{F})\cdot \mathbf{n}\, dS = \oint_C \mathbf{F}\cdot \mathbf{T}\, ds = -\frac{\sqrt{2}}{6}$.

15. Parametrize the first surface:
$$\mathbf{r}: D \mapsto \mathbb{R}^3, \quad \mathbf{r}(u,v) = \langle u, v, \sqrt{4-u^2-v^2}\rangle,$$
where $D = \{(u,v): u^2+v^2 = 4\}$.
$$\iint_S (\nabla \times \mathbf{F})\cdot \mathbf{n}\, dS$$
$$= \int_{-2}^{2}\int_{-\sqrt{4-v^2}}^{\sqrt{4-v^2}} \langle 2v, -2u, 1\rangle$$
$$\cdot \langle \tfrac{u}{\sqrt{4-u^2-v^2}}, \tfrac{v}{\sqrt{4-u^2-v^2}}, 1\rangle\, du\, dv$$
$$= \int_{-2}^{2}\int_{-\sqrt{4-v^2}}^{\sqrt{4-v^2}} du\, dv = 4\pi \text{ (the area of a circle of radius 2)}.$$

Parametrize the second surface:
$$\mathbf{r}: D \to \mathbb{R}^3, \quad \mathbf{r}(u,v) = \langle u, v, 2-\sqrt{u^2+v^2}\rangle,$$
where $D = \{(u,v): u^2+v^2 \le 4\}$.
$$\iint_S (\nabla \times \mathbf{F})\cdot \mathbf{n}\, dS$$
$$= \int_{-2}^{2}\int_{-\sqrt{4-v^2}}^{\sqrt{4-v^2}} \langle 2v, -2u, 1\rangle$$
$$\cdot \langle u(u^2+v^2)^{-1/2}, v(u^2+v^2)^{-1/2}, 1\rangle\, du\, dv$$
$$= \int_{-2}^{2}\int_{-\sqrt{4-v^2}}^{\sqrt{4-v^2}} du\, dv = 4\pi.$$

Parametrize the third surface:
$$\mathbf{r}: D \mapsto \mathbb{R}^3, \quad \mathbf{r}(u,v) = \langle u, v, 4-u^2-v^2\rangle,$$
where $D = \{(u,v): u^2+v^2 \le 4\}$;
$$\iint_S (\nabla \times \mathbf{F})\cdot \mathbf{n}\, dS$$
$$= \int_{-2}^{2}\int_{-\sqrt{4-v^2}}^{\sqrt{4-v^2}} \langle 2v, -2u, 1\rangle \cdot \langle 2u, 2v, 1\rangle\, du\, dv$$
$$= \int_{-2}^{2}\int_{-\sqrt{4-v^2}}^{\sqrt{4-v^2}} du\, dv = 4\pi.$$

Section 17.3

1. $\iint_D 6 - 4\, dx\, dy = 32\pi$.

3. $\int_0^1 \int_0^1 8x + e^y\, dy\, dx = 3 + e$.

5. $\int_0^2 \int_1^2 (-2x - 2y)\, dy\, dx = -10$.

7. $\oint_C P\, dx + Q\, dy = \iint_D \frac{\partial Q}{\partial x} - \frac{\partial P}{\partial Y}\, dy\, dx = -\frac{17}{10}$.

9. $\oint_C P\, dx + Q\, dy = \iint_D \frac{\partial Q}{\partial x} - \frac{\partial P}{\partial y}\, dy\, dx = -2.\bar{6}$.

11. $\iint_D \frac{\partial Q}{\partial x} - \frac{\partial P}{\partial y}\, dy\, dx \approx 0.03216$.

13. Area
$$= \tfrac{1}{2} \oint_{C_2}(x\, dy - y\, dx) - \oint_{C_1}(x\, dy - y\, dx) = 4\pi.$$

15. Area $= -\tfrac{1}{2} \oint_C x\, dy - y\, dx = 27\sqrt{3}$.

17. $\oint_C x\,dy = \int_0^{2\pi} \langle 0, a\cos t\rangle \cdot \langle -a\sin t, b\cos t\rangle\,dt$

$= \int_0^{2\pi} ab\cos^2 t\,dt = \int_0^{2\pi} ab\frac{(1+\cos 2t)}{2}\,dt$

$= \frac{ab}{2}\left(t + \frac{\sin 2t}{2}\right)\Big]_0^{2\pi} = \frac{ab}{2}(2\pi) = \pi ab.$

$-\oint_C y\,dx = -\int_0^{2\pi} \langle b\sin t, 0\rangle \cdot \langle -a\sin t, b\cos t\rangle\,dt$

$= \int_0^{2\pi} ab\sin^2 t\,dt = ab\int_0^{2\pi} \frac{1-\cos 2t}{2}\,dt$

$= \frac{ab}{2}\left(t - \frac{\sin 2t}{2}\right)\Big]_0^{2\pi} = \frac{ab}{2}(2\pi) = \pi ab.$

Section 17.4

1. $\iiint_R (2 - 2 + 1)\,dV = \frac{1}{6}.$

3. $\iiint_R 2 + 3 + 4\,dV = 27\pi.$

5. $\iiint_R \nabla \cdot \mathbf{F}\,dV = \iint_S \mathbf{F} \cdot \mathbf{n}\,dS = -\frac{4}{3}.$

7. $\iiint_R \nabla \cdot \mathbf{F}\,dV = \iint_S \mathbf{F} \cdot \mathbf{n}\,dS = 3\pi.$

9. flux $= \iint_S \mathbf{F} \cdot \mathbf{n}\,dS$

$= \iiint_{R_1} \nabla \cdot \mathbf{F}\,dV + \iiint_{R_2} \nabla \cdot \mathbf{F}\,dV$

$= \frac{17\pi}{3}$ where R_1 is the entire region above the paraboloid inside the sphere and R_2 is the portion that lies below the plane $z = 1$..

11. $\iint_S \mathbf{F} \cdot \mathbf{n}\,dS = \iiint_R \nabla \cdot \mathbf{F}\,dV = \frac{10\pi}{3}.$

A141

Appendices

Appendix 1. TRIANGLE TRIGONOMETRY

1. $\frac{5}{13}$.

3. $\frac{13}{12}$.

5. $\frac{5}{12}$.

7. $\frac{\pi}{5} = 36°$.
$\sin(\frac{\pi}{5}) \approx .5878,$ $\qquad \cos(\frac{\pi}{5}) \approx .8090,$
$\tan(\frac{\pi}{5}) \approx .7265,$ $\qquad \sec(\frac{\pi}{5}) \approx 1.2361,$
$\csc(\frac{\pi}{5}) \approx 1.7013,$ $\qquad \cot(\frac{\pi}{5}) \approx 1.3764.$

9. $-\frac{5\pi}{4} = -225°$.
$\sin(-\frac{5\pi}{4}) \approx .7071,$ $\qquad \cos(-\frac{5\pi}{4}) \approx -.7071,$
$\tan(-\frac{5\pi}{4}) = -1,$ $\qquad \sec(-\frac{5\pi}{4}) \approx -1.4142,$
$\csc(-\frac{5\pi}{4}) \approx 1.4142,$ $\qquad \cot(-\frac{5\pi}{4}) = -1.$

11. $\frac{8\pi}{3} = 480°$.
$\sin(\frac{8\pi}{3}) \approx .8660,$ $\qquad \cos(\frac{8\pi}{3}) = -.5,$
$\tan(\frac{8\pi}{3}) \approx -1.7321,$ $\qquad \sec(\frac{8\pi}{3}) = -2,$
$\csc(\frac{8\pi}{3}) \approx 1.1547,$ $\qquad \cot(\frac{8\pi}{3}) \approx -.5774.$

13. $100° = \frac{5\pi}{9}$.
$\sin(100°) \approx .9848,$ $\qquad \cos(100°) \approx -.1736,$
$\tan(100°) \approx -5.6713,$ $\qquad \sec(100°) \approx -5.7588,$
$\csc(100°) \approx 1.0154,$ $\qquad \cot(100°) \approx -.1763.$

15. $270° = \frac{3\pi}{2}$.
$\sin(270°) = -1,$ $\qquad \cos(270°) = 0,$
$\tan(270°)$ undef., $\qquad \sec(270°)$ undef.,
$\csc(270°) = -1,$ $\qquad \cot(270°) = 0.$

17. $-270° = -\frac{3\pi}{2}$.
$\sin(-270°) = 1,$ $\qquad \cos(-270°) = 0,$
$\tan(-270°)$ undef., $\qquad \sec(-270°)$ undef.,
$\csc(-270°) = 1,$ $\qquad \cot(-270°) = 0.$

19. angle $C = 100°$,
side $a \approx 4.077$ cm, $\qquad$ side $b \approx 5.026$ cm.

21. No triangle possible with given data.

23. side $c \approx 3.6$ inches,
angle $A \approx 33.7°,$ $\qquad$ angle $B \approx 56.3°.$

25. ≈ 76.84 miles.

27. $\approx 61°$.

29. ≈ 807 meters.

Appendix 2. TECHNIQUES OF INTEGRATION

1. $\sin x - \frac{1}{3}\sin^3 x + C.$

3. $\frac{1}{5}\cos^5 x - \frac{1}{3}\cos^3 x + C.$

5. $\frac{1}{4}\tan^4 x + \frac{1}{6}\tan^6 x + C.$

7. $\frac{1}{5}\tan^5 x - \frac{1}{3}\tan^3 x + \tan x - x + C.$

9. $\tan x - \cot x + C.$

11. $\frac{3x}{8} + \frac{\sin 2x}{4} + \frac{\sin 4x}{32} + C.$

13. $-\frac{1}{5}\cot^5 x - \frac{1}{7}\cot^7 x + C.$

15. $\frac{-1}{1 + \tan x} + C.$

17. $2\arcsin(\frac{x}{2}) - \frac{1}{2}x\sqrt{4 - x^2} + C.$

19. $\frac{\sqrt{x^2 - 25}}{25x} + C.$

21. $\frac{-x}{\sqrt{x^2 - 1}} + C.$

23. $\frac{9}{4}\arcsin(\frac{2x}{3}) + \frac{x}{2}\sqrt{9 - 4x^2} + C.$

25. $\frac{1}{243}(9x^2 + 49)^{3/2} - \frac{49}{81}(9x^2 + 49)^{1/2} + C.$

27. $\frac{1}{2}\arctan\left(\frac{x - 2}{2}\right) + C.$

29. $-2\sqrt{9 - 8x - x^2} - 5\arcsin\left(\frac{x + 4}{5}\right) + C.$

31. $\frac{1}{2}\left(\arctan(x + 2) + \frac{x + 2}{x^2 + 4x + 5}\right) + C.$

33. $\frac{2}{3\sqrt{7}}\arctan\left(\frac{4x - 3}{3\sqrt{7}}\right) + C.$

Appendix 3. METHOD OF PARTIAL FRACTIONS

1. $3\ln|x| + 2\ln|x-4| + C$,
or $\ln|x|^3(x-4)^2 + C$.

3. $6\ln|x-1| + 5/(x-1) + C$.

5. $2\ln|x| - \ln|x-2| + 4\ln|x+2| + C$,
or $\ln\dfrac{x^2(x+2)^4}{|x-2|} + C$.

7. $5\ln|x| - \dfrac{2}{x} + \dfrac{3}{2x^2} - \dfrac{1}{3x^3} + 4\ln|x+3| + C$.

9. $3\ln|x+5| + \ln(x^2+4) + \frac{1}{2}\arctan(x/2) + C$
or $\ln\left((x^2+4)|x+5|^3\right) + \frac{1}{2}\arctan(x/2) + C$.

11. $\ln(x^2+1) - \dfrac{4}{x^2+1} + C$.

13. $\dfrac{x^3}{3} - 9x - \dfrac{1}{9x} - \dfrac{\ln(x^2+9)}{2}$
$+ \dfrac{728}{27}\arctan\left(\dfrac{x}{3}\right) + C$.

15. $2\ln|x-4| + 2\ln|x+1| - \frac{3}{2}(x+1)^{-2} + C$.

Appendix 4. POLAR COORDINATES

For 1-16, coordinates are approximate.
1. $(1.5, 2.5981)$.
3. $(1.5, -2.5981)$.
5. $(-\pi, 0)$.
7. $(0, 0)$.
9. $(5, 53.130°)$; $(5, .92730)$.
11. $(4.2857, -49.4°)$; $(4.2857, -.86220)$.
13. $(4.4429, 45°)$; $(4.4429, \frac{\pi}{4})$.
15. $(1000, 90°)$; $(1000, \frac{\pi}{2})$.

17. a is the radius of the circle and $(a, 0)$ is the center of the circle.

19. $(2a, 0)$ is the furthest right-hand edge of the cardioid. $(0, \pm a)$ are the y-intercepts.

21. $x = a$ is an asymptote. $(a \pm 1, 0)$ are the x-intercepts. For $0 < a \leq 1$, $(0, 0)$ is also a point on the graph.

23. $(a \pm 1, 0)$ are x-intercepts. If $a \geq 1$, $(0, 0)$ is also a point on the graph. (0 ± 1) are y-intercepts.

25. $(a, 0)$ is the x-intercept. $x = a$ is an asymptote.

27. $e = \frac{1}{3}$: ellipse, $e = 3$: hyperbola.

29. The answers will be the same.

31. The outer tip of each leaf is a distance of a from the origin.

33. Using $\cos(\theta)$ puts the first leaf on the positive x-axis. Using $\sin(\theta)$ puts the center of the first leaf in the direction $\theta = \frac{\pi}{2}k$.

35. If r is an even function of θ, the graphs are identical. If r is an odd function of θ, the graphs are symmetric across the origin.

37. $\sin(\pi-\theta) = \sin(\theta)$, so for any equation with θ only in terms of $\sin(\theta)$ or $\csc(\theta)$, the new and old graphs are identical. $\cos(\pi - \theta) = -\cos\theta$, so for any equation with θ only in terms of $\cos(\theta)$ or $\sec(\theta)$, the new graph is a reflection of the old graph across the origin.

39. If r is an odd function of θ, the new graph is the same as the old graph. If r is an even function of θ, the new graph is a reflection of the old graph across the origin.

41. The new graph is a clockwise rotation of the old graph an angle of $\frac{\pi}{2}$.

43. $(x^2 + y^2)^2 - 3x^2 y = 0$;
$x(t) = 3\sin(t)\cos^3(t)$,
$y(t) = 3\sin^2(t)\cos^2(t)$.

45. $x^3 + y^2 x - 5y^2 = 0$;
$x(t) = 5\sin^2(t)$,
$y(t) = 5\sin^2(t)\tan(t)$.

47. $(x-1)^2(x^2 + y^2) - 49x^2 = 0$;
$x(t) = 1 - 7\cos(t)$,
$y(t) = \tan(t) - 7\sin(t)$.

49. $(x^2 + y^2 - x)^2 - 9(x^2 + y^2) = 0$;
$x(t) = (-3 + \cos(t))\cos(t)$,
$y(t) = (-3 + \cos(t))\sin(t)$.

51. ≈ 2.96. **53.** ≈ 1.68.

55. ≈ 4.84. **57.** ≈ 0.21.

59. ≈ 0.02. **61.** $\frac{3\pi}{2} \approx 4.71$.

63. $\frac{3\pi}{2} \approx 4.71$. **65.** $\frac{3\pi}{2} \approx 4.71$.

67. $\frac{3\pi}{2} \approx 4.71$. **69.** ≈ 0.54.

71. ≈ 0.54. **73.** $\frac{\pi}{8} \approx 0.39$.

75. $\frac{\pi}{12} \approx 0.26$. **77.** $\frac{\pi}{12} \approx 0.26$.

79. $\frac{\pi a^2}{4k}$.

Appendix 5. COMPLEX NUMBERS

1. $z \approx 8.06 \text{ cis } 5.23$.

3. $2 - 7i$.

5. $-34 + 8i$.

7. With $z_1 = 6 - 5i \approx 7.81 \text{ cis } 5.59$,
and $z_2 = -4 - 2i \approx 4.47 \text{ cis } 3.61$,
so $z_1 z_2 \approx 34.91 \text{ cis } 9.20$.

9. $4 \text{ cis}(-30°)$.

11. The roots are:
$4 \text{ cis}(20°)$, $4 \text{ cis}(60°)$, $4 \text{ cis}(100°)$,
$4 \text{ cis}(140°)$, $4 \text{ cis}(180°)$, $4 \text{ cis}(220°)$,
$4 \text{ cis}(260°)$, $4 \text{ cis}(300°)$, $4 \text{ cis}(340°)$.

The real root is $4 \text{ cis}(180°) = -4$.

The conjugate pairs of roots are:
$4 \text{ cis}(60°)$ and $4 \text{ cis}(300°)$;
$4 \text{ cis}(100°)$ and $4 \text{ cis}(260°)$;
$4 \text{ cis}(140°)$ and $4 \text{ cis}(220°)$;
$4 \text{ cis}(20°)$ and $4 \text{ cis}(340°)$.

13. 2.

15. $z_1 \approx .71 + 2.12i$, $z_2 \approx -.71 - 2.12i$.

17. $z_1 = 2$, $z_2 = 2i$, $z_3 = -2$, $z_4 = -2i$.

19. $z_1 \approx 1.08 + 1.87i$,
$z_2 \approx -2.15$, $z_3 \approx 1.08 - 1.87i$.

Appendix 6. TAYLOR'S FORMULA

1. $x \approx -0.89361$.

3. $x \approx 0.73887$.

5. $x \approx 1.89772$.

7. $x \approx -0.77015$.

9. $x \approx 0.21886$.

11. $R_N(1.2) = \dfrac{f^{(N+1)}(c)}{(N+1)!}(1.2 - 1)$
is the error term. Now, since
$f^{(N+1)}(c) = \dfrac{(-1)^N N!}{c^N}$ with $1 \le c \le 1.2$,
so $|R_N(1.2)| = \left|\dfrac{(-1)^N N!}{c^N(N+1)!}(.2)\right| \le \dfrac{.2}{N+1}$.
When $N = 199$, $\dfrac{.2}{N+1} = \dfrac{.2}{200} = .001$,
and $P_{199}(x) = \sum_{n=1}^{199}(-1)^{n+1}\dfrac{(x-1)^n}{n}$.

$P_{199}(1.2) = \sum_{n=1}^{199}(-1)^{n+1}\dfrac{(.2)^n}{n} \approx .18232$
approximates $\ln(1.2)$ within $.001$.

$P(.2) = \sum_{n=1}^{\infty}(-1)^{n+1}\dfrac{(.2)^n}{n}$ is an alternating series, so the error in $P_N(x)$ is less than or equal to $\dfrac{(.2)^{N+1}}{N+1}$. For $N = 3$, $\dfrac{(.2)^4}{4} = .0004 < .001$.
Hence, $P_3(.2) = .2 - \dfrac{(.2)^2}{2} + \dfrac{(.2)^3}{3} = .182\overline{6}$
approximates $\ln(1.2)$ within $.001$ as well.

13. $R_N(1) = \dfrac{f^{(N+1)}(c)}{(N+1)!}(1 - 0)$ is error term.
$f^{(N+1)}(c) = e^c$, where $0 < c < 1$, so
$|R_N(1)| = \left|\dfrac{e^c}{(N+1)!}\right| < \dfrac{3}{(N+1)!}$.
When $N = 6$, $\dfrac{3}{7!} \approx .0005952 < .001$, and
$P_6(x) = 1 + x + \dfrac{x^2}{2!} + \dfrac{x^3}{3!} + \dfrac{x^4}{4!} + \dfrac{x^5}{5!} + \dfrac{x^6}{6!}$.
$P_6(1) = 1 + 1 + \dfrac{1}{2} + \dfrac{1}{6} + \dfrac{1}{24} + \dfrac{1}{120} + \dfrac{1}{720}$
$= 2.7180\overline{5}$ approximates e within $.001$.

DIFFERENTIATION PRACTICE

1. $\dfrac{ds}{dt} = 8t + 4.$

3. $\dfrac{dM}{dw} = \dfrac{1}{2}w^{-1/2} + 2w^{-3/2} - \dfrac{9}{2}w^{-5/2}.$

5. $\dfrac{dy}{dx} = 12x + 5x^{-2} - \dfrac{4}{3}x^{-5/3}.$

7. $\dfrac{dN}{dt} = \dfrac{2}{3}t^{-1/3} + \dfrac{1}{3}t^{-4/3}.$

9. $\dfrac{dy}{dr} = 10(5r - 4).$

11. $\dfrac{dy}{dx} = \dfrac{1}{\sqrt{2x}}.$

13. $\dfrac{dy}{dt} = \dfrac{3}{\sqrt{6t + 5}}.$

15. $\dfrac{dy}{dz} = \dfrac{4z + 4}{3(2z^2 + 4z + 8)^{2/3}}.$

17. $\dfrac{dy}{dx} = 8x^{1/3} - 2x^{-3/2}.$

19. $\dfrac{dy}{dx} = \dfrac{2}{3}t^{-1/3} + \dfrac{3}{2}t^{-5/2}.$

21. $\dfrac{dy}{dx} = 7(x^5 - 4x + 8)^6 (5x^4 - 4).$

23. $\dfrac{dL}{dw} = 4(w^4 - 8w^2 + 15)^3 (4w^3 - 16w).$

25. $\dfrac{dy}{dx} = 4x(3x^2 - 1)^3.$

27. $\dfrac{dT}{ds} = -6(2s^{-4} + 3s^{-2} + 2)^{-7}(-8s^{-5} - 6s^{-3}).$

29. $\dfrac{dy}{dx} = \dfrac{1}{3}(3 + 2x^{-2} + 4x^{-3})^{-2/3}(-4x^{-3} - 12x^{-4}).$

31. $\dfrac{dy}{dx} = 4(7x + (x^2 + 6)^{1/2})^3 (7 + x(x^2 + 6)^{-1/2}).$

33. $\dfrac{dA}{dz} = \dfrac{1}{2}(1 + (1 + 2z)^{1/2})^{-1/2}(1 + 2z)^{-1/2}.$

35. $\dfrac{dA}{ds} = \dfrac{s}{2}(s^2 + 9)^{-3/4}(4s + 5)^4 + 16(s^2 + 9)^{1/4}(4s + 5)^3.$

37. $\dfrac{dy}{dx} = 7(x^2 + 1)(5x^2 + 1).$

39. $\dfrac{dy}{dx} = (2x - 5)^{-1/2}.$

41. $\dfrac{ds}{dt} = \dfrac{1}{2}(t^2 + t + 1)^{-1/2}(2t + 1).$

43. $\dfrac{dy}{dx} = 7 + x(x^2 + 6)^{-1/2}.$

45. $\dfrac{dx}{dz} = \dfrac{1}{2}(z^2 + (z^2 + 9)^{1/2})^{-1/2} \cdot (2z + z(z^2 + 9)^{-1/2}).$

47. $\dfrac{dy}{dx} = 18x((x^2 - 6)^3 + 1)^2 (x^2 - 6)^2.$

49. $\dfrac{dy}{dx} = \dfrac{5x^{3/2}}{2(x + 1)^{7/2}}.$

51. $\dfrac{dy}{dx} = \dfrac{2(3x^2 - 5)(62x)}{(2x^2 + 7)^3}.$

53. $\dfrac{dy}{dx} = 2(6x - 7)^2 (8x^2 + 9)(168x^2 - 112x + 81).$

55. $\dfrac{dx}{dy} = \dfrac{3}{2}(y^2 - 2)^{-1/4}(25y^2 + 2y - 20).$

57. $\dfrac{dy}{dx} = 3(x^6 + 1)^4 (3x + 2)^2 (33x^6 + 20x^5 + 3).$

59. $\dfrac{dy}{dx} = -\dfrac{2}{3}(7y - 2)^{-3}(2y + 1)^{-1/3}(28y + 25).$

61. $\dfrac{dA}{dr} = \dfrac{1}{2}r^{-1/2}(r + 1)^{-1/2}(r + 2)^{-1/2}(3r^2 + 6r + 2).$

63. $\dfrac{dy}{dx} = \dfrac{1}{4}(3x^3 - x + 1)^{-3/4}(x^2 - 6)^2 (81x^4 - 79x^2 + 24x + 6).$

65. $\dfrac{dy}{dx} = 20x^{3/2} + \dfrac{21}{2}x^{5/2}.$

67. $\dfrac{dy}{dx} = \dfrac{12}{5}(3x + 5)^{-1/5}.$

69. $\dfrac{dy}{dx} = \dfrac{(-2x^{-3} + (2/3)x^{-2/3})(3x^3 - x + 1)}{2(3x^3 - x + 1)^{3/2}(x^{-2} + 2x^{1/3})^{1/2}} - \dfrac{(x^{-2} + 2x^{1/3})(9x^2 - 1)}{2(3x^3 - x + 1)^{3/2}(x^{-2} + 2x^{1/3})^{1/2}}.$

71. $\dfrac{dy}{du} = \dfrac{-53}{2(7u - 9)^{3/2}(2u + 5)^{1/2}}.$

73. $\dfrac{dy}{dx} = \dfrac{27z^2 - 60z + 5}{(6z + 1)^4}.$

75. $\dfrac{dR}{dw} = \dfrac{(2w - 4)(w + 1)(w + 3) - (w - 1)(w - 3)(2w + 4)}{((w + 1)(w + 3))^2}.$

77. $\dfrac{ds}{dt} = \dfrac{-48t}{(9t^2 + 16)^{5/3}}.$

79. $\dfrac{dy}{dx} = \dfrac{7(x^2 - 1)(3x^3 + 50x^2 + 9x - 10)}{(3x + 10)^5}.$

81. $f'(x) = \dfrac{-6x^2 - 2x + 3}{e^{x^2}}$.

83. $f'(x) = \dfrac{10}{3}\tan^{2/3}(x^2)\sec^2(x^2)$.

85. $f'(x) = 24x\sec^3(4x^2 - 8)\tan(4x^2 - 8)$.

87. $f'(x) = 36x(3x^2 - 5)^2 \tan(3x^2 - 5)^3 \cdot \sec^2(3x^2 - 5)^3$.

89. $f'(x) = -4\sin^3(\csc x)\cos(\csc x)(\csc x \cot x)$.

91. $f'(x) = 3e^x \tan^2(e^x) \sec^2(e^x)$.

93. $f'(x) = \dfrac{\sec^2(\sqrt{x})}{2\sqrt{x}}$.

95. $f'(x) = 7(e^x + e^{3x})^6(e^x + 3e^{3x})$.

97. $f'(x) = \cos\!\left(\tfrac{1}{x}\right) + \tfrac{1}{x}\sin\!\left(\tfrac{1}{x}\right)$.

99. $f'(x) = -\dfrac{3}{2x^2}\sin^{1/2}\!\left(\dfrac{1}{x}\right)\cos\!\left(\dfrac{1}{x}\right)e^{\cos x}$
$\quad -\sin^{3/2}\!\left(\dfrac{1}{x}\right)e^{\cos x}(\sin x)$.

INTEGRATION PRACTICE

1. $\dfrac{x^2}{2}\arcsin x - \dfrac{1}{4}\arcsin x + \dfrac{x}{4}\sqrt{1 - x^2} + C$.

3. $\dfrac{1}{6}\sin^3(2x) - \dfrac{1}{10}\sin^5(2x) + C$.

5. $\dfrac{x}{25\sqrt{x^2 + 25}} + C$.

7. $2\ln|x - 1| - \ln|x| - \dfrac{x}{(x-1)^2} + C$.

9. $-\sqrt{4 + 4x - x^2} + 2\arcsin\!\left(\dfrac{x - 2}{\sqrt{8}}\right) + C$.

11. $\dfrac{1}{3}e^{2x}(2\sin(3x) - 3\cos(3x)) + C$.

13. $-\sqrt{4 - x^2} + C$.

15. $2\arctan(x^{1/2}) + C$.

17. $\dfrac{10x\sin(5x) - (25x^2 - 2)\cos(5x)}{125} + C$.

19. $\dfrac{2}{3}(1 + e^x)^{3/2} + C$.

21. $\dfrac{1}{3}\tan^3 x + C$.

23. $-\dfrac{1}{4}(8 - x^3)^{4/3} + C$.

25. $\dfrac{1}{2}e^{2x} - e^x + \ln(1 + e^x) + C$.

27. $\dfrac{1}{3}(16 - x^2)^{3/2} - 16(16 - x^2)^{1/2} + C$.

29. $x\arctan(5x) - (1/10)\ln|1 + 25x^2| + C$.

31. $\dfrac{1}{\sqrt{5}}\ln\left|\sqrt{5}x + \sqrt{7 + 5x^2}\right| + C$.

33. $\dfrac{1}{5}(x^2 - 25)^{5/2} + \dfrac{25}{3}(x^2 - 25)^{3/2} + C$.

35. $-\dfrac{1}{4}x^2 e^{-4x} - \dfrac{1}{8}xe^{-4x} - \dfrac{1}{32}e^{-4x} + C$.

37. $\dfrac{1}{7}\cos(7x) + C$.

39. $x^3 \sin x + 3x^2 \cos x - 6x\sin x$
$\quad -6\cos x + \sin x + C$.

41. $24x - \dfrac{10}{3}\ln|\sin(3x)| - \dfrac{1}{3}\cot(3x) + C$.

43. $-2\sqrt{1 + \cos x} + C$.

45. $(1/3)\sec^3 x - \sec x + C$.

47. $(1/4)x^4 - 2x^2 + 4\ln|x| + C$.

49. $\dfrac{3}{64}(2x + 3)^{8/3} - \dfrac{9}{20}(2x + 3)^{5/3}$
$\quad + \dfrac{27}{16}(2x + 3)^{2/3} + C$.

Index

‖v‖ (norm of vector), 710–11

absolute extrema, over closed
    domains, 891–94
absolutely converging series,
    661–65
  ratio test for, 663–64, 666
  root test for, 664–65, 666
acceleration, 693
acceleration vectors, 765
  curvature and, 797
  osculating plan and, 790
addition
  associative law for, 705, 726
  commutative law for, 705, 726
  of matrices, 724–25, 726
  of vectors, 701, 705, 706
additive identity law for 0, for
    vectors, 705
additive inverse law, for vectors,
    705
additive inverses, of vectors, 702
additive property, for multiple
    integrals, 916
algebra
  of matrices, 723–25
  of vectors, 701–6
alternating harmonic sequences,
    621, 622, 629
alternating harmonic series, 643,
    644, 661, 664
alternating sequences, 621
alternating series test for
    convergence, 653–55,
    665
analytic reasoning, for verifying
    vector properties, 705
anti-commutative cross products,
    of vectors, 734
antiderivatives
  for position vector functions,
    775–76
  potentials as, 1025–27
Archimedean property of real
    numbers, 636, 643
arc length, 777–80
  curvature and, 795–96
  defined, 778
  line integrals of scalar fields
    with respect to, 988–90
  of piece-wise smooth curves,
    986
  reparametrization by, 781

  of smooth curves, 986
  speed with respect to, 782
  velocity with respect to, 782
arc length element, 986
arc length function, 780–83, 986
  derivative of, 986
area. *See also* surface area
  calculating using Green's
    theorem, 1046–47
arithmetic sequences, 623–24
associative law of addition
  for matrices, 726
  for vectors, 705
average rate of change, 846, 866
average velocity vector, 762

best linear approximation, 845,
    859–61
best quadratic approximation,
    881–84
binormal vectors, unit, 790
bounded sequences, 629–30,
    631–32
bowl down elliptic paraboloids,
    835–36, 837
bowl up elliptic paraboloids, 835,
    837

calculus, fundamental theorems
    of, 780
Cantor, Georg, 645
Cantor's middle-third set, 645–46
capped surfaces, 1039–41
Cartesian plane, 699
  curves in, 805
chain rule, 772–73, 780, 794
  gradient form of, 870–72
  for total derivatives, 979–83
chaotic sequences, 630
circles, 809
circulation, along a curve, 992
closed and bounded intervals, 891
closed curves, 987
  piece-wise conservative vector
    fields, 1022–23
  positively oriented, 1033–34
closed domains, absolute extrema
    over, 891–94
closed-form, for sequences, 679
closed surfaces, 1033
  oriented, 1033
cofactors, expansion by, 729

column vectors, 727
commutative law of addition
  for matrices, 726
  for vectors, 705
comparison test for series
    convergence, 657–59,
    666
components, of vectors, 695–96,
    705
  relative to a direction, 716–19
  terminal point and, 750
conditionally converging series,
    661–62
cones, elliptic, 810
conic sections, 808–9
conservative fields, 1022–25,
    1027–29
  Stokes' theorem and, 1039–40
constant differences, in arithmetic
    sequences, 623–24
constant position functions,
    derivatives of, 768
constant ratio
  in geometric sequences, 624–26
  in geometric series, 640
constant sequences, 621, 623
constant unit speed, 780
constant vectors, 776
continuity
  of multivariable functions,
    855–56
  of position functions, 759
continuously differentiable
    functions, 880–81
continuous multivariable
    functions, 855–56
  critical points, 886–88
  order of integration, 909
  relative or local maxima or
    minima, 886
  stationary critical points, 887
contours
  approximating with slope fields,
    828–29
  defined, 820
  maps, 820–22
  plots, 820–24
contractions, 685
convergence
  absolute, 661–65
  conditional, 661–62
  interval of, 671–74
  radius of, 671–74
  of sequences, 626–29, 643

convergence *(cont.)*
  of sequences *(cont.)*
    bounded, 631–32
    speed of, 628–29
    to limits, 626–27
  of series, 639, 641–42, 646–55, 670–74
convergence tests for series, 646–55, 656–66
  alternating series test, 653–55, 665–66
  comparison test, 657–59, 666
  integral test, 648–52, 665
  limit comparison test, 659–61, 666
  Nth term test, 647–48, 659, 665
  ratio test, 663–64, 666, 670, 672–73
  root test, 664–65, 666, 671–72
coordinate functions, 749
coordinate systems
  cylindrical, 937–39, 940–41, 942–44, 999–1000, 1002
  left-handed, 738
  polar, 927–30
  rectangular, 697, 938–41
  for representing vectors, 695–96
  right-handed, 738
  spherical, 939–42, 944–46, 999–1000, 1002
critical points, 886–88, 896
  defined, 886
  stationary, 887–88
crosshairs, 818
cross product rule, 770–72
  order of functions in, 772
cross products, of vectors, 733
  anti-commutative, 734
  as area of parallelogram, 736
  curvature and, 797
  geometric interpretations of, 735–39
  of parallel vectors, 734
  squared norm of, 736
cross-sections, 809
  area functions, 908
  quadratic functions, 837–38
curl
  del notation, 969–70
  line integrals and, 1030–31
  physical interpretation of, 966–67
  representing with matrix, 965
  Stokes' theorem, 1034–36, 1039, 1040
  of vector fields, 964–67
Curry, Haskel, 819
Currying, 819
curvature, 792–98
  arc length and, 795–96
  computing, 793–95, 796
  defined, 793
  of a graph, 799
  radius of, 793–94
curvature vector, 793
curves, 749–800, 985–87
  capping curves, 1039–41
  in Cartesian plane, 805
  circulation along, 992
  closed, 987
  differential calculus of, 758–65
  level, 820–23, 828–29, 873–74
  measuring rate of change along, 870–72
  normal vectors to, 785–91
  parametrized, 751–54, 781, 987, 998
    smooth, 781
    in space, 1001–2
    theorem, 994–95
  piece-wise smooth, 986
  in a plane, curvature of, 793–95, 796
  simple, 987
  smooth, 760, 986
    parametrized, 781
  in space, 752–56
    computing curvature of, 796–98
    parametrized, 1001–2
    torsion of, 790–91
  surfaces capping, 1039–41
  visualizing vector-valued functions with, 750
cylinders
  elliptic, 810
  infinite, 808
cylindrical coordinates, 937–39
  parametrizing image curves in, 1002
  parametrizing surfaces with, 999–1000
  spherical coordinates and, 940–41
  triple integrals in, 942–44
cylindrical shells, volume of, 943–44
cylindrical surfaces, graphing, 807–8

definite integrals, 776–77, 985
  defined, 776
  as limit, 619
  physical interpretation of, 903–4
  value of, 903
degrees of freedom, 998
del notation
  for divergence and curl of a vector field, 967–68
  for gradient of a scalar field, 864
  scalar fields, 969–70
density, triple integral and, 913
derivatives. *See also* partial derivatives; total derivatives
  of arc length function, 986
  of constant position functions, 768
  directional, 865–70, 872–73
  geometric interpretation, 845
  higher-order partial, 878–84
  as limit, 619
  partial, 845, 846–51, 878–84
  physical interpretation, 845
  of position functions, 758, 760–63
    graphical interpretation of, 761–63
  of scalar fields, 960
  for scalar-valued functions, 845
  second derivative test for functions of two variables, 888–91
  total, 845, 854–61, 979–83
  of vector fields, 960–70
  of vectors
    higher-order, 764–65
    linearity properties of, 768
    properties of, 768–73
determinants
  calculating, 730
  of Jacobian matrix, 974–75
  of square matrices, 728–30
differentiable multivariable functions, 856, 880–81
discriminants for, 889–90
differentiable position functions, 760
differentiable scalar fields, del notation, 969–70
differential calculus
  of curves, 758–65
  of multivariable functions, 845–900
differentiation
  implicit, 828
  of power series function representation, 675
directed line segments
  equivalence class of, 695
  for representing vectors, 694–95
directional derivatives, 865–68
  along a path, 872–73
  defined, 866, 867
  direction of fastest change, 868–70
direction angles, of vectors, 722
direction cosines, of vectors, 721–22
directions, 693
direction vectors, 741–44, 760

INDEX

discriminants
  for differentiable functions, 889–90
  of quadratic polynomials, 838–40, 889
displacement vectors, 761
distance traveled, 778
distributive laws for scalars
  first, 705
  second, 705
divergence
  negative, 962–63
  physical interpretation of, 962–64
  positive, 962–63
  of sequences, 626, 629–30
  of series, 639, 641–42, 643, 646–55
  of vector fields, 961–64
divergence theorem, 1049–51
  verifying, 1050–51
division, scalar, 703–4
domains, of multivariable functions, 803
dot product rule, 770, 771
dot products, of vectors, 708–19
  analytic description of, 713–14
  calculating scalar component from, 718
  calculating vector component from, 718
  defined, 708
  divergence as, 967
  geometric description of, 713–15, 732–33
  gradients, 865
  of nonzero perpendicular vectors, 715
  norms of vectors, 710–11
  standard basis vectors, 712
  theorem, 709–10
  unit vectors, 711–13
  of vector with itself, 713
double integrals, 681, 904–11
  approximating, 948–51
  approximating with Monte Carlo method, 953
  calculating iteratively, 906–8
  calculating over general regions, 915–25
  calculating over rectangular regions, 905–6
  calculating by slicing method, 917–21
  changing order of integration, 908–9, 923–25
  defined, 904
  Green's theorem and, 1043
  interpreting as signed volume, 909–11
  Jacobian matrices and, 975–79

  in polar coordinates, 927–30
  properties of, 916
  subdividing region of integration, 921–23
  value of, 906
  volume and, 904, 908–11
double Riemann sums, 905–6

ellipses, 808–13
  calculating area of, using Green's theorem, 1046
ellipsoids, 809
elliptic cones, 810
elliptic cylinders, 810
elliptic paraboloids, 811
  bowl down, 835–36, 837
  bowl up, 835, 837
equivalence classes, of directed line segments, 695
Euler's constant, 645
expansion by minors (cofactors), 729
extrema
  absolute, over closed domains, 891–94
  under constraints, 895–900
  in contour plots, 822–23
  critical points, 886–88
  of multivariable functions, 885–94
extreme value theorem, for multivariable functions, 891

Fibonacci sequences, 681
first-degree Taylor polynomial approximation, 859–61
first fundamental theorem of calculus, 1021
first-order partial derivatives, 879, 880
fishnet, 817
fixed points, of functions, 683
fixed-point theorem, 686–88
flow lines, for visualizing velocity vectors, 960
fluids
  curl, 966–67
  incompressible, 964
  velocity field of, 959–60, 966, 992
flux, 1016–18
  Gauss' theorem, 1049–51
force, 693
force fields, 958–59
  conservative, 1023
  line integrals for, 991
formulas
  harmonic sequences, 622

  sequences, 622–23, 624–25
function graphs, parametrizing, 998–99
functions. *See also* multivariable functions
  continuous at a point, 855–56
  fixed point of, 683
  iterating, 683, 684–85
  power series, 619, 668–77
  vector-valued, 749
fundamental theorem for line integrals, 1027–29
fundamental theorems of calculus
  first, 1021
  second, 1021
fundamental theorems of vector calculus, 1021–51
  divergence theorem (Gauss' theorem), 1049–51
  Green's theorem, 1043–47
  line integrals, 1027–29
  Stokes' theorem, 1033–41
fundamental vector product, 1003–7
  finding surface area with, 1009, 1010
  forming principal unit normal vector from, 1014

gas, velocity field of, 962–63
Gauss' theorem, 1049–51
general equation, of a line, 740
geometric problems in space, solving with vectors, 732–39
geometric reasoning, for verifying vector properties, 705–6
geometric sequences, 624–26, 630
geometric series, 640–43, 644, 666
  theorem, 640
gradients, 864–74
  calculating, 864–65
  chain rule, 870–72
  directional derivatives, 865–68, 872–73
  direction of fastest change, 868–70
  level curves and surfaces and, 873–74
  rate of change along a path, 870–72
  tangent plane equation, 874–76
  vector fields defined by, 865
gradient vectors, 845
graphical representation
  of derivatives, 761–63
  of image curves, 761
graphing
  curvature of, 799
  cylindrical surfaces, 807–8

graphing *(cont.)*
   iteration process, 684–86
   linear functions, 829–33
   multivariable functions, 804–5, 815–29
   parametrized surfaces, 1001–2
   planes, 807
   quadratic functions, 834–42
   quadratic surfaces, 808–13
   of scalar-valued functions, 804–5
   spheres, 806
Green's theorem, 1043–47
   calculating area using, 1046–47
   verification of, 1044–46

harmonic sequences, 621, 625–26, 631
   formulas describing, 622
harmonic series, 643, 646
helix
   parametrization, 782
   in space curves, 753, 756
higher-order derivatives, 764–65
higher-order partial derivatives, 878–84
horizontally simple regions, 918–19
horizontal planes, 832
horizontal tangent planes, 887
hyperbolas, 809, 810–12
hyperbolic paraboloids, 812
hyperboloid of one sheet, 811
hyperboloid of two sheets, 811

identity matrix, 723
image curves, 751
   graphical interpretation of, 761
   of moving objects, 758
   for parametrization in cylindrical coordinates, 1002
   for parametrization in spherical coordinates, 1002
   parametrizing tangent line to, 763–64
implicit differentiation, 828
improper integrals, comparing series to, 648
incompressible vector fields, 964, 968
indefinite integrals, 775–76
indexes, of sequences, 621, 623
inductive sequences, 680
inertia, moment of, 936
infinite cylinders, 808
infinite processes, 619–91
   defined, 619
   iteration, 679–89
   power series, 668–77

   sequences, 619–34
   series, 636–44, 646–55, 656–6
infinite sets, vs. sequences, 623
infinite trough, 808
initial points
   vectors, 750
   of vectors, 694
initial position vectors, 741–44
initial seed, for iterative processes, 681, 683
inner product, 708
instantaneous direction of movement, 762
instantaneous rate of change, 846, 847, 850–51, 867
   derivative and, 845
   fastest, 859–60
   measuring with Jacobian matrix, 973
   of volume per unit of volume, 962
instantaneous speed, 762
instantaneous velocity vector, 762
integrals. *See also* definite integrals; double integrals; line integrals; triple integrals
   approximating with Monte Carlo method, 952–54
   indefinite, 775–76
   path, 985
   surface, 1008–18
integral test for series convergence, 648–52, 665
integration, 775–83
   changing order of, 908–9, 923–25
   region of, subdividing, 921–23
intercepts, in graph of plane, 807
intermediate value theorem, 687
intervals
   closed and bounded, 891
   of convergence, 671–74
   of validity, for Maclaurin series, 676–77
irrotational vector fields, 967, 968
iteration, 679–89
   of contractions, 685
   of double integrals, 906–8
   graphing, 684–86
   Newton's method, 688–89
iterative sequences, 681–82

Jacobian matrix, 972–83
   defined, 972
   determinant as scale factor, 974–75
   in double integration, 975–79
   trace of, 973
Jacobian scale factor, 976–77

   triple integration and, 984–85

Lagrange multiplier theorem, 895–900
Laplacian, of scalar fields, 968–70
left-handed coordinate system, 738
left-hand rule, 738
Leibniz notation, chain rule, 773
level curves, 820–23
   gradient and, 873–74
   plotting, 828–29
level surfaces, 823–24
   gradient and, 873–74
limit comparison test, for series, 659–61, 666
limit properties, of sequences, 632–34
limits
   of multivariable functions, 855
   of position functions, 758–59
   sequence convergence to, 620, 626–27, 632–34
linear functions, 829–33, 834
   contractions of, 685
   fixed point of, 683
   graphing, 829–33
   of n variables, 833
   of one variable, 829
   point-slope form of, 832
   Taylor form of, 832–33
   of three variables, 833
   of two variables, 829–33
linearity properties, of derivatives, 768
linearly dependent vectors, 735
linearly independent vectors, 735
line integrals, 985–95
   arc length, 985–87
   curl and, 1030–31
   curves, 985–87
   defined, 991
   fundamental theorems for, 1021–31
   Green's theorem and, 1046–47
   notations for, 992–94
   orientation effects on, 994–95
   of scalar fields with respect to arc length, 988–90
   Stokes' theorem, 1034–36
   of vector fields, 990–92
lines
   general equation for, 740
   parametric equations of, 740–43
   parametrized, 750–54, 988–89, 993–94
   vector parametric form of, 741–44
locally linear functions, 856
local maxima, 886, 888

INDEX

in contour plots, 822
local minima, 886, 888
  in contour plots, 822
local slope, of slice function, 849–50

machine graphics, visualizing space curves with, 754–56
Maclaurin series, 675–77
magnitude, 693
main diagonal, in matrices, 723
mass, surface, 1012–13
matrices
  adding and subtracting, 724–26
  algebra of, 723–26
  computing vector cross products with, 733
  defined, 603, 723
  equal, 723
  expansion by minors, 729
  identity, 723
  Jacobian, 972–83
  main diagonal in, 723
  multiplication of, 725
  real, 723
  for representing curl, 965
  scalar multiplication of, 724, 726
  square, 723, 726, 728–30
  trace of, 973
  vectors and, 723–30
  zero, 723
matrix algebra, 723–26
  properties of, 726
matrix product, chain rule and, 981
maxima
  relative or local, 822, 886, 890
  strict, 838–40
minima
  relative or local, 882, 886, 888, 890
  strict, 838–40, 842
minors
  defined, 729
  expansion by, 729
mixed second-order partial derivatives, 879
Möbius band, 1015
moment of inertia, 936
monotonically decreasing sequences, 631
monotonically increasing bounded sequences, 649
monotonically increasing bounded series, 657
monotonically increasing sequences, 631
monotonic sequences, 630–32
Monte Carlo method, for approximating integrals, 952–54
multiple integrals, 904–14
  double integrals, 904–11
  numerical techniques for, 948–54
  properties of, 916
  triple integrals, 911–14
multiplication
  of matrices, 725, 726
  scalar, of vectors, 703–4
multivariable functions, 801–42
  best linear approximations of, 859–61
  continuity of, 855–56
  contour plots of, 820–23
  defined, 801
  differentiable at a point, 856
  differential calculus of, 845–900
  domains of, 803
  examples, 802–13
  extreme value theorem for, 891
  finding extrema of, 885–94
  gradients of, 864–74
  graphing, 804–5, 815–29
  integral calculus of, 903–55
  limits of, 855
  linear, 829–33
  locally linear, 856
  notation, 803–5
  partial derivatives of, 845, 846–51
  perspective, 824–25
  quadratic functions, 834–43
  total derivatives of, 845, 854–61
  visualizing and interpreting, 815–29
  wireframe plotting of, 825–26
n-dimensional space, representing vectors in, 698–99
negative volume, double integrals, 909–10
net vector change in position, 777
net volume, double integrals, 909–10
Newton's method, iterative function for, 688–89
non-orientable surfaces, 1015
normal vectors, 785
  parametrized surfaces and, 1003
  principal unit, 785–88
  to a curve, 785–91
  to a plane, 744–45
  unit, 785
norms, of vectors ($\|v\|$), 710–11
notation
  del, 864, 967–68
  line integrals, 992–94
  multivariable functions, 803–5
  partial derivatives, 847
scalars, 694, 749
sequences, 621
series, 638
summation, 638, 640
vectors, 694, 749
Nth term test, 647–48, 659, 665
numerical techniques
  for general regions of integration, 951–52
  multiple integrals for, 948–54

1-space, 698
one-to-one parametrization, 1033
optimization, 885–94
order of integration, changing, 923–25
orientation
  of line integrals, 994–95
  preserving, 994–95
  reversing, 995
  of parametrized curves, 751–54, 987
oriented closed surfaces, 1033
oriented surfaces, 1014–15
origins, in quadric surfaces, 812
orthogonal projection, 717
orthogonal vector components, 719
orthogonal vectors, 715–16, 873–74
osculating plane, 789–90
outward flux, 1017–18

parabolas, 809, 811, 812
  slice functions as, 816
paraboloids
  elliptic, 811, 835–36, 837
  hyperbolic, 812
  of revolution, 817
paradox, 637
parallelograms, approximating surface area with, 1009
parallel vectors, cross products of, 734
parametric equations of lines, 740–43
parametric equations of planes, 743–45
parametrization, one-to-one, 1033
parametrized curves, 750–50, 987, 990
  orientation of, 751–54, 987
  smooth, by arc length, 781
  in space, 1001–2
  theorems, 994–95
parametrized lines, 988–89, 993–94
  tangent, 763–64
parametrized surfaces, 998–1007

parametrized surfaces *(cont.)*
  function graphs, 998–99
  graphing, 1001–2
  parametrizing with cylindrical or spherical coordinates, 999–1000
  simple, 1033
  surface area of, 1008–12
  surface integrals, 1008–18
partial derivatives, 845, 846–51
  calculating, 847–49
  defined, 846–47
  first-order, 879, 880
  geometrical interpretation, 849–50
  higher-order, 878–84
  notation, 847
  physical interpretation, 850–51
  second-order, 878–84
  third-order, 880–81
partial differentiation, rules for, 847–48
partial sums, series as limit of, 638
path integrals, 985
perpendicular vectors, 715
perspective, in multivariable functions, 824–25
piece-wise smooth curves, 986
  arc length of, 986
  conservative vector fields, 1022–23
Pisa, Leonardo di, 681
plane curves, computing curvature of, 796
planes
  graphing, 807
  Green's theorem and, 1043–47
  horizontal, 832
  intersections with surfaces, 837–38
  linear functions, 829–33
  normal vectors to, 744–45, 874
  osculating, 789–90
  parametric equations of, 743–45
  vertical, 832
point-slope form of linear functions, 832
point of validity, for Taylor series, 676
polar coordinates, double integrals in, 927–30
polynomial functions, terms of, 619
position functions, 759–56
  chain rule, 773
  continuity of, 759
  derivatives of, 758, 760–63
  differentiable, 760
  limits of, 758–59

for parametrized curves, 752–53, 998
for parametrized surfaces, 998
position vector functions, antiderivatives for, 775–76
position vectors
  derivatives of, 768–73
  product rules, 769
positively oriented closed curves, 1033–34
potentials, 1025–27
  as antiderivatives, 1025–27
  of conservative vector fields, 1027–29
  defined, 1025
potential scalar fields, 1025–27
power series, 619, 668–77
  convergence of, 670–76
  defined, 668–69
  evaluating, 669–70
  form of, 669
  functions defined by, 674–75
preservation, of line integral orientation, 994–95
principal unit normal vector, 785–88, 794, 1014–18
  acceleration vector and, 790
  Green's theorem and, 1043
  inward, 1033
  outward, 1033
  Stokes' theorem and, 1033
principal views, 754
product rules
  cross product rule, 770–72
  dot product rule, 770, 771
  position vectors, 769
  for real-valued functions, 769
  scalar, 769–70
  for vector-valued functions, 769
p-series, 666
Pythagorean theorem, 802
  computing vector norms with, 710–11

quadratic functions, 834–43
  cross-sections, 837–38
  discriminants, 838–40
  graphing, 834–42
  slicing, 836–38
  Taylor form for, 840–42
quadratic polynomials, discriminants of, 889
quadric surfaces, 808–13, 834
  defined, 808
  graphing, 808–13
  origins in, 812

radius of convergence, 671–74
radius of curvature, 793–94

computing, 795
rate of change
  average, 846, 866
  instantaneous, 845, 846, 847, 850–51, 859–60, 867, 962, 973
  total, 858
ratio test, for series convergence, 663–64, 666, 670, 672–73
rays, vectors vs., 694
real matrices, 723
real numbers, Archimedean property of, 636, 643
real-valued functions
  chain rule for, 772–73
  derivatives of constant functions, 768
  product rules for, 769
  Riemann sums for, 903
  total distance traveled, 778
rectangular coordinates
  cylindrical coordinates and, 938–39
  for representing vectors, 697
  spherical coordinates and, 939–41
recursive sequences, 680–82
regions
  closed and bounded, 891
  describing with polar coordinates, 927–30
  double integrals, 915–25
  horizontally simple, 918–19
  numerical techniques for, 951–52
  rectangular, double integrals over, 905–6
  subdividing, 921–23
  triple integrals, 932–35
  vertically simple, 918–19
relative maxima, 886
  theorem, 890
relative minima, 886
  theorem, 890
remainder, for Taylor series, 676
reparametrization by arc length, 781
resultant vectors, 701
reversal, of line integral orientation, 995
Riemann sums, 903, 948
  approximating wires with, 988
  defined, 776
  double, 905–6
  limiting values for, 903
  triple, 912, 913
right-handed coordinate system, 738
right-hand rule, 737–38, 1030
rise over run, 830, 833, 859

INDEX

root test for series convergence, 664–65, 666, 671–72
row vectors, 727

saddle points, 812, 826, 835, 837, 888
   theorems, 839, 890
saddles, 812, 835, 837, 838
scalar components, relative to a direction, 716–18
scalar division, 703–4
scalar fields
   defined, 801, 802
   del notation, 969–70
   derivatives of, 960
   Laplacian of, 968–70
   line integrals of, with respect to arc length, 988–90
   surface integrals of, 1012–14
scalar multiplication
   identity law for 1, 705
   of matrices, 724, 726
   property of 0, 705
   property of −1, 705
   vector norms and, 711
   of vectors, 703–4
scalar product, 708, 709. *See also* dot products
scalar product rules, 769–70
scalars (scalar quantities), 693
   functions, total distance traveled, 778
   notation, 694, 749
   as one-dimensional vectors, 698
   properties of (theorem), 705
scalar-valued functions, 801
   derivatives for, 845
   of two variables, graphing, 804–5
scale factor, Jacobian determinant as, 974–75
second-degree Taylor polynomial approximation, 883–84, 888
second derivative test, for functions of two variables, 888–91
second fundamental theorem of calculus, 1021
second-order partial derivatives, 878–84
sequences, 619–34
   alternating, 621
   alternating harmonic, 621, 622, 629
   arithmetic, 623–24
   bounded, 629–30, 631–32
   chaotic, 630
   closed-form description, 679
   constant, 621, 623

converging, 626–29, 631–32, 642, 643
   divergence of, 626, 629–30
   Fibonacci sequences, 681
   formulas, 622–23, 624–25
   geometric, 624–26, 630
   harmonic, 621, 622, 625–26, 631, 643
   index of, 621
   inductive, 680
   vs. infinite sets, 623
   iterative, 681–82
   limit properties of, 632–34
   limits of, 620
   monotonic, 630–32
   monotonically increasing bounded, 649
   notation, 621
   partial sums, diverging, 642
   recursive, 680–82
   starting index value, 623
   terms of, 620–21, 642
   theorem, 632
   unbounded, 629–30
series
   absolutely convergent, 661–65
   alternating harmonic, 643, 644, 661, 664
   conditional convergent, 661–62
   convergence tests for, 646–55, 656–66
   converging, 639, 641–42, 646–55, 670–74
   defined, 619, 637–39
   diverging, 639, 641–42, 643, 646–55, 670
   geometric, 640–43, 644, 666
   harmonic, 643, 646
   as limit of partial sums, 638
   Maclaurin series, 675–77
   monotonically increasing bounded, 657
   notation, 638
   power, 668–77
   properties of, 644
   p-series, 666
   sum of, 619, 636, 638, 639
   Taylor series, 619, 675–77
   telescoping, 639, 666
signed volume, double integrals interpreted as, 909–11
simple curves, 987
   piece-wise smooth, 1044, 1046
simple parametrized surfaces, 1033
singular points, 1003
slice functions
   local slope of, 849–50
   with more than two inputs, 817–20
   with two inputs, 815–18

slicing
   for calculating double integrals, 907–8, 917–21
   contour plots, 820–23
   quadratic functions, 836–38
slope
   derivative and, 845
   local, of slice function, 849–50
slope fields, approximating contours with, 828–29
slope-intercept equation, 740–41
slope with respect to x, 830–32
slope with respect to y, 830–32
smooth curves, 760, 986
   arc length of, 986
   parametrizing, by arc length, 781
   piece-wise, conservative vector fields, 1022–23
smooth points, 1003
smooth surfaces, parametrization of, 1003–7
space curves, 752–56
   computing curvature of, 796–98
   principal views of, 754
   torsion of, 790–91
   visualizing with machine graphics, 754–56
space-time, 698
speed
   defined, 693
   derivative of arc length function as, 986
   with respect to arc length, 782
   of sequence convergence, 628–29
sphere of radius, 941
spheres
   cross-sections of, 809
   equations of, 806, 809
   graphing, 806
   surface area of, 1011–12
spherical coordinates, 939–42
   parametrizing image curves with, 1002
   parametrizing surfaces with, 999–1000
   triple integrals in, 944–46
square matrices, 723, 726
   determinants of, 728–30
standard basis vectors, 712
stationary critical points, 887–88
Stokes, George G., 1034
Stokes' theorem, 1033–41
   verifying, 1037–40
strictly monotonically decreasing sequences, 631
strictly monotonically increasing sequences, 631
strict maximums, 838–40
   theorem, 839

strict minimums, 838–40, 842
  theorem, 839
subtraction
  of matrices, 724–25
  of vectors, 702–3
summation notation, 638, 640
sums, of series, 619, 636, 638, 639
surface area. *See also* area
  approximating with parallelograms, 1009
  finding with fundamental vector product, 1009, 1010
  of parametrized surfaces, 1008–12
  of spheres, 1011–12
surface area elements, 1010
surface integrals, 1008–18
  defined, 1008
  divergence theorem, 1050–51
  Green's theorem and, 1043
  of scalar fields, 1012–14
  Stokes' theorem, 1034–36, 1038, 1040–41
  of vector fields, 1015–18
surface mass, 1012–13
surfaces
  closed, 1033
  cylindrical, 807–8
  describing, 805–6
  intersections with planes, 837–38
  level, 823–24, 873–74
  non-orientable, 1015
  oriented, 1014–15
  parametrized, 998–1018, 1033
  quadric, 808–13, 834

tangent lines, 763–64
  parametrizing, 763–64
tangent planes
  best linear approximation and, 859–60
  equation of, 874–75
  horizontal, 887
  parametrized surfaces and, 1003–7
tangent vectors, unit, 764, 794
Taylor form
  of linear functions, 832–33
  for quadratic functions, 840–42
Taylor polynomials, 845
  approximating with higher-order partial derivatives, 878
  first-degree approximations, 859–61
  second-degree approximations, 883–84, 888
Taylor series, 619, 675–77
  point of validity for, 676

  remainder for, 676
telescoping series, 639, 666
terminal points, of vectors, 694, 750
terms, of sequences, 620–21, 623–26
theorems
  divergence theorem, 1049–51
  first fundamental theorem of calculus, 1021
  fixed-point theorem, 686–88
  fundamental theorems of vector calculus, 1021–51
  Gauss' theorem, 1049–51
  geometric series, 640
  Green's theorem, 1043–47
  intermediate value theorem, 687
  second fundamental theorem of calculus, 1021
  sequences, 632
  Stokes' theorem, 1033–41
third-order partial derivatives, 880–81
three-dimensional space, representing vectors in, 697–98
time
  as fourth dimension, 698
  in vector-valued functions, 750
torsion
  absolute value of, 791
  of space curves, 790–91
total derivatives, 845, 854–61
  calculating, 857–58
  chain rule for, 979–83
  defined, 856–57
  interpreting, 858–59
  properties of, 980–83
  of vector fields, 972–83
total distance traveled, 778
total rate of change, 858
trace, of a matrix, 973
triangle law, 701
triple integrals, 911–14
  calculating, 913–14
  in cylindrical coordinates, 942–44
  Jacobian scale factor and, 984–85
  over general regions, 932–35
  physical interpretation of, 913
  properties of, 916
  in spherical coordinates, 944–46
triple Riemann sums, 912, 913
triple scalar products, 734–35, 739
trough, 836, 838
  infinite, 808
two-dimensional vector fields, Green's theorem and, 1043–47

unbounded sequences, 629–30
uniform density, triple integrals and, 913
unit binormal vectors, 790
unit circle, parametrized, 752
unit normal vectors, 785, 786
unit tangent vectors, 763–64, 785, 787, 788, 794
  acceleration vector and, 790
  curvature and, 793
unit vectors, 711–13, 718, 786, 788

validity
  intervals of, for Maclaurin series, 676–77
  point of, for Taylor series, 676
variable density, triple integral and, 913
vector components, 695–96, 705
  orthogonal, 719
  relative to a direction, 716–19
  terminal point and, 750
vector fields
  conservative, 1022–25, 1027–29
  continuous, 958
  curl of, 964–67, 1030–31
  defined, 957–58
  defined by gradients, 865
  derivatives of, 960–70
  differentiable, 958
  divergence of, 961–64
  flux value in, 1016–18
  force fields, 958–59
  incompressible, 968
  incompressible at a point, 964
  incompressible at every point, 964
  irrotational, 968
  irrotational at a point, 967
  irrotational at every point, 967
  line integrals of, 990–92
  Stokes' theorem, 1034–36
  surface integrals of, 1015–18
  total derivative of, 972–83
  two-dimensional, Green's theorem and, 1043–47
  velocity fields, 958–59
  visualizing, 959–60
vector parametric form, of lines, 741, 742
vector quantity, 693. *See also* vectors
vectors, 693–748, 957–1018
  acceleration, 765
  addition of, 701, 705, 706
  additive inverses of, 702
  algebra of, 701–6
  analytic properties of, 695–96, 704–5

average velocity, 762
column, 727
computing angle between, 715–16
cross products of, 733
defined, 693
direction, 741–44, 760
direction angles, 722
direction cosines of, 721–22
displacement, 761
dot products of, 708–19
fundamental theorems, 1021–51
geometric representation of, 694–95, 732–39
gradient, 845
initial points of, 694, 750
initial position, 741–44
instantaneous velocity, 762
linearly dependent, 735
linearly independent, 735
matrices and, 723–30
net vector change in position, 777
normal, 744–45, 785–91, 874, 1003
notation for, 693, 749
orthogonal, 715–16, 873–74
parallel, cross products of, 734
perpendicular, 715
principal unit normal, 785–88
properties of (theorem), 705–6, 709–10, 734
rays vs., 694
representing in 3-dimensional space, 697–98
representing in n-dimensional space, 698–99
resultant, 701
row, 727
scalar division of, 703–4
scalar multiplication of, 703–4
standard basis, 712
subtraction of, 702–3
terminal points of, 694, 750
unit, 711–13, 718, 786, 788
unit binormal, 790
unit normal, 785
unit tangent, 763–64, 785, 787, 788, 794
velocity, 762–63, 776–77, 787, 788, 797, 958, 960, 991
written as linear combinations, 712
zero, 695, 778
vector-valued functions
cross product rule, 770–72
defined, 749
dot product rule, 770, 771
limits of, 758–59
linearity properties of derivative, 768
position functions, 759–56
product rules for, 769
2- or 3-dimensional, 750
vector values, given by gradients, 865
velocity
defined, 693
with respect to arc length, 782

velocity vector fields, 958–59
curl expressed in terms of line integrals for, 1030–31
line integrals for, 992
velocity vectors, 762–63, 787, 788, 958
average, 762
curvature and, 797
function, 776–77
instantaneous, 762
line integrals and, 991
visualizing with flow lines, 960
vertical line test, 805
vertically simple regions, double integrals, 918–19
vertical planes, 832
volume
divergence theorem, 1050
double integrals and, 904, 908–11
triple integrals, 934–35

wireframe plotting, of multivariable functions, 825–26
wires
approximating with Riemann sums, 988
mass of, 988–89

Zeno's paradox, 636–37, 640
zero matrices, 723
zero vector, 695, 778

57. $\int \dfrac{\sqrt{a+bu}}{u}\,du = 2\sqrt{a+bu} + a\int \dfrac{du}{u\sqrt{a+bu}}$

58. $\int \dfrac{\sqrt{a+bu}}{u^2}\,du = -\dfrac{\sqrt{a+bu}}{u} + \dfrac{b}{2}\int \dfrac{du}{u\sqrt{a+bu}}$

59. $\int u^n \sqrt{a+bu}\,du = \dfrac{2u^n(a+bu)^{3/2}}{b(2n+3)} - \dfrac{2na}{b(2n+3)}\int \dfrac{u^{n-1}}{\sqrt{a+bu}}\,du$

60. $\int \dfrac{u^n\,du}{\sqrt{a+bu}} = \dfrac{2u^n\sqrt{a+bu}}{b(2n+1)} - \dfrac{2na}{b(2n+1)}\int \dfrac{u^{n-1}\,du}{\sqrt{a+bu}}$

61. $\int \dfrac{du}{u^n\sqrt{a+bu}} = -\dfrac{\sqrt{a+bu}}{a(n-1)u^{n-1}} - \dfrac{b(2n-3)}{2a(n-1)}\int \dfrac{du}{u^{n-1}\sqrt{a+bu}}$

62. $\int \dfrac{du}{u\sqrt{a+bu}} = \dfrac{1}{\sqrt{a}}\ln\left|\dfrac{\sqrt{a+bu}-\sqrt{a}}{\sqrt{a+bu}+\sqrt{a}}\right| + C,\ \text{(if } a > 0\text{)};\quad = \dfrac{2}{\sqrt{-a}}\arctan\sqrt{\dfrac{a+bu}{-a}} + C,\ \text{(if } a < 0\text{)}$

Trigonometric Forms

63. $\int \sin^2 u\,du = \dfrac{1}{2}u - \dfrac{1}{4}\sin 2u + C$

64. $\int \cos^2 u\,du = \dfrac{1}{2}u + \dfrac{1}{4}\sin 2u + C$

65. $\int \tan^2 u\,du = \tan u - u + C$

66. $\int \cot^2 u\,du = -\cot u - u + C$

67. $\int \sin^3 u\,du = -\dfrac{1}{3}(2 + \sin^2 u)\cos u + C$

68. $\int \cos^3 u\,du = \dfrac{1}{3}(2 + \cos^2 u)\sin u + C$

69. $\int \tan^3 u\,du = \dfrac{1}{2}\tan^2 u + \ln|\cos u| + C$

70. $\int \cot^3 u\,du = -\dfrac{1}{2}\cot^2 u - \ln|\sin u| + C$

71. $\int \sec^3 u\,du = \dfrac{1}{2}\sec u \tan u + \dfrac{1}{2}\ln|\sec u + \tan u| + C$

72. $\int \csc^3 u\,du = -\dfrac{1}{2}\csc u \cot u + \dfrac{1}{2}\ln|\csc u - \cot u| + C$

73. $\int \sin^n u\,du = -\dfrac{1}{n}\sin^{n-1} u \cos u + \dfrac{n-1}{n}\int \sin^{n-2} u\,du$

74. $\int \cos^n u\,du = \dfrac{1}{n}\cos^{n-1} u \sin u + \dfrac{n-1}{n}\int \cos^{n-2} u\,du$

75. $\int \tan^n u\,du = \dfrac{1}{n-1}\tan^{n-1} u - \int \tan^{n-2} u\,du$

76. $\int \cot^n u\,du = \dfrac{-1}{n-1}\cot^{n-1} u - \int \cot^{n-2} u\,du$

77. $\int \sec^n u\,du = \dfrac{1}{n-1}\tan u \sec^{n-2} u + \dfrac{n-2}{n-1}\int \sec^{n-2} u\,du$

78. $\int \csc^n u\,du = \dfrac{-1}{n-1}\cot u \csc^{n-2} u + \dfrac{n-2}{n-1}\int \csc^{n-2} u\,du$

79. $\int \sin au \sin bu\,du = \dfrac{\sin(a-b)u}{2(a-b)} - \dfrac{\sin(a+b)u}{2(a+b)} + C$

80. $\int \cos au \cos bu\,du = \dfrac{\sin(a-b)u}{2(a-b)} + \dfrac{\sin(a+b)u}{2(a+b)} + C$

81. $\int \sin au \cos bu\,du = -\dfrac{\cos(a-b)u}{2(a-b)} - \dfrac{\cos(a+b)u}{2(a+b)} + C$

82. $\int u \sin u\,du = \sin u - u\cos u + C$

83. $\int u \cos u\,du = \cos u + u\sin u + C$

84. $\int u^n \sin u\,du = -u^n \cos u + n\int u^{n-1}\cos u\,du$

85. $\int u^n \cos u\,du = u^n \sin u - n\int u^{n-1}\sin u\,du$

86. $\int \sin^n u \cos^m u\,du = -\dfrac{\sin^{n-1} u \cos^{m+1} u}{n+m} + \dfrac{n-1}{n+m}\int \sin^{n-2} u \cos^m u\,du$

$= \dfrac{\sin^{n+1} u \cos^{m-1} u}{n+m} + \dfrac{m-1}{n+m}\int \sin^n u \cos^{m-2} u\,du$

(continued on next page)

Inverse Trigonometric Forms

87. $\int \arcsin u \, du = u \arcsin u + \sqrt{1-u^2} + C$

88. $\int \arccos u \, du = u \arccos u - \sqrt{1-u^2} + C$

89. $\int \arctan u \, du = u \arctan u - \frac{1}{2} \ln(1+u^2) + C$

90. $\int u \arctan u \, du = \frac{u^2+1}{2} \arctan u - \frac{u}{2} + C$

91. $\int u \arcsin u \, du = \frac{2u^2-1}{4} \arcsin u + \frac{u\sqrt{1-u^2}}{4} + C$

92. $\int u \arccos u \, du = \frac{2u^2-1}{4} \arccos u - \frac{u\sqrt{1-u^2}}{4} + C$

93. $\int u^n \arcsin u \, du = \frac{1}{n+1}\left[u^{n+1} \arcsin u - \int \frac{u^{n+1} du}{\sqrt{1-u^2}}\right], \quad n \neq -1$

94. $\int u^n \arccos u \, du = \frac{1}{n+1}\left[u^{n+1} \arccos u + \int \frac{u^{n+1} du}{\sqrt{1-u^2}}\right], \quad n \neq -1$

95. $\int u^n \arctan u \, du = \frac{1}{n+1}\left[u^{n+1} \arctan u - \int \frac{u^{n+1} du}{1+u^2}\right], \quad n \neq -1$

Exponential and Logarithmic Forms

96. $\int u e^{au} du = \frac{1}{a^2}(au-1)e^{au} + C$

97. $\int u^n e^{au} du = \frac{1}{a} u^n e^{au} - \frac{n}{a} \int u^{n-1} e^{au} du$

98. $\int e^{au} \sin bu \, du = \frac{e^{au}}{a^2+b^2}(a \sin bu - b \cos bu) + C$

99. $\int e^{au} \cos bu \, du = \frac{e^{au}}{a^2+b^2}(a \cos bu + b \sin bu) + C$

100. $\int \ln u \, du = u \ln u - u + C$

101. $\int u^n \ln u \, du = \frac{u^{n+1}}{(n+1)^2}[(n+1) \ln u - 1] + C$

102. $\int \frac{1}{u \ln u} du = \ln |\ln u| + C$

Hyperbolic Forms

103. $\int \sinh u \, du = \cosh u + C$

104. $\int \cosh u \, du = \sinh u + C$

105. $\int \tanh u \, du = \ln \cosh u + C$

106. $\int \coth u \, du = \ln |\sinh u| + C$

107. $\int \operatorname{sech} u \, du = \arctan(\sinh u) + C$

108. $\int \operatorname{csch} u \, du = \ln |\tanh \frac{1}{2} u| + C$

109. $\int \operatorname{sech}^2 u \, du = \tanh u + C$

110. $\int \operatorname{csch}^2 u \, du = -\coth u + C$

111. $\int \operatorname{sech} u \tanh u \, du = -\operatorname{sech} u + C$

112. $\int \operatorname{csch} u \coth u \, du = -\operatorname{csch} u + C$

Forms involving $2au - u^2$

113. $\int \frac{\sqrt{2au-u^2}}{u} du = \sqrt{2au-u^2} + a \arccos\left(\frac{a-u}{a}\right) + C$

114. $\int \frac{\sqrt{2au-u^2}}{u^2} du = -\frac{2\sqrt{2au-u^2}}{u} - \arccos\left(\frac{a-u}{a}\right) + C$

115. $\int \frac{du}{\sqrt{2au-u^2}} = \arccos\left(\frac{a-u}{a}\right) + C$

116. $\int \frac{du}{u\sqrt{2au-u^2}} = -\frac{\sqrt{2ua-u^2}}{au} + C$

117. $\int \frac{u \, du}{\sqrt{2au-u^2}} = -\sqrt{2au-u^2} + a \arccos\left(\frac{a-u}{a}\right) + C$

118. $\int \sqrt{2au-u^2} \, du = \frac{u-a}{2}\sqrt{2au-u^2} + \frac{a^2}{2} \arccos\left(\frac{a-u}{a}\right) + C$

119. $\int u\sqrt{2au-u^2} \, du = \frac{2u^2-au-3a^2}{6}\sqrt{2au-u^2} + \frac{a^3}{2} \arccos\left(\frac{a-u}{a}\right) + C$

120. $\int \frac{u^2 \, du}{\sqrt{2au-u^2}} = -\frac{(u+3a)}{2}\sqrt{2au-u^2} + \frac{3a^2}{2} \arccos\left(\frac{a-u}{a}\right) + C$